Foundations of Reinforcement Learning with Applications in Finance

Foundations of Reinforcement Learning with Applications in Finance aims to demystify Reinforcement Learning, and to make it a practically useful tool for those studying and working in applied areas — especially finance.

Reinforcement Learning is emerging as a powerful technique for solving a variety of complex problems across industries that involve Sequential Optimal Decisioning under Uncertainty. Its penetration in high-profile problems like self-driving cars, robotics, and strategy games points to a future where Reinforcement Learning algorithms will have decisioning abilities far superior to humans. But when it comes getting educated in this area, there seems to be a reluctance to jump right in, because Reinforcement Learning appears to have acquired a reputation for being mysterious and technically challenging.

This book strives to impart a lucid and insightful understanding of the topic by emphasizing the foundational mathematics and implementing models and algorithms in well-designed Python code, along with robust coverage of several financial trading problems that can be solved with Reinforcement Learning. This book has been created after years of iterative experimentation on the pedagogy of these topics while being taught to university students as well as industry practitioners.

Features

- Focus on the foundational theory underpinning Reinforcement Learning and software design of the corresponding models and algorithms
- Suitable as a primary text for courses in Reinforcement Learning, but also as supplementary reading for applied/financial mathematics, programming, and other related courses
- Suitable for a professional audience of quantitative analysts or data scientists
- Blends theory/mathematics, programming/algorithms and real-world financial nuances while always striving to maintain simplicity and to build intuitive understanding
- To access the code base for this book, please go to: https://github.com/TikhonJelvis/RL-book.

Foundations of Reinforcement Learning with Applications in Finance

Ashwin Rao
Tikhon Jelvis

CRC Press
Taylor & Francis Group
Boca Raton London New York

CRC Press is an imprint of the
Taylor & Francis Group, an **informa** business

A CHAPMAN & HALL BOOK

First edition published 2023
by CRC Press
6000 Broken Sound Parkway NW, Suite 300, Boca Raton, FL 33487-2742

and by CRC Press
4 Park Square, Milton Park, Abingdon, Oxon, OX14 4RN

Library of Congress Cataloging-in-Publication Data

Names: Rao, Ashwin, author. | Jelvis, Tikhon, author.
Title: Foundations of reinforcement learning with applications in finance /
Ashwin Rao, Tikhon Jelvis.
Description: 1 Edition. | Boca Raton, FL : Chapman & Hall, CRC Press,
[2023] | Includes bibliographical references and index.
Identifiers: LCCN 2022030498 (print) | LCCN 2022030499 (ebook) | ISBN
9781032124124 (hardback) | ISBN 9781032134291 (paperback) | ISBN
9781003229193 (ebook)
Subjects: LCSH: Finance--Study and teaching. | Reinforcement learning.
Classification: LCC HG152 .R36 2023 (print) | LCC HG152 (ebook) | DDC
332.076--dc23/eng/20220928
LC record available at https://lccn.loc.gov/2022030498
LC ebook record available at https://lccn.loc.gov/2022030499

ISBN: 978-1-032-12412-4 (hbk)
ISBN: 978-1-032-13429-1 (pbk)
ISBN: 978-1-003-22919-3 (ebk)

DOI: 10.1201/ 9781003229193

Typeset in TeXGyrePagella font
by KnowledgeWorks Global Ltd

Publisher's note: This book has been prepared from camera-ready copy provided by the authors.

To access the code base for this book, please go to: https://github.com/TikhonJelvis/RL-book.

Contents

MODULE III **Reinforcement Learning Algorithms**

MODULE IV **Finishing Touches**

Preface

We (Ashwin and Tikhon) have spent all of our educational and professional lives operating in the intersection of Mathematics and Computer Science—Ashwin for more than three decades and Tikhon for more than a decade. During these periods, we've commonly held two obsessions. The first is to bring together the (strangely) disparate worlds of Mathematics and Software Development. The second is to focus on the pedagogy of various topics across Mathematics and Computer Science. Fundamentally, this book was born out of a deep desire to release our twin obsessions so as to not just educate the next generation of scientists and engineers but to also present some new and creative ways of teaching technically challenging topics in ways that are easy to absorb and retain.

Apart from these common obsessions, each of us has developed some expertise in a few topics that come together in this book. Ashwin's undergraduate and doctoral education was in Computer Science, with specializations in Algorithms, Discrete Mathematics and Abstract Algebra. He then spent more than two decades of his career (across the Finance and Retail industries) in the realm of Computational Mathematics, recently focused on Machine Learning and Optimization. In his role as an Adjunct Professor at Stanford University, Ashwin specializes in Reinforcement Learning and Mathematical Finance. The content of this book is essentially an expansion of the content of the course CME 241 he teaches at Stanford. Tikhon's education is in Computer Science, and he has specialized in Software Design, with an emphasis on treating software design as mathematical specification of "what to do" versus computational mechanics of "how to do". This is a powerful way of developing software, particularly for mathematical applications, significantly improving readability, modularity and correctness. This leads to code that naturally and clearly reflects mathematical structures, thus blurring the artificial lines between Mathematics and Programming. He has also championed the philosophy of leveraging programming as a powerful way to learn mathematical concepts. Ashwin has been greatly influenced by Tikhon on this philosophy and both of us have been quite successful in imparting our students with deep understanding of a variety of mathematical topics by using programming as a powerful pedagogical tool.

In fact, the key distinguishing feature of this book is to promote learning through an appropriate blend of A) intuitive understanding of the concepts, B) mathematical rigor and C) programming of the models and algorithms (with sound software design principles that reflect the mathematical specification). We've found this unique approach to teaching facilitates strong retention of the concepts because students are *active learners* when they code everything they are learning, in a manner that reflects the mathematical concepts. We have strived to create a healthy balance between content accessibility and intuition development on one hand versus technical rigor and completeness on the other hand. Throughout the book, we provide proper mathematical notation, theorems (and sometimes formal proofs) as well as well-designed working code for various models and algorithms. But we have always accompanied this formalism with intuition development using simple examples and appropriate visualization.

We want to highlight that this book emphasizes the *foundational components* of Reinforcement Learning—Markov Decision Processes, Bellman Equations, Fixed-Points,

Dynamic Programming, Function Approximation, Sampling, Experience-Replay, Batch Methods, Value-based versus Policy-based Learning, balancing Exploration and Exploitation, blending Planning and Learning etc. So although we have covered several key algorithms in this book, we do not dwell on specifics of the algorithms—rather, we emphasize the core principles and always allow for various types of flexibility in tweaking those algorithms (our investment in modular software design of the algorithms facilitates this flexibility). Likewise, we have kept the content of the financial applications fairly basic, emphasizing the core ideas, and developing working code for simplified versions of these financial applications. Getting these financial applications to be effective in practice is a much more ambitious endeavor—we don't attempt that in this book, but we highlight what it would take to make it work in practice. The theme of this book is understanding of core concepts rather than addressing all the nuances (and frictions) one typically encounters in the real world. The financial content in this book is a significant fraction of the broader topic of Mathematical Finance, and we hope that this book provides the side benefit of a fairly quick yet robust education in the key topics of Portfolio Management, Derivatives Pricing and Order-Book Trading.

We were introduced to modern Reinforcement Learning by the works of Richard Sutton, including his seminal book with Andrew Barto. There are several other works by other authors, many with more mathematical detail, but we found Sutton's works much easier to learn this topic. Our book tends to follow Sutton's less rigorous but more intuitive approach, but we provide a bit more mathematical formalism/detail, and we use precise working code instead of the typical psuedo-code found in textbooks on these topics. We have also been greatly influenced by David Silver's excellent RL lectures series at University College London that is available on youtube. We have strived to follow the structure of David Silver's lecture series, typically augmenting it with more detail. So it pays to emphasize that the content of this book is not our original work. Rather, our contribution is to present content that is widely and publicly available in a manner that is easier to learn (particularly due to our augmented approach of "learning by coding"). Likewise, the financial content is not our original work—it is based on standard material on Mathematical Finance and based on a few papers that treat the Financial problems as Stochastic Control problems. However, we found the presentation in these papers not easy to understand for the typical student. Moreover, some of these papers did not explicitly model these problems as Markov Decision Processes, and some of them did not consider Reinforcement Learning as an option to solve these problems. So we presented the content in these papers in more detail, specifically with clearer notation and explanations, and with working Python code. It's interesting to note that Ashwin worked on some of these finance problems during his time at Goldman Sachs and Morgan Stanley, but at that time, these problems were not viewed from the lens of Stochastic Control. While designing the content of CME 241 at Stanford, Ashwin realized that several problems from his past finance career can be cast as Markov Decision Processes, which led him to the above-mentioned papers, which in turn led to the content creation for CME 241, that then extended into the financial content in this book. There are several Appendices in this book to succinctly provide appropriate pre-requisite mathematical/financial content. We have strived to provide references throughout the chapters and appendices to enable curious students to learn each topic in more depth. However, we are not exhaustive in our list of references because typically each of our references tends to be fairly exhaustive in the papers/books it in turn references.

We have many people to thank—those who provided the support and encouragement for us to write this book. Ashwin would like to thank his managers Paritosh Desai and Mike McNamara at Target Corporation, as well as all of the faculty and staff he works with at Stanford University, for providing a wonderful environment and excellent support for CME 241, notably George Papanicolau, Kay Giesecke, Peter Glynn, Gianluca Iaccarino,

Indira Choudhury and Jess Galvez. Ashwin would also like to thank all his students who implicitly proofread the contents, and his course assistants Sven Lerner and Jeff Gu. Tikhon would like to thank Paritosh Desai for help exploring the software design and concepts used in the book as well as his immediate team at Target for teaching him about how Markov Decision Processes apply in the real world.

Author Biographies

Ashwin Rao is the Chief Science Officer of Wayfair, an e-commerce company where he and his team develop mathematical models and algorithms for supply-chain and logistics, merchandising, marketing, search, personalization, pricing and customer service. Ashwin is also an Adjunct Professor at Stanford University, focusing his research and teaching in the area of Stochastic Control, particularly Reinforcement Learning algorithms with applications in Finance and Retail. Previously, Ashwin was a Managing Director at Morgan Stanley and a Trading Strategist at Goldman Sachs. Ashwin holds a Bachelor's degree in Computer Science and Engineering from IIT-Bombay and a Ph.D. in Computer Science from the University of Southern California, where he specialized in Algorithms Theory and Abstract Algebra.

Tikhon Jelvis is a programmer who specializes in bringing ideas from programming languages and functional programming to machine learning and data science. He has developed inventory optimization, simulation and demand forecasting systems as a Principal Scientist at Target and is a speaker and open-source contributor in the Haskell community, where he serves on the board of directors for Haskell.org.

Summary of Notation

$\mathbb{Z}$	Set of integers
$\mathbb{Z}^+$	Set of positive integers, i.e., $\{1, 2, 3, \ldots\}$
$\mathbb{Z}_{\geq 0}$	Set of non-negative integers, i.e., $\{0, 1, 2, 3, \ldots\}$
$\mathbb{R}$	Set of real numbers
$\mathbb{R}^+$	Set of positive real numbers
$\mathbb{R}_{\geq 0}$	Set of non-negative real numbers
$\log(x)$	*Natural Logarithm* (to the base e) of x
$|x|$	*Absolute Value* of x
$sign(x)$	$+1$ if $x > 0$, -1 if $x < 0$, 0 if $x = 0$
$[a, b]$	Set of real numbers that are $\geq a$ and $\leq b$. The notation $x \in [a, b]$ is shorthand for $x \in \mathbb{R}$ and $a \leq x \leq b$
$[a, b)$	Set of real numbers that are $\geq a$ and $< b$. The notation $x \in [a, b)$ is shorthand for $x \in \mathbb{R}$ and $a \leq x < b$
$(a, b]$	Set of real numbers that are $> a$ and $leqb$. The notation $x \in (a, b]$ is shorthand for $x \in \mathbb{R}$ and $a < x \leq b$
$\emptyset$	The Empty Set (Null Set)
$\sum_{i=1}^{n} a_i$	Sum of terms $a_1, a_2, \ldots, a_n$
$\prod_{i=1}^{n} a_i$	Product of terms $a_1, a_2, \ldots, a_n$
$\approx$	approximately equal to
$x \in \mathcal{X}$	x is an element of the set $\mathcal{X}$
$x \notin \mathcal{X}$	x is not an element of the set $\mathcal{X}$
$\mathcal{X} \cup \mathcal{Y}$	*Union* of the sets $\mathcal{X}$ and $\mathcal{Y}$
$\mathcal{X} \cap \mathcal{Y}$	*Intersection* of the sets $\mathcal{X}$ and $\mathcal{Y}$
$\mathcal{X} - \mathcal{Y}$	*Set Difference* of the sets $\mathcal{X}$ and $\mathcal{Y}$, i.e., the set of elements within the set $\mathcal{X}$ that are not elements of the set $\mathcal{Y}$
$\mathcal{X} \times \mathcal{Y}$	*Cartesian Product* of the sets $\mathcal{X}$ and $\mathcal{Y}$
$\mathcal{X}^k$	For a set $\mathcal{X}$ and an integer $k \geq 1$, this refers to the *Cartesian Product* $\mathcal{X} \times \mathcal{X} \times \ldots \times \mathcal{X}$ with k occurrences of $\mathcal{X}$ in the Cartesian Product (note: $\mathcal{X}^1 = \mathcal{X}$)
$f : X \to Y$	*Function f* with *Domain X* and *Co-domain Y*
f^k	For a function f and an integer $k \geq 0$, this refers to the *function composition* of f with itself, repeated k times. So, $f^k(x)$ is the value $f(f(\ldots f(x) \ldots))$ with k occurrences of f in this function-composition expression (note: $f^1 = f$ and f^0 is the identity function)
f^{-1}	*Inverse function* of a bijective function $f : \mathcal{X} \to \mathcal{Y}$, i.e., for all $x \in \mathcal{X}, f^{-1}(f(x)) = x$ and for all $y \in \mathcal{Y}, f(f^{-1}(y)) = y$
$f'(x_0)$	*Derivative* of the function $f : \mathcal{X} \to \mathbb{R}$ with respect to it's domain variable $x \in \mathcal{X}$, evaluated at $x = x_0$

$f''(x_0)$	*Second Derivative* of the function $f : \mathcal{X} \to \mathbb{R}$ with respect to it's domain variable $x \in \mathcal{X}$, evaluated at $x = x_0$
$\mathbb{P}[X]$	*Probability Density Function* (PDF) of random variable X
$\mathbb{P}[X = x]$	Probability that random variable X takes the value x
$\mathbb{P}[X\|Y]$	*Probability Density Function* (PDF) of random variable X, conditional on the value of random variable Y (i.e., PDF of X expressed as a function of the values of Y)
$\mathbb{P}[X = x\|Y = y]$	Probability that random variable X takes the value x, conditional on random variable Y taking the value y
$\mathbb{E}[X]$	*Expected Value* of random variable X
$\mathbb{E}[X\|Y]$	*Expected Value* of random variable X, conditional on the value of random variable Y (i.e., Expected Value of X expressed as a function of the values of Y)
$\mathbb{E}[X\|Y = y]$	*Expected Value* of random variable X, conditional on random variable Y taking the value y
$x \sim \mathcal{N}(\mu, \sigma^2)$	Random variable x follows a *Normal Distribution* with mean μ and variance σ^2
$x \sim Poisson(\lambda)$	Random variable x follows a *Poisson Distribution* with mean λ
$f(x; \boldsymbol{w})$	Here f refers to a parameterized function with domain $\mathcal{X}$ ($x \in \mathcal{X}$), $\boldsymbol{w}$ refers to the parameters controlling the definition of the function f
$\boldsymbol{v}^T$	*Row-vector* with components equal to the components of the *Column-vector* $\boldsymbol{v}$, i.e., *Transpose* of the *Column-vector* $\boldsymbol{v}$ (by default, we assume vectors are expressed as *Column-vectors*)
$\boldsymbol{A}^T$	*Transpose* of the *matrix* $\boldsymbol{A}$
$\|\boldsymbol{v}\|$	L^2 norm of vector $\boldsymbol{v} \in \mathbb{R}^m$, i.e., if $\boldsymbol{v} = (v_1, v_2, \ldots, v_m)$, then $\|\boldsymbol{v}\| = \sqrt{v_1^2 + v_2^2 + \ldots + v_m^2}$
$\boldsymbol{A}^{-1}$	*Matrix-Inverse* of the *square matrix* $\boldsymbol{A}$
$\boldsymbol{A} \cdot \boldsymbol{B}$	*Matrix-Multiplication* of matrices $\boldsymbol{A}$ and $\boldsymbol{B}$ (note: vector notation $\boldsymbol{v}$ typically refers to a column-vector, i.e., a matrix with 1 column, and so $\boldsymbol{v}^T \cdot \boldsymbol{w}$ is simply the *inner-product* of same-dimensional vectors $\boldsymbol{v}$ and $\boldsymbol{w}$)
$\boldsymbol{I}_m$	$m \times m$ *Identity Matrix*
$\boldsymbol{Diagonal}(\boldsymbol{v})$	$m \times m$ Diagonal Matrix whose elements are the same (also in same order) as the elements of the m-dimensional Vector $\boldsymbol{v}$
$dim(\boldsymbol{v})$	Dimension of a vector $\boldsymbol{v}$
$\mathbb{I}_c$	$\mathbb{I}$ represents the *Indicator function* and $\mathbb{I}_c = 1$ if condition c is True, $= 0$ if c is False
$\arg\max_{x \in \mathcal{X}} f(x)$	This refers to the value of $x \in \mathcal{X}$ that maximizes $f(x)$, i.e., $\max_{x \in \mathcal{X}} f(x) = f(\arg\max_{x \in \mathcal{X}} f(x))$
$\nabla_{\boldsymbol{w}} f(\boldsymbol{w})$	Gradient of the function f with respect to $\boldsymbol{w}$ (note: $\boldsymbol{w}$ could be an arbitrary data structure and this gradient is of the same data type as the data type of $\boldsymbol{w}$)
$x \leftarrow y$	Variable x is assigned (or updated to) the value of y

CHAPTER 1

Overview

1.1 LEARNING REINFORCEMENT LEARNING

Reinforcement Learning (RL) is emerging as a practical, powerful technique for solving a variety of complex business problems across industries that involve Sequential Optimal Decisioning under Uncertainty. Although RL is classified as a branch of Machine Learning (ML), it tends to be viewed and treated quite differently from other branches of ML (Supervised and Unsupervised Learning). Indeed, **RL seems to hold the key to unlocking the promise of AI**—machines that adapt their decisions to vagaries in observed information, while continuously steering towards the optimal outcome. Its penetration in high-profile problems like self-driving cars, robotics and strategy games points to a future where RL algorithms will have decisioning abilities far superior to humans.

But when it comes getting educated in RL, there seems to be a reluctance to jump right in because RL seems to have acquired a reputation of being mysterious and exotic. We often hear even technical people claim that RL involves "advanced math" and "complicated engineering", and so there seems to be a psychological barrier to entry. While real-world RL algorithms and implementations do get fairly elaborate and complicated in overcoming the proverbial last-mile of business problems, the foundations of RL can actually be learned without heavy technical machinery. **A key goal of this book is to demystify RL by finding a balance between A) providing depth of understanding and B) keeping technical content basic.** So now we list the key features of this book that enable this balance:

- Focus on the foundational theory underpinning RL. Our treatment of this theory is based on undergraduate-level Probability, Optimization, Statistics and Linear Algebra. **We emphasize rigorous but simple mathematical notations and formulations in developing the theory**, and encourage you to write out the equations rather than just reading from the book. Occasionally, we invoke some advanced mathematics (e.g., Stochastic Calculus), but the majority of the book is based on easily understandable mathematics. In particular, two basic theory concepts—Bellman Optimality Equation and Generalized Policy Iteration—are emphasized throughout the book as they form the basis of pretty much everything we do in RL, even in the most advanced algorithms.
- Parallel to the mathematical rigor, we bring the concepts to life with simple examples and informal descriptions to help you develop an intuitive understanding of the mathematical concepts. **We drive towards creating appropriate mental models to visualize the concepts**. Often, this involves turning mathematical abstractions into physical examples (emphasizing visual intuition). So we go back and forth between rigor and intuition, between abstractions and visuals, so as to blend them nicely and get the best of both worlds.

DOI: 10.1201/9781003229193-1

- Each time you learn a new mathematical concept or algorithm, we encourage you to write small pieces of code (in Python) that implements the concept/algorithm. As an example, if you just learned a surprising theorem, we'd ask you to write a simulator to simply verify the statement of the theorem. We emphasize this approach not just to bolster the theoretical and intuitive understanding with a hands-on experience, but also because there is a strong emotional effect of seeing expected results emanating from one's code, which in turn promotes long-term retention of the concepts. Most importantly, we avoid messy and complicated ML/RL/BigData tools/packages and stick to bare-bones Python/numpy as these unnecessary tools/packages are huge blockages to core understanding. We believe **coding from scratch and designing the code to reflect the mathematical structure/concepts is the correct approach to truly understand the concepts/algorithms**.

- Lastly, it is important to work with examples that are A) simplified versions of real-world problems in a business domain rich with applications, B) adequately comprehensible without prior business-domain knowledge, C) intellectually interesting and D) sufficiently marketable to employers. We've chosen Financial Trading applications. For each financial problem, we first cover the traditional approaches (including solutions from landmark papers) and then cast the problem in ways that can be solved with RL. **We have made considerable effort to make this book self-contained in terms of the financial knowledge required to navigate these problems**.

1.2 WHAT YOU WILL LEARN FROM THIS BOOK

Here is what you will specifically learn and gain from the book:

- You will learn about the simple but powerful theory of Markov Decision Processes (MDPs)—a framework for Sequential Optimal Decisioning under Uncertainty. You will firmly understand the power of Bellman Equations, which is at the heart of all Dynamic Programming as well as all RL algorithms.

- You will master Dynamic Programming (DP) Algorithms, which are a class of (in the language of AI) Planning Algorithms. You will learn about Policy Iteration, Value Iteration, Backward Induction, Approximate Dynamic Programming and the all-important concept of Generalized Policy Iteration, which lies at the heart of all DP as well as all RL algorithms.

- You will gain a solid understanding of a variety of Reinforcement Learning (RL) Algorithms, starting with the basic algorithms like SARSA and Q-Learning and moving on to several important algorithms that work well in practice, including Gradient Temporal Difference, Deep Q-Networks, Least-Squares Policy Iteration, Policy Gradient, Monte-Carlo Tree Search. You will learn about how to gain advantages in these algorithms with bootstrapping, off-policy learning and deep-neural-networks-based function approximation. You will learn how to balance exploration and exploitation with Multi-Armed Bandits techniques like Upper Confidence Bounds, Thompson Sampling, Gradient Bandits and Information State-Space algorithms. You will also learn how to blend Planning and Learning methodologies, which is very important in practice.

- You will exercise with plenty of "from-scratch" Python implementations of models and algorithms. Throughout the book, we emphasize healthy Python programming practices including interface design, type annotations, functional programming and inheritance-based polymorphism (always ensuring that the programming principles reflect the mathematical principles). The larger take-away from this book will be a rare (and high-in-demand) ability to blend Applied Mathematics concepts with Software Design paradigms.

- You will go deep into important Financial Trading problems, including:
  - (Dynamic) Asset-Allocation to maximize Utility of Consumption
  - Pricing and Hedging of Derivatives in an Incomplete Market
  - Optimal Exercise/Stopping of Path-Dependent American Options
  - Optimal Trade Order Execution (managing Price Impact)
  - Optimal Market-Making (Bid/Ask managing Inventory Risk)

- We treat each of the above problems as MDPs (i.e., Optimal Decisioning formulations), first going over classical/analytical solutions to these problems, then introducing real-world frictions/considerations, and tackling with DP and/or RL.

- As a bonus, we throw in a few applications beyond Finance, including a couple from Supply-Chain and Clearance Pricing in a Retail business.

- We implement a wide range of Algorithms and develop various models in a git code base that we refer to and explain in detail throughout the book. This code base not only provides detailed clarity on the algorithms/models but also serves to educate on healthy programming patterns suitable not just for RL but more generally for any Applied Mathematics work.

- In summary, this book blends Theory/Mathematics, Programming/Algorithms and Real-World Financial Nuances while always keeping things simple and intuitive.

1.3   EXPECTED BACKGROUND TO READ THIS BOOK

There is no shortcut to learning Reinforcement Learning or learning the Financial Applications content. You will need to allocate 100-200 hours of effort to learn this material (assuming you have no prior background in these topics). This extent of effort incorporates the time required to write out the equations/theory as well as the coding of the models/algorithms, while you are making your way through this book. Note that although we have kept the Mathematics, Programming and Financial content fairly basic, this topic is only for technically-inclined readers. Below we outline the technical preparation that is required to follow the material covered in this book.

- Experience with (but not necessarily expertise in) Python is expected, and a good deal of comfort with numpy is required. Note that much of the Python programming in this book is for mathematical modeling and for numerical algorithms, so one doesn't need to know Python from the perspective of building engineering systems or user-interfaces. So you don't need to be a professional software developer/engineer but you need to have a healthy interest in learning Python best practices associated with mathematical modeling, algorithms development and numerical programming (we teach these best practices in this book). We don't use any of the popular (but messy and complicated) Big Data/Machine Learning libraries such as Pandas, PySpark, scikit, Tensorflow, PyTorch, OpenCV, NLTK etc. (all you need to know is numpy).
- Familiarity with git and use of an Integrated Development Environment (IDE), e.g., Pycharm or Emacs (with Python plugins), is recommended, but not required.
- Familiarity with LaTeX for writing equations is recommended but not required (other typesetting tools, or even hand-written math is fine, but LaTeX is a skill that is very valuable if you'd like a future in the general domain of Applied Mathematics).

- You need to be strong in undergraduate-level Probability as it is the most important foundation underpinning RL.
- You will also need to have some preparation in undergraduate-level Numerical Optimization, Statistics, Linear Algebra.
- No background in Finance is required, but a strong appetite for Mathematical Finance is required.

1.4 DECLUTTERING THE JARGON LINKED TO REINFORCEMENT LEARNING

Machine Learning has exploded in the past decade or so, and Reinforcement Learning (treated as a branch of Machine Learning and hence, a branch of A.I.) has surfaced to the mainstream in both academia and in the industry. It is important to understand what Reinforcement Learning aims to solve, rather than the more opaque view of RL as a technique to learn from data. RL aims to solve problems that involve making *Sequential Optimal Decisions under Uncertainty*. Let us break down this jargon so as to develop an intuitive (and high-level) understanding of the features pertaining to the problems RL solves.

Firstly, let us understand the term *Uncertainty*. This means the problems under consideration involve random variables that evolve over time. The technical term for this is *stochastic processes*. We will cover this in detail later in this book, but for now, it's important to recognize that evolution of random variables over time is very common in nature (e.g., weather) and in business (e.g., customer demand or stock prices), but modeling and navigating such random evolutions can be enormously challenging.

The next term is *Optimal Decisions*, which refers to the technical term *Optimization*. This means there is a well-defined quantity to be maximized (the "goal"). The quantity to be maximized might be financial (like investment value or business profitability), or it could be a safety or speed metric (such as health of customers or time to travel), or something more complicated like a blend of multiple objectives rolled into a single objective.

The next term is *Sequential*, which refers to the fact that as we move forward in time, the relevant random variables' values evolve, and the optimal decisions have to be adjusted to the "changing circumstances". Due to this non-static nature of the optimal decisions, the term *Dynamic Decisions* is often used in the literature covering this subject.

Putting together the three notions of (Uncertainty/Stochastic, Optimization, Sequential/Dynamic Decisions), these problems (that RL tackles) have the common feature that one needs to *overpower the uncertainty by persistent steering towards the goal*. This brings us to the term *Control* (in references to *persistent steering*). These problems are often (aptly) characterized by the technical term *Stochastic Control*. So you see that there is indeed a lot of jargon here. All of this jargon will become amply clear after the first few chapters in this book where we develop mathematical formalism to understand these concepts precisely (and also write plenty of code to internalized these concepts). For now, we just wanted to familiarize you with the range of jargon linked to Reinforcement Learning.

This jargon overload is due to the confluence of terms from Control Theory (emerging from Engineering disciplines), from Operations Research, and from Artificial Intelligence (emerging from Computer Science). For simplicity, we prefer to refer to the class of problems RL aims to solve as *stochastic control* problems. Reinforcement Learning is a class of algorithms that are used to solve Stochastic Control problems. We should point out here that there are other disciplines (beyond Control Theory, Operations Research and Artificial Intelligence) with a rich history of developing theory and techniques within the general space of Stochastic Control. Figure 1.1 (a popular image on the internet, first seen by us in

David Silver's RL course slides) illustrates the many faces of Stochastic Control, which has often been referred to as "The many faces of Reinforcement Learning".

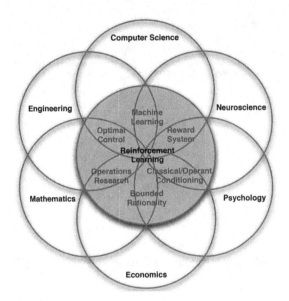

Figure 1.1 **Many Faces of Reinforcement Learning (Image Credit: David Silver's RL Course)**

It is also important to recognize that Reinforcement Learning is considered to be a branch of Machine Learning. While there is no crisp definition for *Machine Learning* (ML), ML generally refers to the broad set of techniques to infer mathematical models/functions by acquiring ("learning") knowledge of patterns and properties in the presented data. In this regard, Reinforcement Learning does fit this definition. However, unlike the other branches of ML (Supervised Learning and Unsupervised Learning), Reinforcement Learning is a lot more ambitious—it not only learns the patterns and properties of the presented data, it also learns about the appropriate behaviors to be exercised (appropriate decisions to be made) so as to drive towards the optimization objective. It is sometimes said that Supervised Learning and Unsupervised learning are about "minimization" (i.e., they minimize the fitting error of a model to the presented data), while Reinforcement Learning is about "maximization" (i.e., RL identifies the suitable decisions to be made to maximize a well-defined objective). Figure 1.2 depicts the in-vogue classification of Machine Learning.

More importantly, the class of problems RL aims to solve can be described with a simple yet powerful mathematical framework known as *Markov Decision Processes* (abbreviated as MDPs). We have an entire chapter dedicated to deep coverage of MDPs, but we provide a quick high-level introduction to MDPs in the next section.

1.5 INTRODUCTION TO THE MARKOV DECISION PROCESS (MDP) FRAMEWORK

The framework of a Markov Decision Process is depicted in Figure 1.3. As the figure indicates, the *Agent* and the *Environment* interact in a time-sequenced loop. The term *Agent* refers to an algorithm (AI algorithm) and the term *Environment* refers to an abstract entity that serves up uncertain outcomes to the Agent. It is important to note that the

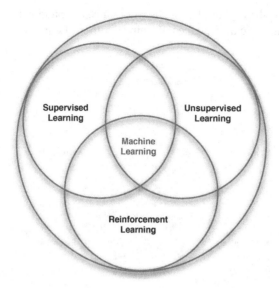

Figure 1.2 **Branches of Machine Learning (Image Credit: David Silver's RL Course)**

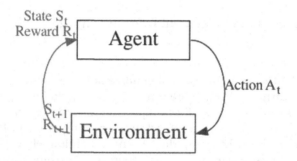

Figure 1.3 **The MDP Framework**

Environment is indeed abstract in this framework and can be used to model all kinds of real-world situations such as the financial market serving up random stock prices or customers of a company serving up random demand or a chess opponent serving up random moves (from the perspective of the Agent), or really anything at all you can imagine that serves up something random at each time step (it is up to us to model an Environment appropriately to fit the MDP framework).

As the figure indicates, at each time step t, the Agent observes an abstract piece of information (which we call *State*) and a numerical (real number) quantity that we call *Reward*. Note that the concept of *State* is indeed completely abstract in this framework, and we can model *State* to be any data type, as complex or elaborate as we'd like. This flexibility in modeling *State* permits us to model all kinds of real-world situations as an MDP. Upon observing a *State* and a *Reward* at time step t, the *Agent* responds by taking an *Action*. Again, the concept of *Action* is completely abstract and is meant to represent an activity performed by an AI algorithm. It could be a purchase or sale of a stock responding to market stock price movements, or it could be movement of inventory from a warehouse to a store in response to large sales at the store, or it could be a chess move in response to the opponent's chess move (opponent is *Environment*), or really anything at all you can imagine that responds to observations (*State* and *Reward*) served by the *Environment*.

Upon receiving an *Action* from the *Agent* at time step t, the *Environment* responds (with time ticking over to $t + 1$) by serving up the next time step's random *State* and random *Reward*. A technical detail (that we shall explain in detail later) is that the *State* is assumed to have the *Markov Property*, which means:

- The next *State*/*Reward* depends only on Current *State* (for a given *Action*).
- The current *State* encapsulates all relevant information from the history of the interaction between the *Agent* and the *Environment*.
- The current *State* is a sufficient statistic of the future (for a given *Action*).

The goal of the *Agent* at any point in time is to maximize the *Expected Sum* of all future *Reward*s by controlling (at each time step) the *Action* as a function of the observed *State* (at that time step). This function from a *State* to *Action* at any time step is known as the *Policy* function. So we say that the agent's job is exercise control by determining the *Optimal Policy* function. Hence, this is a dynamic (i.e., time-sequenced) control system under uncertainty. If the above description was too terse, don't worry—we will explain all of this in great detail in the coming chapters. For now, we just wanted to provide a quick flavor of what the MDP framework looks like. Now we sketch the above description with some (terse) mathematical notation to provide a bit more of the overview of the MDP framework. The following notation is for discrete time steps (continuous time steps notation is analogous, but technically more complicated to describe here):

We denote time steps as $t = 1, 2, 3, \ldots$. Markov State at time t is denoted as $S_t \in \mathcal{S}$, where $\mathcal{S}$ is referred to as the *State Space* (a countable set). Action at time t is denoted as $A_t \in \mathcal{A}$, where $\mathcal{A}$ is referred to as the *Action Space* (a countable set). Reward at time t is denoted as $R_t \in \mathcal{D}$, where $\mathcal{D}$ is a countable subset of $\mathbb{R}$ (representing the numerical feedback served by the Environment, along with the State, at each time step t).

We represent the transition probabilities from one time step to the next with the following notation:

$$p(r, s'|s, a) = \mathbb{P}[(R_{t+1} = r, S_{t+1} = s')|S_t = s, A_t = a]$$

$\gamma \in [0, 1]$ is known as the discount factor used to discount Rewards when accumulating Rewards, as follows:

$$\text{Return } G_t = R_{t+1} + \gamma \cdot R_{t+2} + \gamma^2 \cdot R_{t+3} + \cdots$$

The discount factor γ allows us to model situations where a future reward is less desirable than a current reward of the same quantity.

The goal is to find a *Policy* $\pi : \mathcal{S} \to \mathcal{A}$ that maximizes $\mathbb{E}[G_t|S_t = s]$ for all $s \in \mathcal{S}$. In subsequent chapters, we clarify that the MDP framework actually considers more general policies than described here—policies that are stochastic, i.e., functions that take as input a state and output a probability distribution of actions (rather than a single action). However, for ease of understanding of the core concepts, in this chapter, we stick to deterministic policies $\pi : \mathcal{S} \to \mathcal{A}$.

The intuition here is that the two entities *Agent* and *Environment* interact in a time-sequenced loop wherein the *Environment* serves up next states and rewards based on the transition probability function p and the *Agent* exerts control over the vagaries of p by exercising the policy π in a way that optimizes the Expected "accumulated rewards" (i.e., Expected Return) from any state.

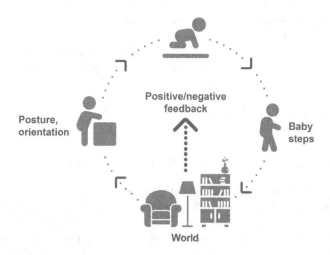

Figure 1.4 Baby Learning MDP

It's worth pointing out that the MDP framework is inspired by how babies (*Agent*) learn to perform tasks (i.e., take *Actions*) in response to the random activities and events (*States* and *Rewards*) they observe as being served up from the world (*Environment*) around them. Figure 1.4 illustrates this—at the bottom of the figure (labeled "World", i.e., *Environment*), we have a room in a house with a vase atop a bookcase. At the top of the figure is a baby (learning *Agent*) on the other side of the room who wants to make her way to the bookcase, reach for the vase, and topple it—doing this efficiently (i.e., in quick time and quietly) would mean a large *Reward* for the baby. At each time step, the baby finds herself in a certain posture (e.g., lying on the floor, or sitting up, or trying to walk etc.) and observes various visuals around the room—her posture and her visuals would constitute the *State* for the baby at each time step. The baby's *Actions* are various options of physical movements to try to get to the other side of the room (assume the baby is still learning how to walk). The baby tries one physical movement, but is unable to move forward with that movement. That would mean a negative *Reward*—the baby quickly learns that this movement is probably not a good idea. Then she tries a different movement, perhaps trying to stand on her feet and start walking. She makes a couple of good steps forward (positive *Rewards*), but then falls down and hurts herself (that would be a big negative *Reward*). So by trial and error, the baby learns about the consequences of different movements (different actions). Eventually, the baby learns that by holding on to the couch, she can walk across, and then when she reaches the bookcase, she learns (again by trial and error) a technique to climb the bookcase that is quick yet quiet (so she doesn't raise her mom's attention). This means the baby learns of the optimal policy (best actions for each of the states she finds herself in) after essentially what is a "trial and error" method of learning what works and what doesn't. This example is essentially generalized in the MDP framework, and the baby's "trial and error" way of learning is essentially a special case of the general technique of Reinforcement Learning.

1.6 REAL-WORLD PROBLEMS THAT FIT THE MDP FRAMEWORK

As you might imagine by now, all kinds of problems in nature and in business (and indeed, in our personal lives) can be modeled as Markov Decision Processes. Here is a sample of such problems:

- Self-driving vehicle (Actions constitute speed/steering to optimize safety/time).
- Game of Chess (Actions constitute moves of the pieces to optimize chances of winning the game).
- Complex Logistical Operations, such as those in a Warehouse (Actions constitute inventory movements to optimize throughput/time).
- Making a humanoid robot walk/run on a difficult terrain (Actions are walking movements to optimize time to destination).
- Manage an investment portfolio (Actions are trades to optimize long-term investment gains).
- Optimal decisions during a football game (Actions are strategic game calls to optimize chances of winning the game).
- Strategy to win an election (Actions constitute political decisions to optimize chances of winning the election).

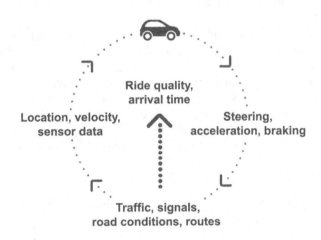

Figure 1.5 **Self-driving Car MDP**

Figure 1.5 illustrates the MDP for a self-driving car. At the top of the figure is the *Agent* (the car's driving algorithm) and at the bottom of the figure is the *Environment* (constituting everything the car faces when driving—other vehicles, traffic signals, road conditions, weather etc.). The *State* consists of the car's location, velocity, and all of the information picked up by the car's sensors/cameras. The *Action* consists of the steering, acceleration and brake. The *Reward* would be a combination of metrics on ride comfort and safety, as well as the negative of each time step (because maximizing the accumulated Reward would then amount to minimizing time taken to reach the destination).

1.7 THE INHERENT DIFFICULTY IN SOLVING MDPS

"Solving" an MDP refers to identifying the optimal policy with an algorithm. This section paints an intuitive picture of why solving a general MDP is fundamentally a hard problem. Often, the challenge is simply that the *State Space* is very large (involving many variables) or complex (elaborate data structure), and hence, is computationally intractable. Likewise, sometimes the *Action Space* can be quite large or complex.

But the main reason for why solving an MDP is inherently difficult is the fact that there is no direct feedback on what the "correct" Action is for a given State. What we mean by that is that unlike a supervised learning framework, the MDP framework doesn't give us anything other than a *Reward* feedback to indicate if an Action is the *right one* or not. A large Reward might encourage the Agent, but it's not clear if one just got lucky with the large Reward or if there could be an even larger Reward if the Agent tries the Action again. The linkage between Actions and Rewards is further complicated by the fact that there is time-sequenced complexity in an MDP, meaning an Action can influence future States, which in turn influences future Actions. Consequently, we sometimes find that Actions can have delayed consequences, i.e., the Rewards for a good Action might come after many time steps (e.g., in a game of Chess, a brilliant move leads to a win after several further moves).

The other problem one encounters in real-world situations is that the Agent often doesn't know the *model* of the environment. By model, we are referring to the probabilities of state-transitions and rewards (the function p we defined above). This means the Agent has to simultaneously learn the model (from the real-world data stream) and solve for the optimal policy.

Lastly, when there are many actions, the Agent needs to try them all to check if there are some hidden gems (great actions that haven't been tried yet), which in turn means one could end up wasting effort on "duds" (bad actions). So the agent has to find the balance between *exploitation* (retrying actions that have yielded good rewards so far) and *exploration* (trying actions that have either not been tried enough or not been tried at all).

All of this seems to indicate that we don't have much hope in solving MDPs in a reliable and efficient manner. But it turns out that with some clever mathematics, we can indeed make some good inroads. We outline the core idea of this "clever mathematics" in the next section.

1.8 VALUE FUNCTION, BELLMAN EQUATIONS, DYNAMIC PROGRAMMING AND RL

Perhaps the most important concept we want to highlight in this entire book is the idea of a *Value Function* and how we can compute it in an efficient manner with either *Planning* or *Learning* algorithms. The Value Function $V^\pi : \mathcal{S} \to \mathbb{R}$ for a given (fixed) policy π is defined as:

$$V^\pi(s) = \mathbb{E}_{\pi,p}[G_t | S_t = s] \text{ for all } s \in \mathcal{S}$$

The intuitive way to understand Value Function is that it tells us how much "accumulated future reward" (i.e., *Return*) we expect to obtain from a given state. The randomness under the expectation comes from the uncertain future states and rewards the Agent is going to see (based on the function p), upon taking future actions prescribed by the policy π. The key in evaluating the Value Function for a given policy is that it can be expressed recursively, in terms of the Value Function for the next time step's states. In other words,

$$V^\pi(s) = \sum_{r,s'} p(r, s' | s, \pi(s)) \cdot (r + \gamma \cdot V^\pi(s')) \text{ for all } s \in \mathcal{S}$$

This equation says that when the Agent follows a deterministic policy π, in a given state s, it takes an action $a = \pi(s)$, then sees a random next state s' and a random reward r, so $V^\pi(s)$ can be broken into the expectation of r (immediate next step's expected reward) and the remainder of the future expected accumulated rewards (which can be written in terms of the expectation of $V^\pi(s')$). We won't get into the details of how to solve this recursive formulation in this chapter (will cover this in great detail in future chapters), but it's important for you to recognize for now that this recursive formulation is the key to evaluating the Value Function for a given policy.

However, evaluating the Value Function for a given policy is not the end goal—it is simply a means to the end goal of evaluating the *Optimal Value Function* (from which we obtain the *Optimal Policy*). The Optimal Value Function $V^* : \mathcal{S} \to \mathbb{R}$ is defined as:

$$V^*(s) = \max_\pi V^\pi(s) \text{ for all } s \in \mathcal{S}$$

The good news is that even the Optimal Value Function can be expressed recursively, as follows:

$$V^*(s) = \max_a \sum_{r,s'} p(r, s'|s, a) \cdot (r + \gamma \cdot V^*(s')) \text{ for all } s \in \mathcal{S}$$

Furthermore, we can prove that there exists an Optimal Policy π^* achieving $V^*(s)$ for all $s \in \mathcal{S}$ (the proof is constructive, which gives a simple method to obtain the function π^* from the function V^*). Specifically, this means that the Value Function obtained by following the optimal policy π^* is the same as the Optimal Value Function V^*, i.e.,

$$V^{\pi^*}(s) = V^*(s) \text{ for all } s \in \mathcal{S}$$

There is a bit of terminology here to get familiar with. The problem of calculating $V^\pi(s)$ (Value Function for a give policy) is known as the *Prediction* problem (since this amounts to statistical estimation of the expected returns from any given state when following a policy π). The problem of calculating the Optimal Value Function V^* (and hence, Optimal Policy π^*), is known as the *Control* problem (since this requires steering of the policy such that we obtain the maximum expected return from any state). Solving the Prediction problem is typically a stepping stone toward solving the (harder) problem of Control. These recursive equations for V^π and V^* are known as the (famous) Bellman Equations (which you will hear a lot about in future chapters). In a continuous-time formulation, the Bellman Equation is referred to as the famous *Hamilton-Jacobi-Bellman (HJB)* equation.

The algorithms to solve the prediction and control problems based on Bellman equations are broadly classified as:

- Dynamic Programming, a class of (in the language of A.I.) *Planning* algorithms.
- Reinforcement Learning, a class of (in the language of A.I.) *Learning* algorithms.

Now let's talk a bit about the difference between Dynamic Programming and Reinforcement Learning algorithms. Dynamic Programming algorithms (which we cover a lot of in this book) assume that the agent knows of the transition probabilities p and the algorithm takes advantage of the knowledge of those probabilities (leveraging the Bellman Equation to efficiently calculate the Value Function). Dynamic Programming algorithms are considered to be *Planning* and not *Learning* (in the language of A.I.) because the algorithm doesn't need to interact with the Environment and doesn't need to learn from the (states, rewards) data stream coming from the Environment. Rather, armed with the transition probabilities, the algorithm can reason about future probabilistic outcomes and perform the requisite optimization calculation to infer the Optimal Policy. So it *plans* its path to success, rather than *learning* about how to succeed.

However, in typical real-world situations, one doesn't really know the transition probabilities p. This is the realm of Reinforcement Learning (RL). RL algorithms interact with the Environment, learn with each new (state, reward) pair received from the Environment, and incrementally figure out the Optimal Value Function (with the "trial and error" approach that we outlined earlier). However, note that the Environment interaction could be *real* interaction or *simulated* interaction. In the latter case, we do have a model of the transitions but the structure of the model is so complex that we only have access to samples of the next state and reward (rather than an explicit representation of the probabilities). This is known as a *Sampling Model* of the Environment. With access to such a sampling model of the environment (e.g., a robot learning on a simulated terrain), we can employ the same RL algorithm that we would have used when interacting with a real environment (e.g., a robot learning on an actual terrain). In fact, most RL algorithms in practice learn from sampling models of the environment. As we explained earlier, RL is essentially a "trial and error" learning approach and hence, is quite laborious and fundamentally inefficient. The recent progress in RL is coming from more efficient ways of learning the Optimal Value Function, and better ways of approximating the Optimal Value Function. One of the key challenges for RL in the future is to identify better ways of finding the balance between "exploration" and "exploitation" of actions. In any case, one of the key reasons RL has started doing well lately is due to the assistance it has obtained from Deep Learning (typically Deep Neural Networks are used to approximate the Value Function and/or to approximate the Policy Function). RL with such deep learning approximations is known by the catchy modern term *Deep RL*.

We believe the current promise of A.I. is dependent on the success of Deep RL. The next decade will be exciting as RL research will likely yield improved algorithms, and its pairing with Deep Learning will hopefully enable us to solve fairly complex real-world stochastic control problems.

1.9 OUTLINE OF CHAPTERS

The chapters in this book are organized into 4 modules as follows:

- Module I: Processes and Planning Algorithms (Chapters 3 – 6)
- Module II: Modeling Financial Applications (Chapters 7 – 10)
- Module III: Reinforcement Learning Algorithms (Chapters 11 – 14)
- Module IV: Finishing Touches (Chapters 15 – 17)

Before covering the contents of the chapters in these 4 modules, the book starts with 2 preliminary chapters. The first of these preliminary chapters is *this chapter* (the one you are reading) which serves as an *Overview*, covering the pedagogical aspects of learning RL (and more generally Applied Mathematics), outline of the learnings to be acquired from this book, background required to read this book, a high-level overview of Stochastic Control, MDP, Value Function, Bellman Equation, DP and RL, and finally, the outline of chapters in this book. The second preliminary chapter is called *Programming and Design*. Since this book makes heavy use of Python code for developing mathematical models and for algorithms implementations, we cover the requisite Python background (specifically the design paradigms we use in our Python code) in this chapter. To be clear, this chapter is not a full Python tutorial—the reader is expected to have some background in Python already. It is a tutorial of some key techniques and practices in Python (that many readers of this book might not be accustomed to) that we use heavily in this book and that are also highly relevant to programming in the broader area of Applied Mathematics. We cover *Classes, Inheritance, Abstract Classes, Static Types, Generic Programming, Dataclasses, Immutability, First-class Functions, Lambdas, Iterators* and *Generators*.

The remaining chapters in this book are organized in the 4 modules we listed above.

1.9.1 Module I: Processes and Planning Algorithms

The first module of this book covers the theory of Markov Decision Processes (MDP), Dynamic Programming (DP) and Approximate Dynamics Programming (ADP) across Chapters 3 – 6.

In order to understand the MDP framework, we start with the foundations of *Markov Processes* (sometimes referred to as Markov Chains) in Chapter 3. Markov Processes do not have any Rewards or Actions, they only have states and states transitions. We believe spending a lot of time on this simplified framework of Markov Processes is excellent preparation before getting to MDPs. Chapter 3 then builds upon Markov Processes to include the concept of Reward (but not Action)—the inclusion of Reward yields a framework known as Markov Reward Process. With Markov Reward Processes, we can talk about Value Functions and Bellman Equation, which serve as great preparation for understanding Value Function and Bellman Equation later in the context of MDPs. Chapter 3 motivates these concepts with examples of stock prices and with a simple inventory example that serves first as a Markov Process and then as a Markov Reward Process. There is also a significant amount of programming content in this chapter to develop comfort as well as depth in these concepts.

Chapter 4 on *Markov Decision Processes* lays the foundational theory underpinning RL—the framework for representing problems dealing with sequential optimal decisioning under uncertainty (Markov Decision Process). You will learn about the relationship between Markov Decision Processes and Markov Reward Processes, about the Value Function and the Bellman Equations. Again, there is a considerable amount of programming exercises in this chapter. The heavy investment in this theory together with hands-on programming will put you in a highly advantaged position to learn the following chapters in a very clear and speedy manner.

Chapter 5 on *Dynamic Programming* covers the Planning technique of Dynamic Programming (DP), which is an important class of foundational algorithms that can be an alternative to RL if the MDP is not too large or too complex. Also, learning these algorithms provides important foundations to be able to understand subsequent RL algorithms more deeply. You will learn about several important DP algorithms by the end of the chapter, and you will learn about why DP gets difficult in practice which draws you to the motivation behind RL. Again, we cover plenty of programming exercises that are quick to implement and will aid considerably in internalizing the concepts. Finally, we emphasize a special algorithm—Backward Induction—for solving finite-horizon Markov Decision Processes, which is the setting for the financial applications we cover in this book.

The Dynamic Programming algorithms covered in Chapter 5 suffer from the two so-called curses: Curse of Dimensionality and Curse of Modeling. These curses can be cured with a combination of sampling and function approximation. Module III covers the sampling cure (using Reinforcement Learning). Chapter 6 on *Function Approximation and Approximate Dynamic Programming* covers the topic of function approximation and shows how an intermediate cure—Approximate Dynamic Programming (function approximation without sampling)—is often quite viable and can be suitable for some problems. As part of this chapter, we implement linear function approximation and approximation with deep neural networks (forward and back propagation algorithms), so we can use these approximations in Approximate Dynamic Programming algorithms and later also in RL.

1.9.2 Module II: Modeling Financial Applications

The second module of this book covers the background on Utility Theory and 5 financial applications of Stochastic Control across Chapters 7 – 10.

We begin this module with Chapter 7 on *Utility Theory* which covers a very important Economics concept that is a pre-requisite for most of the Financial Applications we cover in subsequent chapters. This is the concept of risk-aversion (i.e., how people want to be compensated for taking risk) and the related concepts of risk-premium and Utility functions. The remaining chapters in this module cover not only the 5 financial applications, but also great detail on how to model them as MDPs, develop DP/ADP algorithms to solve them, and write plenty of code to implement the algorithms, which helps internalize the learnings quite well. Note that in practice these financial applications can get fairly complex and DP/ADP algorithms don't quite scale, which means we need to tap into RL algorithms to solve them. So we revisit these financial applications in Module III when we cover RL algorithms.

Chapter 8 is titled *Dynamic Asset Allocation and Consumption*. This chapter covers the first of the 5 Financial Applications. This problem is about how to adjust the allocation of one's wealth to various investment choices in response to changes in financial markets. The problem also involves how much wealth to consume in each interval over one's lifetime so as to obtain the best utility from wealth consumption. Hence, it is the joint problem of (dynamic) allocation of wealth to financial assets and appropriate consumption of one's wealth over a period of time. This problem is best understood in the context of Merton's landmark paper in 1969 where he stated and solved this problem. This chapter is mainly focused on the mathematical derivation of Merton's solution of this problem with Dynamic Programming. You will also learn how to solve the asset allocation problem in a simple setting with Approximate Backward Induction (an ADP algorithm covered in Chapter 6).

Chapter 9 covers a very important topic in Mathematical Finance: *Pricing and Hedging of Derivatives*. Full and rigorous coverage of derivatives pricing and hedging is a fairly elaborate and advanced topic, and beyond the scope of this book. But we have provided a way to understand the theory by considering a very simple setting—that of a single-period with discrete outcomes and no provision for rebalancing of the hedges, that is typical in the general theory. Following the coverage of the foundational theory, we cover the problem of optimal exercise of American Options and the problem of optimal pricing/hedging of derivatives in an *incomplete market* (both problems are modeled as MDPs). In this chapter, you will learn about some highly important financial foundations such as the concepts of arbitrage, replication, market completeness, and the all-important risk-neutral measure. You will learn the proofs of the two fundamental theorems of asset pricing in this simple setting. We also provide an overview of the general theory (beyond this simple setting). Next, you will learn how to model the general problem of optimal stopping as an MDP and specialize to the case of optimal exercise of American Options. You will learn how to use Backward Induction (a DP algorithm we learned in Chapter 5) to solve this problem when the state space is not too big. Finally, you will learn how to price/hedge derivatives incorporating real-world frictions by modeling this problem as an MDP. By the end of this chapter, you would have developed significant expertise in pricing and hedging complex derivatives, a skill that is in high demand in the finance industry.

Chapter 10 on *Order-Book Algorithms* covers the remaining two Financial Applications, pertaining to the world of Algorithmic Trading. The current practice in Algorithmic Trading is to employ techniques that are rules-based and heuristic. However, Algorithmic Trading is quickly transforming into Machine Learning-based Algorithms. In this chapter, you will be first introduced to the mechanics of trade order placements (market orders and limit orders), and then introduced to a very important real-world problem—how to submit a large-sized market order by splitting the shares to be transacted and timing the splits optimally in order to overcome "price impact" and gain maximum proceeds. You will learn about the classical methods based on Dynamic Programming. Next, you will learn about the market frictions and the need to tackle them with RL. In the second half of this chapter, we cover the Algorithmic-Trading twin of the Optimal Execution problem—that of a market-maker having to submit dynamically-changing bid and ask limit orders so she can make maximum gains. You will learn about how market-makers (a big and thriving industry) operate. Then you will learn about how to formulate this problem as an MDP. We will do a thorough coverage of the classical Dynamic Programming solution by Avellaneda and Stoikov. Finally, you will be exposed to the real-world nuances of this problem, and hence, the need to tackle with a market-calibrated simulator and RL.

1.9.3 Module III: Reinforcement Learning Algorithms

The third module of this book covers Reinforcement Learning algorithms across Chapters 11 – 14.

Chapter 11 on *Monte-Carlo and Temporal-Difference for Prediction* starts a new phase in this book—our entry into the world of RL algorithms. To understand the basics of RL, we start this chapter by restricting the RL problem to a very simple one—one where the state space is small and manageable as a table enumeration (known as tabular RL) and one where we only have to calculate the Value Function for a Fixed Policy (this problem is known as the Prediction problem, versus the optimization problem which is known as the Control problem). The restriction to Tabular Prediction is important because it makes it much easier to understand the core concepts of Monte-Carlo (MC) and Temporal-Difference (TD) in this simplified setting. The later part of this chapter extends Tabular Prediction to Prediction with Function Approximation (leveraging the function approximation foundations we develop in Chapter 6 in the context of ADP). The remaining chapters will build upon this chapter by adding more complexity and more nuances, while retaining much of the key core concepts developed in this chapter. As ever, you will learn by coding plenty of MC and TD algorithms from scratch.

Chapter 12 on *Monte-Carlo and Temporal-Difference for Control* makes the natural extension from Prediction to Control, while initially remaining in the tabular setting. The investments made in understanding the core concepts of MC and TD in Chapter 11 bear fruit here as important Control Algorithms such as SARSA and Q-learning can now be learned with enormous clarity. In this chapter, we implement both SARSA and Q-Learning from scratch in Python. This chapter also introduces a very important concept for the future success of RL in the real world: off-policy learning (Q-Learning is the simplest off-policy learning algorithm, and it has had good success in various applications). The later part of this chapter extends Tabular Control to Control with Function Approximation (leveraging the function approximation foundations we develop in Chapter 6).

Chapter 13 on *Batch RL, Experience-Replay, DQN, LSPI, Gradient TD* moves on from basic and more traditional RL algorithms to recent innovations in RL. We start this chapter with the important ideas of *Batch RL* and *Experience-Replay*, which makes more efficient use of data by storing data as it comes and re-using it in batches throughout the learning process of the algorithm. We emphasize the innovative Deep Q-Networks algorithm that leverages Experience-Replay and Deep Neural Networks for Q-Learning. We also emphasize a simple but important Batch RL technique using linear function approximation—the algorithm for Prediction is known as Least-Squares Temporal Difference and its extension to solve the Control problem is known as Least-Squares Policy Iteration. We discuss these algorithms in the context of one of the Financial Applications from Module II. In the later part of this chapter, we provide deeper insights into the core mathematics underpinning RL algorithms (back to the basics of Bellman Equation). Understanding *Value Function Geometry* will place you in a highly advantaged situation in terms of truly understanding what is it that makes some Algorithms succeed in certain situations and fail in other situations. This chapter also explains how to break out of the so-called Deadly Triad (when bootstrapping, function approximation and off-policy are employed together, RL algorithms tend to fail). The state-of-the-art Gradient TD Algorithm resists the deadly triad, and we dive deep into its inner workings to understand how and why.

Chapter 14 on *Policy Gradient Algorithms* introduces a very different class of RL algorithms that are based on improving the policy using the gradient of the policy function approximation (rather than the usual policy improvement based on explicit argmax on Q-Value Function). When action spaces are large or continuous, Policy Gradient tends to be the only option and so, this chapter is useful to overcome many real-world challenges (including those in many financial applications) where the action space is indeed large. You will learn about the mathematical proof of the elegant Policy Gradient Theorem and implement a few Policy Gradient Algorithms from scratch. You will learn about state-of-the-art Actor-Critic methods and a couple of specialized algorithms that have worked well in practice. Lastly, you will also learn about Evolutionary Strategies, an algorithm that looks quite similar to Policy Gradient Algorithms, but is technically not an RL Algorithm. However, learning about Evolutionary Strategies is important because some real-world applications, including Financial Applications, can indeed be tackled well with Evolutionary Strategies.

1.9.4   Module IV: Finishing Touches

The fourth module of this book (Chapters 15 – 17) provides the finishing touches—covering the miscellaneous topic of balancing exploration versus exploitation, showing how to blend together the Planning approach of Module I with the Learning approach of Module III, and providing perspectives on practical nuances in the real world.

Chapter 15 on *Multi-Armed Bandits: Exploration versus Exploitation* is a deep-dive into the topic of balancing exploration and exploitation, a topic of great importance in RL algorithms. Exploration versus Exploitation is best understood in the simpler setting of the Multi-Armed Bandit (MAB) problem. You will learn about various state-of-the-art MAB algorithms, implement them in Python, and draw various graphs to understand how they perform versus each other in various problem settings. You will then be exposed to Contextual Bandits, which is a popular approach in optimizing choices of Advertisement placements. Finally, you will learn how to apply the MAB algorithms within RL.

Chapter 16 on *Blending Learning and Planning* brings the various pieces of Planning and Learning concepts learned in this book together. You will learn that in practice, one needs to be creative about blending planning and learning concepts (a technique known as Model-based RL). In practice, many problems are indeed tackled using Model-based RL. You will also get familiar with an algorithm (Monte Carlo Tree Search) that was highly popularized when it solved the Game of GO, a problem that was thought to be insurmountable by present AI technology.

Chapter 17 is the concluding chapter on *Summary and Real-World Considerations*. The purpose of this chapter is two-fold: Firstly, to summarize the key learnings from this book, and secondly, to provide some commentary on how to take the learnings from this book into practice (to solve real-world problems). We specifically focus on the challenges one faces in the real-world—modeling difficulties, problem-size difficulties, operational challenges, data challenges (access, cleaning, organization), and also change-management challenges as one shifts an enterprise from legacy systems to an AI system. We also provide some guidance on how to go about building an end-to-end system based on RL.

1.9.5   Short Appendix Chapters

Finally, we have 7 short Appendix chapters at the end of this book. The first appendix is on *Moment Generating Functions* and its use in various calculations across this book. The second appendix is on *Portfolio Theory* covering the mathematical foundations of balancing return versus risk in portfolios and the much-celebrated Capital Asset Pricing Model (CAPM). The third appendix covers the basics of *Stochastic Calculus* as we need some of this theory (*Ito Integral, Ito's Lemma* etc.) in the derivations in a couple of the chapters in Module II. The fourth appendix is on the *HJB Equation*, which is a key part of the derivation of the closed-form solutions for 2 of the 5 financial applications we cover in Module II. The fifth appendix covers the derivation of the famous *Black-Scholes Equation* (and its solution for Call/Put Options). The sixth appendix is a technical perspective of *Function Approximations as Affine Spaces*, which helps develop a deeper mathematical understanding of function approximations. The seventh appendix covers the formulas for bayesian updates to conjugate prior distributions for the parameters of Gaussian and Bernoulli data distributions.

CHAPTER 2

Programming and Design

Programming is creative work with few constraints: imagine something and you can probably build it—in *many* different ways. Liberating and gratifying, but also challenging. Just like starting a novel from a blank page or a painting from a blank canvas, a new program is so open that it's a bit intimidating. Where do you start? What will the system look like? How will you get it *right*? How do you split your problem up? How do you prevent your code from evolving into a complete mess?

There's no easy answer. Programming is inherently iterative—we rarely get the right design at first, but we can always edit code and refactor over time. But iteration itself is not enough; just like a painter needs technique and composition, a programmer needs patterns and design.

Existing teaching resources tend to deemphasize programming techniques and design. Theory-heavy and algorithms-heavy books show models and algorithms as self-contained procedures written in pseudocode, without the broader context (and corresponding design considerations) of a real codebase. Newer AI/ML materials sometimes take a different tack and provide real code examples using industry-strength frameworks, but rarely touch on software design questions.

In this book, we take a third approach. Starting *from scratch*, we build a Python framework that reflects the key ideas and algorithms we cover in this book. The abstractions we define map to the key concepts we introduce; how we structure the code maps to the relationships between those concepts.

Unlike the pseudocode approach, we do not implement algorithms in a vacuum; rather, each algorithm builds on abstractions introduced earlier in the book. By starting from scratch (rather than using an existing ML framework) we keep the code reasonably simple, without needing to worry about specific examples going out of date. We can focus on the concepts important to this book while teaching programming and design *in situ*, demonstrating an intentional approach to code design.

2.1 CODE DESIGN

How can we take a complex domain like reinforcement learning and turn it into code that is easy to understand, debug and extend? How can we split this problem into manageable pieces? How do those pieces interact?

There is no simple single answer to these questions. No two programming challenges are identical and the same challenge has many reasonable solutions. A solid design will not be completely clear up-front; it helps to have a clear direction in mind, but expect to revisit specific decisions over time. That's exactly what we did with the code for this book: we had a vision for a Python Reinforcement Learning framework that matched the topics we present, but as we wrote more and more of the book, we revised the framework code as we came up with better ideas or found new requirements our previous design did not cover.

We might have no easy answers, but we do have patterns and principles that—in our experience—consistently produce quality code. Taken together, these ideas form a philosophy of code design oriented around defining and combining **abstractions** that reflect how we think about our domain. Since code itself can point to specific design ideas and capabilities, there's a feedback loop: expanding the programming abstractions we've designed can help us find new algorithms and functionality, improving our understanding of the domain.

Just what *is* an abstraction? An appropriately abstract question! An abstraction is a "compound idea": a single concept that combines multiple separate ideas into one. We can combine ideas along two axes:

- We can *compose* different concepts together, thinking about how they behave as one unit. A car engine has thousands of parts that interact in complex ways, but we can think about it as a single object for most purposes.
- We can *unify* different concepts by identifying how they are similar. Different breeds of dogs might look totally different, but we can think of all of them as dogs.

The human mind can only handle so many distinct ideas at a time—we have an inherently limited working memory. A rather simplified model is that we only have a handful of "slots" in working memory, and we simply can't track more independent thoughts at the same time. The way we overcome this limitation is by coming up with *new* ideas (new *abstractions*) that combine multiple concepts into one.

We want to organize code around abstractions for the same reason that we use abstractions to understand more complex ideas. How do you understand code? Do you run the program in your head? That's a natural starting point and it works for simple programs, but it quickly becomes difficult and then impossible. A computer doesn't have working-memory limitations and can run *billions* of instructions a second that we can't possibly keep up with. The computer doesn't need structure or abstraction in the code it runs, but we need it to have any hope of writing or understanding anything beyond the simplest of programs. Abstractions in our code group information and logic so that we can think about rich concepts rather than tracking every single bit of information and every single instruction separately.

The details may differ, but designing code around abstractions that correspond to a solid mental model of the domain works well in any area and with any programming language. It might take some extra up-front thought but, done well, this style of design pays dividends. Our goal is to write code that makes life easier *for ourselves*; this helps for everything from "one-off" experimental code through software engineering efforts with large teams.

2.2 ENVIRONMENT SETUP

You can follow along with all of the examples in this book by getting a copy of the RL framework from GitHub[1] and setting up a dedicated Python 3 environment for the code.

The Python code depends on several Python libraries. Once you have a copy of the code repository, you can create an environment with the right libraries by running a few shell commands.

---

[1] If you are not familiar with Git and GitHub, look through GitHub's Getting Started documentation.

First, move to the directory with the codebase:

```
cd rl-book
```

Then, create and activate a Python virtual environment[2]:

```
python3 -m venv .venv
source .venv/bin/activate
```

You only need to create the environment once, but you will need to activate it every time that you want to work on the code from a new shell. Once the environment is activated, you can install the right versions of each Python dependency:

```
pip install -r requirements.txt
```

To access the framework itself, you need to install it in editable mode (-e):

```
pip install -e .
```

Once the environment is set up, you can confirm that it works by running the frameworks automated tests:

```
python -m unittest discover
```

If everything installed correctly, you should see an "OK" message on the last line of the output after running this command.

2.3   CLASSES AND INTERFACES

What does designing clean abstractions actually entail? There are always two parts to answering this question:

1. Understanding the domain concept that you are modeling.
2. Figuring out how to express that concept with features and patterns provided by your programming language.

Let's jump into an extended example to see exactly what this means. One of the key building blocks for Reinforcement Learning—all of statistics and machine learning, really—is Probability. How are we going to handle uncertainty and randomness in our code?

One approach would be to keep Probability implicit. Whenever we have a random variable, we could call a function and get a random result (technically referred to as a *sample*). If we were writing a Monopoly game with two six-sided dice, we would define it like this:

```
from random import randint
def six_sided()
    return randint(1, 6)
def roll_dice():
    return six_sided() + six_sided()
```

This works, but it's pretty limited. We can't do anything except get one outcome at a time. More importantly, this only captures a slice of how we *think* about Probability: there's *randomness* but we never even mentioned probability distributions (referred to as simply *distributions* for the rest of this chapter). We have outcomes and we have a function we

---

[2]A Python "virtual environment" is a way to manage Python dependencies on a per-project basis. Having a different environment for different Python projects lets each project have its own version of Python libraries, which avoids problems when one project needs an older version of a library and another project needs a newer version.

can call repeatedly, but there's no way to tell that function apart from a function that has nothing to do with Probability but just happens to return an integer.

How can we write code to get the expected value of a distribution? If we have a parametric distribution (e.g., a distribution like Poisson or Gaussian, characterized by parameters), can we get the parameters out if we need them?

Since distributions are implicit in the code, the *intentions* of the code aren't clear and it is hard to write code that generalizes over distributions. Distributions are absolutely crucial for machine learning, so this is not a great starting point.

2.3.1 A Distribution Interface

To address these problems, let's define an abstraction for probability distributions.

How do we represent a distribution in code? What can we *do* with distributions? That depends on exactly what kind of distribution we're working with. If we know something about the structure of a distribution—perhaps it's a Poisson distribution where $\lambda = 5$, perhaps it's an empirical distribution with set probabilities for each outcome—we could do quite a bit: produce an exact Probability Distribution Function (PDF) or Cumulative Distribution Function (CDF), calculate expectations and do various operations efficiently. But that isn't the case for all the distributions we work with! What if the distribution comes from a complicated simulation? At the extreme, we might not be able to do anything except draw samples from the distribution.

Sampling is the least common denominator. We can sample distributions where we don't know enough to do anything else, and we can sample distributions where we know the exact form and parameters. Any abstraction we start with for a probability distribution needs to cover sampling, and any abstraction that requires more than just sampling will not let us handle all the distributions we care about.

In Python, we can express this idea with a class:

```python
from abc import ABC, abstractmethod
class Distribution(ABC):
    @abstractmethod
    def sample(self):
        pass
```

This class defines an **interface**: a definition of what we require for something to qualify as a distribution. Any kind of distribution we implement in the future will be able to, at minimum, generate samples; when we write functions that sample distributions, they can require their inputs to inherit from `Distribution`.

The class itself does not actually implement `sample`. `Distribution` captures the *abstract concept* of distributions that we can sample, but we would need to specify a specific distribution to actually sample anything. To reflect this in Python, we've made `Distribution` an **abstract base class** (ABC), with `sample` as an *abstract* method—a method without an implementation. Abstract classes and abstract methods are features that Python provides to help us define interfaces for abstractions. We can define the `Distribution` class to structure the rest of our probability distribution code before we define any specific distributions.

2.3.2 A Concrete Distribution

Now that we have an interface, what do we do with it? An interface can be approached from two sides:

- Something that **requires** the interface. This will be code that uses operations specified in the interface and work with *any* value that satisfies those requirements.

- Something that **provides** the interface. This will be some value that supports the operations specified in the interface.

If we have some code that requires an interface and some other code that satisfies the interface, we know that we can put the two together and get something that works—even if the two sides were written without any knowledge or reference to each other. The interface manages how the two sides interact.

To use our `Distribution` class, we can start by providing a **concrete class**[3] that implements the interface. Let's say that we wanted to model dice—perhaps for a game of D&D or Monopoly. We could do this by defining a `Die` class that represents an n-sided die and inherits from `Distribution`:

```
import random
class Die(Distribution):
    def __init__(self, sides):
        self.sides = sides

    def sample(self):
        return random.randint(1, self.sides)
six_sided = Die(6)
def roll_dice():
    return six_sided.sample() + six_sided.sample()
```

This version of `roll_dice` has exactly the same behavior as `roll_dice` in the previous section, but it took a bunch of extra code to get there. What was the point?

The key difference is that we now have a value that represents the *distribution* of rolling a die, not just the outcome of a roll. The code is easier to understand—when we come across a `Die` object, the meaning and intention behind it is clear—and it gives us a place to add additional die-specific functionality. For example, it would be useful for debugging if we could print not just the *outcome* of rolling a die but the die itself—otherwise, how would we know if we rolled a die with the right number of sides for the given situation?

If we were using a function to represent our die, printing it would not be useful:

```
>>> print(six_sided)
<function six_sided at 0x7f00ea3e3040>
```

That said, the `Die` class we've defined so far isn't much better:

```
>>> print(Die(6))
<__main__.Die object at 0x7ff6bcadc190>
```

With a class—and unlike a function—we can fix this. Python lets us change some of the built-in behavior of objects by overriding special methods. To change how the class is printed, we can override `__repr__`:[4]

```
class Die(Distribution):
    ...
    def __repr__(self):
        return f"Die(sides={self.sides})"
```

---

[3]In this context, a concrete class is any class that is not an abstract class. More generally, "concrete" is the opposite of "abstract"—when an abstraction can represent multiple more specific concepts, we call any of the specific concepts "concrete".

[4]Our definition of `__repr__` used a Python feature called an "f-string". Introduced in Python 3.6, f-strings make it easier to inject Python values into strings. By putting an f in front of a string literal, we can include a Python value in a string: `f"{1 + 1}"` gives us the string `"2"`.

Much better:

```
>>> print(Die(6))
Die(sides=6)
```

This seems small but makes debugging *much* easier, especially as the codebase gets larger and more complex.

2.3.2.1   Dataclasses

The Die class we wrote is intentionally simple. Our die is defined by a single property: the number of sides it has. The __init__ method takes the number of sides as an input and puts it into an *attribute* of the class; once a Die object is created, there is no reason to change this value—if we need a die with a different number of sides, we can just create a new object. Abstractions do not have to be complex to be useful.

Unfortunately, some of the default behavior of Python classes isn't well-suited to simple classes. We've already seen that we need to override __repr__ to get useful behavior, but that's not the only default that's inconvenient. Python's default way to compare objects for equality—the __eq__ method—uses the is operator, which means it compares objects *by identity*. This makes sense for classes in general which can change over time, but it is a poor fit for simple abstraction like Die. Two Die objects with the same number of sides have the same behavior and represent the same probability distribution, but with the default version of __eq__, two Die objects declared separately will never be equal:

```
>>> six_sided = Die(6)
>>> six_sided == six_sided
True
>>> six_sided == Die(6)
False
>>> Die(6) == Die(6)
False
```

This behavior is inconvenient and confusing, the sort of edge-case that leads to hard-to-spot bugs. Just like we overrode __repr__, we can fix this by overriding __eq__:

```
def __eq__(self, other):
    return self.sides == other.sides
```

This fixes the weird behavior we saw earlier:

```
>>> Die(6) == Die(6)
True
```

However, this simple implementation will lead to errors if we use == to compare a Die with a non-Die value:

```
>>> Die(6) == None
Traceback (most recent call last):
  File "<stdin>", line 1, in <module>
  File ".../rl/chapter1/probability.py", line 18, in __eq__
    return self.sides == other.sides
AttributeError: 'NoneType' object has no attribute 'sides'
```

We generally won't be comparing values of different types with ==—for None, Die(6) is None would be more idiomatic—but the usual expectation in Python is that == on different types will return False rather than raising an exception. We can fix by explicitly checking the type of other:

```
def __eq__(self, other):
    if isinstance(other, Die):
        return self.sides == other.sides
    return False
```

```
>>> Die(6) == None
False
```

Most of the classes we will define in the rest of the book follow this same pattern—they're defined by a small number of parameters, all that __init__ does is set a few attributes and they need custom __repr__ and __eq__ methods. Manually defining __init__, __repr__ and __eq__ for every single class isn't *too* bad—the definitions are entirely systematic—but it carries some real costs:

- Extra code without important content makes it harder to *read* and *navigate* through a codebase.
- It's easy for mistakes to sneak in. For example, if you add an attribute to a class but forget to add it to its __eq__ method, you won't get an error—== will just ignore that attribute. Unless you have tests that explicitly check how == handles your new attribute, this oversight can sneak through and lead to weird behavior in code that uses your class.
- Frankly, writing these methods by hand is just *tedious*.

Luckily, Python 3.7 introduced a feature that fixes all of these problems: **dataclasses**. The dataclasses module provides a decorator[5] that lets us write a class that behaves like Die without needing to manually implement __init__, __repr__ or __eq__. We still have access to "normal" class features like inheritance ((Distribution)) and custom methods (sample):

```
from dataclasses import dataclass

@dataclass
class Die(Distribution):
    sides: int

    def sample(self):
        return random.randint(1, self.sides)
```

This version of Die has the exact behavior we want in a way that's easier to write and—more importantly—*far* easier to read. For comparison, here's the code we would have needed *without* dataclasses:

```
class Die(Distribution):
    def __init__(self, sides):
        self.sides = sides

    def __repr__(self):
        return f"Die(sides={self.sides})"

    def __eq__(self, other):
        if isinstance(other, Die):
            return self.sides == other.sides
        return False

    def sample(self):
        return random.randint(1, self.sides)
```

---

[5]Python decorators are modifiers that can be applied to class, function and method definitions. A decorator is written *above* the definition that it applies to, starting with a @ symbol. Examples include abstractmethod—which we saw earlier—and dataclass.

As you can imagine, the difference would be even starker for classes with more attributes!

Dataclasses provide such a useful foundation for classes in Python that the *majority* of the classes we define in this book are dataclasses—we use dataclasses unless we have a *specific* reason not to.

2.3.2.2 Immutability

Once we've created a `Die` object, it does not make sense to change its number of sides—if we need a distribution for a different die, we can create a new object instead. If we change the `sides` of a `Die` object in one part of our code, it will also change in every other part of the codebase that uses that object, in ways that are hard to track. Even if the change made sense in one place, chances are it is not expected in other parts of the code. Changing state can create invisible connections between seemingly separate parts of the codebase, which becomes hard to mentally track. A sure recipe for bugs!

Normally, we avoid this kind of problem in Python purely by convention: nothing *stops* us from changing `sides` on a `Die` object, but we know not to do that. This is doable, but hardly ideal; just like it is better to rely on seatbelts rather than pure driver skill, it is better to have the language prevent us from doing the wrong thing than relying on pure convention. Normal Python classes don't have a convenient way to stop attributes from changing, but luckily dataclasses do:

```
@dataclass(frozen=True)
class Die(Distribution):
    ...
```

With `frozen=True`, attempting to change `sides` will raise an exception:

```
>>> d = Die(6)
>>> d.sides = 10
Traceback (most recent call last):
  File "<stdin>", line 1, in <module>
  File "<string>", line 4, in __setattr__
dataclasses.FrozenInstanceError: cannot assign to field 'sides'
```

An object that we cannot change is called **immutable**. Instead of changing the object *in place*, we can return a fresh copy with the attribute changed; `dataclasses` provides a `replace` function that makes this easy:

```
>>> import dataclasses
>>> d6 = Die(6)
>>> d20 = dataclasses.replace(d6, sides=20)
>>> d20
Die(sides=20)
```

This example is a bit convoluted—with such a simple object, we would just write `d20 = Die(20)`—but `dataclasses.replace` becomes a lot more useful with more complex objects that have multiple attributes.

Returning a fresh copy of data rather than modifying in place is a common pattern in Python libraries. For example, the majority of Pandas operations—like `drop` or `fillna`—return a *copy* of the dataframe rather than modifying the dataframe in place. These methods have an `inplace` argument as an option, but this leads to enough confusing behavior that the Pandas team is currently deliberating on deprecating `inplace` altogether.

Apart from helping prevent odd behavior and bugs, `frozen=True` has an important bonus: we can use immutable objects as dictionary keys and set elements. Without `frozen=True`, we would get a `TypeError` because non-frozen dataclasses do not implement `__hash__`:

```
>>> d = Die(6)
>>> {d : "abc"}
Traceback (most recent call last):
  File "<stdin>", line 1, in <module>
TypeError: unhashable type: 'Die'
>>> {d}
Traceback (most recent call last):
  File "<stdin>", line 1, in <module>
TypeError: unhashable type: 'Die'
```

With frozen=True, dictionaries and sets work as expected:

```
>>> d = Die(6)
>>> {d : "abc"}
{Die(sides=6): 'abc'}
>>> {d}
{Die(sides=6)}
```

Immutable dataclass objects act like plain data—not too different from strings and ints. In this book, we follow the same practice with frozen=True as we do with dataclasses in general: we set frozen=True unless there is a specific reason not to.

2.3.3 Checking Types

A die has to have an int number of sides—0.5 sides or "foo" sides simply doesn't make sense. Python will not stop us from *trying* Die("foo"), but we would get a TypeError if we tried sampling it:

```
>>> foo = Die("foo")
>>> foo.sample()
Traceback (most recent call last):
  File "<stdin>", line 1, in <module>
  File ".../rl/chapter1/probability.py", line 37, in sample
    return random.randint(1, self.sides)
  File ".../lib/python3.8/random.py", line 248, in randint
    return self.randrange(a, b+1)
TypeError: can only concatenate str (not "int") to str
```

The types of an object's attributes are a useful indicator of how the object should be used. Python's dataclasses let us use **type annotations** (also known as "type hints") to specify the type of each attribute:

```
@dataclass(frozen=True)
class Die(Distribution):
    sides: int
```

In normal Python, these type annotations exist primarily for documentation—a user can see the types of each attribute at a glance, but the language does not raise an error when an object is created with the wrong types in an attribute. External tools—Integrated Development Environments (IDEs) and typecheckers—can catch type mismatches in annotated Python code without running the code. With a type-aware editor, Die("foo") would be underlined with an error message:

> Argument of type "Literal['foo']" cannot be assigned to parameter "sides" of type "int" in function "__init__"
>
> "Literal['foo']" is incompatible with "int" [reportGeneralTypeIssues]

This particular message comes from **pyright** running over the language server protocol (LSP), but Python has a number of different typecheckers available[6].

---

[6]Python has a number of external typecheckers, including:

Instead of needing to call `sample` to see an error—which we then have to carefully read to track back to the source of the mistake—the mistake is highlighted for us without even needing to run the code.

2.3.3.1 Static Typing

Being able to find type mismatches *without running code* is called **static typing**. Some languages (e.g., Java, C++) require *all* code to be statically typed; Python does not. In fact, Python started out as a **dynamically typed** language with no type annotations. Type errors would only come up when the code containing the error was run.

Python is still *primarily* a dynamically typed language—type annotations are optional in most places and there is no built-in checking for annotations. In the `Die("foo")` example, we only got an error when we ran code that passed `sides` into a function that *required* an int (`random.randint`). We can get static checking with external tools, but even then it remains *optional*—even statically checked Python code runs dynamic type checks, and we can freely mix statically checked and "normal" Python. Optional static typing on top of a dynamically typed languages is called **gradual typing** because we can incrementally add static types to an existing dynamically typed codebase.

Dataclass attributes are not the only place where knowing types is useful; it would also be handy for function parameters, return values and variables. Python supports *optional* annotations on all of these; dataclasses are the only language construct where annotations are *required*. To help mix annotated and unannotated code, typecheckers will report mismatches in code that is explicitly annotated but will usually not try to guess types for unannotated code.

How would we add type annotations to our example code? So far, we've defined two classes:

- `Distribution`, an abstract class defining interfaces for probability distributions in general
- `Die`, a concrete class for the distribution of an n-sided die

We've already annotated the `sides` in `Die` has to be an `int`. We also know that the *outcome* of a die roll is an `int`. We can annotate this by adding `-> int` after `def sample(...)`:

```
@dataclass(frozen=True)
class Die(Distribution):
    sides: int

    def sample(self) -> int:
        return random.randint(1, self.sides)
```

Other kinds of concrete distributions would have other sorts of outcomes. A coin flip would either be `"heads"` or `"tails"`; a normal distribution would produce a `float`. Type annotations are particularly important when writing code for any kind of mathematical modeling because we ensure that the type specifications in our code correspond clearly to the precise specification of sets in our mathematical (notational) description.

- mypy
- pyright
- pytype
- pyre

The PyCharm IDE also has a propriety typchecker built-in.

These tools can be run from the command line or integrated into editors. Different checkers *mostly* overlap in functionality and coverage but have slight differences in the sort of errors they detect and the style of error messages they generate.

2.3.4   Type Variables

Annotating `sample` for the `Die` class was straightforward: the outcome of a die roll is always a number, so `sample` always returns `int`. But what type would `sample` in general have? The `Distribution` class defines an interface for *any* distribution, which means it needs to cover *any* type of outcomes. For our first version of the `Distribution` class, we didn't annotate anything for `sample`:

```
class Distribution(ABC):
    @abstractmethod
    def sample(self):
        pass
```

This works—annotations are optional—but it can get confusing: some code we write will work for any kind of distribution, some code needs distributions that return numbers, other code will need something else... In every instance `sample` better return *something*, but that isn't explicitly annotated. When we leave out annotations our code will still work, but our editor or IDE will not catch as many mistakes.

The difficulty here is that different kinds of distributions—different *implementations of the* `Distribution` *interface*—will return different types from `sample`. To deal with this, we need **type variables**: variables that stand in for *some* type that can be different in different contexts. Type variables are also known as "generics" because they let us write classes that generically work for any type.

To add annotations to the abstract `Distribution` class, we will need to define a type variable for the outcomes of the distribution, then tell Python that `Distribution` is "generic" in that type:

```
from typing import Generic, TypeVar
# A type variable named "A"
A = TypeVar("A")

# Distribution is "generic in A"
class Distribution(ABC, Generic[A]):
    # Sampling must produce a value of type A
    @abstractmethod
    def sample(self) -> A:
        pass
```

In this code, we've defined a type variable `A`[7] and specified that `Distribution` uses `A` by inheriting from `Generic[A]`. We can now write type annotations for distributions *with specific types of outcomes*: for example, `Die` would be an instance of `Distribution[int]` since the outcome of a die roll is always an `int`. We can make this explicit in the class definition:

```
class Die(Distribution[int]):
    ...
```

This lets us write specialized functions that only work with certain kinds of distributions. Let's say we wanted to write a function that approximated the expected value of a distribution by sampling repeatedly and calculating the mean. This function works for distributions that have numeric outcomes—`float` or `int`—but not other kinds of distributions (How would we calculate an average for a random name?). We can annotate this explicitly by using `Distribution[float]`:[8]

---

[7] Traditionally, type variables have one-letter capitalized names—although it's perfectly fine to use full words if that would make the code clearer.

[8] The `float` type in Python *also* covers `int`, so we can pass a `Distribution[int]` anywhere that a `Distribution[float]` is required.

```
import statistics
def expected_value(d: Distribution[float], n: int = 100) -> float:
    return statistics.mean(d.sample() for _ in range(n))
```

With this function:

- `expected_value(Die(6))` would be fine
- `expected_value(RandomName())` (where `RandomName` is a `Distribution[str]`) would be a type error

Using `expected_value` on a distribution with non-numeric outcomes would raise a type error at runtime. Having this highlighted in the editor can save us time—we see the mistake right away, rather than waiting for tests to run—and will catch the problem even if our test suite doesn't.

2.3.5 Functionality

So far, we've covered two abstractions for working with probability distributions:

- `Distribution`: an abstract class that defines the *interface* for probability distributions
- `Die`: a distribution for rolling fair n-sided dice

This is an illustrative example, but it doesn't let us do much. If all we needed were n-sided dice, a separate `Distribution` class would be overkill. Abstractions are a means for managing complexity, but any abstraction we define also adds some complexity to a codebase itself—it's one more concept for a programmer to learn and understand. It's always worth considering whether the added complexity from defining and using an abstraction is worth the benefit. How does the abstraction help us understand the code? What kind of mistakes does it prevent—and what kind of mistakes does it *encourage*? What kind of added functionality does it give us? If we don't have sufficiently solid answers to these questions, we should consider leaving the abstraction out.

If all we cared about were dice, `Distribution` wouldn't carry its weight. Reinforcement Learning, though, involves both a wide range of specific distributions—any given Reinforcement Learning problem can have domain-specific distributions—as well as algorithms that need to work for all of these problems. This gives us two reasons to define a `Distribution` abstraction: `Distribution` will *unify* different applications of Reinforcement Learning and will *generalize* our Reinforcement Learning code to work in different contexts. By programming against a general interface like `Distribution`, our algorithms will be able to work for the different applications we present in the book—and even work for applications we weren't thinking about when we designed the code.

One of the practical advantages of defining general-purpose abstractions in our code is that it gives us a place to add functionality that will work for *any* instance of the abstraction. For example, one of the most common operations for a probability distribution that we can sample is drawing n samples. Of course, we could just write a loop every time we needed to do this:

```
samples = []
for _ in range(100):
    samples += [distribution.sample()]
```

This code is *fine*, but it's not *great*. A `for` loop in Python might be doing pretty much *anything*; it's used for repeating pretty much anything. It's hard to understand what a loop is doing at a glance, so we'd have to carefully read the code to see that it's putting 100 samples in a list. Since this is such a common operation, we can add a method for it instead:

```
class Distribution(ABC, Generic[A]):
    ...
    def sample_n(self, n: int) -> Sequence[A]:
        return [self.sample() for _ in range(n)]
```

The implementation here is different—it's using a **list comprehension**[9] rather than a normal `for` loop—but it's accomplishing the same thing. The more important distinction happens when we *use* the method; instead of needing a `for` loop or list comprehension each time, we can just write:

```
samples = distribution.sample_n(100)
```

The meaning of this line of code—and the programmer's intention behind it—are immediately clear at a glance.

Of course, this example is pretty limited. The list comprehension to build a list with 100 samples is a bit more complicated than just calling `sample_n(100)`, but not by much—it's still perfectly readable at a glance. This pattern of implementing general-purpose functions on our abstractions becomes a lot more useful as the functions themselves become more complicated.

However, there is another advantage to defining methods like `sample_n`: some kinds of distributions might have more efficient or more accurate ways to implement the same logic. If that's the case, we would override `sample_n` to use the better implementation. Code that uses `sample_n` would automatically benefit; code that used a loop or comprehension instead would not. For example, this happens if we implement a distribution by wrapping a function from numpy's random module:

```
import numpy as np
@dataclass
class Gaussian(Distribution[float]):
    μ: float
    σ: float

    def sample(self) -> float:
        return np.random.normal(loc=self.μ, scale=self.σ)

    def sample_n(self, n: int) -> Sequence[float]:
        return np.random.normal(loc=self.μ, scale=self.σ, size=n)
```

numpy is optimized for array operations, which means that there is an up-front cost to

---

[9]List comprehensions are a Python feature to build lists by looping over something. The simplest pattern is the same as writing a `for` loop:

```
foo = [expr for x in xs]
# is the same as:
foo = []
for x in xs:
    foo += [expr]
```

List comprehensions can combine multiple lists, acting like nested `for` loops:

```
>>> [(x, y) for x in range(3) for y in range(2)]
[(0, 0), (0, 1), (1, 0), (1, 1), (2, 0), (2, 1)]
```

They can also have `if` clauses to only keep elements that match a condition:

```
>>> [x for x in range(10) if x % 2 == 0]
[0, 2, 4, 6, 8]
```

Some combination of `for` and `if` clauses can let us build surprisingly complicated lists! Comprehensions will often be easier to read than loops—a loop could be doing *anything*, a comprehension is always creating a list—but it's always a judgement call. A couple of nested for loops might be easier to read than a sufficiently convoluted comprehension!

calling `numpy.random.normal` *the first time*, but it can quickly generate additional samples after that. The performance impact is significant[10]:

```
>>> d = Gaussian(μ=0, σ=1)
>>> timeit.timeit(lambda: [d.sample() for _ in range(100)])
293.33819171399955
>>> timeit.timeit(lambda: d.sample_n(100))
5.566364272999635
```

That's a 53× difference!

The code for `Distribution` and several concrete classes implementing the `Distribution` interface (including `Gaussian`) is in the file rl/distribution.py.

2.4 ABSTRACTING OVER COMPUTATION

So far, we've seen how we can build up a programming model for our domain by defining interfaces (like `Distribution`) and writing classes (like `Die`) that implement these interfaces. Classes and interfaces give us a way to model the "things" in our domain, but, in an area like Reinforcement Learning, "things" aren't enough: we also want some way to abstract over the *actions* that we're taking or the computation that we're performing.

Classes do give us one way to model behavior: **methods**. A common analogy is that objects act as "nouns" and methods act as "verbs"—methods are the actions we can take with an object. This is a useful capability that lets us abstract over doing the same kind of action on different sorts of objects. Our `sample_n` method, for example, with its default implementation, gives us two things:

1. If we implement a new type of distribution with a custom sample method, we get `sample_n` for free for that distribution.
2. If we implement a new type of distribution that has a way to get n samples faster than calling sample n times, we can override the method to use the faster algorithm.

If we made `sample_n` a normal function we could get 1. but not 2.; if we left `sample_n` as an abstract method, we'd get 2. but not 1.. Having a non-abstract method on the abstract class gives us the best of both worlds.

So if methods are our "verbs", what else do we need? While methods abstract over actions, they do this *indirectly*—we can talk about objects as standalone values, but we can only use methods *on* objects, with no way to talk about computation itself. Stretching the metaphor with grammar, it's like having verbs without infinitives or gerunds—we'd be able to talk about "somebody skiing", but not about "skiing" itself or somebody "planning to ski"!

In this world, "nouns" (objects) are **first-class citizens** but "verbs" (methods) aren't. What it takes to be a "first-class" value in a programming language is a fuzzy concept; a reasonable litmus test is whether we can pass a value to a function or store it in a data structure. We can do this with objects, but it's not clear what this would mean for methods.

2.4.1 First-Class Functions

If we didn't have a first-class way to talk about actions (as opposed to objects), one way we could work around this would be to represent functions *as* objects with a single method. We'd be able to pass them around just like normal values and, when we needed to actually

---

[10]This code uses the `timeit` module from Python's standard library, which provides a simple way to benchmark small bits of Python code. By default, it measures how long it takes to execute 1,000,000 iterations of the function in seconds, so the two examples here took 0.293 ms and 0.006 ms, respectively.

perform the action or computation, we'd just call the object's method. This solution shows us that it makes sense to have a first-class way to work with actions, but it requires an extra layer of abstraction (an object just to have a single method) which doesn't add anything substantial on its own while making our intentions less clear.

Luckily, we don't have to resort to a one-method object pattern in Python because Python has **first-class functions**: functions are already values that we can pass around and store, without needing a separate wrapper object.

First-class functions give us a new way to abstract over computation. Methods let us talk about the same kind of behavior for different objects; first-class functions let us do something *with* different actions. A simple example might be repeating the same action n times. Without an abstraction, we might do this with a for loop:

```
for _ in range(10):
    do_something()
```

Instead of writing a loop each time, we could factor this logic into a function that took n *and* do_something as arguments:

```
def repeat(action: Callable, n: int):
    for _ in range(n):
        action()
repeat(do_something, 10)
```

repeat takes action and n as arguments, then calls action n times. action has the type Callable which, in Python, covers functions as well as any other objects you can call with the f() syntax. We can also specify the return type and arguments a Callable should have; if we wanted the type of a function that took an int and a str as input and returned a bool, we would write Callable[[int, str], bool].

The version with the repeat function makes our intentions clear in the code. A for loop can do many different things, while repeat will always just repeat. It's not a big difference in this case—the for loop version is sufficiently easy to read that it's not a big impediment—but it becomes more important with complex or domain-specific logic.

Let's take a look at the expected_value function we defined earlier:

```
def expected_value(d: Distribution[float], n: int) -> float:
    return statistics.mean(d.sample() for _ in range(n))
```

We had to restrict this function to Distribution[float] because it only makes sense to take an expectation of a numeric outcome. But what if we have something else like a coin flip? We would still like some way to understand the expectation of the distribution, but to make that meaningful we'd need to have some mapping from outcomes ("heads" or "tails") to numbers (For example, we could say "heads" is 1 and "tails" is 0.). We could do this by taking our Coin distribution, converting it to a Distribution[float] and calling expected_value on that, but this might be inefficient and it's certainly awkward. An alternative would be to provide the mapping as an argument to the expected_value function:

```
def expected_value(
    d: Distribution[A],
    f: Callable[[A], float],
    n: int
) -> float:
    return statistics.mean(f(d.sample()) for _ in range(n))
```

The implementation of expected_value has barely changed—it's the same mean calculation as previously, except we apply f to each outcome. This small change, however, has made the function far more flexible: we can now call `expected_value` on *any* sort of `Distribution`, not just `Distribution[float]`.

Going back to our coin example, we could use it like this:

```python
def payoff(coin: Coin) -> float:
    return 1.0 if coin == "heads" else 0.0
expected_value(coin_flip, payoff)
```

The `payoff` function maps outcomes to numbers and then we calculate the expected value using that mapping.

We'll see first-class functions used in a number of places throughout the book; the key idea to remember is that *functions are values* that we can pass around or store just like any other object.

2.4.1.1 Lambdas

`payoff` itself is a pretty reasonable function: it has a clear name and works as a standalone concept. Often, though, we want to use a first-class function in some specific context where giving the function a name is not needed or even distracting. Even in cases with reasonable names like `payoff`, it might not be worth introducing an extra named function if it will only be used in one place.

Luckily, Python gives us an alternative: `lambda`. Lambdas are function literals. We can write `3.0` and get a number without giving it a name, and we can write a `lambda` expression to get a function without giving it a name. Here's the same example as with the `payoff` function but using a `lambda` instead:

```python
expected_value(coin_flip, lambda coin: 1.0 if coin == "heads" else 0.0)
```

The `lambda` expression here behaves exactly the same as `def payoff` did in the earlier version. Note how the lambda as a single expression with no `return`—if you ever need multiple statements in a function, you'll have to use a `def` instead of a `lambda`. In practice, lambdas are great for functions whose bodies are *short* expressions, but anything that's too long or complicated will read better as a standalone function `def`.

2.4.2 Iterative Algorithms

First-class functions give us an abstraction over *individual* computations: we can pass functions around, give them inputs and get outputs, but the computation between the input and the output is a complete black box. The caller of the function has no control over what happens inside the function. This limitation can be a real problem!

One common scenario in Reinforcement Learning—and other areas in numerical programming—is algorithms that *iteratively converge* to the correct result. We can run the algorithm repeatedly to get more and more accurate results, but the improvements with each iteration get progressively smaller. For example, we can approximate the square root of a by starting with some initial guess x_0 and repeatedly calculating x_{n+1}:

$$x_{n+1} = \frac{x_n + \frac{a}{x_n}}{2}$$

At each iteration, x_{n+1} gets closer to the right answer by smaller and smaller steps. At some point the change from x_n to x_{n+1} becomes small enough that we decide to stop. In Python, this logic might look something like this:

```
def sqrt(a: float) -> float:
    x = a / 2 # initial guess
    while abs(x_n - x) > 0.01:
        x_n = (x + (a / x)) / 2
    return x_n
```

The hard coded 0.01 in the while loop should be suspicious. How do we know that 0.01 is the right stopping condition? How do we decide when to stop at all?

The trick with this question is that we *can't* know when to stop when we're implementing a general-purpose function because the level of precision we need will depend on what the result is used for! It's the *caller* of the function that knows when to stop, not the *author*.

The first improvement we can make is to turn the 0.01 into an extra parameter:

```
def sqrt(a: float, threshold: float) -> float:
    x = a / 2 # initial guess
    while abs(x_n - x) > threshold:
        x_n = (x + (a / x)) / 2
    return x_n
```

This is a definite improvement over a literal 0.01, but it's still limited. We've provided an extra parameter for how the function behaves, but the control is still fundamentally with the function. The caller of the function might want to stop before the method converges if it's taking too many iterations or too much time, but there's no way to do that by changing the threshold parameter. We could provide additional parameters for all of these, but we'd quickly end up with the logic for how to stop iteration requiring a lot more code and complexity than the iterative algorithm itself! Even that wouldn't be enough; if the function isn't behaving as expected in some specific application, we might want to print out intermediate values or graph the convergence over time—so should we include additional control parameters *for that*?

Then what do we do when we have n other iterative algorithms? Do we copy-paste the same stopping logic and parameters into each one? We'd end up with a lot of redundant code!

2.4.2.1 Iterators and Generators

This friction points to a conceptual distinction that our code does not support: *what happens at each iteration* is logically separate from *how we do the iteration*, but the two are fully coupled in our implementation. We need some way to abstract over iteration in some way that lets us separate *producing* values iteratively from *consuming* them.

Luckily, Python provides powerful facilities for doing exactly this: **iterators** and **generators**. Iterators give us a way of *consuming* values and generators give us a way of *producing* values.

You might not realize it, but chances are your Python code uses iterators all the time. Python's for loop uses an iterator under the hood to get each value it's looping over—this is how for loops work for lists, dictionaries, sets, ranges and even custom types. Try it out:

```
for x in [3, 2, 1]: print(x)
for x in {3, 2, 1}: print(x)
for x in range(3): print(x)
```

Note how the iterator for the set ({3, 2, 1}) prints 1 2 3 rather than 3 2 1—sets do not preserve the order in which elements are added, so they iterate over elements in some kind of internally defined order instead.

When we iterate over a dictionary, we will print the *keys* rather than the *values* because that is the default iterator. To get values or key-value pairs we'd need to use the values and `items` methods respectively, each of which returns a different kind of iterator over the dictionary.

```
d = {'a': 1, 'b': 2, 'c': 3}
for k in d: print(k)
for v in d.values(): print(v)
for k, v in d.items(): print(k, v)
```

In each of these three cases we're still looping over the same dictionary, we just get a different view each time—iterators give us the flexibility of iterating over the same structure in different ways.

Iterators aren't just for loops: they give us a first-class abstraction for iteration. We can pass them into functions; for example, Python's `list` function can convert any iterator into a list. This is handy when we want to see the elements of specialized iterators if the iterator itself does not print out its values:

```
>>> range(5)
range(0, 5)
>>> list(range(5))
[0, 1, 2, 3, 4]
```

Since iterators are first-class values, we can also write general-purpose iterator functions. The Python standard library has a set of operations like this in the `itertools` module; for example, `itertools.takewhile` lets us stop iterating as soon as some condition stops holding:

```
>>> elements = [1, 3, 2, 5, 3]
>>> list(itertools.takewhile(lambda x: x < 5, elements))
[1, 3, 2]
```

Note how we converted the result of `takewhile` into a list—without that, we'd see that `takewhile` returns some kind of opaque internal object that implements that iterator specifically. This works fine—we can use the `takewhile` object anywhere we could use any other iterator—but it looks a bit odd in the Python interpreter:

```
>>> itertools.takewhile(lambda x: x < 5, elements)
<itertools.takewhile object at 0x7f8e3baefb00>
```

Now that we've seen a few examples of how we can *use* iterators, how do we define our own? In the most general sense, a Python `Iterator` is any object that implements a `__next__()` method, but implementing iterators this way is pretty awkward. Luckily, Python has a more convenient way to create an iterator by creating a *generator* using the `yield` keyword. `yield` acts similar to `return` from a function, except instead of stopping the function altogether, it outputs the yielded value to an iterator and pauses the function until the yielded element is consumed by the caller.

This is a bit of an abstract description, so let's look at how this would apply to our `sqrt` function. Instead of looping and stopping based on some condition, we'll write a version of `sqrt` that returns an iterator with each iteration of the algorithm as a value:

```
def sqrt(a: float) -> Iterator[float]:
    x = a / 2 # initial guess
    while True:
        x = (x + (a / x)) / 2
        yield x
```

With this version, we update x at each iteration and then `yield` the updated value. Instead of getting a single value, the caller of the function gets an iterator that contains an infinite number of iterations; it is up to the caller to decide how many iterations to evaluate and when to stop. The sqrt function itself has an infinite loop, but this isn't a problem because execution of the function pauses at each yield which lets the caller of the function stop it whenever they want.

To do 10 iterations of the sqrt algorithm, we could use itertools.islice:

```
>>> iterations = list(itertools.islice(sqrt(25), 10))
>>> iterations[-1]
5.0
```

A fixed number of iterations can be useful for exploration, but we probably want the threshold-based convergence logic we had earlier. Since we now have a first-class abstraction for iteration, we can write a general-purpose converge function that takes an iterator and returns a version of that same iterator that stops as soon as two values are sufficiently close. Python 3.10 and later provides itertools.pairwise, which makes the code pretty simple:

```
def converge(values: Iterator[float], threshold: float) -> Iterator[float]:
    for a, b in itertools.pairwise(values):
        yield a

        if abs(a - b) < threshold:
            break
```

For older versions of Python, we'd have to implement our version of pairwise as well:

```
def pairwise(values: Iterator[A]) -> Iterator[Tuple[A, A]]:
    a = next(values, None)
    if a is None:
        return

    for b in values:
        yield (a, b)
        a = b
```

Both of these follow a common pattern with iterators: each function takes an iterator as an input *and returns an iterator as an output*. This doesn't always have to be the case, but we get a major advantage when it is: iterator → iterator operations *compose*. We can get relatively complex behavior by starting with an iterator (like our sqrt example) then applying *multiple* operations to it. For example, somebody calling sqrt might want to converge at some threshold but, just in case the algorithm gets stuck for some reason, also have a hard stop at 10,000 iterations. We don't need to write a new version of sqrt or even converge to do this; instead, we can use converge *with* itertools.islice:

```
results = converge(sqrt(n), 0.001)
capped_results = itertools.islice(results, 10000)
```

This is a powerful programming style because we can write and test each operation—sqrt, converge, islice—in isolation and get complex behavior by combining them in the right way. If we were writing the same logic *without* iterators, we would need a single loop that calculated each step of sqrt, checked for convergence *and* kept a counter to stop after 10,000 steps—and we'd need to replicate this pattern for every single such algorithm!

Iterators and generators will come up all throughout this book because they provide a programming abstraction for *processes*, making them a great foundation for the *mathematical* processes that underlie Reinforcement Learning.

2.5 KEY TAKEAWAYS FROM THIS CHAPTER

- The code in this book is designed not only to illustrate the topics and algorithms covered in each chapter but also to demonstrate some general principles of code design. Paying attention to code design gives us a codebase that's easier to understand, debug and extend.
- Code design is centered around abstractions. An abstraction is a code construct that combines separate pieces of information or unifies distinct concepts into one. Programming languages like Python provide different tools for expressing abstractions, but the goal is always the same: make the code easier to understand.
- Classes and interfaces (abstract classes in Python) define abstractions over data. Multiple implementations of the same concept can have the same interface (like `Distribution`), and code written using that interface will work for objects of *any* compatible class. Traditional Python classes have some inconvenient limitations which can be avoided by using dataclasses (classes with the `@dataclass` annotation).
- Python has type annotations that are required for dataclasses but are also useful for describing interfaces in functions and methods. Additional tools like `mypy` or PyCharm can use these type annotations to catch errors without needing to run the code.
- Functions are first-class values in Python, meaning that they can be stored in variables and passed to other functions as arguments. Classes abstract over data; functions abstract over computation.
- Iterators abstract over *iteration*: computations that happen in sequence, producing a value after each iteration. Reinforcement learning focuses primarily on iterative algorithms, so iterators become one of the key abstractions for working with different reinforcement learning algorithms.

I

Processes and Planning Algorithms

CHAPTER 3

Markov Processes

This book is about "Sequential Decisioning under Sequential Uncertainty". In this chapter, we will ignore the "sequential decisioning" aspect and focus just on the "sequential uncertainty" aspect.

3.1 THE CONCEPT OF *STATE* IN A PROCESS

For a gentle introduction to the concept of *State*, we start with an informal notion of the terms *Process* and *State* (this will be formalized later in this chapter). Informally, think of a *Process* as producing a sequence of random outcomes at discrete time steps that we'll index by a time variable $t = 0, 1, 2, \ldots$. The random outcomes produced by a Process might be key financial/trading/business metrics one cares about, such as prices of financial derivatives or the value of a portfolio held by an investor. To understand and reason about the evolution of these random outcomes of a Process, it is beneficial to focus on the internal representation of the Process at each point in time t, that is fundamentally responsible for driving the outcomes produced by the Process. We refer to this internal representation of the Process at time t as the (random) *State* of the Process at time t and denote it as S_t. Specifically, we are interested in the probability of the next State S_{t+1}, given the present State S_t and the past States $S_0, S_1, \ldots, S_{t-1}$, i.e., $\mathbb{P}[S_{t+1}|S_t, S_{t-1}, \ldots, S_0]$. So to clarify, we distinguish between the internal representation (*State*) and the output (outcomes) of the Process. The *State* could be any data type—it could be something as simple as the daily closing price of a single stock, or it could be something quite elaborate like the number of shares of each publicly traded stock held by each bank in the U.S., as noted at the end of each week.

3.2 UNDERSTANDING MARKOV PROPERTY FROM STOCK PRICE EXAMPLES

We will be learning about Markov Processes in this chapter, and these Processes have *States* that possess a property known as the *Markov Property*. So we will now learn about the *Markov Property of States*. Let us develop some intuition for this property with some examples of random evolution of stock prices over time.

To aid with the intuition, let us pretend that stock prices take on only integer values and that it's acceptable to have zero or negative stock prices. Let us denote the stock price at time t as X_t. Let us assume that from time step t to the next time step $t + 1$, the stock price can either go up by 1 or go down by 1, i.e., the only two outcomes for X_{t+1} are $X_t + 1$ or $X_t - 1$. To understand the random evolution of the stock prices in time, we just need to quantify the probability of an up-move $\mathbb{P}[X_{t+1} = X_t + 1]$ since the probability of a down-move $\mathbb{P}[X_{t+1} = X_t - 1] = 1 - \mathbb{P}[X_{t+1} = X_t + 1]$. We will consider 3 different processes

DOI: 10.1201/9781003229193-3

for the evolution of stock prices. These 3 processes will prescribe $\mathbb{P}[X_{t+1} = X_t + 1]$ in 3 different ways.

Process 1:

$$\mathbb{P}[X_{t+1} = X_t + 1] = \frac{1}{1 + e^{-\alpha_1(L-X_t)}}$$

where L is an arbitrary reference level and $\alpha_1 \in \mathbb{R}_{\geq 0}$ is a "pull strength" parameter. Note that this probability is defined as a logistic function of $L - X_t$ with the steepness of the logistic function controlled by the parameter α_1 (see Figure 3.1)

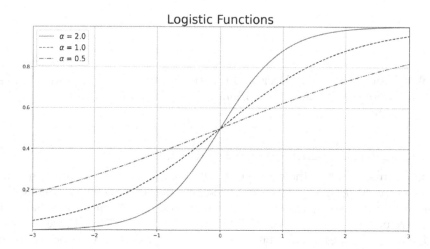

Figure 3.1 **Logistic Curves**

The way to interpret this logistic function of $L - X_t$ is: if X_t is greater than the reference level L (making $\mathbb{P}[X_{t+1} = X_t + 1] < 0.5$), then there is more of a down-pull than an up-pull. Likewise, if X_t is less than L, then there is more of an up-pull. The extent of the pull is controlled by the magnitude of the parameter α_1. We refer to this behavior as *mean-reverting behavior*, meaning the stock price tends to revert to the "mean" (i.e., to the reference level L).

We can model the State $S_t = X_t$ and note that the probabilities of the next State S_{t+1} depend only on the current State S_t and not on the previous States $S_0, S_1, \ldots, S_{t-1}$. Informally, we phrase this property as: "The future is independent of the past given the present". Formally, we can describe this property of the states as:

$$\mathbb{P}[S_{t+1}|S_t, S_{t-1}, \ldots, S_0] = \mathbb{P}[S_{t+1}|S_t] \text{ for all } t \geq 0$$

This is a highly desirable property since it helps make the mathematics of such processes much easier and the computations much more tractable. We call this the *Markov Property* of States, or simply that these are *Markov States*.

Let us now code this up. Firstly, we create a dataclass to represent the dynamics of this process. As you can see in the code below, the dataclass `Process1` contains two attributes `level_param: int` and `alpha1: float = 0.25` to represent L and α_1, respectively. It contains the method `up_prob` to calculate $\mathbb{P}[X_{t+1} = X_t + 1]$ and the method `next_state`, which samples from a Bernoulli distribution (whose probability is obtained from the method `up_prob`) and creates the next state S_{t+1} from the current state S_t. Also, note the nested dataclass `State` meant to represent the state of Process 1 (its only attribute `price: int` reflects the fact that the state consists of only the current price, which is an integer).

```
import numpy as np
from dataclasses import dataclass
@dataclass
class Process1:
    @dataclass
    class State:
        price: int

    level_param: int  # level to which price mean-reverts
    alpha1: float = 0.25  # strength of mean-reversion (non-negative value)

    def up_prob(self, state: State) -> float:
        return 1. / (1 + np.exp(-self.alpha1 * (self.level_param - state.price)))

    def next_state(self, state: State) -> State:
        up_move: int = np.random.binomial(1, self.up_prob(state), 1)[0]
        return Process1.State(price=state.price + up_move * 2 - 1)
```

Next, we write a simple simulator using Python's generator functionality (using `yield`) as follows:

```
def simulation(process, start_state):
    state = start_state
    while True:
        yield state
        state = process.next_state(state)
```

Now we can use this simulator function to generate sampling traces. In the following code, we generate num_traces number of sampling traces over time_steps number of time steps starting from a price X_0 of start_price. The use of Python's generator feature lets us do this "lazily" (on-demand) using the `itertools.islice` function.

```
import itertools
def process1_price_traces(
    start_price: int,
    level_param: int,
    alpha1: float,
    time_steps: int,
    num_traces: int
) -> np.ndarray:
    process = Process1(level_param=level_param, alpha1=alpha1)
    start_state = Process1.State(price=start_price)
    return np.vstack([
        np.fromiter((s.price for s in itertools.islice(
            simulation(process, start_state),
            time_steps + 1
        )), float) for _ in range(num_traces)])
```

The entire code is in the file rl/chapter2/stock_price_simulations.py. We encourage you to play with this code with different start_price, level_param, alpha1, time_steps, num_traces. You can plot graphs of sampling traces of the stock price or graphs of the terminal distributions of the stock price at various time points (both of these plotting functions are made available for you in this code).

Now let us consider a different process.

Process 2:

$$\mathbb{P}[X_{t+1} = X_t + 1] = \begin{cases} 0.5(1 - \alpha_2(X_t - X_{t-1})) & \text{if } t > 0 \\ 0.5 & \text{if } t = 0 \end{cases}$$

where α_2 is a "pull strength" parameter in the closed interval $[0, 1]$. The intuition here is that the direction of the next move $X_{t+1} - X_t$ is biased in the reverse direction of the previous move $X_t - X_{t-1}$, and the extent of the bias is controlled by the parameter α_2.

We note that if we model the state S_t as X_t, we won't satisfy the Markov Property because the probabilities of X_{t+1} depend on not just X_t but also on X_{t-1}. However, we can perform a little trick here and create an augmented state S_t consisting of the pair $(X_t, X_t - X_{t-1})$. In case $t = 0$, the state S_0 can assume the value $(X_0, Null)$ where $Null$ is just a symbol denoting the fact that there have been no stock price movements thus far. With the state S_t as this pair $(X_t, X_t - X_{t-1})$, we can see that the Markov Property is indeed satisfied:

$$\mathbb{P}[(X_{t+1}, X_{t+1} - X_t)|(X_t, X_t - X_{t-1}), (X_{t-1}, X_{t-1} - X_{t-2}), \ldots, (X_0, Null)]$$

$$= \mathbb{P}[(X_{t+1}, X_{t+1} - X_t)|(X_t, X_t - X_{t-1})] = 0.5(1 - \alpha_2(X_{t+1} - X_t)(X_t - X_{t-1}))$$

It would be natural to wonder why the state doesn't comprise of simply $X_t - X_{t-1}$—in other words, why is X_t also required to be part of the state. It is true that knowledge of simply $X_t - X_{t-1}$ fully determines the probabilities of $X_{t+1} - X_t$. So if we set the state to be simply $X_t - X_{t-1}$ at any time step t, then we do get a Markov Process with just two states $+1$ and -1 (with probability transitions between them). However, this simple Markov Process doesn't tell us what the value of stock price X_t is by looking at the state $X_t - X_{t-1}$ at time t. In this application, we not only care about Markovian state transition probabilities, but we also care about knowledge of stock price at any time t from knowledge of the state at time t. Hence, we model the state as the pair (X_t, X_{t-1}).

Note that if we had modeled the state S_t as the entire stock price history $(X_0, X_1, \ldots, X_t)$, then the Markov Property would be satisfied trivially. The Markov Property would also be satisfied if we had modeled the state S_t as the pair (X_t, X_{t-1}) for $t > 0$ and S_0 as $(X_0, Null)$. However, our choice of $S_t := (X_t, X_t - X_{t-1})$ is the "simplest/minimal" internal representation. In fact, in this entire book, our endeavor in modeling states for various processes is to ensure the Markov Property with the "simplest/minimal" representation for the state. The corresponding dataclass for Process 2 is shown below:

```
handy_map: Mapping[Optional[bool], int] = {True: -1, False: 1, None: 0}
@dataclass
class Process2:
    @dataclass
    class State:
        price: int
        is_prev_move_up: Optional[bool]

    alpha2: float = 0.75  # strength of reverse-pull (value in [0,1])

    def up_prob(self, state: State) -> float:
        return 0.5 * (1 + self.alpha2 * handy_map[state.is_prev_move_up])

    def next_state(self, state: State) -> State:
        up_move: int = np.random.binomial(1, self.up_prob(state), 1)[0]
        return Process2.State(
            price=state.price + up_move * 2 - 1,
            is_prev_move_up=bool(up_move)
        )
```

The code for generation of sampling traces of the stock price is almost identical to the code we wrote for Process 1.

```
def process2_price_traces(
    start_price: int,
    alpha2: float,
    time_steps: int,
    num_traces: int
) -> np.ndarray:
    process = Process2(alpha2=alpha2)
```

```
start_state = Process2.State(price=start_price, is_prev_move_up=None)
return np.vstack([
    np.fromiter((s.price for s in itertools.islice(
        simulation(process, start_state),
        time_steps + 1
    )), float) for _ in range(num_traces)])
```

Now let us look at a more complicated process.

Process 3: This is an extension of Process 2, where the probability of next movement depends not only on the last movement but on all past movements. Specifically, it depends on the number of past up-moves (call it $U_t = \sum_{i=1}^{t} \max(X_i - X_{i-1}, 0)$) relative to the number of past down-moves (call it $D_t = \sum_{i=1}^{t} \max(X_{i-1} - X_i, 0)$) in the following manner:

$$\mathbb{P}[X_{t+1} = X_t + 1] = \begin{cases} \frac{1}{1+(\frac{U_t+D_t}{D_t}-1)^{\alpha_3}} & \text{if } t > 0 \\ 0.5 & \text{if } t = 0 \end{cases}$$

where $\alpha_3 \in \mathbb{R}_{\geq 0}$ is a "pull strength" parameter. Let us view the above probability expression as:

$$f(\frac{D_t}{U_t + D_t}; \alpha_3)$$

where $f : [0,1] \to [0,1]$ is a sigmoid-shaped function

$$f(x; \alpha) = \frac{1}{1 + (\frac{1}{x} - 1)^{\alpha}}$$

whose steepness at $x = 0.5$ is controlled by the parameter α (note: values of $\alpha < 1$ will produce an inverse sigmoid as seen in Figure 3.2, which shows unit-sigmoid functions f for different values of α).

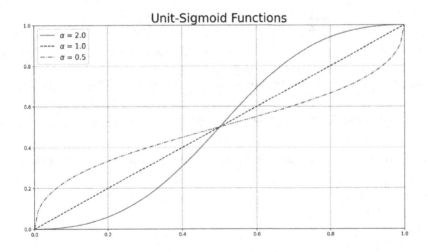

Figure 3.2 **Unit-Sigmoid Curves**

The probability of next up-movement is fundamentally dependent on the quantity $\frac{D_t}{U_t+D_t}$ (the function f simply serves to control the extent of the "reverse pull"). $\frac{D_t}{U_t+D_t}$ is the fraction of past time steps when there was a down-move. So, if number of down-moves in history are greater than number of up-moves in history, then there will be more of an up-pull than a down-pull for the next price movement $X_{t+1} - X_t$ (likewise, the other way round when $U_t > D_t$). The extent of this "reverse pull" is controlled by the "pull strength" parameter α_3 (governed by the sigmoid-shaped function f).

Again, note that if we model the state S_t as X_t, we won't satisfy the Markov Property because the probabilities of next state $S_{t+1} = X_{t+1}$ depends on the entire history of stock price moves and not just on the current state $S_t = X_t$. However, we can again do something clever and create a compact enough state S_t consisting of simply the pair (U_t, D_t). With this representation for the state S_t, the Markov Property is indeed satisfied:

$$\mathbb{P}[(U_{t+1}, D_{t+1})|(U_t, D_t), (U_{t-1}, D_{t-1}), \ldots, (U_0, D_0)] = \mathbb{P}[(U_{t+1}, D_{t+1})|(U_t, D_t)]$$

$$= \begin{cases} f(\frac{D_t}{U_t+D_t}; \alpha_3) & \text{if } U_{t+1} = U_t + 1, D_{t+1} = D_t \\ f(\frac{U_t}{U_t+D_t}; \alpha_3) & \text{if } U_{t+1} = U_t, D_{t+1} = D_t + 1 \end{cases}$$

It is important to note that unlike Processes 1 and 2, the stock price X_t is actually not part of the state S_t in Process 3. This is because U_t and D_t together contain sufficient information to capture the stock price X_t (since $X_t = X_0 + U_t - D_t$, and noting that X_0 is provided as a constant).

The corresponding dataclass for Process 3 is shown below:

```python
@dataclass
class Process3:
    @dataclass
    class State:
        num_up_moves: int
        num_down_moves: int

    alpha3: float = 1.0  # strength of reverse-pull (non-negative value)

    def up_prob(self, state: State) -> float:
        total = state.num_up_moves + state.num_down_moves
        if total == 0:
            return 0.5
        elif state.num_down_moves == 0:
            return state.num_down_moves ** self.alpha3
        else:
            return 1. / (1 + (total / state.num_down_moves - 1) ** self.alpha3)

    def next_state(self, state: State) -> State:
        up_move: int = np.random.binomial(1, self.up_prob(state), 1)[0]
        return Process3.State(
            num_up_moves=state.num_up_moves + up_move,
            num_down_moves=state.num_down_moves + 1 - up_move
        )
```

The code for generation of sampling traces of the stock price is shown below:

```python
def process3_price_traces(
    start_price: int,
    alpha3: float,
    time_steps: int,
    num_traces: int
) -> np.ndarray:
    process = Process3(alpha3=alpha3)
    start_state = Process3.State(num_up_moves=0, num_down_moves=0)
    return np.vstack([
        np.fromiter((start_price + s.num_up_moves - s.num_down_moves
                     for s in itertools.islice(simulation(process, start_state),
                                               time_steps + 1)), float)
        for _ in range(num_traces)])
```

As suggested for Process 1, you can plot graphs of sampling traces of the stock price, or plot graphs of the probability distributions of the stock price at various terminal time points T for Processes 2 and 3, by playing with the code in rl/chapter2/stock_price_simulations.py.

Figure 3.3 shows a single sampling trace of stock prices for each of the 3 processes. Figure 3.4 shows the probability distribution of the stock price at terminal time $T = 100$ over 1000 traces.

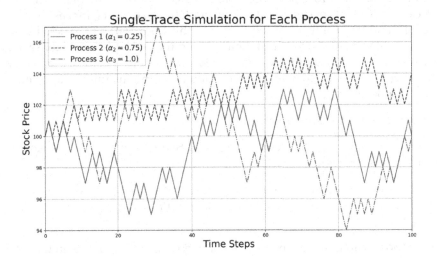

Figure 3.3 **Single Sampling Trace**

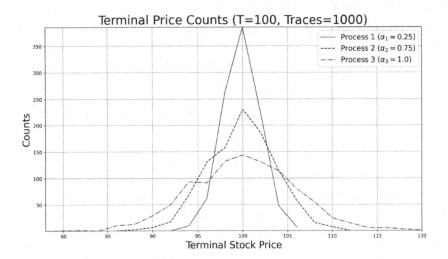

Figure 3.4 **Terminal Distribution**

Having developed the intuition for the Markov Property of States, we are now ready to formalize the notion of Markov Processes (some of the literature refers to Markov Processes as Markov Chains, but we will stick with the term Markov Processes).

3.3 FORMAL DEFINITIONS FOR MARKOV PROCESSES

Our formal definitions in this book will be restricted to Discrete-Time Markov Processes, where time moves forward in discrete time steps $t = 0, 1, 2, \ldots$. Also for ease of exposition, our formal definitions in this book will be restricted to sets of states that are countable. A countable set can be either a finite set or an infinite set of the same cardinality as the set of natural numbers, i.e., a set that is enumerable. This book will cover examples of continuous-time Markov Processes, where time is a continuous variable [1]. This book will also cover examples of sets of states that are uncountable [2]. However, for ease of exposition, our definitions and development of the theory in this book will be restricted to discrete-time and countable sets of states. The definitions and theory can be analogously extended to continuous-time or to uncountable sets of states (we request you to self-adjust the definitions and theory accordingly when you encounter continuous-time or uncountable sets of states in this book).

Definition 3.3.1. A *Markov Process* consists of:

- A countable set of states $\mathcal{S}$ (known as the State Space) and a set $\mathcal{T} \subseteq \mathcal{S}$ (known as the set of Terminal States).

- A time-indexed sequence of random states $S_t \in \mathcal{S}$ for time steps $t = 0, 1, 2, \ldots$ with each state transition satisfying the Markov Property: $\mathbb{P}[S_{t+1}|S_t, S_{t-1}, \ldots, S_0] = \mathbb{P}[S_{t+1}|S_t]$ for all $t \geq 0$.

- Termination: If an outcome for S_T (for some time step T) is a state in the set $\mathcal{T}$, then this sequence outcome terminates at time step T.

We refer to $\mathbb{P}[S_{t+1}|S_t]$ as the transition probabilities for time t.

Definition 3.3.2. A *Time-Homogeneous Markov Process* is a Markov Process with the additional property that $\mathbb{P}[S_{t+1}|S_t]$ is independent of t.

This means the dynamics of a Time-Homogeneous Markov Process can be fully specified with the function

$$\mathcal{P} : (\mathcal{S} - \mathcal{T}) \times \mathcal{S} \to [0, 1]$$

defined as:

$$\mathcal{P}(s, s') = \mathbb{P}[S_{t+1} = s'|S_t = s] \text{ for time steps } t = 0, 1, 2, \ldots, \text{ for all } s \in \mathcal{S} - \mathcal{T}, s' \in \mathcal{S}$$

such that

$$\sum_{s' \in \mathcal{S}} \mathcal{P}(s, s') = 1 \text{ for all } s \in \mathcal{S} - \mathcal{T}$$

We refer to the function $\mathcal{P}$ as the transition probability function of the Time-Homogeneous Markov Process, with the first argument to $\mathcal{P}$ to be thought of as the *source state* and the second argument as the *destination state*.

Note that the arguments to $\mathcal{P}$ in the above specification are devoid of the time index t (hence, the term *Time-Homogeneous* which means "time-invariant"). Moreover, note that a Markov Process that is not time-homogeneous can be converted to a Time-Homogeneous Markov Process by augmenting all states with the time index t. This means if the original

---

[1] Markov Processes in continuous-time often go by the name *Stochastic Processes*, and their calculus goes by the name *Stochastic Calculus* (see Appendix C).

[2] Uncountable sets are those with cardinality larger than the set of natural numbers, e.g., the set of real numbers, which are not enumerable.

state space of a Markov Process that is not time-homogeneous is $\mathcal{S}$, then the state space of the corresponding Time-Homogeneous Markov Process is $\mathbb{Z}_{\geq 0} \times \mathcal{S}$ (where $\mathbb{Z}_{\geq 0}$ denotes the domain of the time index). This is because each time step has its own unique set of (augmented) states, which means the entire set of states in $\mathbb{Z}_{\geq 0} \times \mathcal{S}$ can be covered by time-invariant transition probabilities, thus qualifying as a Time-Homogeneous Markov Process. Therefore, henceforth, any time we say *Markov Process,* assume we are referring to a *Discrete-Time, Time-Homogeneous Markov Process with a Countable State Space* (unless explicitly specified otherwise), which in turn will be characterized by the transition probability function $\mathcal{P}$. Note that each of the 3 stock price processes we covered are examples of a (Time-Homogeneous) Markov Process, even without requiring augmenting the state with the time index.

The classical definitions and theory of Markov Processes model "termination" with the idea of *Absorbing States*. A state s is called an absorbing state if $\mathcal{P}(s, s) = 1$. This means, once we reach an absorbing state, we are "trapped" there, hence capturing the notion of "termination". So the classical definitions and theory of Markov Processes typically don't include an explicit specification of states as terminal and non-terminal. However, when we get to Markov Reward Processes and Markov Decision Processes (frameworks that are extensions of Markov Processes), we will need to explicitly specify states as terminal and non-terminal states rather than model the notion of termination with absorbing states. So, for consistency in definitions and in the development of the theory, we are going with a framework where states in a Markov Process are explicitly specified as terminal or non-terminal states. We won't consider an absorbing state as a terminal state as the Markov Process keeps moving forward in time forever when it gets to an absorbing state. We will refer to $\mathcal{S} - \mathcal{T}$ as the set of Non-Terminal States $\mathcal{N}$ (and we will refer to a state in $\mathcal{N}$ as a non-terminal state). The sequence $S_0, S_1, S_2, \ldots$ terminates at time step $t = T$ if $S_T \in \mathcal{T}$.

3.3.1 Starting States

Now it's natural to ask the question: How do we "start" the Markov Process (in the stock price examples, this was the notion of the start state)? More generally, we'd like to specify a probability distribution of start states so we can perform simulations and (let's say) compute the probability distribution of states at specific future time steps. While this is a relevant question, we'd like to separate the following two specifications:

- Specification of the transition probability function $\mathcal{P}$.
- Specification of the probability distribution of start states (denote this as $\mu : \mathcal{N} \to [0, 1]$).

We say that a Markov Process is fully specified by $\mathcal{P}$ in the sense that this gives us the transition probabilities that govern the complete dynamics of the Markov Process. A way to understand this is to relate specification of $\mathcal{P}$ to the specification of movement rules in a game (such as chess or monopoly). These games are specified with a finite (in fact, fairly compact) set of rules that is easy for a newbie to the game to understand. However, when we want to *actually play* the game, we need to specify the starting position (one could start these games at arbitrary, but legal, starting positions and not just at some canonical starting position). The specification of starting positions is basically the specification of μ. Given μ together with $\mathcal{P}$ enables us to generate sampling traces of the Markov Process (i.e., *play* games like chess or monopoly). These sampling traces typically result in a wide range of outcomes due to sampling and long-running of the Markov Process (versus compact specification of transition probabilities). These sampling traces enable us to answer questions such as probability distribution of states at specific future time steps or expected time of first occurrence of a specific state etc., given a certain starting probability distribution μ.

Thinking about the separation between specifying the rules of the game versus actually playing the game helps us understand the need to separate the notion of dynamics specification $\mathcal{P}$ (fundamental to the time-homogeneous character of the Markov Process) and the notion of starting distribution μ (required to perform sampling traces). Hence, the separation of concerns between $\mathcal{P}$ and μ is key to the conceptualization of Markov Processes. Likewise, we separate the concerns in our code design as well, as evidenced by how we separated the next_state method in the Process dataclasses and the simulation function.

3.3.2 Terminal States

Games are examples of Markov Processes that terminate at specific states (based on rules for winning or losing the game). In general, in a Markov Process, termination might occur after a variable number of time steps (like in the games examples), or like we will see in many financial application examples, termination might occur after a fixed number of time steps, or like in the stock price examples we saw earlier, there is, in fact, no termination.

If all random sequences of states (sampling traces) reach a terminal state, then we say that these random sequences of the Markov Process are *Episodic* (otherwise, we call these sequences as *Continuing*). The notion of episodic sequences is important in Reinforcement Learning since some Reinforcement Learning algorithms require episodic sequences.

When we cover some of the financial applications later in this book, we will find that the Markov Process terminates after a fixed number of time steps, say T. In these applications, the time index t is part of the state representation, each state with time index $t = T$ is labeled a terminal state, and all states with time index $t < T$ will transition to states with time index $t + 1$.

Now we are ready to write some code for Markov Processes, where we illustrate how to specify that certain states are terminal states.

3.3.3 Markov Process Implementation

The first thing we do is to create separate classes for non-terminal states $\mathcal{N}$ and terminal states $\mathcal{T}$, with an abstract base class for all states $\mathcal{S}$ (terminal or non-terminal). In the code below, the abstract base class (ABC) State represents $\mathcal{S}$. The class State is parameterized by a generic type (TypeVar('S')) representing a generic state space Generic[S].

The concrete class Terminal represents $\mathcal{T}$ and the concrete class NonTerminal represents $\mathcal{N}$. The method on_non_terminal will prove to be very beneficial in the implementation of various algorithms we shall be writing for Markov Processes and also for Markov Reward Processes and Markov Decision Processes (which are extensions of Markov Processes). The method on_non_terminal enables us to calculate a value for all states in $\mathcal{S}$ even though the calculation is defined only for all non-terminal states $\mathcal{N}$. The argument f to on_non_terminal defines this value through a function from $\mathcal{N}$ to an arbitrary value-type X. The argument default provides the default value for terminal states $\mathcal{T}$ so that on_non_terminal can be used on any object in State (i.e. for any state in $\mathcal{S}$, terminal or non-terminal). As an example, let's say you want to calculate the expected number of states one would traverse after a certain state and before hitting a terminal state. Clearly, this calculation is well-defined for non-terminal states, and the function f would implement this by either some kind of analytical method or by sampling state-transition sequences and averaging the counts of non-terminal states traversed across those sequences. By defining (defaulting) this value to be 0 for terminal states, we can then invoke such a calculation for all states $\mathcal{S}$, terminal or non-terminal, and embed this calculation in an algorithm without worrying about special handing in the code for the edge case of being a terminal state.

```
from abc import ABC
from dataclasses import dataclass
from typing import Generic, Callable, TypeVar
S = TypeVar('S')
X = TypeVar('X')
class State(ABC, Generic[S]):
    state: S

    def on_non_terminal(
        self,
        f: Callable[[NonTerminal[S]], X],
        default: X
    ) -> X:
        if isinstance(self, NonTerminal):
            return f(self)
        else:
            return default
@dataclass(frozen=True)
class Terminal(State[S]):
    state: S

@dataclass(frozen=True)
class NonTerminal(State[S]):
    state: S
```

Now we are ready to write a class to represent Markov Processes. We create an abstract class `MarkovProcess` parameterized by a generic type (`TypeVar('S')`) representing a generic state space `Generic[S]`. The abstract class has an abstract method called `transition` that is meant to specify the transition probability distribution of next states, given a current non-terminal state. We know that `transition` is well-defined only for non-terminal states, and hence, its argument is clearly type-annotated as `NonTerminal[S]`. The return type of `transition` is `Distribution[State[S]]`, which, as we know from Chapter 2, represents the probability distribution of next states. We also have a method `simulate` that enables us to generate an `Iterable` (generator) of sampled states, given as input a `start_state_distribution: Distribution[NonTerminal[S]]` (from which we sample the starting state). The sampling of next states relies on the implementation of the `sample` method for the `Distribution[State[S]]` object produced by the `transition` method.

Here's the full body of the abstract class `MarkovProcess`[3]:

```
from abc import abstractmethod
from rl.distribution import Distribution
from typing import Iterable
class MarkovProcess(ABC, Generic[S]):
    @abstractmethod
    def transition(self, state: NonTerminal[S]) -> Distribution[State[S]]:
        pass

    def simulate(
        self,
        start_state_distribution: Distribution[NonTerminal[S]]
    ) -> Iterable[State[S]]:
        state: State[S] = start_state_distribution.sample()
        yield state

        while isinstance(state, NonTerminal):
            state = self.transition(state).sample()
            yield state
```

---

[3]`MarkovProcess` is defined in the file rl/markov_process.py.

3.4 STOCK PRICE EXAMPLES MODELED AS MARKOV PROCESSES

So if you have a mathematical specification of the transition probabilities of a Markov Process, all you need to do is to create a concrete class that implements the interface of the abstract class MarkovProcess (specifically by implementing the abstract method transition) in a manner that captures your mathematical specification of the transition probabilities. Let us write this for the case of Process 3 (the 3rd example of stock price transitions we covered earlier). We name the concrete class as StockPriceMP3. Note that the generic state space S is now replaced with a specific state space represented by the type @dataclass StateMP3. The code should be self-explanatory since we implemented this process as a standalone in the previous section. Note the use of the Categorical distribution in the transition method to capture the 2-outcomes probability distribution of next states (for movements up or down).

```python
from rl.distribution import Categorical
from rl.gen_utils.common_funcs import get_unit_sigmoid_func

@dataclass
class StateMP3:
    num_up_moves: int
    num_down_moves: int

@dataclass
class StockPriceMP3(MarkovProcess[StateMP3]):

    alpha3: float = 1.0  # strength of reverse-pull (non-negative value)

    def up_prob(self, state: StateMP3) -> float:
        total = state.num_up_moves + state.num_down_moves
        return get_unit_sigmoid_func(self.alpha3)(
            state.num_down_moves / total
        ) if total else 0.5

    def transition(
        self,
        state: NonTerminal[StateMP3]
    ) -> Categorical[State[StateMP3]]:
        up_p = self.up_prob(state.state)
        return Categorical({
            NonTerminal(StateMP3(
                state.state.num_up_moves + 1, state.state.num_down_moves
            )): up_p,
            NonTerminal(StateMP3(
                state.state.num_up_moves, state.state.num_down_moves + 1
            )): 1 - up_p
        })
```

To generate sampling traces, we write the following function:

```python
from rl.distribution import Constant
import numpy as np

def process3_price_traces(
    start_price: int,
    alpha3: float,
    time_steps: int,
    num_traces: int
) -> np.ndarray:
    mp = StockPriceMP3(alpha3=alpha3)
    start_state_distribution = Constant(
        NonTerminal(StateMP3(num_up_moves=0, num_down_moves=0))
    )
    return np.vstack([np.fromiter(
        (start_price + s.state.num_up_moves - s.state.num_down_moves for s in
         itertools.islice(
            mp.simulate(start_state_distribution),
            time_steps + 1
        )),
```

```
    float
) for _ in range(num_traces)])
```

We leave it to you as an exercise to similarly implement Stock Price Processes 1 and 2 that we had covered earlier. The complete code along with the driver to set input parameters, run all 3 processes and create plots is in the file rl/chapter2/stock_price_mp.py. We encourage you to change the input parameters in __main__ and get an intuitive feel for how the simulation results vary with the changes in parameters.

3.5   FINITE MARKOV PROCESSES

Now let us consider Markov Processes with a finite state space. So we can represent the state space as $\mathcal{S} = \{s_1, s_2, \ldots, s_n\}$. Assume the set of non-terminal states $\mathcal{N}$ has $m \leq n$ states. Let us refer to Markov Processes with finite state spaces as Finite Markov Processes. Since Finite Markov Processes are a subclass of Markov Processes, it would make sense to create a concrete class FiniteMarkovProcess that implements the interface of the abstract class MarkovProcess (specifically implement the abstract method transition). But first, let's think about the data structure required to specify an instance of a FiniteMarkovProcess (i.e., the data structure we'd pass to the __init__ method of FiniteMarkovProcess). One choice is a $m \times n$ 2D numpy array representation, i.e., matrix elements representing transition probabilities

$$\mathcal{P} : \mathcal{N} \times \mathcal{S} \to [0, 1]$$

However, we often find that this matrix is sparse—transitions from source states land at small sets of destination states. So we'd like a sparse representation, and we can accomplish this by conceptualizing $\mathcal{P}$ in an equivalent curried form[4] as follows:

$$\mathcal{N} \to (\mathcal{S} \to [0, 1])$$

With this curried view, we can represent the outer $\to$ as a map (in Python, as a dictionary of type Mapping) whose keys are the non-terminal states $\mathcal{N}$, and each non-terminal-state key maps to a FiniteDistribution[S] type that represents the inner $\to$, i.e. a finite probability distribution of the next states transitioned to from a given non-terminal state.

Note that the FiniteDistribution[S] will only contain the set of states transitioned to with non-zero probability. To make things concrete, here's a toy Markov Process data structure example of a city with highly unpredictable weather outcomes from one day to the next (note: Categorical type inherits from FiniteDistribution type in the code at rl/distribution.py):

```
from rl.distribution import Categorical
{
  "Rain": Categorical({"Rain": 0.3, "Nice": 0.7}),
  "Snow": Categorical({"Rain": 0.4, "Snow": 0.6}),
  "Nice": Categorical({"Rain": 0.2, "Snow": 0.3})
}
```

It is common to view this Markov Process representation as a directed graph, as depicted in Figure 3.5. The nodes are the states and the directed edges are the probabilistic state transitions, with the transition probabilities labeled on them.

Our goal now is to define a FiniteMarkovProcess class that is a concrete class implementation of the abstract class MarkovProcess. This requires us to wrap the states in

---

[4]Currying is the technique of converting a function that takes multiple arguments into a sequence of functions that each takes a single argument, as illustrated above for the $\mathcal{P}$ function.

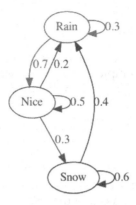

Figure 3.5 Weather Markov Process

the keys/values of the `FiniteMarkovProcess` dictionary with the appropriate `Terminal` or `NonTerminal` wrapping. Let's create an alias called `Transition` for this wrapped dictionary data structure since we will use this wrapped data structure often:

```
from typing import Mapping
from rl.distribution import FiniteDistribution

Transition = Mapping[NonTerminal[S], FiniteDistribution[State[S]]]
```

To create a `Transition` data type from the above example of the weather Markov Process, we'd need to wrap each of the "Rain", "Snow" and "Nice" strings with `NonTerminal`.

Now we are ready to write the code for the `FiniteMarkovProcess` class.[5] The `__init__` method (constructor) takes as argument a `transition_map` whose type is similar to `Transition[S]` except that we use the `S` type directly in the `Mapping` representation instead of `NonTerminal[S]` or `State[S]` (this is convenient for users to specify their Markov Process in a succinct `Mapping` representation without the burden of wrapping each `S` with a `NonTerminal[S]` or `Terminal[S]`). The dictionary we created above for the weather Markov Process can be used as the `transition_map` argument. However, this means the `__init__` method needs to wrap the specified `S` states as `NonTerminal[S]` or `Terminal[S]` when creating the attribute `self.transition_map`. We also have an attribute `self.non_terminal_states: Sequence[NonTerminal[S]]` that is an ordered sequence of the non-terminal states. We implement the `transition` method by simply returning the `FiniteDistribution[State[S]]` the given `state: NonTerminal[S]` maps to in the attribute `self.transition_map: Transition[S]`. Note that along with the `transition` method, we have implemented the `__repr__` method for a well-formatted display of `self.transition_map`.

```
from typing import Sequence
from rl.distribution import FiniteDistribution, Categorical

class FiniteMarkovProcess(MarkovProcess[S]):
    non_terminal_states: Sequence[NonTerminal[S]]
    transition_map: Transition[S]

    def __init__(self, transition_map: Mapping[S, FiniteDistribution[S]]):
        non_terminals: Set[S] = set(transition_map.keys())
        self.transition_map = {
            NonTerminal(s): Categorical(
                {(NonTerminal(s1) if s1 in non_terminals else Terminal(s1)): p
                 for s1, p in v}
```

---

[5]`FiniteMarkovProcess` is defined in the file rl/markov_process.py.

```
        ) for s, v in transition_map.items()
    }
    self.non_terminal_states = list(self.transition_map.keys())
def __repr__(self) -> str:
    display = ""

    for s, d in self.transition_map.items():
        display += f"From State {s.state}:\n"
        for s1, p in d:
            opt = (
                "Terminal State" if isinstance(s1, Terminal) else "State"
            )
            display += f"  To {opt} {s1.state} with Probability {p:.3f}\n"
    return display

def transition(self, state: NonTerminal[S]) -> FiniteDistribution[State[S]]:
    return self.transition_map[state]
```

3.6 SIMPLE INVENTORY EXAMPLE

To help conceptualize Finite Markov Processes, let us consider a simple example of changes in inventory at a store. Assume you are the store manager and that you are tasked with controlling the ordering of inventory from a supplier. Let us focus on the inventory of a particular type of bicycle. Assume that each day there is random (non-negative integer) demand for the bicycle with the probabilities of demand following a Poisson distribution (with Poisson parameter $\lambda \in \mathbb{R}_{\geq 0}$), i.e. demand i for each $i = 0, 1, 2, \ldots$ occurs with probability

$$f(i) = \frac{e^{-\lambda}\lambda^i}{i!}$$

Denote $F : \mathbb{Z}_{\geq 0} \to [0, 1]$ as the Poisson cumulative probability distribution function, i.e.,

$$F(i) = \sum_{j=0}^{i} f(j)$$

Assume you have storage capacity for at most $C \in \mathbb{Z}_{\geq 0}$ bicycles in your store. Each evening at 6pm when your store closes, you have the choice to order a certain number of bicycles from your supplier (including the option to not order any bicycles on a given day). The ordered bicycles will arrive 36 hours later (at 6am the day after the day after you order—we refer to this as *delivery lead time* of 36 hours). Denote the *State* at 6pm store-closing each day as (α, β), where α is the inventory in the store (referred to as On-Hand Inventory at 6pm) and β is the inventory on a truck from the supplier (that you had ordered the previous day) that will arrive in your store the next morning at 6am (β is referred to as On-Order Inventory at 6pm). Due to your storage capacity constraint of at most C bicycles, your ordering policy is to order $C - (\alpha + \beta)$ if $\alpha + \beta < C$ and to not order if $\alpha + \beta \geq C$. The precise sequence of events in a 24-hour cycle is:

- Observe the (α, β) *State* at 6pm store-closing (call this state S_t).
- Immediately order according to the ordering policy described above.
- Receive bicycles at 6am if you had ordered 36 hours ago.
- Open the store at 8am.
- Experience random demand from customers according to demand probabilities stated above (number of bicycles sold for the day will be the minimum of demand on the day and inventory at store opening on the day).

- Close the store at 6pm and observe the state (this state is S_{t+1}).

If we let this process run for a while, in steady-state, we ensure that $\alpha + \beta \le C$. So to model this process as a Finite Markov Process, we shall only consider the steady-state (finite) set of states

$$\mathcal{S} = \{(\alpha, \beta) | \alpha \in \mathbb{Z}_{\ge 0}, \beta \in \mathbb{Z}_{\ge 0}, 0 \le \alpha + \beta \le C\}$$

So restricting ourselves to this finite set of states, our order quantity equals $C - (\alpha + \beta)$ when the state is (α, β).

If the current state S_t is (α, β), there are only $\alpha + \beta + 1$ possible next states S_{t+1} as follows:

$$(\alpha + \beta - i, C - (\alpha + \beta)) \text{ for } i = 0, 1, \dots, \alpha + \beta$$

with transition probabilities governed by the Poisson probabilities of demand as follows:

$$\mathcal{P}((\alpha, \beta), (\alpha + \beta - i, C - (\alpha + \beta))) = f(i) \text{ for } 0 \le i \le \alpha + \beta - 1$$

$$\mathcal{P}((\alpha, \beta), (0, C - (\alpha + \beta))) = \sum_{j=\alpha+\beta}^{\infty} f(j) = 1 - F(\alpha + \beta - 1)$$

Note that the next state's (S_{t+1}) On-Hand can be zero resulting from any of infinite possible demand outcomes greater than or equal to $\alpha + \beta$.

So we are now ready to write code for this simple inventory example as a Markov Process. All we have to do is to create a derived class inherited from `FiniteMarkovProcess` and write a method to construct the `transition_map: Transition`. Note that the generic state type `S` is replaced here with the `@dataclass InventoryState` consisting of the pair of On-Hand and On-Order inventory quantities comprising the state of this Finite Markov Process.

```
from rl.distribution import Categorical
from scipy.stats import poisson

@dataclass(frozen=True)
class InventoryState:
    on_hand: int
    on_order: int

    def inventory_position(self) -> int:
        return self.on_hand + self.on_order

class SimpleInventoryMPFinite(FiniteMarkovProcess[InventoryState]):
    def __init__(
        self,
        capacity: int,
        poisson_lambda: float
    ):
        self.capacity: int = capacity
        self.poisson_lambda: float = poisson_lambda

        self.poisson_distr = poisson(poisson_lambda)
        super().__init__(self.get_transition_map())

    def get_transition_map(self) -> \
            Mapping[InventoryState, FiniteDistribution[InventoryState]]:
        d: Dict[InventoryState, Categorical[InventoryState]] = {}
        for alpha in range(self.capacity + 1):
            for beta in range(self.capacity + 1 - alpha):
                state = InventoryState(alpha, beta)
                ip = state.inventory_position()
                beta1 = self.capacity - ip
                state_probs_map: Mapping[InventoryState, float] = {
```

```
                InventoryState(ip - i, beta1):
                (self.poisson_distr.pmf(i) if i < ip else
                 1 - self.poisson_distr.cdf(ip - 1))
                for i in range(ip + 1)
        }
        d[InventoryState(alpha, beta)] = Categorical(state_probs_map)
    return d
```

Let us utilize the __repr__ method written previously to view the transition probabilities for the simple case of $C = 2$ and $\lambda = 1.0$ (this code is in the file rl/chapter2/simple_inventory_mp.py)

```
user_capacity = 2
user_poisson_lambda = 1.0

si_mp = SimpleInventoryMPFinite(
    capacity=user_capacity,
    poisson_lambda=user_poisson_lambda
)

print(si_mp)
```

The output we get is nicely displayed as:

```
From State InventoryState(on_hand=0, on_order=0):
  To State InventoryState(on_hand=0, on_order=2) with Probability 1.000
From State InventoryState(on_hand=0, on_order=1):
  To State InventoryState(on_hand=1, on_order=1) with Probability 0.368
  To State InventoryState(on_hand=0, on_order=1) with Probability 0.632
From State InventoryState(on_hand=0, on_order=2):
  To State InventoryState(on_hand=2, on_order=0) with Probability 0.368
  To State InventoryState(on_hand=1, on_order=0) with Probability 0.368
  To State InventoryState(on_hand=0, on_order=0) with Probability 0.264
From State InventoryState(on_hand=1, on_order=0):
  To State InventoryState(on_hand=1, on_order=1) with Probability 0.368
  To State InventoryState(on_hand=0, on_order=1) with Probability 0.632
From State InventoryState(on_hand=1, on_order=1):
  To State InventoryState(on_hand=2, on_order=0) with Probability 0.368
  To State InventoryState(on_hand=1, on_order=0) with Probability 0.368
  To State InventoryState(on_hand=0, on_order=0) with Probability 0.264
From State InventoryState(on_hand=2, on_order=0):
  To State InventoryState(on_hand=2, on_order=0) with Probability 0.368
  To State InventoryState(on_hand=1, on_order=0) with Probability 0.368
  To State InventoryState(on_hand=0, on_order=0) with Probability 0.264
```

For a graphical view of this Markov Process, see Figure 3.6. The nodes are the states, labeled with their corresponding α and β values. The directed edges are the probabilistic state transitions from 6pm on a day to 6pm on the next day, with the transition probabilities labeled on them.

We can perform a number of interesting experiments and calculations with this simple Markov Process, and we encourage you to play with this code (specifically varying the capacity C and Poisson mean λ) to run simulations and probabilistic calculations of natural curiosity for a store owner.

There is a rich and interesting theory for Markov Processes. However, we won't go into this theory as our coverage of Markov Processes so far is a sufficient building block to take us to the incremental topics of Markov Reward Processes and Markov Decision Processes. However, before we move on, we'd like to show just a glimpse of the rich theory with the

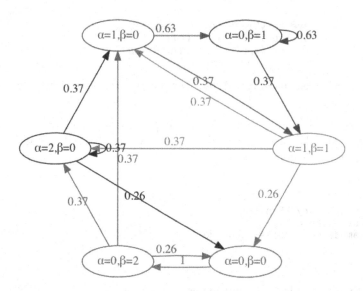

Figure 3.6 **Simple Inventory Markov Process**

calculation of *Stationary Probabilities* and apply it to the case of the above simple inventory Markov Process.

3.7 STATIONARY DISTRIBUTION OF A MARKOV PROCESS

Definition 3.7.1. The *Stationary Distribution* of a (Discrete-Time, Time-Homogeneous) Markov Process with state space $\mathcal{S} = \mathcal{N}$ and transition probability function $\mathcal{P} : \mathcal{N} \times \mathcal{N} \to [0, 1]$ is a probability distribution function $\pi : \mathcal{N} \to [0, 1]$ such that:

$$\pi(s) = \sum_{s' \in \mathcal{N}} \pi(s) \cdot \mathcal{P}(s, s') \text{ for all } s \in \mathcal{N}$$

The intuitive view of the stationary distribution π is that (under specific conditions we are not listing here) if we let the Markov Process run forever, then in the long run, the states occur at specific time steps with relative frequencies (probabilities) given by a distribution π that is independent of the time step. The probability of occurrence of a specific state s at a time step (asymptotically far out in the future) should be equal to the sum-product of probabilities of occurrence of all the states at the previous time step and the transition probabilities from those states to s. But since the states' occurrence probabilities are invariant in time, the π distribution for the previous time step is the same as the π distribution for the time step we considered. This argument holds for all states s, and that is exactly the statement of the definition of *Stationary Distribution* formalized above.

If we specialize this definition of *Stationary Distribution* to Finite-States, Discrete-Time, Time-Homogeneous Markov Processes with state space $\mathcal{S} = \{s_1, s_2, \ldots, s_n\} = \mathcal{N}$, then we can express the Stationary Distribution π as follows:

$$\pi(s_j) = \sum_{i=1}^{n} \pi(s_i) \cdot \mathcal{P}(s_i, s_j) \text{ for all } j = 1, 2, \ldots n$$

Below we use bold-face notation to represent functions as vectors and matrices (since we assume finite states). So, $\boldsymbol{\pi}$ is a column vector of length n and $\mathcal{P}$ is the $n \times n$ transition probability matrix (rows are source states, columns are destination states with each row summing to 1). Then, the statement of the above definition can be succinctly expressed as:

$$\pi^T = \pi^T \cdot \mathcal{P}$$

which can be re-written as:

$$\mathcal{P}^T \cdot \pi = \pi$$

But this is simply saying that π is an eigenvector of $\mathcal{P}^T$ with eigenvalue of 1. So then, it should be easy to obtain the stationary distribution π from an eigenvectors and eigenvalues calculation of $\mathcal{P}^T$.

Let us write code to compute the stationary distribution. We shall add two methods in the `FiniteMarkovProcess` class, one for setting up the transition probability matrix $\mathcal{P}$ (`get_transition_matrix` method) and another to calculate the stationary distribution π (`get_stationary_distribution`) from this transition probability matrix. Note that $\mathcal{P}$ is restricted to $\mathcal{N} \times \mathcal{N} \to [0,1]$ (rather than $\mathcal{N} \times \mathcal{S} \to [0,1]$) because these probability transitions suffice for all the calculations we will be performing for Finite Markov Processes. Here's the code for the two methods (the full code for `FiniteMarkovProcess` is in the file `rl/markov_process.py`):

```
import numpy as np
from rl.distribution import FiniteDistribution, Categorical
    def get_transition_matrix(self) -> np.ndarray:
        sz = len(self.non_terminal_states)
        mat = np.zeros((sz, sz))

        for i, s1 in enumerate(self.non_terminal_states):
            for j, s2 in enumerate(self.non_terminal_states):
                mat[i, j] = self.transition(s1).probability(s2)

        return mat

    def get_stationary_distribution(self) -> FiniteDistribution[S]:
        eig_vals, eig_vecs = np.linalg.eig(self.get_transition_matrix().T)
        index_of_first_unit_eig_val = np.where(
            np.abs(eig_vals - 1) < 1e-8)[0][0]
        eig_vec_of_unit_eig_val = np.real(
            eig_vecs[:, index_of_first_unit_eig_val])
        return Categorical({
            self.non_terminal_states[i].state: ev
            for i, ev in enumerate(eig_vec_of_unit_eig_val /
                                   sum(eig_vec_of_unit_eig_val))
        })
```

We skip the theory that tells us about the conditions under which a stationary distribution is well-defined, or the conditions under which there is a unique stationary distribution. Instead, we just go ahead with this calculation here assuming this Markov Process satisfies those conditions (it does!). So, we simply seek the index of the `eig_vals` vector with eigenvalue equal to 1 (accounting for floating-point error). Next, we pull out the column of the `eig_vecs` matrix at the `eig_vals` index calculated above and convert it into a real-valued vector (eigenvectors/eigenvalues calculations are, in general, complex numbers calculations—see the reference for the `np.linalg.eig` function). So this gives us the real-valued eigenvector with eigenvalue equal to 1. Finally, we have to normalize the eigenvector, so its values add up to 1 (since we want probabilities) and return the probabilities as a `Categorical` distribution).

Running this code for the simple case of capacity $C = 2$ and Poisson mean $\lambda = 1.0$ (instance of `SimpleInventoryMPFinite`) produces the following output for the stationary distribution π:

```
{InventoryState(on_hand=0, on_order=0): 0.117,
 InventoryState(on_hand=0, on_order=1): 0.279,
 InventoryState(on_hand=0, on_order=2): 0.117,
 InventoryState(on_hand=1, on_order=0): 0.162,
 InventoryState(on_hand=1, on_order=1): 0.162,
 InventoryState(on_hand=2, on_order=0): 0.162}
```

This tells us that On-Hand of 0 and On-Order of 1 is the state occurring most frequently (28% of the time) when the system is played out indefinitely.

Let us summarize the 3 different representations we've covered:

- Functional Representation: as given by the transition method, i.e., given a non-terminal state, the transition method returns a probability distribution of next states. This representation is valuable when performing simulations by sampling the next state from the returned probability distribution of next states. This is applicable to the general case of Markov Processes (including infinite state spaces).
- Sparse Data Structure Representation: as given by transition map: Transition, which is convenient for compact storage and useful for visualization (e.g., __repr__ method display or as a directed graph figure). This is applicable only to Finite Markov Processes.
- Dense Data Structure Representation: as given by the get_transition_matrix 2D numpy array, which is useful for performing linear algebra that is often required to calculate mathematical properties of the process (e.g., to calculate the stationary distribution). This is applicable only to Finite Markov Processes.

Now we are ready to move to our next topic of *Markov Reward Processes*. We'd like to finish this section by stating that the Markov Property owes its name to a mathematician from a century ago—Andrey Markov. Although the Markov Property seems like a simple enough concept, the concept has had profound implications on our ability to compute or reason with systems involving time-sequenced uncertainty in practice. There are several good books to learn more about Markov Processes—we recommend the book by Paul Gagniuc (Gagniuc 2017).

3.8 FORMALISM OF MARKOV REWARD PROCESSES

As we've said earlier, the reason we covered Markov Processes is that we want to make our way to Markov Decision Processes (the framework for Reinforcement Learning algorithms) by adding incremental features to Markov Processes. Now we cover an intermediate framework between Markov Processes and Markov Decision Processes, known as Markov Reward Processes. We essentially just include the notion of a numerical *reward* to a Markov Process each time we transition from one state to the next. These rewards are random, and all we need to do is to specify the probability distributions of these rewards as we make state transitions.

The main purpose of Markov Reward Processes is to calculate how much reward we would accumulate (in expectation, from each of the non-terminal states) if we let the process run indefinitely, bearing in mind that future rewards need to be discounted appropriately (otherwise, the sum of rewards could blow up to ∞). In order to solve the problem of calculating expected accumulative rewards from each non-terminal state, we will first set up some formalism for Markov Reward Processes, develop some (elegant) theory on calculating rewards accumulation, write plenty of code (based on the theory), and apply the theory and code to the simple inventory example (which we will embellish with rewards equal to negative of the costs incurred at the store).

Definition 3.8.1. A *Markov Reward Process* is a Markov Process, along with a time-indexed sequence of *Reward* random variables $R_t \in \mathcal{D}$ (a countable subset of $\mathbb{R}$) for time steps $t = 1, 2, \ldots$, satisfying the Markov Property (including Rewards): $\mathbb{P}[(R_{t+1}, S_{t+1})|S_t, S_{t-1}, \ldots, S_0] = \mathbb{P}[(R_{t+1}, S_{t+1})|S_t]$ for all $t \geq 0$.

It pays to emphasize again (like we emphasized for Markov Processes), that the definitions and theory of Markov Reward Processes we cover (by default) are for discrete-time, for countable state spaces and countable set of pairs of next state and reward transitions (with the knowledge that the definitions and theory are analogously extensible to continuous-time and uncountable spaces/transitions). In the more general case, where states or rewards are uncountable, the same concepts apply except that the mathematical formalism needs to be more detailed and more careful. Specifically, we'd end up with integrals instead of summations, and probability density functions (for continuous probability distributions) instead of probability mass functions (for discrete probability distributions). For ease of notation and, more importantly, for ease of understanding of the core concepts (without being distracted by heavy mathematical formalism), we've chosen to stay with discrete-time, countable $\mathcal{S}$ and countable $\mathcal{D}$ (by default). However, there will be examples of Markov Reward Processes in this book involving continuous-time and uncountable $\mathcal{S}$ and $\mathcal{D}$ (please adjust the definitions and formulas accordingly).

We refer to $\mathbb{P}[(R_{t+1}, S_{t+1})|S_t]$ as the transition probabilities of the Markov Reward Process for time t.

Since we commonly assume Time-Homogeneity of Markov Processes, we shall also (by default) assume Time-Homogeneity for Markov Reward Processes, i.e., $\mathbb{P}[(R_{t+1}, S_{t+1})|S_t]$ is independent of t.

With the default assumption of time-homogeneity, the transition probabilities of a Markov Reward Process can be expressed as a transition probability function:

$$\mathcal{P}_R : \mathcal{N} \times \mathcal{D} \times \mathcal{S} \to [0, 1]$$

defined as:

$$\mathcal{P}_R(s, r, s') = \mathbb{P}[(R_{t+1} = r, S_{t+1} = s')|S_t = s] \text{ for time steps } t = 0, 1, 2, \ldots,$$

for all $s \in \mathcal{N}, r \in \mathcal{D}, s' \in \mathcal{S}$, such that $\sum_{s' \in \mathcal{S}} \sum_{r \in \mathcal{D}} \mathcal{P}_R(s, r, s') = 1$ for all $s \in \mathcal{N}$

The subsection on *Start States* we had covered for Markov Processes naturally applies to Markov Reward Processes as well. So we won't repeat the section here, rather we simply highlight that when it comes to simulations, we need a separate specification of the probability distribution of start states. Also, by inheriting from our framework of Markov Processes, we model the notion of a "process termination" by explicitly specifying states as terminal states or non-terminal states. The sequence $S_0, R_1, S_1, R_2, S_2, \ldots$ terminates at time step $t = T$ if $S_T \in \mathcal{T}$, with R_T being the final reward in the sequence.

If all random sequences of states in a Markov Reward Process terminate, we refer to it as *episodic* sequences (otherwise, we refer to it as *continuing* sequences).

Let's write some code that captures this formalism. We create a derived `@abstractclass` `MarkovRewardProcess` that inherits from the `@abstractclass` `MarkovProcess`. Analogous to `MarkovProcess`'s `transition` method (that represents $\mathcal{P}$), `MarkovRewardProcess` has an abstract method `transition_reward` that represents $\mathcal{P}_R$. Note that the return type of `transition_reward` is `Distribution[Tuple[State[S], float]]`, representing the probability distribution of (next state, reward) pairs transitioned to.

Also, analogous to `MarkovProcess`'s `simulate` method, `MarkovRewardProcess` has the method `simulate_reward`, which generates a stream of `TransitionStep[S]` objects. Each

TransitionStep[S] object consists of a 3-tuple: (state, next state, reward) representing the sampled transitions within the generated sampling trace. Here's the actual code:

```
@dataclass(frozen=True)
class TransitionStep(Generic[S]):
    state: NonTerminal[S]
    next_state: State[S]
    reward: float

class MarkovRewardProcess(MarkovProcess[S]):

    @abstractmethod
    def transition_reward(self, state: NonTerminal[S])\
            -> Distribution[Tuple[State[S], float]]:
        pass

    def simulate_reward(
        self,
        start_state_distribution: Distribution[NonTerminal[S]]
    ) -> Iterable[TransitionStep[S]]:
        state: State[S] = start_state_distribution.sample()
        reward: float = 0.

        while isinstance(state, NonTerminal):
            next_distribution = self.transition_reward(state)

            next_state, reward = next_distribution.sample()
            yield TransitionStep(state, next_state, reward)

            state = next_state
```

So the idea is that if someone wants to model a Markov Reward Process, they'd simply have to create a concrete class that implements the interface of the abstract class MarkovRewardProcess (specifically implement the abstract method transition_reward). But note that the abstract method transition of MarkovProcess also needs to be implemented to make the whole thing concrete. However, we don't have to implement it in the concrete class implementing the interface of MarkovRewardProcess—in fact, we can implement it in the MarkovRewardProcess class itself by tapping the method transition_reward. Here's the code for the transition method in MarkovRewardProcess:

```
from rl.distribution import Distribution, SampledDistribution

    def transition(self, state: NonTerminal[S]) -> Distribution[State[S]]:
        distribution = self.transition_reward(state)

        def next_state(distribution=distribution):
            next_s, _ = distribution.sample()
            return next_s

        return SampledDistribution(next_state)
```

Note that since the transition_reward method is abstract in MarkovRewardProcess[6], the only thing the transition method can do is to tap into the sample method of the abstract Distribution object produced by transition_reward and return a SampledDistribution.

Now let us develop some more theory. Given a specification of $\mathcal{P}_R$, we can extract:

- The transition probability function $\mathcal{P} : \mathcal{N} \times \mathcal{S} \to [0, 1]$ of the implicit Markov Process defined as:

$$\mathcal{P}(s, s') = \sum_{r \in \mathcal{D}} \mathcal{P}_R(s, r, s')$$

- The reward transition function:

$$\mathcal{R}_T : \mathcal{N} \times \mathcal{S} \to \mathbb{R}$$

---

[6]The full definition of MarkovRewardProcess is in the file rl/markov_process.py.

defined as:

$$\mathcal{R}_T(s, s') = \mathbb{E}[R_{t+1}|S_{t+1} = s', S_t = s] = \sum_{r \in \mathcal{D}} \frac{\mathcal{P}_R(s, r, s')}{\mathcal{P}(s, s')} \cdot r = \sum_{r \in \mathcal{D}} \frac{\mathcal{P}_R(s, r, s')}{\sum_{r \in \mathcal{D}} \mathcal{P}_R(s, r, s')} \cdot r$$

The Rewards specification of most Markov Reward Processes we encounter in practice can be directly expressed as the reward transition function $\mathcal{R}_T$ (versus the more general specification of $\mathcal{P}_R$). Lastly, we want to highlight that we can transform either of $\mathcal{P}_R$ or $\mathcal{R}_T$ into a "more compact" reward function that is sufficient to perform key calculations involving Markov Reward Processes. This reward function

$$\mathcal{R} : \mathcal{N} \to \mathbb{R}$$

is defined as:

$$\mathcal{R}(s) = \mathbb{E}[R_{t+1}|S_t = s] = \sum_{s' \in \mathcal{S}} \mathcal{P}(s, s') \cdot \mathcal{R}_T(s, s') = \sum_{s' \in \mathcal{S}} \sum_{r \in \mathcal{D}} \mathcal{P}_R(s, r, s') \cdot r$$

We've created a bit of notational clutter here. So it would be a good idea for you to take a few minutes to pause, reflect and internalize the differences between $\mathcal{P}_R$, $\mathcal{P}$ (of the implicit Markov Process), $\mathcal{R}_T$ and $\mathcal{R}$. This notation will analogously re-appear when we learn about Markov Decision Processes in Chapter 4. Moreover, this notation will be used considerably in the rest of the book, so it pays to get comfortable with their semantics.

3.9 SIMPLE INVENTORY EXAMPLE AS A MARKOV REWARD PROCESS

Now we return to the simple inventory example and embellish it with a reward structure to turn it into a Markov Reward Process (business costs will be modeled as negative rewards). Let us assume that your store business incurs two types of costs:

- Holding cost of h for each bicycle that remains in your store overnight. Think of this as "interest on inventory"—each day your bicycle remains unsold, you lose the opportunity to gain interest on the cash you paid to buy the bicycle. Holding cost also includes the cost of upkeep of inventory.
- Stockout cost of p for each unit of "missed demand", i.e., for each customer wanting to buy a bicycle that you could not satisfy with available inventory, e.g., if 3 customers show up during the day wanting to buy a bicycle each, and you have only 1 bicycle at 8am (store opening time), then you lost two units of demand, incurring a cost of $2p$. Think of the cost of p per unit as the lost revenue plus disappointment for the customer. Typically $p \gg h$.

Let us go through the precise sequence of events, now with the incorporation of rewards, in each 24-hour cycle:

- Observe the (α, β) *State* at 6pm store-closing (call this state S_t).
- Immediately order according to the ordering policy given by: Order quantity = $\max(C - (\alpha + \beta), 0)$.
- Record any overnight holding cost incurred as described above.
- Receive bicycles at 6am if you had ordered 36 hours ago.
- Open the store at 8am.
- Experience random demand from customers according to the specified Poisson probabilities (Poisson mean = λ).
- Record any stockout cost due to missed demand as described above.

- Close the store at 6pm, register the reward R_{t+1} as the negative sum of overnight holding cost and the day's stockout cost, and observe the state (this state is S_{t+1}).

Since the customer demand on any day can be an infinite set of possibilities (Poisson distribution over the entire range of non-negative integers), we have an infinite set of pairs of next state and reward we could transition to from a given current state. Let's see what the probabilities of each of these transitions looks like. For a given current state $S_t := (\alpha, \beta)$, if customer demand for the day is i, then the next state S_{t+1} is:

$$(\max(\alpha + \beta - i, 0), \max(C - (\alpha + \beta), 0))$$

and the reward R_{t+1} is:

$$-h \cdot \alpha - p \cdot \max(i - (\alpha + \beta), 0)$$

Note that the overnight holding cost applies to each unit of on-hand inventory at store closing ($= \alpha$) and the stockout cost applies only to any units of "missed demand" ($= \max(i - (\alpha + \beta), 0)$). Since two different values of demand $i \in \mathbb{Z}_{\geq 0}$ do not collide on any unique pair (s', r) of next state and reward, we can express the transition probability function $\mathcal{P}_R$ for this Simple Inventory Example as a Markov Reward Process as:

$$\mathcal{P}_R((\alpha, \beta), -h \cdot \alpha - p \cdot \max(i - (\alpha + \beta), 0), (\max(\alpha + \beta - i, 0), \max(C - (\alpha + \beta), 0)))$$

$$= \frac{e^{-\lambda}\lambda^i}{i!} \text{ for all } i = 0, 1, 2, \ldots$$

Now let's write some code to implement this simple inventory example as a Markov Reward Process as described above. All we have to do is to create a concrete class implementing the interface of the abstract class `MarkovRewardProcess` (specifically implement the abstract method `transition_reward`). The code below in `transition_reward` method in class `SimpleInventoryMRP` samples the customer demand from a Poisson distribution, uses the above formulas for the pair of next state and reward as a function of the customer demand sample, and returns an instance of `SampledDistribution`. Note that the generic state type `S` is replaced here with the `@dataclass InventoryState` to represent a state of this Markov Reward Process, comprising of the On-Hand and On-Order inventory quantities.

```python
from rl.distribution import SampledDistribution
import numpy as np

@dataclass(frozen=True)
class InventoryState:
    on_hand: int
    on_order: int

    def inventory_position(self) -> int:
        return self.on_hand + self.on_order

class SimpleInventoryMRP(MarkovRewardProcess[InventoryState]):
    def __init__(
        self,
        capacity: int,
        poisson_lambda: float,
        holding_cost: float,
        stockout_cost: float
    ):
        self.capacity = capacity
        self.poisson_lambda: float = poisson_lambda
        self.holding_cost: float = holding_cost
        self.stockout_cost: float = stockout_cost

    def transition_reward(
```

```
        self,
        state: NonTerminal[InventoryState]
) -> SampledDistribution[Tuple[State[InventoryState], float]]:
    def sample_next_state_reward(state=state) ->\
            Tuple[State[InventoryState], float]:
        demand_sample: int = np.random.poisson(self.poisson_lambda)
        ip: int = state.state.inventory_position()
        next_state: InventoryState = InventoryState(
            max(ip - demand_sample, 0),
            max(self.capacity - ip, 0)
        )
        reward: float = - self.holding_cost * state.state.on_hand\
            - self.stockout_cost * max(demand_sample - ip, 0)
        return NonTerminal(next_state), reward

    return SampledDistribution(sample_next_state_reward)
```

The above code can be found in the file rl/chapter2/simple_inventory_mrp.py. We leave it as an exercise for you to use the simulate_reward method inherited by SimpleInventoryMRP to perform simulations and analyze the statistics produced from the sampling traces.

3.10   FINITE MARKOV REWARD PROCESSES

Certain calculations for Markov Reward Processes can be performed easily if:

- the state space is finite ($S = \{s_1, s_2, \ldots, s_n\}$), and
- the set of unique pairs of next state and reward transitions from each of the states in $\mathcal{N}$ is finite

If we satisfy the above two characteristics, we refer to the Markov Reward Process as a Finite Markov Reward Process. So let us write some code for a Finite Markov Reward Process. We create a concrete class FiniteMarkovRewardProcess that primarily inherits from FiniteMarkovProcess (a concrete class) and secondarily implements the interface of the abstract class MarkovRewardProcess. Our first task is to think about the data structure required to specify an instance of FiniteMarkovRewardProcess (i.e., the data structure we'd pass to the __init__ method of FiniteMarkovRewardProcess). Analogous to how we curried $\mathcal{P}$ for a Markov Process as $\mathcal{N} \rightarrow (S \rightarrow [0,1])$ (where $S = \{s_1, s_2, \ldots, s_n\}$ and $\mathcal{N}$ has $m \leq n$ states), here we curry $\mathcal{P}_R$ as:

$$\mathcal{N} \rightarrow (S \times \mathcal{D} \rightarrow [0,1])$$

Since S is finite and since the set of unique pairs of next state and reward transitions is also finite, this leads to the analog of the Transition data type for the case of Finite Markov Reward Processes (named RewardTransition) as follows:

```
StateReward = FiniteDistribution[Tuple[State[S], float]]
RewardTransition = Mapping[NonTerminal[S], StateReward[S]]
```

The FiniteMarkovRewardProcess class has 3 responsibilities:

- It needs to accept as input to __init__ a Mapping type similar to RewardTransition using simply S instead of NonTerminal[S] or State[S] in order to make it convenient for the user to specify a FiniteRewardProcess as a succinct dictionary, without being encumbered with wrapping S as NonTerminal[S] or Terminal[S] types. This means the __init__ method (constructor) needs to appropriately wrap S as NonTerminal[S] or Terminal[S] types to create the attribute transition_reward_map: RewardTransition[S]. Also, the __init__ method needs to create a transition_map: Transition[S] (extracted from the input to __init__) in order to instantiate its concrete parent class FiniteMarkovProcess.

- It needs to implement the `transition_reward` method analogous to the implementation of the `transition` method in `FiniteMarkovProcess`
- It needs to compute the reward fuction $\mathcal{R} : \mathcal{N} \to \mathbb{R}$ from the transition probability function $\mathcal{P}_R$ (i.e. from `self.transition_reward_map: RewardTransition`) based on the expectation calculation we specified above (as mentioned earlier, $\mathcal{R}$ is key to the relevant calculations we shall soon be performing on Finite Markov Reward Processes). To perform further calculations with the reward function $\mathcal{R}$, we need to produce it as a 1-dimensional numpy array (i.e., a vector) attribute of the class (we name it as `reward_function_vec`).

Here's the code that fulfills the above three responsibilities[7]:

```python
import numpy as np
from rl.distribution import FiniteDistribution, Categorical
from collections import defaultdict
from typing import Mapping, Tuple, Dict, Set

class FiniteMarkovRewardProcess(FiniteMarkovProcess[S],
                                MarkovRewardProcess[S]):

    transition_reward_map: RewardTransition[S]
    reward_function_vec: np.ndarray

    def __init__(
        self,
        transition_reward_map: Mapping[S, FiniteDistribution[Tuple[S, float]]]
    ):
        transition_map: Dict[S, FiniteDistribution[S]] = {}

        for state, trans in transition_reward_map.items():
            probabilities: Dict[S, float] = defaultdict(float)
            for (next_state, _), probability in trans:
                probabilities[next_state] += probability

            transition_map[state] = Categorical(probabilities)

        super().__init__(transition_map)

        nt: Set[S] = set(transition_reward_map.keys())
        self.transition_reward_map = {
            NonTerminal(s): Categorical(
                {(NonTerminal(s1) if s1 in nt else Terminal(s1), r): p
                 for (s1, r), p in v}
            ) for s, v in transition_reward_map.items()
        }

        self.reward_function_vec = np.array([
            sum(probability * reward for (_, reward), probability in
                self.transition_reward_map[state])
            for state in self.non_terminal_states
        ])

    def transition_reward(self, state: NonTerminal[S]) -> StateReward[S]:
        return self.transition_reward_map[state]
```

3.11 SIMPLE INVENTORY EXAMPLE AS A FINITE MARKOV REWARD PROCESS

Now we'd like to model the simple inventory example as a Finite Markov Reward Process so we can take advantage of the algorithms that apply to Finite Markov Reward Processes. As we've noted previously, our ordering policy ensures that in steady-state, the sum of On-Hand (denote as α) and On-Order (denote as β) won't exceed the capacity C. So we constrain the set of states such that this condition is satisfied: $0 \leq \alpha + \beta \leq C$ (i.e., finite

---

[7]The code for `FiniteMarkovRewardProcess` (and more) is in rl/markov_process.py.

number of states). Although the set of states is finite, there are an infinite number of pairs of next state and reward outcomes possible from any given current state. This is because there are an infinite set of possibilities of customer demand on any given day (resulting in infinite possibilities of stockout cost, i.e., negative reward, on any day). To qualify as a Finite Markov Reward Process, we'll need to model in a manner such that we have a finite set of pairs of next state and reward outcomes from a given current state. So what we'll do is that instead of considering (S_{t+1}, R_{t+1}) as the pair of next state and reward, we model the pair of next state and reward to instead be $(S_{t+1}, \mathbb{E}[R_{t+1}|(S_t, S_{t+1})])$ (we know $\mathcal{P}_R$ due to the Poisson probabilities of customer demand, so we can actually calculate this conditional expectation of reward). So given a state s, the pairs of next state and reward would be: $(s', \mathcal{R}_T(s, s'))$ for all the s' we transition to from s. Since the set of possible next states s' are finite, these newly-modeled rewards associated with the transitions $(\mathcal{R}_T(s, s'))$ are also finite and hence, the set of pairs of next state and reward from any current state are also finite. Note that this creative alteration of the reward definition is purely to reduce this Markov Reward Process into a Finite Markov Reward Process. Let's now work out the calculation of the reward transition function $\mathcal{R}_T$.

When the next state's (S_{t+1}) On-Hand is greater than zero, it means all of the day's demand was satisfied with inventory that was available at store-opening $(= \alpha + \beta)$, and hence, each of these next states S_{t+1} correspond to no stockout cost and only an overnight holding cost of $h\alpha$. Therefore,

$$\mathcal{R}_T((\alpha, \beta), (\alpha + \beta - i, C - (\alpha + \beta))) = -h\alpha \text{ for } 0 \leq i \leq \alpha + \beta - 1$$

When next state's (S_{t+1}) On-Hand is equal to zero, there are two possibilities:

1. The demand for the day was exactly $\alpha + \beta$, meaning all demand was satisifed with available store inventory (so no stockout cost and only overnight holding cost), or
2. The demand for the day was strictly greater than $\alpha + \beta$, meaning there's some stockout cost in addition to overnight holding cost. The exact stockout cost is an expectation calculation involving the number of units of missed demand under the corresponding Poisson probabilities of demand exceeding $\alpha + \beta$.

This calculation is shown below:

$$\mathcal{R}_T((\alpha, \beta), (0, C - (\alpha + \beta))) = -h\alpha - p(\sum_{j=\alpha+\beta+1}^{\infty} f(j) \cdot (j - (\alpha + \beta)))$$

$$= -h\alpha - p(\lambda(1 - F(\alpha + \beta - 1)) - (\alpha + \beta)(1 - F(\alpha + \beta)))$$

So now we have a specification of $\mathcal{R}_T$, but when it comes to our coding interface, we are expected to specify $\mathcal{P}_R$ as that is the interface through which we create a FiniteMarkovRewardProcess. Fear not—a specification of $\mathcal{P}_R$ is easy once we have a specification of $\mathcal{R}_T$. We simply create 4-tuples (s, r, s', p) for all $s \in \mathcal{N}, s' \in \mathcal{S}$ such that $r = \mathcal{R}_T(s, s')$ and $p = \mathcal{P}(s, s')$ (we know $\mathcal{P}$ along with $\mathcal{R}_T$), and the set of all these 4-tuples (for all $s \in \mathcal{N}, s' \in \mathcal{S}$) constitute the specification of $\mathcal{P}_R$, i.e., $\mathcal{P}_R(s, r, s') = p$. This turns our reward-definition-altered mathematical model of a Finite Markov Reward Process into a programming model of the FiniteMarkovRewardProcess class. This reward-definition-altered model enables us to gain from the fact that we can leverage the algorithms we'll be writing for Finite Markov Reward Processes (including some simple and elegant linear-algebra-solver-based solutions). The downside of this reward-definition-altered model is that it prevents us from generating samples of the specific rewards encountered when transitioning from one state to another (because we no longer capture the probabilities of individual reward outcomes). Note that we can indeed generate sampling traces, but each transition step in the sampling trace will only show us the "mean reward" (specifically, the expected reward conditioned on current state and next state).

In fact, most Markov Processes you'd encounter in practice can be modeled as a combination of $\mathcal{R}_T$ and $\mathcal{P}$, and you'd simply follow the above $\mathcal{R}_T$ to $\mathcal{P}_R$ representation transformation drill to present this information in the form of $\mathcal{P}_R$ to instantiate a `FiniteMarkovRewardProcess`. We designed the interface to accept $\mathcal{P}_R$ as input since that is the most general interface for specifying Markov Reward Processes.

So now let's write some code for the simple inventory example as a Finite Markov Reward Process as described above. All we have to do is to create a derived class inherited from `FiniteMarkovRewardProcess` and write a method to construct the `transition_reward_map` input to the constructor (`__init__`) of FiniteMarkovRewardProcess (i.e., $\mathcal{P}_R$). Note that the generic state type S is replaced here with the @dataclass InventoryState to represent the inventory state, comprising of the On-Hand and On-Order inventory quantities.

```python
from scipy.stats import poisson
@dataclass(frozen=True)
class InventoryState:
    on_hand: int
    on_order: int

    def inventory_position(self) -> int:
        return self.on_hand + self.on_order

class SimpleInventoryMRPFinite(FiniteMarkovRewardProcess[InventoryState]):
    def __init__(
        self,
        capacity: int,
        poisson_lambda: float,
        holding_cost: float,
        stockout_cost: float
    ):
        self.capacity: int = capacity
        self.poisson_lambda: float = poisson_lambda
        self.holding_cost: float = holding_cost
        self.stockout_cost: float = stockout_cost

        self.poisson_distr = poisson(poisson_lambda)
        super().__init__(self.get_transition_reward_map())

    def get_transition_reward_map(self) -> \
            Mapping[
                InventoryState,
                FiniteDistribution[Tuple[InventoryState, float]]
            ]:
        d: Dict[InventoryState, Categorical[Tuple[InventoryState, float]]] = {}
        for alpha in range(self.capacity + 1):
            for beta in range(self.capacity + 1 - alpha):
                state = InventoryState(alpha, beta)
                ip = state.inventory_position()
                beta1 = self.capacity - ip
                base_reward = - self.holding_cost * state.on_hand
                sr_probs_map: Dict[Tuple[InventoryState, float], float] =\
                    {(InventoryState(ip - i, beta1), base_reward):
                        self.poisson_distr.pmf(i) for i in range(ip)}
                probability = 1 - self.poisson_distr.cdf(ip - 1)
                reward = base_reward - self.stockout_cost *\
                    (probability * (self.poisson_lambda - ip) +
                        ip * self.poisson_distr.pmf(ip))
                sr_probs_map[(InventoryState(0, beta1), reward)] = probability
                d[state] = Categorical(sr_probs_map)
        return d
```

The above code is in the file rl/chapter2/simple_inventory_mrp.py). We encourage you to play with the inputs to `SimpleInventoryMRPFinite` in `__main__` and view the transition probabilities and rewards of the constructed Finite Markov Reward Process.

3.12 VALUE FUNCTION OF A MARKOV REWARD PROCESS

Now we are ready to formally define the main problem involving Markov Reward Processes. As we've said earlier, we'd like to compute the "expected accumulated rewards" from any non-terminal state. However, if we simply add up the rewards in a sampling trace following time step t as $\sum_{i=t+1}^{\infty} R_i = R_{t+1} + R_{t+2} + \ldots$, the sum would often diverge to infinity. So we allow for rewards accumulation to be done with a discount factor $\gamma \in [0, 1]$: We define the (random) *Return* G_t as the "discounted accumulation of future rewards" following time step t. Formally,

$$G_t = \sum_{i=t+1}^{\infty} \gamma^{i-t-1} \cdot R_i = R_{t+1} + \gamma \cdot R_{t+2} + \gamma^2 \cdot R_{t+3} + \ldots$$

We use the above definition of *Return* even for a terminating sequence (say terminating at $t = T$, i.e., $S_T \in \mathcal{T}$), by treating $R_i = 0$ for all $i > T$.

Note that γ can range from a value of 0 on one extreme (called "myopic") to a value of 1 on another extreme (called "far-sighted"). "Myopic" means the Return is the same as Reward (no accumulation of future Rewards in the Return). With "far-sighted" ($\gamma = 1$), the Return calculation can diverge for continuing (non-terminating) Markov Reward Processes but "far-sighted" is indeed applicable for episodic Markov Reward Processes (where all random sequences of the process terminate). Apart from the Return divergence consideration, $\gamma < 1$ helps algorithms become more tractable (as we shall see later when we get to Reinforcement Learning). We should also point out that the reason to have $\gamma < 1$ is not just for mathematical convenience or computational tractability—there are valid modeling reasons to discount Rewards when accumulating to a Return. When Reward is modeled as a financial quantity (revenues, costs, profits etc.), as will be the case in most financial applications, it makes sense to incorporate time-value-of-money which is a fundamental concept in Economics/Finance that says there is greater benefit in receiving a dollar now versus later (which is the economic reason why interest is paid or earned). So it is common to set γ to be the discounting based on the prevailing interest rate ($\gamma = \frac{1}{1+r}$, where r is the interest rate over a single time step). Another technical reason for setting $\gamma < 1$ is that our models often don't fully capture future uncertainty and so, discounting with γ acts to undermine future rewards that might not be accurate (due to future uncertainty modeling limitations). Lastly, from an AI perspective, if we want to build machines that acts like humans, psychologists have indeed demonstrated that human/animal behavior prefers immediate reward over future reward.

Note that we are (as usual) assuming the fact that the Markov Reward Process is time-homogeneous (time-invariant probabilities of state transitions and rewards).

As you might imagine now, we'd want to identify non-terminal states with large expected returns and those with small expected returns. This, in fact, is the main problem involving a Markov Reward Process—to compute the "Expected Return" associated with each non-terminal state in the Markov Reward Process. Formally, we are interested in computing the *Value Function*

$$V : \mathcal{N} \to \mathbb{R}$$

defined as:

$$V(s) = \mathbb{E}[G_t | S_t = s] \text{ for all } s \in \mathcal{N}, \text{ for all } t = 0, 1, 2, \ldots$$

For the rest of the book, we will assume that whenever we are talking about a Value Function, the discount factor γ is appropriate to ensure that the Expected Return from each state is finite.

Now we show a creative piece of mathematics due to Richard Bellman. Bellman noted (Bellman 1957b) that the Value Function has a recursive structure. Specifically,

$$
\begin{aligned}
V(s) &= \mathbb{E}[R_{t+1}|S_t = s] + \gamma \cdot \mathbb{E}[R_{t+2}|S_t = s] + \gamma^2 \cdot \mathbb{E}[R_{t+3}|S_t = s] + \dots \\
&= \mathcal{R}(s) + \gamma \cdot \sum_{s' \in \mathcal{N}} \mathbb{P}[S_{t+1} = s'|S_t = s] \cdot \mathbb{E}[R_{t+2}|S_{t+1} = s'] \\
&\quad + \gamma^2 \cdot \sum_{s' \in \mathcal{N}} \mathbb{P}[S_{t+1} = s'|S_t = s] \sum_{s'' \in \mathcal{N}} \mathbb{P}[S_{t+2} = s''|S_{t+1} = s'] \cdot \mathbb{E}[R_{t+3}|S_{t+2} = s''] \\
&\quad + \dots \\
&= \mathcal{R}(s) + \gamma \cdot \sum_{s' \in \mathcal{N}} \mathcal{P}(s,s') \cdot \mathcal{R}(s') + \gamma^2 \cdot \sum_{s' \in \mathcal{N}} \mathcal{P}(s,s') \sum_{s'' \in \mathcal{N}} \mathcal{P}(s',s'') \cdot \mathcal{R}(s'') + \dots \\
&= \mathcal{R}(s) + \gamma \cdot \sum_{s' \in \mathcal{N}} \mathcal{P}(s,s') \cdot \left(\mathcal{R}(s') + \gamma \cdot \sum_{s'' \in \mathcal{N}} \mathcal{P}(s',s'') \cdot \mathcal{R}(s'') + \dots\right) \\
&= \mathcal{R}(s) + \gamma \cdot \sum_{s' \in \mathcal{N}} \mathcal{P}(s,s') \cdot V(s') \text{ for all } s \in \mathcal{N}
\end{aligned}
$$

$$(3.1)$$

Note that although the transitions to random states $s', s'', \dots$ are in the state space of $\mathcal{S}$ rather than $\mathcal{N}$, the right-hand-side above sums over states $s', s'', \dots$ only in $\mathcal{N}$ because transitions to terminal states (in $\mathcal{T} = \mathcal{S} - \mathcal{N}$) don't contribute any reward beyond the rewards produced *before reaching* the terminal state.

We refer to this recursive equation (3.1) for the Value Function as the Bellman Equation for Markov Reward Processes. Figure 3.7 is a convenient visualization aid of this important equation. In the rest of the book, we will depict quite a few of these type of state-transition visualizations to aid with creating mental models of key concepts.

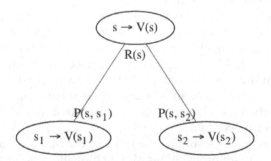

Figure 3.7 **Visualization of MRP Bellman Equation**

For the case of Finite Markov Reward Processes, assume $\mathcal{S} = \{s_1, s_2, \dots, s_n\}$ and assume $\mathcal{N}$ has $m \leq n$ states. Below we use bold-face notation to represent functions as column vectors and matrices since we have finite states/transitions. So, $\boldsymbol{V}$ is a column vector of length m, $\boldsymbol{\mathcal{P}}$ is an $m \times m$ matrix, and $\boldsymbol{\mathcal{R}}$ is a column vector of length m (rows/columns corresponding to states in $\mathcal{N}$), so we can express the above equation in vector and matrix notation as follows:

$$V = \mathcal{R} + \gamma \mathcal{P} \cdot V$$

Therefore,

$$\Rightarrow V = (I_m - \gamma \mathcal{P})^{-1} \cdot \mathcal{R} \tag{3.2}$$

where I_m is the $m \times m$ identity matrix.

Let us write some code to implement the calculation of Equation (3.2). In the FiniteMarkovRewardProcess class, we implement the method get_value_function_vec that performs the above calculation for the Value Function V in terms of the reward function $\mathcal{R}$ and the transition probability function $\mathcal{P}$ of the implicit Markov Process. The Value Function V is produced as a 1D numpy array (i.e. a vector). Here's the code:

```
def get_value_function_vec(self, gamma: float) -> np.ndarray:
    return np.linalg.solve(
        np.eye(len(self.non_terminal_states)) -
        gamma * self.get_transition_matrix(),
        self.reward_function_vec
    )
```

Invoking this get_value_function_vec method on SimpleInventoryMRPFinite for the simple case of capacity $C = 2$, Poisson mean $\lambda = 1.0$, holding cost $h = 1.0$, stockout cost $p = 10.0$, and discount factor $\gamma = 0.9$ yields the following result:

```
{NonTerminal(state=InventoryState(on_hand=0, on_order=0)): -35.511,
 NonTerminal(state=InventoryState(on_hand=0, on_order=1)): -27.932,
 NonTerminal(state=InventoryState(on_hand=0, on_order=2)): -28.345,
 NonTerminal(state=InventoryState(on_hand=1, on_order=0)): -28.932,
 NonTerminal(state=InventoryState(on_hand=1, on_order=1)): -29.345,
 NonTerminal(state=InventoryState(on_hand=2, on_order=0)): -30.345}
```

The corresponding values of the attribute reward_function_vec (i.e., $\mathcal{R}$) are:

```
{NonTerminal(state=InventoryState(on_hand=1, on_order=0)): -3.325,
 NonTerminal(state=InventoryState(on_hand=0, on_order=0)): -10.0,
 NonTerminal(state=InventoryState(on_hand=0, on_order=1)): -2.325,
 NonTerminal(state=InventoryState(on_hand=0, on_order=2)): -0.274,
 NonTerminal(state=InventoryState(on_hand=1, on_order=1)): -1.274,
 NonTerminal(state=InventoryState(on_hand=2, on_order=0)): -2.274}
```

This tells us that On-Hand of 0 and On-Order of 2 has the highest expected reward. However, the Value Function is highest for On-Hand of 0 and On-Order of 1.

This computation for the Value Function works if the state space is not too large (the size of the square linear system of equations is equal to number of non-terminal states). When the state space is large, this direct method of solving a linear system of equations won't scale, and we have to resort to numerical methods to solve the recursive Bellman Equation. This is the topic of Dynamic Programming and Reinforcement Learning algorithms that we shall learn in this book.

3.13   SUMMARY OF KEY LEARNINGS FROM THIS CHAPTER

Before we end this chapter, we'd like to highlight the two highly important concepts we learnt in this chapter:

- Markov Property: A concept that enables us to reason effectively and compute efficiently in practical systems involving sequential uncertainty.

- Bellman Equation: A mathematical insight that enables us to express the Value Function recursively—this equation (and its Optimality version covered in Chapter 4) is, in fact, the core idea within all Dynamic Programming and Reinforcement Learning algorithms.

Markov Decision Processes

We've said before that this book is about "sequential decisioning" under "sequential uncertainty". In Chapter 3, we covered the "sequential uncertainty" aspect with the framework of Markov Processes, and we extended the framework to also incorporate the notion of uncertain "Reward" each time we make a state transition—we called this extended framework Markov Reward Processes. However, this framework had no notion of "sequential decisioning". In this chapter, we further extend the framework of Markov Reward Processes to incorporate the notion of "sequential decisioning", formally known as Markov Decision Processes. Before we step into the formalism of Markov Decision Processes, let us develop some intuition and motivation for the need to have such a framework—to handle sequential decisioning. Let's do this by re-visiting the simple inventory example we covered in Chapter 3.

4.1 SIMPLE INVENTORY EXAMPLE: HOW MUCH TO ORDER?

When we covered the simple inventory example in Chapter 3 as a Markov Reward Process, the ordering policy was:

$$\theta = \max(C - (\alpha + \beta), 0)$$

where $\theta \in \mathbb{Z}_{\geq 0}$ is the order quantity, $C \in \mathbb{Z}_{\geq 0}$ is the space capacity (in bicycle units) at the store, α is the On-Hand Inventory and β is the On-Order Inventory ((α, β) comprising the *State*). We calculated the Value Function for the Markov Reward Process that results from following this policy. Now we ask the question: Is this Value Function good enough? More importantly, we ask the question: Can we improve this Value Function by following a different ordering policy? Perhaps by ordering less than that implied by the above formula for θ? This leads to the natural question—Can we identify the ordering policy that yields the *Optimal* Value Function (one with the highest expected returns, i.e., lowest expected accumulated costs, from each state)? Let us get an intuitive sense for this optimization problem by considering a concrete example.

Assume that instead of bicycles, we want to control the inventory of a specific type of toothpaste in the store. Assume you have space for 20 units of toothpaste on the shelf assigned to the toothpaste (assume there is no space in the backroom of the store). Asssume that customer demand follows a Poisson distribution with Poisson parameter $\lambda = 3.0$. At 6pm store-closing each evening, when you observe the *State* as (α, β), you now have a choice of ordering a quantity of toothpastes from any of the following values for the order quantity $\theta : \{0, 1, \ldots, \max(20 - (\alpha + \beta), 0)\}$. Let's say at Monday 6pm store-closing, $\alpha = 4$ and $\beta = 3$. So, you have a choice of order quantities from among the integers in the range of 0 to $(20 - (4 + 3) = 13)$ (i.e., 14 choices). Previously, in the Markov Reward Process model, you

DOI: 10.1201/9781003229193-4

would have ordered 13 units on Monday store-closing. This means on Wednesday morning at 6am, a truck would have arrived with 13 units of the toothpaste. If you sold say 2 units of the toothpaste on Tuesday, then on Wednesday 8am at store-opening, you'd have $4 + 3 - 2 + 13 = 18$ units of toothpaste on your shelf. If you keep following this policy, you'd typically have almost a full shelf at store-opening each day, which covers almost a week worth of expected demand for the toothpaste. This means your risk of going out-of-stock on the toothpaste is extremely low, but you'd be incurring considerable holding cost (you'd have close to a full shelf of toothpastes sitting around almost each night). So as a store manager, you'd be thinking—"I can lower my costs by ordering less than that prescribed by the formula of $20 - (\alpha + \beta)$". But how much less? If you order too little, you'd start the day with too little inventory and might risk going out-of-stock. That's a risk you are highly uncomfortable with since the stockout cost per unit of missed demand (we called it p) is typically much higher than the holding cost per unit (we called it h). So you'd rather "err" on the side of having more inventory. But how much more? We also need to factor in the fact that the 36-hour lead time means a large order incurs large holding costs *two days later*. Most importantly, to find this right balance in terms of a precise mathematical optimization of the Value Function, we'd have to factor in the uncertainty of demand (based on daily Poisson probabilities) in our calculations. Now this gives you a flavor of the problem of sequential decisioning (each day you have to decide how much to order) under sequential uncertainty.

To deal with the "decisioning" aspect, we will introduce the notion of *Action* to complement the previously introduced notions of *State* and *Reward*. In the inventory example, the order quantity is our *Action*. After observing the *State*, we choose from among a set of Actions (in this case, we choose from within the set $\{0, 1, \ldots, \max(C - (\alpha + \beta), 0)\}$). We note that the Action we take upon observing a state affects the next day's state. This is because the next day's On-Order is exactly equal to today's order quantity (i.e., today's action). This in turn might affect our next day's action since the action (order quantity) is typically a function of the state (On-Hand and On-Order inventory). Also note that the Action we take on a given day will influence the Rewards after a couple of days (i.e. after the order arrives). It may affect our holding cost adversely if we had ordered too much or it may affect our stockout cost adversely if we had ordered too little and then experienced high demand.

4.2 THE DIFFICULTY OF SEQUENTIAL DECISIONING UNDER UNCERTAINTY

This simple inventory example has given us a peek into the world of Markov Decision Processes, which in general, have two distinct (and inter-dependent) high-level features:

- At each time step t, an *Action* A_t is picked (from among a specified choice of actions) upon observing the *State* S_t.
- Given an observed *State* S_t and a performed *Action* A_t, the probabilities of the state and reward of the next time step (S_{t+1} and R_{t+1}) are in general a function of not just the state S_t, but also of the action A_t.

We are tasked with maximizing the *Expected Return* from each state (i.e., maximizing the Value Function). This seems like a pretty hard problem in the general case because there is a cyclic interplay between:

- actions depending on state, on one hand, and

- next state/reward probabilities depending on action (and state) on the other hand.

There is also the challenge that actions might have delayed consequences on rewards, and it's not clear how to disentangle the effects of actions from different time steps on a future reward. So without direct correspondence between actions and rewards, how can we control the actions so as to maximize expected accumulated rewards? To answer this question, we will need to set up some notation and theory. Before we formally define the Markov Decision Process framework and its associated (elegant) theory, let us set up a bit of terminology.

Using the language of AI, we say that at each time step t, the *Agent* (the algorithm we design) observes the state S_t, after which the Agent performs action A_t, after which the *Environment* (upon seeing S_t and A_t) produces a random pair: the next state state S_{t+1} and the next reward R_{t+1}, after which the *Agent* oberves this next state S_{t+1}, and the cycle repeats (until we reach a terminal state). This cyclic interplay is depicted in Figure 4.1. Note that time ticks over from t to $t+1$ when the environment sees the state S_t and action A_t.

The MDP framework was formalized in a paper by Richard Bellman (Bellman 1957a) and the MDP theory was developed further in Richard Bellman's book named *Dynamic Programming* (Bellman 1957b) and in Ronald Howard's book named *Dynamic Programming and Markov Processes* (Howard 1960).

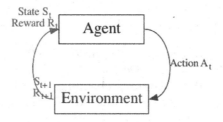

Figure 4.1 **Markov Decision Process**

4.3   FORMAL DEFINITION OF A MARKOV DECISION PROCESS

Similar to the definitions of Markov Processes and Markov Reward Processes, for ease of exposition, the definitions and theory of Markov Decision Processes below will be for discrete-time, for countable state spaces and countable set of pairs of next state and reward transitions (with the knowledge that the definitions and theory are analogously extensible to continuous-time and uncountable spaces, which we shall indeed encounter later in this book).

Definition 4.3.1. A *Markov Decision Process* comprises of:

- A countable set of states $\mathcal{S}$ (known as the State Space), a set $\mathcal{T} \subseteq \mathcal{S}$ (known as the set of Terminal States), and a countable set of actions $\mathcal{A}$ (known as the Action Space).

- A time-indexed sequence of environment-generated random states $S_t \in \mathcal{S}$ for time steps $t = 0, 1, 2, \ldots$, a time-indexed sequence of environment-generated *Reward* random variables $R_t \in \mathcal{D}$ (a countable subset of $\mathbb{R}$) for time steps $t = 1, 2, \ldots$, and a time-indexed sequence of agent-controllable actions $A_t \in \mathcal{A}$ for time steps $t = 0, 1, 2, \ldots$. (Sometimes we restrict the set of actions allowable from specific states, in which case, we abuse the $\mathcal{A}$ notation to refer to a function whose domain is $\mathcal{N}$ and range is $\mathcal{A}$, and we say that the set of actions allowable from a state $s \in \mathcal{N}$ is $\mathcal{A}(s)$.)

- Markov Property:

$$\mathbb{P}[(R_{t+1}, S_{t+1})|(S_t, A_t, S_{t-1}, A_{t-1}, \ldots, S_0, A_0)] = \mathbb{P}[(R_{t+1}, S_{t+1})|(S_t, A_t)] \text{ for all } t \geq 0$$

- Termination: If an outcome for S_T (for some time step T) is a state in the set $\mathcal{T}$, then this sequence outcome terminates at time step T.

As in the case of Markov Reward Processes, we denote the set of non-terminal states $\mathcal{S} - \mathcal{T}$ as $\mathcal{N}$ and refer to any state in $\mathcal{N}$ as a non-terminal state. The sequence:

$$S_0, A_0, R_1, S_1, A_1, R_1, S_2, \ldots$$

terminates at time step T if $S_T \in \mathcal{T}$ (i.e., the final reward is R_T and the final action is A_{T-1}).

In the more general case, where states or rewards are uncountable, the same concepts apply except that the mathematical formalism needs to be more detailed and more careful. Specifically, we'd end up with integrals instead of summations, and probability density functions (for continuous probability distributions) instead of probability mass functions (for discrete probability distributions). For ease of notation and more importantly, for ease of understanding of the core concepts (without being distracted by heavy mathematical formalism), we've chosen to stay with discrete-time, countable $\mathcal{S}$, countable $\mathcal{A}$ and countable $\mathcal{D}$ (by default). However, there will be examples of Markov Decision Processes in this book involving continuous-time and uncountable $\mathcal{S}$, $\mathcal{A}$ and $\mathcal{D}$ (please adjust the definitions and formulas accordingly).

We refer to $\mathbb{P}[(R_{t+1}, S_{t+1})|(S_t, A_t)]$ as the transition probabilities of the Markov Decision Process for time t.

As in the case of Markov Processes and Markov Reward Processes, we shall (by default) assume Time-Homogeneity for Markov Decision Processes, i.e., $\mathbb{P}[(R_{t+1}, S_{t+1})|(S_t, A_t)]$ is independent of t. This means the transition probabilities of a Markov Decision Process can, in the most general case, be expressed as a state-reward transition probability function:

$$\mathcal{P}_R : \mathcal{N} \times \mathcal{A} \times \mathcal{D} \times \mathcal{S} \to [0, 1]$$

defined as:

$$\mathcal{P}_R(s, a, r, s') = \mathbb{P}[(R_{t+1} = r, S_{t+1} = s')|(S_t = s, A_t = a)]$$

for time steps $t = 0, 1, 2, \ldots$, for all $s \in \mathcal{N}, a \in \mathcal{A}, r \in \mathcal{D}, s' \in \mathcal{N}$ such that:

$$\sum_{s' \in \mathcal{S}} \sum_{r \in \mathcal{D}} \mathcal{P}_R(s, a, r, s') = 1 \text{ for all } s \in \mathcal{N}, a \in \mathcal{A}$$

Henceforth, any time we say Markov Decision Process, assume we are referring to a Discrete-Time, Time-Homogeneous Markov Decision Process with countable spaces and countable transitions (unless explicitly specified otherwise), which in turn can be characterized by the state-reward transition probability function $\mathcal{P}_R$. Given a specification of $\mathcal{P}_R$, we can construct:

- The state transition probability function

$$\mathcal{P} : \mathcal{N} \times \mathcal{A} \times \mathcal{S} \to [0, 1]$$

    defined as:

$$\mathcal{P}(s, a, s') = \sum_{r \in \mathcal{D}} \mathcal{P}_R(s, a, r, s')$$

- The reward transition function:

$$\mathcal{R}_T : \mathcal{N} \times \mathcal{A} \times \mathcal{S} \to \mathbb{R}$$

defined as:

$$\mathcal{R}_T(s, a, s') = \mathbb{E}[R_{t+1}|(S_{t+1} = s', S_t = s, A_t = a)]$$
$$= \sum_{r \in \mathcal{D}} \frac{\mathcal{P}_R(s, a, r, s')}{\mathcal{P}(s, a, s')} \cdot r$$
$$= \sum_{r \in \mathcal{D}} \frac{\mathcal{P}_R(s, a, r, s')}{\sum_{r \in \mathcal{D}} \mathcal{P}_R(s, a, r, s')} \cdot r$$

The Rewards specification of most Markov Decision Processes we encounter in practice can be directly expressed as the reward transition function $\mathcal{R}_T$ (versus the more general specification of $\mathcal{P}_R$). Lastly, we want to highlight that we can transform either of $\mathcal{P}_R$ or $\mathcal{R}_T$ into a "more compact" reward function that is sufficient to perform key calculations involving Markov Decision Processes. This reward function

$$\mathcal{R} : \mathcal{N} \times \mathcal{A} \to \mathbb{R}$$

is defined as:

$$\mathcal{R}(s, a) = \mathbb{E}[R_{t+1}|(S_t = s, A_t = a)]$$
$$= \sum_{s' \in \mathcal{S}} \mathcal{P}(s, a, s') \cdot \mathcal{R}_T(s, a, s')$$
$$= \sum_{s' \in \mathcal{S}} \sum_{r \in \mathcal{D}} \mathcal{P}_R(s, a, r, s') \cdot r$$

4.4 POLICY

Having understood the dynamics of a Markov Decision Process, we now move on to the specification of the *Agent*'s actions as a function of the current state. In the general case, we assume that the Agent will perform a random action A_t, according to a probability distribution that is a function of the current state S_t. We refer to this function as a *Policy*. Formally, a *Policy* is a function

$$\pi : \mathcal{N} \times \mathcal{A} \to [0, 1]$$

defined as:

$$\pi(s, a) = \mathbb{P}[A_t = a|S_t = s] \text{ for time steps } t = 0, 1, 2, \ldots, \text{ for all } s \in \mathcal{N}, a \in \mathcal{A}$$

such that:

$$\sum_{a \in \mathcal{A}} \pi(s, a) = 1 \text{ for all } s \in \mathcal{N}$$

Note that the definition above assumes that a Policy is Markovian, i.e., the action probabilities depend only on the current state and not the history. The definition above also assumes that a Policy is *Stationary*, i.e., $\mathbb{P}[A_t = a|S_t = s]$ is invariant in time t. If we do encounter a situation where the policy would need to depend on the time t, we'll simply include t to be part of the state, which would make the Policy stationary (albeit at the cost of state-space bloat and hence, computational cost).

When we have a policy such that the action probability distribution for each state is concentrated on a single action, we refer to it as a deterministic policy. Formally, a deterministic policy $\pi_D : \mathcal{N} \to \mathcal{A}$ has the property that for all $s \in \mathcal{N}$,

$$\pi(s, \pi_D(s)) = 1 \text{ and } \pi(s, a) = 0 \text{ for all } a \in \mathcal{A} \text{ with } a \neq \pi_D(s)$$

So we shall denote deterministic policies simply as the function π_D. We shall refer to non-deterministic policies as stochastic policies (the word stochastic reflecting the fact that the agent will perform a random action according to the probability distribution specified by π). So when we use the notation π, assume that we are dealing with a stochastic (i.e., non-deterministic) policy and when we use the notation π_D, assume that we are dealing with a deterministic policy.

Let's write some code to get a grip on the concept of Policy. We start with the design of an abstract class called `Policy` that represents a general Policy, as we have articulated above. The only method it contains is an abstract method `act` that accepts as input a `state`: `NonTerminal[S]` (as seen before in the classes `MarkovProcess` and `MarkovRewardProcess`, S is a generic type to represent a generic state space) and produces as output a `Distribution[A]` representing the probability distribution of the random action as a function of the input non-terminal state. Note that A represents a generic type to represent a generic action space.

```
A = TypeVar('A')
S = TypeVar('S')

class Policy(ABC, Generic[S, A]):

    @abstractmethod
    def act(self, state: NonTerminal[S]) -> Distribution[A]:
        pass
```

Next, we implement a class for deterministic policies.

```
@dataclass(frozen=True)
class DeterministicPolicy(Policy[S, A]):
    action_for: Callable[[S], A]

    def act(self, state: NonTerminal[S]) -> Constant[A]:
        return Constant(self.action_for(state.state))
```

We will often encounter policies that assign equal probabilities to all actions, from each non-terminal state. We implement this class of policies as follows:

```
from rl.distribution import Choose

@dataclass(frozen=True)
class UniformPolicy(Policy[S, A]):
    valid_actions: Callable[[S], Iterable[A]]

    def act(self, state: NonTerminal[S]) -> Choose[A]:
        return Choose(self.valid_actions(state.state))
```

The above code is in the file rl/policy.py.

Now let's write some code to create some concrete policies for an example we are familiar with—the simple inventory example. We first create a concrete class `SimpleInventoryDeterministicPolicy` for deterministic inventory replenishment policies that is a derived class of `DeterministicPolicy`. Note that the generic state type S is replaced here with the class `InventoryState` that represents a state in the inventory example, comprising of the On-Hand and On-Order inventory quantities. Also note that the generic action type A is replaced here with the `int` type since in this example, the action is the quantity of inventory to be ordered at store-closing (which is an integer quantity). Invoking the `act` method of `SimpleInventoryDeterministicPolicy` runs the following deterministic policy:

$$\pi_D((\alpha, \beta)) = \max(r - (\alpha + \beta), 0)$$

where r is a parameter representing the "reorder point" (meaning, we order only when the inventory position falls below the "reorder point"), α is the On-Hand Inventory at store-closing, β is the On-Order Inventory at store-closing, and inventory position is equal to $\alpha + \beta$. In Chapter 3, we set the reorder point to be equal to the store capacity C.

```python
from rl.distribution import Constant
@dataclass(frozen=True)
class InventoryState:
    on_hand: int
    on_order: int

    def inventory_position(self) -> int:
        return self.on_hand + self.on_order
class SimpleInventoryDeterministicPolicy(
        DeterministicPolicy[InventoryState, int]
):
    def __init__(self, reorder_point: int):
        self.reorder_point: int = reorder_point

        def action_for(s: InventoryState) -> int:
            return max(self.reorder_point - s.inventory_position(), 0)

        super().__init__(action_for)
```

We can instantiate a specific deterministic policy with a reorder point of say 8 as:

```python
si_dp = SimpleInventoryDeterministicPolicy(reorder_point=8)
```

Now let's write some code to create stochastic policies for the inventory example. We create a concrete class `SimpleInventoryStochasticPolicy` that implements the interface of the abstract class `Policy` (specifically implements the abstract method `act`). The code in `act` implements a stochastic policy as a `SampledDistribution[int]` driven by a sampling of the Poison distribution for the reorder point. Specifically, the reorder point r is treated as a Poisson random variable with a specified mean (of say $\lambda \in \mathbb{R}_{\geq 0}$). We sample a value of the reorder point r from this Poisson distribution (with mean λ). Then, we create a sample order quantity (*action*) $\theta \in \mathbb{Z}_{\geq 0}$ defined as:

$$\theta = \max(r - (\alpha + \beta), 0)$$

```python
import numpy as np
from rl.distribution import SampledDistribution

class SimpleInventoryStochasticPolicy(Policy[InventoryState, int]):
    def __init__(self, reorder_point_poisson_mean: float):
        self.reorder_point_poisson_mean: float = reorder_point_poisson_mean

    def act(self, state: NonTerminal[InventoryState]) -> \
            SampledDistribution[int]:
        def action_func(state=state) -> int:
            reorder_point_sample: int = \
                np.random.poisson(self.reorder_point_poisson_mean)
            return max(
                reorder_point_sample - state.state.inventory_position(),
                0
            )
        return SampledDistribution(action_func)
```

We can instantiate a specific stochastic policy with a reorder point Poisson distribution mean of say 8.0 as:

```python
si_sp = SimpleInventoryStochasticPolicy(reorder_point_poisson_mean=8.0)
```

We will revisit the simple inventory example in a bit after we cover the code for Markov Decision Processes, when we'll show how to simulate the Markov Decision Process for this simple inventory example, with the agent running a deterministic policy. But before we move on to the code design for Markov Decision Processes (to accompany the above implementation of Policies), we need to cover an important insight linking Markov Decision Processes, Policies and Markov Reward Processes.

4.5 [MARKOV DECISION PROCESS, POLICY] := MARKOV REWARD PROCESS

This section has an important insight—that if we evaluate a Markov Decision Process (MDP) with a fixed policy π (in general, with a fixed stochastic policy π), we get the Markov Reward Process (MRP) that is *implied* by the combination of the MDP and the policy π. Let's clarify this with notational precision. But first we need to point out that we have some notation clashes between MDP and MRP. We used $\mathcal{P}_R$ to denote the transition probability function of the MRP as well as to denote the state-reward transition probability function of the MDP. We used $\mathcal{P}$ to denote the transition probability function of the Markov Process implicit in the MRP as well as to denote the state transition probability function of the MDP. We used $\mathcal{R}_T$ to denote the reward transition function of the MRP as well as to denote the reward transition function of the MDP. We used $\mathcal{R}$ to denote the reward function of the MRP as well as to denote the reward function of the MDP. We can resolve these notation clashes by noting the arguments to $\mathcal{P}_R, \mathcal{P}, \mathcal{R}_T$ and $\mathcal{R}$, but to be extra-clear, we'll put a superscript of π to each of the functions $\mathcal{P}_R, \mathcal{P}, \mathcal{R}_T$ and $\mathcal{R}$ of the π-implied MRP so as to distinguish between these functions for the MDP versus the π-implied MRP.

Let's say we are given a fixed policy π and an MDP specified by its state-reward transition probability function $\mathcal{P}_R$. Then the transition probability function $\mathcal{P}_R^\pi$ of the MRP implied by the evaluation of the MDP with the policy π is defined as:

$$\mathcal{P}_R^\pi(s, r, s') = \sum_{a \in \mathcal{A}} \pi(s, a) \cdot \mathcal{P}_R(s, a, r, s')$$

Likewise,

$$\mathcal{P}^\pi(s, s') = \sum_{a \in \mathcal{A}} \pi(s, a) \cdot \mathcal{P}(s, a, s')$$

$$\mathcal{R}_T^\pi(s, s') = \sum_{a \in \mathcal{A}} \pi(s, a) \cdot \mathcal{R}_T(s, a, s')$$

$$\mathcal{R}^\pi(s) = \sum_{a \in \mathcal{A}} \pi(s, a) \cdot \mathcal{R}(s, a)$$

So any time we talk about an MDP evaluated with a fixed policy, you should know that we are effectively talking about the implied MRP. This insight is now going to be key in the design of our code to represent Markov Decision Processes.

We create an abstract class called `MarkovDecisionProcess` (code shown below) with two abstract methods—`step` and `actions`. The `step` method is key: it is meant to specify the distribution of pairs of next state and reward, given a non-terminal state and action. The `actions` method's interface specifies that it takes as input a `state: NonTerminal[S]` and produces as output an `Iterable[A]` to represent the set of actions allowable for the input `state` (since the set of actions can be potentially infinite—in which case we'd have to return an `Iterator[A]`—the return type is fairly generic, i.e., `Iterable[A]`).

The `apply_policy` method takes as input a `policy: Policy[S, A]` and returns a `MarkovRewardProcess` representing the implied MRP. Let's understand the code in `apply_policy`: Firstly, we construct a `class RewardProcess` that implements the abstract method `transition_reward` of `MarkovRewardProcess`. `transition_reward` takes as input a `state: NonTerminal[S]`, creates `actions: Distribution[A]` by applying the given `policy` on `state`, and finally uses the `apply` method of `Distribution` to transform `actions: Distribution[A]` into a `Distribution[Tuple[State[S], float]]` (distribution of (next state, reward) pairs) using the abstract method `step`.

We also write the `simulate_actions` method that is analogous to the `simulate_reward` method we had written for `MarkovRewardProcess` for generating a sampling trace. In this case, each step in the sampling trace involves sampling an action from the given `policy` and then sampling the pair of next state and reward, given the `state` and sampled `action`. Each generated `TransitionStep` object consists of the 4-tuple: (state, action, next state, reward). Here's the actual code:

```python
from rl.distribution import Distribution

@dataclass(frozen=True)
class TransitionStep(Generic[S, A]):
    state: NonTerminal[S]
    action: A
    next_state: State[S]
    reward: float

class MarkovDecisionProcess(ABC, Generic[S, A]):
    @abstractmethod
    def actions(self, state: NonTerminal[S]) -> Iterable[A]:
        pass

    @abstractmethod
    def step(
        self,
        state: NonTerminal[S],
        action: A
    ) -> Distribution[Tuple[State[S], float]]:
        pass

    def apply_policy(self, policy: Policy[S, A]) -> MarkovRewardProcess[S]:
        mdp = self

        class RewardProcess(MarkovRewardProcess[S]):
            def transition_reward(
                self,
                state: NonTerminal[S]
            ) -> Distribution[Tuple[State[S], float]]:
                actions: Distribution[A] = policy.act(state)
                return actions.apply(lambda a: mdp.step(state, a))

        return RewardProcess()

    def simulate_actions(
        self,
        start_states: Distribution[NonTerminal[S]],
        policy: Policy[S, A]
    ) -> Iterable[TransitionStep[S, A]]:
        state: State[S] = start_states.sample()

        while isinstance(state, NonTerminal):
            action_distribution = policy.act(state)

            action = action_distribution.sample()
            next_distribution = self.step(state, action)

            next_state, reward = next_distribution.sample()
            yield TransitionStep(state, action, next_state, reward)
            state = next_state
```

The above code is in the file rl/markov_decision_process.py.

4.6 SIMPLE INVENTORY EXAMPLE WITH UNLIMITED CAPACITY (INFINITE STATE/ACTION SPACE)

Now we come back to our simple inventory example. Unlike previous situations of this example, here we assume that there is no space capacity constraint on toothpaste. This means we have a choice of ordering any (unlimited) non-negative integer quantity of toothpaste units. Therefore, the action space is infinite. Also, since the order quantity shows up as On-Order the next day and as delivered inventory the day after the next day, the On-Hand and On-Order quantities are also unbounded. Hence, the state space is infinite. Due to the infinite state and action spaces, we won't be able to take advantage of the so-called "Tabular Dynamic Programming Algorithms" we will cover in Chapter 5 (algorithms that are meant for finite state and action spaces). There is still significant value in modeling infinite MDPs of this type because we can perform simulations (by sampling from an infinite space). Simulations are valuable not just to explore various properties and metrics relevant in the real-world problem modeled with an MDP, but simulations also enable us to design approximate algorithms to calculate Value Functions for given policies as well as Optimal Value Functions (which is the ultimate purpose of modeling MDPs).

We will cover details on these approximate algorithms later in the book—for now, it's important for you to simply get familiar with how to model infinite MDPs of this type. This infinite-space inventory example serves as a great learning for an introduction to modeling an infinite (but countable) MDP.

We create a concrete class `SimpleInventoryMDPNoCap` that implements the abstract class `MarkovDecisionProcess` (specifically implements abstract methods `step` and `actions`). The attributes `poisson_lambda`, `holding_cost` and `stockout_cost` have the same semantics as what we had covered for Markov Reward Processes in Chapter 3 (`SimpleInventoryMRP`). The `step` method takes as input a `state: NonTerminal[InventoryState]` and an `order: int` (representing the MDP action). We sample from the Poisson probability distribution of customer demand (calling it `demand_sample: int`). Using `order: int` and `demand_sample: int`, we obtain a sample of the pair of `next_state: InventoryState` and `reward: float`. This sample pair is returned as a `SampledDistribution` object. The above sampling dynamics effectively describe the MDP in terms of this `step` method. The `actions` method returns an `Iterator[int]`, an infinite generator of non-negative integers to represent the fact that the action space (order quantities) for any state comprise of all non-negative integers.

```python
import itertools
import numpy as np
from rl.distribution import SampledDistribution

@dataclass(frozen=True)
class SimpleInventoryMDPNoCap(MarkovDecisionProcess[InventoryState, int]):
    poisson_lambda: float
    holding_cost: float
    stockout_cost: float

    def step(
        self,
        state: NonTerminal[InventoryState],
        order: int
    ) -> SampledDistribution[Tuple[State[InventoryState], float]]:

        def sample_next_state_reward(
            state=state,
            order=order
        ) -> Tuple[State[InventoryState], float]:
            demand_sample: int = np.random.poisson(self.poisson_lambda)
            ip: int = state.state.inventory_position()
            next_state: InventoryState = InventoryState(
                max(ip - demand_sample, 0),
                order
            )
            reward: float = - self.holding_cost * state.state.on_hand\
                - self.stockout_cost * max(demand_sample - ip, 0)
            return NonTerminal(next_state), reward

        return SampledDistribution(sample_next_state_reward)

    def actions(self, state: NonTerminal[InventoryState]) -> Iterator[int]:
        return itertools.count(start=0, step=1)
```

We leave it to you as an exercise to run various simulations of the MRP implied by the deterministic and stochastic policy instances we had created earlier (the above code is in the file rl/chapter3/simple_inventory_mdp_nocap.py). See the method `fraction_of_days_oos` in this file as an example of a simulation to calculate the percentage of days when we'd be unable to satisfy some customer demand for toothpaste due to too little inventory at store-opening (naturally, the higher the re-order point in the policy, the lesser the percentage of days when we'd be Out-of-Stock). This kind of simulation exercise helps build intuition on the tradeoffs we have to make between having too little inventory versus having too much inventory (holding costs versus stockout costs)—essentially leading to our ultimate goal of determining the Optimal Policy (more on this later).

4.7   FINITE MARKOV DECISION PROCESSES

Certain calculations for Markov Decision Processes can be performed easily if:

- the state space is finite ($\mathcal{S} = \{s_1, s_2, \ldots, s_n\}$),
- the action space $\mathcal{A}(s)$ is finite for each $s \in \mathcal{N}$,
- the set of unique pairs of next state and reward transitions from each pair of current non-terminal state and action is finite.

If we satisfy the above three characteristics, we refer to the Markov Decision Process as a Finite Markov Decision Process. Let us write some code for a Finite Markov Decision Process. We create a concrete class `FiniteMarkovDecisionProcess` that implements the interface of the abstract class `MarkovDecisionProcess` (specifically implements the abstract methods `step` and the `actions`). Our first task is to think about the data structure required to specify an instance of `FiniteMarkovDecisionProcess` (i.e., the data structure we'd pass to the `__init__` method of `FiniteMarkovDecisionProcess`). Analogous to how we curried $\mathcal{P}_R$ for a Markov Reward Process as $\mathcal{N} \to (\mathcal{S} \times \mathcal{D} \to [0,1])$ (where $\mathcal{S} = \{s_1, s_2, \ldots, s_n\}$ and $\mathcal{N}$ has $m \leq n$ states), here we curry $\mathcal{P}_R$ for the MDP as:

$$\mathcal{N} \to (\mathcal{A} \to (\mathcal{S} \times \mathcal{D} \to [0,1]))$$

Since $\mathcal{S}$ is finite, $\mathcal{A}$ is finite, and the set of next state and reward transitions for each pair of current state and action is also finite, we can represent $\mathcal{P}_R$ as a data structure of type `StateActionMapping[S, A]` as shown below:

```
StateReward = FiniteDistribution[Tuple[State[S], float]]
ActionMapping = Mapping[A, StateReward[S]]
StateActionMapping = Mapping[NonTerminal[S], ActionMapping[A, S]]
```

The constructor (`__init__` method) of `FiniteMarkovDecisionProcess` takes as input `mapping` which is essentially of the same structure as `StateActionMapping[S, A]`, except that the `Mapping` is specified in terms of S rather than `NonTerminal[S]` or `State[S]` so as to make it easy for a user to specify a `FiniteMarkovDecisionProcess` without the overhead of wrapping S in `NonTerminal[S]` or `Terminal[S]`. But this means `__init__` need to do the wrapping to construct the attribute `self.mapping: StateActionMapping[S, A]`. This represents the complete structure of the Finite MDP—it maps each non-terminal state to an action map, and it maps each action in each action map to a finite probability distribution of pairs of next state and reward (essentially the structure of the $\mathcal{P}_R$ function). Along with the attribute `self.mapping`, we also have an attribute `non_terminal_states: Sequence[NonTerminal[S]]` that is an ordered sequence of non-terminal states. Now let's consider the implementation of the abstract method `step` of `MarkovDecisionProcess`. It takes as input a `state: NonTerminal[S]` and an `action: A`. `self.mapping[state][action]` gives us an object of type

FiniteDistribution[Tuple[State[S], float]] which represents a finite probability distribution of pairs of next state and reward, which is exactly what we want to return. This satisfies the responsibility of FiniteMarkovDecisionProcess in terms of implementing the abstract method step of the abstract class MarkovDecisionProcess. The other abstract method to implement is the actions method which produces an Iterable on the allowed actions $\mathcal{A}(s)$ for a given $s \in \mathcal{N}$ by invoking self.mapping[state].keys(). The __repr__ method shown below is quite straightforward.

```python
from rl.distribution import FiniteDistribution, SampledDistribution
class FiniteMarkovDecisionProcess(MarkovDecisionProcess[S, A]):
    mapping: StateActionMapping[S, A]
    non_terminal_states: Sequence[NonTerminal[S]]
    def __init__(
        self,
        mapping: Mapping[S, Mapping[A, FiniteDistribution[Tuple[S, float]]]]
    ):
        non_terminals: Set[S] = set(mapping.keys())
        self.mapping = {NonTerminal(s): {a: Categorical(
            {(NonTerminal(s1) if s1 in non_terminals else Terminal(s1), r): p
                for (s1, r), p in v}
        ) for a, v in d.items()} for s, d in mapping.items()}
        self.non_terminal_states = list(self.mapping.keys())
    def __repr__(self) -> str:
        display = ""
        for s, d in self.mapping.items():
            display += f"From State {s.state}:\n"
            for a, d1 in d.items():
                display += f"  With Action {a}:\n"
                for (s1, r), p in d1:
                    opt = "Terminal " if isinstance(s1, Terminal) else ""
                    display += f"    To [{opt}State {s1.state} and "\
                        + f"Reward {r:.3f}] with Probability {p:.3f}\n"
        return display
    def step(self, state: NonTerminal[S], action: A) -> StateReward[S]:
        action_map: ActionMapping[A, S] = self.mapping[state]
        return action_map[action]
    def actions(self, state: NonTerminal[S]) -> Iterable[A]:
        return self.mapping[state].keys()
```

Now that we've implemented a finite MDP, let's implement a finite policy that maps each non-terminal state to a probability distribution over a finite set of actions. So we create a concrete class @dataclass FinitePolicy that implements the interface of the abstract class Policy (specifically implements the abstract method act). An instance of FinitePolicy is specified with the attribute self.policy_map: Mapping[S, FiniteDistribution[A]] since this type captures the structure of the $\pi : \mathcal{N} \times \mathcal{A} \to [0, 1]$ function in the curried form

$$\mathcal{N} \to (\mathcal{A} \to [0, 1])$$

for the case of finite $\mathcal{S}$ and finite $\mathcal{A}$. The act method is straightforward. We also implement a __repr__ method for pretty-printing of self.policy_map.

```python
@dataclass(frozen=True)
class FinitePolicy(Policy[S, A]):
    policy_map: Mapping[S, FiniteDistribution[A]]
    def __repr__(self) -> str:
        display = ""
        for s, d in self.policy_map.items():
            display += f"For State {s}:\n"
            for a, p in d:
```

```
            display += f"  Do Action {a} with Probability {p:.3f}\n"
        return display
    def act(self, state: NonTerminal[S]) -> FiniteDistribution[A]:
        return self.policy_map[state.state]
```

Let's also implement a finite deterministic policy as a derived class of `FinitePolicy`.

```
class FiniteDeterministicPolicy(FinitePolicy[S, A]):
    action_for: Mapping[S, A]

    def __init__(self, action_for: Mapping[S, A]):
        self.action_for = action_for
        super().__init__(policy_map={s: Constant(a) for s, a in
                                     self.action_for.items()})

    def __repr__(self) -> str:
        display = ""
        for s, a in self.action_for.items():
            display += f"For State {s}: Do Action {a}\n"
        return display
```

Armed with a `FinitePolicy` class, we can now write a method `apply_finite_policy` in `FiniteMarkovDecisionProcess` that takes as input a `policy: FinitePolicy[S, A]` and returns a `FiniteMarkovRewardProcess[S]` by processing the finite structures of both of the MDP and the Policy, and producing a finite structure of the implied MRP.

```
from collections import defaultdict
from rl.distribution import FiniteDistribution, Categorical

    def apply_finite_policy(self, policy: FinitePolicy[S, A])\
            -> FiniteMarkovRewardProcess[S]:

        transition_mapping: Dict[S, FiniteDistribution[Tuple[S, float]]] = {}

        for state in self.mapping:
            action_map: ActionMapping[A, S] = self.mapping[state]
            outcomes: DefaultDict[Tuple[S, float], float]\
                = defaultdict(float)
            actions = policy.act(state)
            for action, p_action in actions:
                for (s1, r), p in action_map[action]:
                    outcomes[(s1.state, r)] += p_action * p

            transition_mapping[state.state] = Categorical(outcomes)

        return FiniteMarkovRewardProcess(transition_mapping)
```

The code for `FiniteMarkovRewardProcess` is in rl/markov_decision_process.py and the code for `FinitePolicy` and `FiniteDeterministicPolicy` is in rl/policy.py.

4.8 SIMPLE INVENTORY EXAMPLE AS A FINITE MARKOV DECISION PROCESS

Now we'd like to model the simple inventory example as a Finite Markov Decision Process so we can take advantage of the algorithms specifically for Finite Markov Decision Processes. To enable finite states and finite actions, we now re-introduce the constraint of space capacity C and apply the restriction that the order quantity (action) cannot exceed $C - (\alpha + \beta)$, where α is the On-Hand component of the State and β is the On-Order component of the State. Thus, the action space for any given state $(\alpha, \beta) \in \mathcal{S}$ is finite. Next, note that this ordering policy ensures that in steady-state, the sum of On-Hand and On-Order will not exceed the capacity C. So we constrain the set of states to be the steady-state set of finite states

$$\mathcal{S} = \{(\alpha, \beta) | \alpha \in \mathbb{Z}_{\geq 0}, \beta \in \mathbb{Z}_{\geq 0}, 0 \leq \alpha + \beta \leq C\}$$

Although the set of states is finite, there are an infinite number of pairs of next state and reward outcomes possible from any given pair of current state and action. This is because there are an infinite set of possibilities of customer demand on any given day (resulting in infinite possibilities of stockout cost, i.e., negative reward, on any day). To qualify as a Finite Markov Decision Process, we need to model in a manner such that we have a finite set of pairs of next state and reward outcomes from any given pair of current state and action. So what we do is that instead of considering (S_{t+1}, R_{t+1}) as the pair of next state and reward, we model the pair of next state and reward to instead be $(S_{t+1}, \mathbb{E}[R_{t+1}|(S_t, S_{t+1}, A_t)])$ (we know $\mathcal{P}_R$ due to the Poisson probabilities of customer demand, so we can actually calculate this conditional expectation of reward). So given a state s and action a, the pairs of next state and reward would be: $(s', \mathcal{R}_T(s, a, s'))$ for all the s' we transition to from (s, a). Since the set of possible next states s' are finite, these newly-modeled rewards associated with the transitions $(\mathcal{R}_T(s, a, s'))$ are also finite and hence, the set of pairs of next state and reward from any pair of current state and action are also finite. Note that this creative alteration of the reward definition is purely to reduce this Markov Decision Process into a Finite Markov Decision Process. Let's now work out the calculation of the reward transition function $\mathcal{R}_T$.

When the next state's (S_{t+1}) On-Hand is greater than zero, it means all of the day's demand was satisfied with inventory that was available at store-opening ($= \alpha + \beta$), and hence, each of these next states S_{t+1} correspond to no stockout cost and only an overnight holding cost of $h\alpha$. Therefore, for all α, β (with $0 \le \alpha + \beta \le C$) and for all order quantity (action) θ (with $0 \le \theta \le C - (\alpha + \beta)$):

$$\mathcal{R}_T((\alpha, \beta), \theta, (\alpha + \beta - i, \theta)) = -h\alpha \text{ for } 0 \le i \le \alpha + \beta - 1$$

When next state's (S_{t+1}) On-Hand is equal to zero, there are two possibilities:

1. The demand for the day was exactly $\alpha + \beta$, meaning all demand was satisifed with available store inventory (so no stockout cost and only overnight holding cost), or
2. The demand for the day was strictly greater than $\alpha + \beta$, meaning there's some stockout cost in addition to overnight holding cost. The exact stockout cost is an expectation calculation involving the number of units of missed demand under the corresponding Poisson probabilities of demand exceeding $\alpha + \beta$.

This calculation is shown below:

$$\mathcal{R}_T((\alpha, \beta), \theta, (0, \theta)) = -h\alpha - p(\sum_{j=\alpha+\beta+1}^{\infty} f(j) \cdot (j - (\alpha + \beta)))$$

$$= -h\alpha - p(\lambda(1 - F(\alpha + \beta - 1)) - (\alpha + \beta)(1 - F(\alpha + \beta)))$$

So now we have a specification of $\mathcal{R}_T$, but when it comes to our coding interface, we are expected to specify $\mathcal{P}_R$ as that is the interface through which we create a FiniteMarkovDecisionProcess. Fear not—a specification of $\mathcal{P}_R$ is easy once we have a specification of $\mathcal{R}_T$. We simply create 5-tuples (s, a, r, s', p) for all $s \in \mathcal{N}, s' \in \mathcal{S}, a \in \mathcal{A}$ such that $r = \mathcal{R}_T(s, a, s')$ and $p = \mathcal{P}(s, a, s')$ (we know $\mathcal{P}$ along with $\mathcal{R}_T$), and the set of all these 5-tuples (for all $s \in \mathcal{N}, s' \in \mathcal{S}, a \in \mathcal{A}$) constitute the specification of $\mathcal{P}_R$, i.e., $\mathcal{P}_R(s, a, r, s') = p$. This turns our reward-definition-altered mathematical model of a Finite Markov Decision Process into a programming model of the FiniteMarkovDecisionProcess class. This reward-definition-altered model enables us to gain from the fact that we can leverage the algorithms we'll be writing for Finite Markov Decision Processes (specifically, the classical Dynamic Programming algorithms—covered in Chapter 5). The downside of this reward-definition-altered model is that it prevents us from generating sampling traces of the specific rewards encountered when transitioning from one state to another (because we no longer capture the probabilities of individual reward outcomes). Note that we can indeed perform simulations, but each transition step in the sampling trace will only show us the "mean reward" (specifically, the expected reward conditioned on current state, action and next state).

In fact, most Markov Processes you'd encounter in practice can be modeled as a combination of $\mathcal{R}_T$ and $\mathcal{P}$, and you'd simply follow the above $\mathcal{R}_T$ to $\mathcal{P}_R$ representation transformation drill to present this information in the form of $\mathcal{P}_R$ to instantiate a FiniteMarkovDecisionProcess. We designed the interface to accept $\mathcal{P}_R$ as input since that is the most general interface for specifying Markov Decision Processes.

So now let's write some code for the simple inventory example as a Finite Markov Decision Process as described above. All we have to do is to create a derived class inherited from FiniteMarkovDecisionProcess and write a method to construct the mapping (i.e., $\mathcal{P}_R$) that the __init__ constuctor of FiniteMarkovRewardProcess requires as input. Note that the generic state type S is replaced here with the @dataclass InventoryState to represent the inventory state, comprising of the On-Hand and On-Order inventory quantities, and the generic action type A is replaced here with int to represent the order quantity.

```python
from scipy.stats import poisson
from rl.distribution import Categorical
InvOrderMapping = Mapping[
    InventoryState,
    Mapping[int, Categorical[Tuple[InventoryState, float]]]
]
class SimpleInventoryMDPCap(FiniteMarkovDecisionProcess[InventoryState, int]):
    def __init__(
        self,
        capacity: int,
        poisson_lambda: float,
        holding_cost: float,
        stockout_cost: float
    ):
        self.capacity: int = capacity
        self.poisson_lambda: float = poisson_lambda
        self.holding_cost: float = holding_cost
        self.stockout_cost: float = stockout_cost

        self.poisson_distr = poisson(poisson_lambda)
        super().__init__(self.get_action_transition_reward_map())
    def get_action_transition_reward_map(self) -> InvOrderMapping:
        d: Dict[InventoryState, Dict[int, Categorical[Tuple[InventoryState,
                                                            float]]]] = {}

        for alpha in range(self.capacity + 1):
            for beta in range(self.capacity + 1 - alpha):
                state: InventoryState = InventoryState(alpha, beta)
                ip: int = state.inventory_position()
                base_reward: float = - self.holding_cost * alpha
                d1: Dict[int, Categorical[Tuple[InventoryState, float]]] = {}

                for order in range(self.capacity - ip + 1):
                    sr_probs_dict: Dict[Tuple[InventoryState, float], float] =\
                        {(InventoryState(ip - i, order), base_reward):
                        self.poisson_distr.pmf(i) for i in range(ip)}

                    probability: float = 1 - self.poisson_distr.cdf(ip - 1)
                    reward: float = base_reward - self.stockout_cost *\
                        (probability * (self.poisson_lambda - ip) +
                        ip * self.poisson_distr.pmf(ip))
                    sr_probs_dict[(InventoryState(0, order), reward)] = \
                        probability
                    d1[order] = Categorical(sr_probs_dict)
```

```
        d[state] = d1
    return d
```

Now let's test this out with some example inputs (as shown below). We construct an instance of the `SimpleInventoryMDPCap` class with these inputs (named `si_mdp` below), then construct an instance of the `FinitePolicy[InventoryState, int]` class (a deterministic policy, named `fdp` below), and combine them to produce the implied MRP (an instance of the `FiniteMarkovRewardProcess[InventoryState]` class).

```
user_capacity = 2
user_poisson_lambda = 1.0
user_holding_cost = 1.0
user_stockout_cost = 10.0

si_mdp: FiniteMarkovDecisionProcess[InventoryState, int] =\
    SimpleInventoryMDPCap(
        capacity=user_capacity,
        poisson_lambda=user_poisson_lambda,
        holding_cost=user_holding_cost,
        stockout_cost=user_stockout_cost
    )

fdp: FiniteDeterministicPolicy[InventoryState, int] = \
    FiniteDeterministicPolicy(
        {InventoryState(alpha, beta): user_capacity - (alpha + beta)
         for alpha in range(user_capacity + 1)
         for beta in range(user_capacity + 1 - alpha)}
    )

implied_mrp: FiniteMarkovRewardProcess[InventoryState] =\
    si_mdp.apply_finite_policy(fdp)
```

The above code is in the file rl/chapter3/simple_inventory_mdp_cap.py. We encourage you to play with the inputs in __main__, produce the resultant implied MRP, and explore its characteristics (such as its Reward Function and its Value Function).

4.9 MDP VALUE FUNCTION FOR A FIXED POLICY

Now we are ready to talk about the Value Function for an MDP evaluated with a fixed policy π (also known as the MDP *Prediction* problem). The term *Prediction* refers to the fact that this problem is about forecasting the expected future returns when the agent follows a specific policy. Just like in the case of MRP, we define the Return G_t at time step t for an MDP as:

$$G_t = \sum_{i=t+1}^{\infty} \gamma^{i-t-1} \cdot R_i = R_{t+1} + \gamma \cdot R_{t+2} + \gamma^2 \cdot R_{t+3} + \dots$$

where $\gamma \in [0, 1]$ is a specified discount factor.

We use the above definition of *Return* even for a terminating sequence (say terminating at $t = T$, i.e., $S_T \in \mathcal{T}$), by treating $R_i = 0$ for all $i > T$.

The Value Function for an MDP evaluated with a fixed policy π

$$V^{\pi} : \mathcal{N} \to \mathbb{R}$$

is defined as:

$$V^{\pi}(s) = \mathbb{E}_{\pi, \mathcal{P}_R}[G_t | S_t = s] \text{ for all } s \in \mathcal{N}, \text{ for all } t = 0, 1, 2, \dots$$

For the rest of the book, we assume that whenever we are talking about a Value Function, the discount factor γ is appropriate to ensure that the Expected Return from each state is finite—in particular, $\gamma < 1$ for continuing (non-terminating) MDPs where the Return could otherwise diverge.

We expand $V^\pi(s) = \mathbb{E}_{\pi,\mathcal{P}_R}[G_t|S_t = s]$ as follows:

$$\mathbb{E}_{\pi,\mathcal{P}_R}[R_{t+1}|S_t = s] + \gamma \cdot \mathbb{E}_{\pi,\mathcal{P}_R}[R_{t+2}|S_t = s] + \gamma^2 \cdot \mathbb{E}_{\pi,\mathcal{P}_R}[R_{t+3}|S_t = s] + \ldots$$

$$= \sum_{a \in \mathcal{A}} \pi(s,a) \cdot \mathcal{R}(s,a) + \gamma \cdot \sum_{a \in \mathcal{A}} \pi(s,a) \sum_{s' \in \mathcal{N}} \mathcal{P}(s,a,s') \sum_{a' \in \mathcal{A}} \pi(s',a') \cdot \mathcal{R}(s',a')$$

$$+ \gamma^2 \cdot \sum_{a \in \mathcal{A}} \pi(s,a) \sum_{s' \in \mathcal{N}} \mathcal{P}(s,a',s') \sum_{a' \in \mathcal{A}} \pi(s',a') \sum_{s'' \in \mathcal{N}} \mathcal{P}(s',a'',s'') \sum_{a'' \in \mathcal{A}} \pi(s'',a'') \cdot \mathcal{R}(s'',a'')$$

$$+ \ldots$$

$$= \mathcal{R}^\pi(s) + \gamma \cdot \sum_{s' \in \mathcal{N}} \mathcal{P}^\pi(s,s') \cdot \mathcal{R}^\pi(s') + \gamma^2 \cdot \sum_{s' \in \mathcal{N}} \mathcal{P}^\pi(s,s') \sum_{s'' \in \mathcal{N}} \mathcal{P}^\pi(s',s'') \cdot \mathcal{R}^\pi(s'') + \ldots$$

But from Equation (3.1) in Chapter 3, we know that the last expression above is equal to the π-implied MRP's Value Function for state s. So, the Value Function V^π of an MDP evaluated with a fixed policy π is exactly the same function as the Value Function of the π-implied MRP. So we can apply the MRP Bellman Equation on V^π, i.e.,

$$V^\pi(s) = \mathcal{R}^\pi(s) + \gamma \cdot \sum_{s' \in \mathcal{N}} \mathcal{P}^\pi(s,s') \cdot V^\pi(s')$$

$$= \sum_{a \in \mathcal{A}} \pi(s,a) \cdot \mathcal{R}(s,a) + \gamma \cdot \sum_{a \in \mathcal{A}} \pi(s,a) \sum_{s' \in \mathcal{N}} \mathcal{P}(s,a,s') \cdot V^\pi(s') \qquad (4.1)$$

$$= \sum_{a \in \mathcal{A}} \pi(s,a) \cdot (\mathcal{R}(s,a) + \gamma \cdot \sum_{s' \in \mathcal{N}} \mathcal{P}(s,a,s') \cdot V^\pi(s')) \text{ for all } s \in \mathcal{N}$$

As we saw in Chapter 3, for finite state spaces that are not too large, Equation (4.1) can be solved for V^π (i.e. solution to the MDP *Prediction* problem) with a linear algebra solution (Equation (3.2) from Chapter 3). More generally, Equation (4.1) will be a key equation for the rest of the book in developing various Dynamic Programming and Reinforcement Algorithms for the MDP *Prediction* problem. However, there is another Value Function that's also going to be crucial in developing MDP algorithms—one which maps a (state, action) pair to the expected return originating from the (state, action) pair when evaluated with a fixed policy. This is known as the *Action-Value Function* of an MDP evaluated with a fixed policy π:

$$Q^\pi : \mathcal{N} \times \mathcal{A} \to \mathbb{R}$$

defined as:

$$Q^\pi(s,a) = \mathbb{E}_{\pi,\mathcal{P}_R}[G_t|(S_t = s, A_t = a)] \text{ for all } s \in \mathcal{N}, a \in \mathcal{A}, \text{ for all } t = 0,1,2,\ldots$$

To avoid terminology confusion, we refer to V^π as the *State-Value Function* (albeit often simply abbreviated to *Value Function*) for policy π, to distinguish from the *Action-Value Function* Q^π. The way to interpret $Q^\pi(s,a)$ is that it's the Expected Return from a given non-terminal state s by first taking the action a and subsequently following policy π. With this interpretation of $Q^\pi(s,a)$, we can perceive $V^\pi(s)$ as the "weighted average" of $Q^\pi(s,a)$ (over all possible actions a from a non-terminal state s) with the weights equal to probabilities of action a, given state s (i.e., $\pi(s,a)$). Precisely,

$$V^\pi(s) = \sum_{a \in \mathcal{A}} \pi(s,a) \cdot Q^\pi(s,a) \text{ for all } s \in \mathcal{N} \qquad (4.2)$$

Combining Equation (4.1) and Equation (4.2) yields:

$$Q^{\pi}(s,a) = \mathcal{R}(s,a) + \gamma \cdot \sum_{s' \in \mathcal{N}} \mathcal{P}(s,a,s') \cdot V^{\pi}(s') \text{ for all } s \in \mathcal{N}, a \in \mathcal{A} \qquad (4.3)$$

Combining Equation (4.3) and Equation (4.2) yields:

$$Q^{\pi}(s,a) = \mathcal{R}(s,a) + \gamma \cdot \sum_{s' \in \mathcal{N}} \mathcal{P}(s,a,s') \sum_{a' \in \mathcal{A}} \pi(s',a') \cdot Q^{\pi}(s',a') \text{ for all } s \in \mathcal{N}, a \in \mathcal{A} \qquad (4.4)$$

Equation (4.1) is known as the MDP State-Value Function Bellman Policy Equation (Figure 4.2 serves as a visualization aid for this Equation). Equation (4.4) is known as the MDP Action-Value Function Bellman Policy Equation (Figure 4.3 serves as a visualization aid for this Equation). Note that Equation (4.2) and Equation (4.3) are embedded in Figure 4.2 as well as in Figure 4.3. Equations (4.1), (4.2), (4.3) and (4.4) are collectively known as the MDP Bellman Policy Equations.

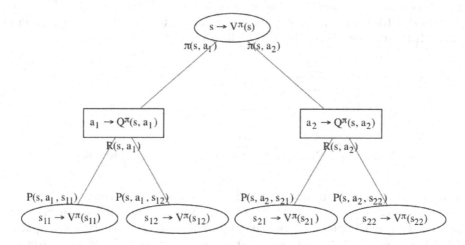

Figure 4.2 **Visualization of MDP State-Value Function Bellman Policy Equation**

For the rest of the book, in these MDP transition figures, we shall always depict states as elliptical-shaped nodes and actions as rectangular-shaped nodes. Notice that transition from a state node to an action node is associated with a probability represented by π and transition from an action node to a state node is associated with a probability represented by $\mathcal{P}$.

Note that for finite MDPs of state space not too large, we can solve the MDP Prediction problem (solving for V^{π} and equivalently, Q^{π}) in a straightforward manner: Given a policy π, we can create the finite MRP implied by π, using the method `apply_policy` in `FiniteMarkovDecisionProcess`, then use the direct linear-algebraic solution that we covered in Chapter 3 to calculate the Value Function of the π-implied MRP. We know that the π-implied MRP's Value Function is the same as the State-Value Function V^{π} of the MDP which can then be used to arrive at the Action-Value Function Q^{π} of the MDP (using Equation (4.3)). For large state spaces, we need to use iterative/numerical methods (Dynamic Programming and Reinforcement Learning algorithms) to solve this Prediction problem (covered later in this book).

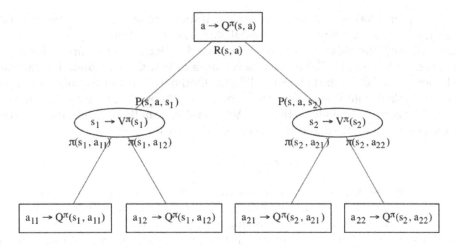

Figure 4.3 Visualization of MDP Action-Value Function Bellman Policy Equation

4.10 OPTIMAL VALUE FUNCTION AND OPTIMAL POLICIES

Finally, we arrive at the main purpose of a Markov Decision Process—to identify a policy (or policies) that would yield the Optimal Value Function (i.e., the best possible *Expected Return* from each of the non-terminal states). We say that a Markov Decision Process is "solved" when we identify its Optimal Value Function (together with its associated Optimal Policy, i.e., a Policy that yields the Optimal Value Function). The problem of identifying the Optimal Value Function and its associated Optimal Policy/Policies is known as the MDP *Control* problem. The term *Control* refers to the fact that this problem involves steering the actions (by iterative modifications of the policy) to drive the Value Function towards Optimality. Formally, the Optimal Value Function

$$V^* : \mathcal{N} \to \mathbb{R}$$

is defined as:

$$V^*(s) = \max_{\pi \in \Pi} V^\pi(s) \text{ for all } s \in \mathcal{N}$$

where Π is the set of stationary (stochastic) policies over the spaces of $\mathcal{N}$ and $\mathcal{A}$.

The way to read the above definition is that for each non-terminal state s, we consider all possible stochastic stationary policies π, and maximize $V^\pi(s)$ across all these choices of π. Note that the maximization over choices of π is done separately for each s, so it's conceivable that different choices of π might maximize $V^\pi(s)$ for different $s \in \mathcal{N}$. Thus, from the above definition of V^*, we can't yet talk about the notion of "An Optimal Policy". So, for now, let's just focus on the notion of Optimal Value Function, as defined above. Note also that we haven't yet talked about how to achieve the above-defined maximization through an algorithm—we have simply *defined* the Optimal Value Function.

Likewise, the Optimal Action-Value Function

$$Q^* : \mathcal{N} \times \mathcal{A} \to \mathbb{R}$$

is defined as:

$$Q^*(s, a) = \max_{\pi \in \Pi} Q^\pi(s, a) \text{ for all } s \in \mathcal{N}, a \in \mathcal{A}$$

V^* is often referred to as the Optimal State-Value Function to distinguish it from the Optimal Action-Value Function Q^* (although, for succinctness, V^* is often also referred to as simply the Optimal Value Function). To be clear, if someone says, Optimal Value Function, by default, they'd be referring to the Optimal State-Value Function V^* (not Q^*).

Much like how the Value Function(s) for a fixed policy have a recursive formulation, Bellman noted (Bellman 1957b) that we can create a recursive formulation for the Optimal Value Function(s). Let us start by unraveling the Optimal State-Value Function $V^*(s)$ for a given non-terminal state s—we consider all possible actions $a \in \mathcal{A}$ we can take from state s, and pick the action a that yields the best Action-Value from thereon, i.e., the action a that yields the best $Q^*(s, a)$. Formally, this gives us the following equation:

$$V^*(s) = \max_{a \in \mathcal{A}} Q^*(s, a) \text{ for all } s \in \mathcal{N} \tag{4.5}$$

Likewise, let's think about what it means to be optimal from a given non-terminal-state and action pair (s, a), i.e, let's unravel $Q^*(s, a)$. First, we get the immediate expected reward $\mathcal{R}(s, a)$. Next, we consider all possible random states $s' \in \mathcal{S}$ we can transition to, and from each of those states which are non-terminal states, we recursively act optimally. Formally, this gives us the following equation:

$$Q^*(s, a) = \mathcal{R}(s, a) + \gamma \cdot \sum_{s' \in \mathcal{N}} \mathcal{P}(s, a, s') \cdot V^*(s') \text{ for all } s \in \mathcal{N}, a \in \mathcal{A} \tag{4.6}$$

Substituting for $Q^*(s, a)$ from Equation (4.6) in Equation (4.5) gives:

$$V^*(s) = \max_{a \in \mathcal{A}} \{\mathcal{R}(s, a) + \gamma \cdot \sum_{s' \in \mathcal{N}} \mathcal{P}(s, a, s') \cdot V^*(s')\} \text{ for all } s \in \mathcal{N} \tag{4.7}$$

Equation (4.7) is known as the MDP State-Value Function Bellman Optimality Equation and is depicted in Figure 4.4 as a visualization aid.

Substituting for $V^*(s)$ from Equation (4.5) in Equation (4.6) gives:

$$Q^*(s, a) = \mathcal{R}(s, a) + \gamma \cdot \sum_{s' \in \mathcal{N}} \mathcal{P}(s, a, s') \cdot \max_{a' \in \mathcal{A}} Q^*(s', a') \text{ for all } s \in \mathcal{N}, a \in \mathcal{A} \tag{4.8}$$

Equation (4.8) is known as the MDP Action-Value Function Bellman Optimality Equation and is depicted in Figure 4.5 as a visualization aid.

Note that Equation (4.5) and Equation (4.6) are embedded in Figure 4.4 as well as in Figure 4.5. Equations (4.7), (4.5), (4.6) and (4.8) are collectively known as the MDP Bellman Optimality Equations. We should highlight that when someone says MDP Bellman Equation or simply Bellman Equation, unless they explicit state otherwise, they'd be referring to the MDP Bellman Optimality Equations (and typically specifically the MDP State-Value Function Bellman Optimality Equation). This is because the MDP Bellman Optimality Equations address the ultimate purpose of Markov Decision Processes—to identify the Optimal Value Function and the associated policy/policies that achieve the Optimal Value Function (i.e., enabling us to solve the MDP *Control* problem).

Again, it pays to emphasize that the Bellman Optimality Equations don't directly give us a recipe to calculate the Optimal Value Function or the policy/policies that achieve the Optimal Value Function—they simply state a powerful mathematical property of the Optimal Value Function that (as we shall see later in this book) helps us come up with algorithms (Dynamic Programming and Reinforcement Learning) to calculate the Optimal Value Function and the associated policy/policies that achieve the Optimal Value Function.

We have been using the phrase "policy/policies that achieve the Optimal Value Function", but we haven't yet provided a clear definition of such a policy (or policies). In fact, as mentioned earlier, it's not clear from the definition of V^* if such a policy (one that

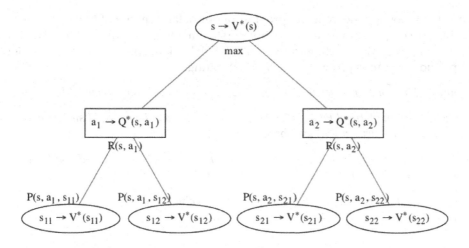

Figure 4.4 Visualization of MDP State-Value Function Bellman Optimality Equation

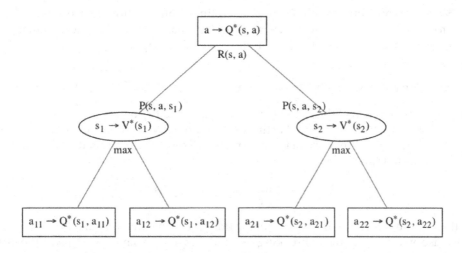

Figure 4.5 Visualization of MDP Action-Value Function Bellman Optimality Equation

would achieve V^*) exists (because it's conceivable that different policies π achieve the maximization of $V^\pi(s)$ for different states $s \in \mathcal{N}$). So instead, we define an *Optimal Policy* $\pi^* : \mathcal{N} \times \mathcal{A} \to [0, 1]$ as one that "dominates" all other policies with respect to the Value Functions for the policies. Formally,

$$\pi^* \in \Pi \text{ is an Optimal Policy if } V^{\pi^*}(s) \geq V^\pi(s) \text{ for all } \pi \in \Pi \text{ and } \textit{for all states } s \in \mathcal{N}$$

The definition of an Optimal Policy π^* says that it is a policy that is "better than or equal to" (on the V^π metric) all other stationary policies *for all* non-terminal states (note that there could be multiple Optimal Policies). Putting this definition together with the definition of the Optimal Value Function V^*, the natural question to then ask is whether there exists an Optimal Policy π^* that maximizes $V^\pi(s)$ *for all* $s \in \mathcal{N}$, i.e., whether there exists a π^* such that $V^*(s) = V^{\pi^*}(s)$ for all $s \in \mathcal{N}$. On the face of it, this seems like a strong statement. However, this answers in the affirmative in most MDP settings of interest. The following theorem and proof is for our default setting of MDP (discrete-time, countable-spaces, time-homogeneous), but the statements and argument themes below apply to various other MDP settings as well. The MDP book by Martin Puterman (Puterman 2014) provides rigorous proofs for a variety of settings.

Theorem 4.10.1. *For any (discrete-time, countable-spaces, time-homogeneous) MDP:*

- *There exists an Optimal Policy $\pi^* \in \Pi$, i.e., there exists a Policy $\pi^* \in \Pi$ such that $V^{\pi^*}(s) \geq V^\pi(s)$ for all policies $\pi \in \Pi$ and for all states $s \in \mathcal{N}$.*

- *All Optimal Policies achieve the Optimal Value Function, i.e. $V^{\pi^*}(s) = V^*(s)$ for all $s \in \mathcal{N}$, for all Optimal Policies π^*.*

- *All Optimal Policies achieve the Optimal Action-Value Function, i.e. $Q^{\pi^*}(s,a) = Q^*(s,a)$ for all $s \in \mathcal{N}$, for all $a \in \mathcal{A}$, for all Optimal Policies π^*.*

Before proceeding with the proof of Theorem (4.10.1), we establish a simple Lemma.

Lemma 4.10.2. *For any two Optimal Policies π_1^* and π_2^*, $V^{\pi_1^*}(s) = V^{\pi_2^*}(s)$ for all $s \in \mathcal{N}$*

Proof. Since π_1^* is an Optimal Policy, from the Optimal Policy definition, we have: $V^{\pi_1^*}(s) \geq V^{\pi_2^*}(s)$ for all $s \in \mathcal{N}$. Likewise, since π_2^* is an Optimal Policy, from the Optimal Policy definition, we have: $V^{\pi_2^*}(s) \geq V^{\pi_1^*}(s)$ for all $s \in \mathcal{N}$. This implies: $V^{\pi_1^*}(s) = V^{\pi_2^*}(s)$ for all $s \in \mathcal{N}$. $\square$

Now we are ready to prove Theorem (4.10.1)

Proof. As a consequence of the above Lemma, all we need to do to prove Theorem (4.10.1) is to establish an Optimal Policy that achieves the Optimal Value Function and the Optimal Action-Value Function. We construct a Deterministic Policy (as a candidate Optimal Policy) $\pi_D^* : \mathcal{N} \to \mathcal{A}$ as follows:

$$\pi_D^*(s) = \arg\max_{a \in \mathcal{A}} Q^*(s,a) \text{ for all } s \in \mathcal{N} \tag{4.9}$$

Note that for any specific s, if two or more actions a achieve the maximization of $Q^*(s,a)$, then we use an arbitrary rule in breaking ties and assigning a single action a as the output of the above $\arg\max$ operation.

First, we show that π_D^* achieves the Optimal Value Functions V^* and Q^*. Since $\pi_D^*(s) = \arg\max_{a \in \mathcal{A}} Q^*(s,a)$ and $V^*(s) = \max_{a \in \mathcal{A}} Q^*(s,a)$ for all $s \in \mathcal{N}$, we can infer for all $s \in \mathcal{N}$ that:

$$V^*(s) = Q^*(s, \pi_D^*(s))$$

This says that we achieve the Optimal Value Function from a given non-terminal state s if we first take the action prescribed by the policy π_D^* (i.e., the action $\pi_D^*(s)$), followed by achieving the Optimal Value Function from each of the next time step's states. But note that each of the next time step's states can achieve the Optimal Value Function by doing the same thing described above ("first take action prescribed by π_D^*, followed by ..."), and so on and so forth for further time step's states. Thus, the Optimal Value Function V^* is achieved if from each non-terminal state, we take the action prescribed by π_D^*. Likewise, the Optimal Action-Value Function Q^* is achieved if from each non-terminal state, we take the action a (argument to Q^*) followed by future actions prescribed by π_D^*. Formally, this says:

$$V^{\pi_D^*}(s) = V^*(s) \text{ for all } s \in \mathcal{N}$$

$$Q^{\pi_D^*}(s,a) = Q^*(s,a) \text{ for all } s \in \mathcal{N}, \text{ for all } a \in \mathcal{A}$$

Finally, we argue that π_D^* is an Optimal Policy. Assume the contradiction (that π_D^* is not an Optimal Policy). Then there exists a policy $\pi \in \Pi$ and a state $s \in \mathcal{N}$ such that $V^\pi(s) > V^{\pi_D^*}(s)$. Since $V^{\pi_D^*}(s) = V^*(s)$, we have: $V^\pi(s) > V^*(s)$ which contradicts the Optimal Value Function Definition: $V^*(s) = \max_{\pi \in \Pi} V^\pi(s)$ for all $s \in \mathcal{N}$. Hence, π_D^* must be an Optimal Policy. $\qquad\qquad\qquad\qquad\qquad\qquad\qquad\qquad\qquad\qquad\qquad\qquad\qquad\qquad\qquad\quad$ □

Equation (4.9) is a key construction that goes hand-in-hand with the Bellman Optimality Equations in designing the various Dynamic Programming and Reinforcement Learning algorithms to solve the MDP Control problem (i.e., to solve for V^*, Q^* and π^*). Lastly, it's important to note that unlike the Prediction problem which has a straightforward linear-algebra-solver for small state spaces, the Control problem is non-linear and so, doesn't have an analogous straightforward linear-algebra-solver. The simplest solutions for the Control problem (even for small state spaces) are the Dynamic Programming algorithms we will cover in Chapter 5.

4.11 VARIANTS AND EXTENSIONS OF MDPS

4.11.1 Size of Spaces and Discrete versus Continuous

Variants of MDPs can be organized by variations in the size and type of:

- State Space
- Action Space
- Time Steps

4.11.1.1 State Space

The definitions we've provided for MRPs and MDPs were for countable (discrete) state spaces. As a special case, we considered finite state spaces since we have pretty straightforward algorithms for exact solution of Prediction and Control problems for finite MDPs (which we shall learn about in Chapter 5). We emphasize finite MDPs because they help you develop a sound understanding of the core concepts and make it easy to program the algorithms (known as "tabular" algorithms since we can represent the MDP in a "table", more specifically a Python data structure like `dict` or `numpy array`). However, these algorithms are practical only if the finite state space is not too large. Unfortunately, in many real-world problems, state spaces are either very large-finite or infinite (sometimes continuous-valued spaces). Large state spaces are unavoidable because phenomena in nature and metrics in business evolve in time due to a complex set of factors and often depend on history. To capture all these factors and to enable the Markov Property, we invariably end up with having to model large state spaces which suffer from two "curses":

- Curse of Dimensionality (size of state space $\mathcal{S}$)
- Curse of Modeling (size/complexity of state-reward transition probabilites $\mathcal{P}_R$)

Curse of Dimensionality is a term coined by Richard Bellman in the context of Dynamic Programming. It refers to the fact that when the number of dimensions in the state space grows, there is an exponential increase in the number of samples required to attain an adequate level of accuracy in algorithms. Consider this simple example (adaptation of an example by Bellman himself)—In a single dimension of space from 0 to 1, 100 evenly spaced sample points suffice to sample the space within a threshold distance of 0.01 between points. An equivalent sampling in 10 dimensions ($[0, 1]^{10}$) within a threshold distance of 0.01 between points will require 10^{20} points. So the 10-dimensional space requires points that are greater by a factor of 10^{18} relative to the points required in single dimension. This explosion in requisite points in the state space is known as the Curse of Dimensionality.

Curse of Modeling refers to the fact that when state spaces are large or when the structure of state-reward transition probabilities is complex, explicit modeling of these transition probabilities is very hard and often impossible (the set of probabilities can go beyond memory or even disk storage space). Even if it's possible to fit the probabilities in available storage space, estimating the actual probability values can be very difficult in complex real-world situations.

To overcome these two curses, we can attempt to contain the state space size with some dimensionality reduction techniques, i.e., including only the most relevant factors in the state representation. Secondly, if future outcomes depend on history, we can include just the past few time steps' values rather than the entire history in the state representation. These savings in state space size are essentially prudent approximations in the state representation. Such state space modeling considerations often require a sound understanding of the real-world problem. Recent advances in unsupervised Machine Learning can also help us contain the state space size. We won't discuss these modeling aspects in detail here—rather, we'd just like to emphasize for now that modeling the state space appropriately is one of the most important skills in real-world Reinforcement Learning, and we will illustrate some of these modeling aspects through a few examples later in this book.

Even after performing these modeling exercises in reducing the state space size, we often still end up with fairly large state spaces (so as to capture sufficient nuances of the real-world problem). We battle these two curses in fundamentally two (complementary) ways:

- Approximation of the Value Function—We create an approximate representation of the Value Function (e.g., by using a supervised learning representation such as a neural network). This permits us to work with an appropriately sampled subset of the state space, infer the Value Function in this state space subset, and interpolate/extrapolate/generalize the Value Function in the remainder of the State Space.
- Sampling from the state-reward transition probabilities $\mathcal{P}_R$—Instead of working with the explicit transition probabilities, we simply use the state-reward sample transitions and employ Reinforcement Learning algorithms to incrementally improve the estimates of the (approximated) Value Function. When state spaces are large, representing explicit transition probabilities is impossible (not enough storage space), and simply sampling from these probability distributions is our only option (and as you shall learn, is surprisingly effective).

This combination of sampling a state space subset, approximation of the Value Function (with deep neural networks), sampling state-reward transitions, and clever Reinforcement Learning algorithms goes a long way in breaking both the curse of dimensionality and curse of modeling. In fact, this combination is a common pattern in the broader field of Applied Mathematics to break these curses. The combination of Sampling and Function Approximation (particularly with the modern advances in Deep Learning) are likely to pave the way for future advances in the broader fields of Real-World AI and Applied Mathematics in general. We recognize that some of this discussion is a bit premature since we haven't even started teaching Reinforcement Learning yet. But we hope that this section provides some high-level perspective and connects the learnings from this chapter to the techniques/algorithms that will come later in this book. We will also remind you of this joint-importance of sampling and function approximation once we get started with Reinforcement Learning algorithms later in this book.

4.11.1.2 Action Space

Similar to state spaces, the definitions we've provided for MDPs were for countable (discrete) action spaces. As a special case, we considered finite action spaces (together with finite state spaces) since we have pretty straightforward algorithms for exact solution of Prediction and Control problems for finite MDPs. As mentioned above, in these algorithms, we represent the MDP in Python data structures like dict or numpy array. However, these finite-MDP algorithms are practical only if the state and action spaces are not too large. In many real-world problems, action spaces do end up as fairly large—either finite-large or infinite (sometimes continuous-valued action spaces). The large size of the action space affects algorithms for MDPs in a couple of ways:

- Large action space makes the representation, estimation and evaluation of the policy π, of the Action-Value function for a policy Q^π and of the Optimal Action-Value function Q^* difficult. We have to resort to function approximation and sampling as ways to overcome the large size of the action space.
- The Bellman Optimality Equation leads to a crucial calculation step in Dynamic Programming and Reinforcement Learning algorithms that involves identifying the action for each non-terminal state that maximizes the Action-Value Function Q. When the action space is large, we cannot afford to evaluate Q for each action for an encountered state (as is done in simple tabular algorithms). Rather, we need to tap into an optimization algorithm to perform the maximization of Q over the action space, for each encountered state. Separately, there is a special class of Reinforcement Learning algorithms called Policy Gradient Algorithms (that we shall later learn about) that are particularly valuable for large action spaces (where other types of Reinforcement Learning algorithms are not efficient and often, simply not an option). However, these techniques to deal with large action spaces require care and attention as they have their own drawbacks (more on this later).

4.11.1.3 Time Steps

The definitions we've provided for MRP and MDP were for discrete time steps. We distinguish discrete time steps as terminating time-steps (known as terminating or episodic MRPs/MDPs) or non-terminating time-steps (known as continuing MRPs/MDPs). We've talked about how the choice of γ matters in these cases ($\gamma = 1$ doesn't work for some continuing MDPs because reward accumulation can blow up to infinity). We won't cover it in this book, but there is an alternative formulation of the Value Function as expected average reward (instead of expected discounted accumulated reward), where we don't discount even for continuing MDPs. We had also mentioned earlier that an alternative to discrete time steps is continuous time steps, which is convenient for analytical tractability.

Sometimes, even if state space and action space components have discrete values (e.g., price of a security traded in fine discrete units, or number of shares of a security bought/sold on a given day), for modeling purposes, we sometimes find it convenient to represent these components as continuous values (i.e., uncountable state space). The advantage of continuous state/action space representation (especially when paired with continuous time) is that we get considerable mathematical benefits from differential calculus as well as from properties of continuous probability distributions (e.g., Gaussian distribution conveniences). In fact, continuous state/action space and continuous time are very popular in Mathematical Finance since some of the groundbreaking work from Mathematical Economics from the 1960s and 1970s—Robert Merton's Portfolio Optimization formulation and solution (Merton 1969) and Black-Scholes' Options Pricing model (Black and

Scholes 1973), to name a couple—are grounded in stochastic calculus[1] which models stock prices/portfolio value as Gaussian evolutions in continuous time (more on this later in the book) and treats trades (buy/sell quantities) as also continuous variables (permitting partial derivatives and tractable partial differential equations).

When all three of state space, action space and time steps are modeled as continuous, the Bellman Optimality Equation we covered in this chapter for countable spaces and discrete-time morphs into a differential calculus formulation and is known as the famous *Hamilton-Jacobi-Bellman* (HJB) equation[2]. The HJB Equation is commonly used to model and solve many problems in engineering, physics, economics and finance. We shall cover a couple of financial applications in this book that have elegant formulations in terms of the HJB equation and equally elegant analytical solutions of the Optimal Value Function and Optimal Policy (tapping into stochastic calculus and differential equations).

4.11.2 Partially-Observable Markov Decision Processes (POMDPs)

You might have noticed in the definition of MDP that there are actually two different notions of state, which we collapsed into a single notion of state. These two notions of state are:

- The internal representation of the environment at each time step t (let's call it $S_t^{(e)}$). This internal representation of the environment is what drives the probabilistic transition to the next time step $t + 1$, producing the random pair of next (environment) state $S_{t+1}^{(e)}$ and reward R_{t+1}.
- The agent state at each time step t (let's call it $S_t^{(a)}$). The agent state is what controls the action A_t the agent takes at time step t, i.e., the agent runs a policy π which is a function of the agent state $S_t^{(a)}$, producing a probability distribution of actions A_t.

In our definition of MDP, note that we implicitly assumed that $S_t^{(e)} = S_t^{(a)}$ at each time step t, and called it the (common) state S_t at time t. Secondly, we assumed that this state S_t is *fully observable* by the agent. To understand *full observability*, let us (first intuitively) understand the concept of *partial observability* in a more generic setting than what we had assumed in the framework for MDP. In this more generic framework, we denote O_t as the information available to the agent from the environment at time step t, as depicted in Figure 4.6. The notion of *partial observability* in this more generic framework is that from the history of observations, actions and rewards up to time step t, the agent does not have full knowledge of the environment state $S_t^{(e)}$. This lack of full knowledge of $S_t^{(e)}$ is known as *partial observability*. *Full observability*, on the other hand, means that the agent can fully construct $S_t^{(e)}$ as a function of the history of observations, actions and rewards up to time step t. Since we have the flexibility to model the exact data structures to represent observations, state and actions in this more generic framework, existence of full observability lets us restructure the observation data at time step t to be $O_t = S_t^{(e)}$. Since we have also assumed $S_t^{(e)} = S_t^{(a)}$, we have:

$$O_t = S_t^{(e)} = S_t^{(a)} \text{ for all time steps } t = 0, 1, 2, \ldots$$

The above statement specialized the framework to that of Markov Decision Processes, which we can now name more precisely as Fully-Observable Markov Decision Processes (when viewed from the lens of the more generic framework described above, that permits partial observability or full observability).

---

[1]Appendix C provides a quick introduction to and overview of Stochastic Calculus.
[2]Appendix D provides a quick introduction to the HJB Equation.

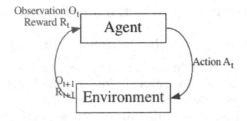

Figure 4.6 **Partially-Observable Markov Decision Process**

In practice, you will often find that the agent doesn't know the true internal representation $(S_t^{(e)})$ of the environment (i.e, partial observability). Think about what it would take to know what drives a stock price from time step t to $t + 1$—the agent would need to have access to pretty much every little detail of trading activity in the entire world, and more!. However, since the MDP framework is simple and convenient, and since we have tractable Dynamic Programming and Reinforcement Learning algorithms to solve MDPs, we often do pretend that $O_t = S_t^{(e)} = S_t^{(a)}$ and carry on with our business of solving the assumed/modeled MDP. Often, this assumption of $O_t = S_t^{(e)} = S_t^{(a)}$ turns out to be a reasonable approximate model of the real-world but there are indeed situations where this assumption is far-fetched. These are situations where we have access to too little information pertaining to the key aspects of the internal state representation $(S_t^{(e)})$ of the environment. It turns out that we have a formal framework for these situations—this framework is known as Partially-Observable Markov Decision Process (POMDP for short). By default, the acronym MDP will refer to a Fully-Observable Markov Decision Process (i.e. corresponding to the MDP definition we have given earlier in this chapter). So let's now define a POMDP.

A POMDP has the usual features of an MDP (discrete-time, countable states, countable actions, countable next state-reward transition probabilities, discount factor, plus assuming time-homogeneity), together with the notion of random observation O_t at each time step t (each observation O_t lies within the Observation Space $\mathcal{O}$) and observation probability function $\mathcal{Z} : \mathcal{S} \times \mathcal{A} \times \mathcal{O} \rightarrow [0, 1]$ defined as:

$$\mathcal{Z}(s', a, o) = \mathbb{P}[O_{t+1} = o|(S_{t+1} = s', A_t = a)]$$

It pays to emphasize that although a POMDP works with the notion of a state S_t, the agent doesn't have knowledge of S_t. It only has knowledge of observation O_t because O_t is the extent of information made available from the environment. The agent will then need to essentially "guess" (probabilistically) what the state S_t might be at each time step t in order to take the action A_t. The agent's goal in a POMDP is the same as that for an MDP: to determine the Optimal Value Function and to identify an Optimal Policy (achieving the Optimal Value Function).

Just like we have the rich theory and algorithms for MDPs, we have the theory and algorithms for POMDPs. POMDP theory is founded on the notion of *belief states*. The informal notion of a belief state is that since the agent doesn't get to see the state S_t (it only sees the observations O_t) at each time step t, the agent needs to keep track of what it thinks the state S_t might be, i.e., it maintains a probability distribution of states S_t conditioned on history. Let's make this a bit more formal.

Let us refer to the history H_t known to the agent at time t as the sequence of data it has collected up to time t. Formally, this data sequence H_t is:

$$(O_0, A_0, R_1, O_1, A_1, R_2, \ldots, O_{t-1}, A_{t-1}, R_t, O_t)$$

A Belief State $b(h)_t$ at time t is a probability distribution over states, conditioned on the history h, i.e.,

$$b(h)_t = (\mathbb{P}[S_t = s_1 | H_t = h], \mathbb{P}[S_t = s_2 | H_t = h], \dots)$$

such that $\sum_{s \in \mathcal{S}} b(h)_t(s) = 1$ for all histories h and for each $t = 0, 1, 2, \dots$.

Since the history H_t satisfies the Markov Property, the belief state $b(h)_t$ satisfies the Markov Property. So we can reduce the POMDP to an MDP M with the set of belief states of the POMDP as the set of states of the MDP M. Note that even if the set of states of the POMDP were finite, the set of states of the MDP M will be infinite (i.e. infinite belief states). We can see that this will almost always end up as a giant MDP M. So although this is useful for theoretical reasoning, practically solving this MDP M is often quite hard computationally. However, specialized techniques have been developed to solve POMDPs but as you might expect, their computational complexity is still quite high. So we end up with a choice when encountering a POMDP—either try to solve it with a POMDP algorithm (computationally inefficient but capturing the reality of the real-world problem) or try to approximate it as an MDP (pretending $O_t = S_t^{(e)} = S_t^{(a)}$) which will likely be computationally more efficient but might be a gross approximation of the real-world problem, which in turn means its effectiveness in practice might be compromised. This is the modeling dilemma we often end up with: what is the right level of detail of real-world factors we need to capture in our model? How do we prevent state spaces from exploding beyond practical computational tractability? The answers to these questions typically have to do with depth of understanding of the nuances of the real-world problem and a trial-and-error process of: formulating the model, solving for the optimal policy, testing the efficacy of this policy in practice (with appropriate measurements to capture real-world metrics), learning about the drawbacks of our model, and iterating back to tweak (or completely change) the model.

Let's consider a classic example of a card game such as Poker or Blackjack as a POMDP where your objective as a player is to identify the optimal policy to maximize your expected return (Optimal Value Function). The observation O_t would be the entire set of information you would have seen up to time step t (or a compressed version of this entire information that suffices for predicting transitions and for taking actions). The state S_t would include, among other things, the set of cards you have, the set of cards your opponents have (which you don't see), and the entire set of exposed as well as unexposed cards not held by players. Thus, the state is only partially observable. With this POMDP structure, we proceed to develop a model of the transition probabilities of next state S_{t+1} and reward R_{t+1}, conditional on current state S_t and current action A_t. We also develop a model of the probabilities of next observation O_{t+1}, conditional on next state S_{t+1} and current action A_t. These probabilities are estimated from data collected from various games (capturing opponent behaviors) and knowledge of the cards-structure of the deck (or decks) used to play the game. Now let's think about what would happen if we modeled this card game as an MDP. We'd no longer have the unseen cards as part of our state. Instead, the state S_t will be limited to the information seen upto time t (i.e., $S_t = O_t$). We can still estimate the transition probabilities, but since it's much harder to estimate in this case, our estimate will likely be quite noisy and nowhere near as reliable as the probability estimates in the POMDP case. The advantage though with modeling it as an MDP is that the algorithm to arrive at the Optimal Value Function/Optimal Policy is a lot more tractable compared to the algorithm for the POMDP model. So it's a tradeoff between the reliability of the probability estimates versus the tractability of the algorithm to solve for the Optimal Value Function/Policy.

The purpose of this subsection on POMDPs is to highlight that by default a lot of problems in the real-world are POMDPs and it can sometimes take quite a bit of domain-knowledge, modeling creativity and real-world experimentation to treat them as MDPs and make the solution to the modeled MDP successful in practice.

The idea of partial observability was introduced in a paper by K.J.Astrom (Åström 1965). To learn more about POMDP theory, we refer you to the POMDP book by Vikram Krishnamurthy (Krishnamurthy 2016).

4.12 SUMMARY OF KEY LEARNINGS FROM THIS CHAPTER

- MDP Bellman Policy Equations
- MDP Bellman Optimality Equations
- Theorem (4.10.1) on the existence of an Optimal Policy, and of each Optimal Policy achieving the Optimal Value Function

CHAPTER 5

Dynamic Programming Algorithms

As a reminder, much of this book is about algorithms to solve the MDP Control problem, i.e., to compute the Optimal Value Function (and an associated Optimal Policy). We will also cover algorithms for the MDP Prediction problem, i.e., to compute the Value Function when the AI agent executes a fixed policy π (which, as we know from Chapter 4, is the same as computing the Value Function of the π-implied MRP). Our typical approach will be to first cover algorithms to solve the Prediction problem before covering algorithms to solve the Control problem—not just because Prediction is a key component in solving the Control problem, but also because it helps understand the key aspects of the techniques employed in the Control algorithm in the simpler setting of Prediction.

5.1 PLANNING VERSUS LEARNING

In this book, we shall look at Prediction and Control from the lens of AI (and we'll specifically use the terminology of AI). We shall distinguish between algorithms that don't have a model of the MDP environment (no access to the $\mathcal{P}_R$ function) versus algorithms that do have a model of the MDP environment (meaning $\mathcal{P}_R$ is available to us either in terms of explicit probability distribution representations or available to us just as a sampling model). The former (algorithms without access to a model) are known as *Learning Algorithms* to reflect the fact that the AI agent will need to interact with the real-world environment (e.g., a robot learning to navigate in an actual forest) and learn the Value Function from data (states encountered, actions taken, rewards observed) it receives through interactions with the environment. The latter (algorithms with access to a model of the MDP environment) are known as *Planning Algorithms* to reflect the fact that the AI agent requires no real-world environment interaction and in fact, projects (with the help of the model) probabilistic scenarios of future states/rewards for various choices of actions, and solves for the requisite Value Function based on the projected outcomes. In both Learning and Planning, the Bellman Equation is the fundamental concept driving the algorithms but the details of the algorithms will typically make them appear fairly different. We will only focus on Planning algorithms in this chapter, and in fact, will only focus on a subclass of Planning algorithms known as Dynamic Programming.

DOI: 10.1201/9781003229193-5

5.2 USAGE OF THE TERM *DYNAMIC PROGRAMMING*

Unfortunately, the term Dynamic Programming tends to be used by different fields in somewhat different ways. So it pays to clarify the history and the current usage of the term. The term *Dynamic Programming* was coined by Richard Bellman himself. Here is the rather interesting story told by Bellman about how and why he coined the term.

> "I spent the Fall quarter (of 1950) at RAND. My first task was to find a name for multistage decision processes. An interesting question is, 'Where did the name, dynamic programming, come from?' The 1950s were not good years for mathematical research. We had a very interesting gentleman in Washington named Wilson. He was Secretary of Defense, and he actually had a pathological fear and hatred of the word, research. I'm not using the term lightly; I'm using it precisely. His face would suffuse, he would turn red, and he would get violent if people used the term, research, in his presence. You can imagine how he felt, then, about the term, mathematical. The RAND Corporation was employed by the Air Force, and the Air Force had Wilson as its boss, essentially. Hence, I felt I had to do something to shield Wilson and the Air Force from the fact that I was really doing mathematics inside the RAND Corporation. What title, what name, could I choose? In the first place I was interested in planning, in decision making, in thinking. But planning, is not a good word for various reasons. I decided therefore to use the word, 'programming.' I wanted to get across the idea that this was dynamic, this was multistage, this was time-varying—I thought, let's kill two birds with one stone. Let's take a word that has an absolutely precise meaning, namely dynamic, in the classical physical sense. It also has a very interesting property as an adjective, and that is it's impossible to use the word, dynamic, in a pejorative sense. Try thinking of some combination that will possibly give it a pejorative meaning. It's impossible. Thus, I thought dynamic programming was a good name. It was something not even a Congressman could object to. So I used it as an umbrella for my activities."

Bellman had coined the term Dynamic Programming to refer to the general theory of MDPs, together with the techniques to solve MDPs (i.e., to solve the Control problem). So the MDP Bellman Optimality Equation was part of this catch-all term *Dynamic Programming*. The core semantic of the term Dynamic Programming was that the Optimal Value Function can be expressed recursively—meaning, to act optimally from a given state, we will need to act optimally from each of the resulting next states (which is the essence of the Bellman Optimality Equation). In fact, Bellman used the term "Principle of Optimality" to refer to this idea of "Optimal Substructure", and articulated it as follows:

> PRINCIPLE OF OPTIMALITY. An optimal policy has the property that whatever the initial state and initial decisions are, the remaining decisions must constitute an optimal policy with regard to the state resulting from the first decisions.

So, you can see that the term Dynamic Programming was not just an algorithm in its original usage. Crucially, Bellman laid out an iterative algorithm to solve for the Optimal Value Function (i.e., to solve the MDP Control problem). Over the course of the next decade, the term Dynamic Programming got associated with (multiple) algorithms to solve the MDP Control problem. The term Dynamic Programming was extended to also refer to algorithms to solve the MDP Prediction problem. Over the next couple of decades, Computer Scientists started referring to the term Dynamic Programming as any algorithm that solves a problem through a recursive formulation as long as the algorithm makes repeated invocations to the solutions of each subproblem (overlapping subproblem structure). A classic such example is the algorithm to compute the Fibonacci sequence by caching the Fibonacci values and re-using those values during the course of the algorithm execution. The algorithm to calculate the shortest path in a graph is another classic example where each shortest (i.e. optimal) path includes sub-paths that are optimal. However, in this book, we won't use the term Dynamic Programming in this broader sense. We will use the term Dynamic Programming to be restricted to algorithms to solve the MDP Prediction and Control problems (even though Bellman originally used it only in the context of Control). More specifically, we will use the term Dynamic Programming in the narrow context of Planning algorithms for problems with the following two specializations:

- The state space is finite, the action space is finite, and the set of pairs of next state and reward (given any pair of current state and action) are also finite.
- We have explicit knowledge of the model probabilities (either in the form of $\mathcal{P}_R$ or in the form of $\mathcal{P}$ and $\mathcal{R}$ separately).

This is the setting of the class `FiniteMarkovDecisionProcess` we had covered in Chapter 4. In this setting, Dynamic Programming algorithms solve the Prediction and Control problems *exactly* (meaning the computed Value Function converges to the true Value Function as the algorithm iterations keep increasing). There are variants of Dynamic Programming algorithms known as Asynchronous Dynamic Programming algorithms, Approximate Dynamic Programming algorithms etc. But without such qualifications, when we use just the term Dynamic Programming, we will be referring to the "classical" iterative algorithms (that we will soon describe) for the above-mentioned setting of the `FiniteMarkovDecisionProcess` class to solve MDP Prediction and Control *exactly*. Even though these classical Dynamical Programming algorithms don't scale to large state/action spaces, they are extremely vital to develop one's core understanding of the key concepts in the more advanced algorithms that will enable us to scale (e.g., the Reinforcement Learning algorithms that we shall introduce in later chapters).

5.3   FIXED-POINT THEORY

We start by covering 3 classical Dynamic Programming algorithms. Each of the 3 algorithms is founded on the Bellman Equations we had covered in Chapter 4. Each of the 3 algorithms is an iterative algorithm where the computed Value Function converges to the true Value Function as the number of iterations approaches infinity. Each of the 3 algorithms is based on the concept of *Fixed-Point* and updates the computed Value Function towards the Fixed-Point (which in this case, is the true Value Function). Fixed-Point is actually a fairly generic and important concept in the broader fields of Pure as well as Applied Mathematics (also important in Theoretical Computer Science), and we believe understanding Fixed-Point theory has many benefits beyond the needs of the subject of this book. Of more relevance is the fact that the Fixed-Point view of Dynamic Programming is the best way to understand Dynamic Programming. We shall not only cover the theory of Dynamic Programming through the Fixed-Point perspective, but we shall also implement Dynamic Programming algorithms in our code based on the Fixed-Point concept. So this section will be a short primer on general Fixed-Point Theory (and implementation in code) before we get to the 3 Dynamic Programming algorithms.

Definition 5.3.1. The Fixed-Point of a function $f : \mathcal{X} \to \mathcal{X}$ (for some arbitrary domain $\mathcal{X}$) is a value $x \in \mathcal{X}$ that satisfies the equation: $x = f(x)$.

Note that for some functions, there will be multiple fixed-points and for some other functions, a fixed-point won't exist. We will be considering functions which have a unique fixed-point (this will be the case for the Dynamic Programming algorithms).

Let's warm up to the above-defined abstract concept of Fixed-Point with a concrete example. Consider the function $f(x) = \cos(x)$ defined for $x \in \mathbb{R}$ (x in radians, to be clear). So we want to solve for an x such that $x = \cos(x)$. Knowing the frequency and amplitude of cosine, we can see that the cosine curve intersects the line $y = x$ at only one point, which should be somewhere between 0 and $\frac{\pi}{2}$. But there is no easy way to solve for this point. Here's an idea: Start with any value $x_0 \in \mathbb{R}$, calculate $x_1 = \cos(x_0)$, then calculate $x_2 = \cos(x_1)$, and so on ..., i.e, $x_{i+1} = \cos(x_i)$ for $i = 0, 1, 2, \ldots$. You will find that x_i and x_{i+1} get closer and closer as i increases, i.e., $|x_{i+1} - x_i| \leq |x_i - x_{i-1}|$ for all $i \geq 1$. So it seems like $\lim_{i \to \infty} x_i = \lim_{i \to \infty} \cos(x_{i-1}) = \lim_{i \to \infty} \cos(x_i)$, which would imply that for large enough i, x_i would serve as an approximation to the solution of the equation $x = \cos(x)$. But why does this method of repeated applications of the function f (no matter what x_0 we start with) work? Why does it not diverge or oscillate? How quickly does it converge? If there were multiple fixed-points, which fixed-point would it converge to (if at all)? Can we characterize a class of functions f for which this method (repeatedly applying f, starting with any arbitrary value of x_0) would work (in terms of solving the equation $x = f(x)$)? These are the questions Fixed-Point theory attempts to answer. Can you think of problems you have solved in the past which fall into this method pattern that we've illustrated above for $f(x) = \cos(x)$? It's likely you have, because most of the root-finding and optimization methods (including multi-variate solvers) are essentially based on the idea of Fixed-Point. If this doesn't sound convincing, consider the simple Newton method:

For a differentiable function $g : \mathbb{R} \to \mathbb{R}$ whose root we want to solve for, the Newton method update rule is:

$$x_{i+1} = x_i - \frac{g(x_i)}{g'(x_i)}$$

Setting $f(x) = x - \frac{g(x)}{g'(x)}$, the update rule is:

$$x_{i+1} = f(x_i)$$

and it solves the equation $x = f(x)$ (solves for the fixed-point of f), i.e., it solves the equation:

$$x = x - \frac{g(x)}{g'(x)} \Rightarrow g(x) = 0$$

Thus, we see the same method pattern as we saw above for $\cos(x)$ (repeated application of a function, starting with any initial value) enables us to solve for the root of g.

More broadly, what we are saying is that if we have a function $f : \mathcal{X} \to \mathcal{X}$ (for some arbitrary domain $\mathcal{X}$), under appropriate conditions (that we will state soon), $f(f(\ldots f(x_0) \ldots))$ converges to a fixed-point of f, i.e., to the solution of the equation $x = f(x)$ (no matter what $x_0 \in \mathcal{X}$ we start with). Now we are ready to state this formally. The statement of the following theorem (due to Stefan Banach) is quite terse, so we will provide plenty of explanation on how to interpret it and how to use it after stating the theorem (we skip the proof of the theorem).

Theorem 5.3.1 (Banach Fixed-Point Theorem). *Let $\mathcal{X}$ be a non-empty set equipped with a complete metric $d : \mathcal{X} \times \mathcal{X} \to \mathbb{R}$. Let $f : \mathcal{X} \to \mathcal{X}$ be such that there exists a $L \in [0, 1)$ such that $d(f(x_1), f(x_2)) \leq L \cdot d(x_1, x_2)$ for all $x_1, x_2 \in \mathcal{X}$ (this property of f is called a contraction, and we refer to f as a contraction function). Then,*

1. *There exists a unique Fixed-Point $x^* \in \mathcal{X}$, i.e.,*

$$x^* = f(x^*)$$

2. *For any $x_0 \in \mathcal{X}$, and sequence $[x_i | i = 0, 1, 2, \ldots]$ defined as $x_{i+1} = f(x_i)$ for all $i = 0, 1, 2, \ldots,$*

$$\lim_{i \to \infty} x_i = x^*$$

3.

$$d(x^*, x_i) \leq \frac{L^i}{1 - L} \cdot d(x_1, x_0)$$

Equivalently,

$$d(x^*, x_{i+1}) \leq \frac{L}{1 - L} \cdot d(x_{i+1}, x_i)$$

$$d(x^*, x_{i+1}) \leq L \cdot d(x^*, x_i)$$

We realize this is quite terse and will now demystify the theorem in a simple, intuitive manner. First, we need to explain what *complete metric* means. Let's start with the term *metric*. A metric is simply a function $d : \mathcal{X} \times \mathcal{X} \to \mathbb{R}$ that satisfies the usual "distance" properties (for any $x_1, x_2, x_3 \in \mathcal{X}$):

1. $d(x_1, x_2) = 0 \Leftrightarrow x_1 = x_2$ (meaning two different points have a distance strictly greater than 0)
2. $d(x_1, x_2) = d(x_2, x_1)$ (meaning distance is directionless)
3. $d(x_1, x_3) \leq d(x_1, x_2) + d(x_2, x_3)$ (meaning the triangle inequality is satisfied)

The term *complete* is a bit of a technical detail on sequences not escaping the set $\mathcal{X}$ (that's required in the proof). Since we won't be doing the proof and since this technical detail is not so important for the intuition, we skip the formal definition of *complete*. A non-empty set $\mathcal{X}$ equipped with the function d (and the technical detail of being *complete*) is known as a complete metric space.

Now we move on to the key concept of *contraction*. A function $f : \mathcal{X} \to \mathcal{X}$ is said to be a contraction function if two points in $\mathcal{X}$ get closer when they are mapped by f (the statement: $d(f(x_1), f(x_2)) \leq L \cdot d(x_1, x_2)$ for all $x_1, x_2 \in \mathcal{X}$, for some $L \in [0, 1)$).

The theorem basically says that for any contraction function f, there is not only a unique fixed-point x^*, one can arrive at x^* by repeated application of f, starting with any initial value $x_0 \in \mathcal{X}$:

$$f(f(\ldots f(x_0) \ldots)) \to x^*$$

We use the notation $f^i : \mathcal{X} \to \mathcal{X}$ for $i = 0, 1, 2, \ldots$ as follows:

$$f^{i+1}(x) = f(f^i(x)) \text{ for all } i = 0, 1, 2, \ldots, \text{ for all } x \in \mathcal{X}$$
$$f^0(x) = x \text{ for all } x \in \mathcal{X}$$

With this notation, the computation of the fixed-point can be expressed as:

$$\lim_{i \to \infty} f^i(x_0) = x^* \text{ for all } x_0 \in \mathcal{X}$$

The algorithm, in iterative form, is:

$$x_{i+1} = f(x_i) \text{ for all } i = 0, 2, \ldots$$

We stop the algorithm when x_i and x_{i+1} are close enough based on the distance-metric d.

Banach Fixed-Point Theorem also gives us a statement on the speed of convergence relating the distance between x^* and any x_i to the distance between any two successive x_i.

This is a powerful theorem. All we need to do is identify the appropriate set $\mathcal{X}$ to work with, identify the appropriate metric d to work with, and ensure that f is indeed a contraction function (with respect to d). This enables us to solve for the fixed-point of f with the above-described iterative process of applying f repeatedly, starting with any arbitrary value of $x_0 \in \mathcal{X}$.

We leave it to you as an exercise to verify that $f(x) = \cos(x)$ is a contraction function in the domain $\mathcal{X} = \mathbb{R}$ with metric d defined as $d(x_1, x_2) = |x_1 - x_2|$. Now let's write some code to implement the fixed-point algorithm we described above. Note that we implement this for any generic type X to represent an arbitrary domain $\mathcal{X}$.

```
X = TypeVar('X')
def iterate(step: Callable[[X], X], start: X) -> Iterator[X]:
    state = start

    while True:
        yield state
        state = step(state)
```

The above function takes as input a function (step: Callable[[X], X]) and a starting value (start: X), and repeatedly applies the function while yielding the values in the form of an Iterator[X], i.e., as a stream of values. This produces an endless stream though. We need a way to specify convergence, i.e., when successive values of the stream are "close enough".

```
def converge(values: Iterator[X], done: Callable[[X, X], bool]) -> Iterator[X]:
    a = next(values, None)
    if a is None:
        return

    yield a

    for b in values:
        yield b
        if done(a, b):
            return

        a = b
```

The above function converge takes as input the generated values from iterate (argument values: Iterator[X]) and a signal to indicate convergence (argument done: Callable[[X, X], bool]), and produces the generated values until done is True. It is the user's responsibility to write the function done and pass it to converge. Now let's use these two functions to solve for $x = \cos(x)$.

```
import numpy as np
x = 0.0
values = converge(
    iterate(lambda y: np.cos(y), x),
    lambda a, b: np.abs(a - b) < 1e-3
)
for i, v in enumerate(values):
    print(f"{i}: {v:.4f}")
```

This prints a trace with the index of the stream and the value at that index as the function cos is repeatedly applied. It terminates when two successive values are within 3 decimal places of each other.

```
0: 0.0000
1: 1.0000
2: 0.5403
3: 0.8576
4: 0.6543
5: 0.7935
6: 0.7014
7: 0.7640
8: 0.7221
9: 0.7504
10: 0.7314
11: 0.7442
12: 0.7356
13: 0.7414
14: 0.7375
15: 0.7401
16: 0.7384
17: 0.7396
18: 0.7388
```

We encourage you to try other starting values (other than the one we have above: $x_0 = 0.0$) and see the trace. We also encourage you to identify other functions f which are contractions in an appropriate metric. The above fixed-point code is in the file rl/iterate.py. In this file, you will find two more functions last and converged to produce the final value of the given iterator when its values converge according to the done function.

5.4 BELLMAN POLICY OPERATOR AND POLICY EVALUATION ALGORITHM

Our first Dynamic Programming algorithm is called *Policy Evaluation*. The Policy Evaluation algorithm solves the problem of calculating the Value Function of a Finite MDP evaluated with a fixed policy π (i.e., the Prediction problem for finite MDPs). We know that this is equivalent to calculating the Value Function of the π-implied Finite MRP. To avoid notation confusion, note that a superscript of π for a symbol means it refers to notation for the π-implied MRP. The precise specification of the Prediction problem is as follows:

Let the states of the MDP (and hence, of the π-implied MRP) be $\mathcal{S} = \{s_1, s_2, \ldots, s_n\}$, and without loss of generality, let $\mathcal{N} = \{s_1, s_2, \ldots, s_m\}$ be the non-terminal states. We are given a fixed policy $\pi : \mathcal{N} \times \mathcal{A} \to [0, 1]$. We are also given the π-implied MRP's transition probability function:

$$\mathcal{P}_R^\pi : \mathcal{N} \times \mathcal{D} \times \mathcal{S} \to [0, 1]$$

in the form of a data structure (since the states are finite, and the pairs of next state and reward transitions from each non-terminal state are also finite). The Prediction problem is to compute the Value Function of the MDP when evaluated with the policy π (equivalently, the Value Function of the π-implied MRP), which we denote as $V^\pi : \mathcal{N} \to \mathbb{R}$.

We know from Chapters 3 and 4 that by extracting (from $\mathcal{P}_R^\pi$) the transition probability function $\mathcal{P}^\pi : \mathcal{N} \times \mathcal{S} \to [0, 1]$ of the implicit Markov Process and the reward function $\mathcal{R}^\pi : \mathcal{N} \to \mathbb{R}$, we can perform the following calculation for the Value Function $V^\pi : \mathcal{N} \to \mathbb{R}$ (expressed as a column vector $V^\pi \in \mathbb{R}^m$) to solve this Prediction problem:

$$V^\pi = (I_m - \gamma \mathcal{P}^\pi)^{-1} \cdot \mathcal{R}^\pi$$

where I_m is the $m \times m$ identity matrix, column vector $\mathcal{R}^\pi \in \mathbb{R}^m$ represents $\mathcal{R}^\pi$, and $\mathcal{P}^\pi$ is an $m \times m$ matrix representing $\mathcal{P}^\pi$ (rows and columns corresponding to the non-terminal states). However, when m is large, this calculation won't scale. So, we look for a numerical algorithm that would solve (for V^π) the following MRP Bellman Equation (for a larger number of finite states).

$$V^\pi = \mathcal{R}^\pi + \gamma \mathcal{P}^\pi \cdot V^\pi$$

We define the *Bellman Policy Operator* $B^\pi : \mathbb{R}^m \to \mathbb{R}^m$ as:

$$B^\pi(V) = \mathcal{R}^\pi + \gamma \mathcal{P}^\pi \cdot V \text{ for any vector } V \text{ in the vector space } \mathbb{R}^m \quad (5.1)$$

So, the MRP Bellman Equation can be expressed as:

$$V^\pi = B^\pi(V^\pi)$$

which means $V^\pi \in \mathbb{R}^m$ is a Fixed-Point of the *Bellman Policy Operator* $B^\pi : \mathbb{R}^m \to \mathbb{R}^m$. Note that the Bellman Policy Operator can be generalized to the case of non-finite MDPs and V^π is still a Fixed-Point for various generalizations of interest. However, since this chapter focuses on developing algorithms for finite MDPs, we will work with the above narrower (Equation (5.1)) definition. Also, for proofs of correctness of the DP algorithms (based on Fixed-Point) in this chapter, we shall assume the discount factor $\gamma < 1$.

Note that B^π is an affine transformation on vectors in $\mathbb{R}^m$ and should be thought of as a generalization of a simple 1-D ($\mathbb{R} \to \mathbb{R}$) affine transformation $y = a + bx$, where the multiplier b is replaced with the matrix $\gamma \mathcal{P}^\pi$ and the shift a is replaced with the column vector $\mathcal{R}^\pi$.

We'd like to come up with a metric for which B^π is a contraction function so we can take advantage of Banach Fixed-Point Theorem and solve this Prediction problem by iterative applications of the Bellman Policy Operator B^π. For any Value Function $V \in \mathbb{R}^m$ (representing $V : \mathcal{N} \to \mathbb{R}$), we shall express the Value for any state $s \in \mathcal{N}$ as $V(s)$.

Our metric $d : \mathbb{R}^m \times \mathbb{R}^m \to \mathbb{R}$ shall be the L^∞ norm defined as:

$$d(X, Y) = \|X - Y\|_\infty = \max_{s \in \mathcal{N}} |(X - Y)(s)|$$

B^π is a contraction function under L^∞ norm because for all $X, Y \in \mathbb{R}^m$,

$$\max_{s \in \mathcal{N}} |(B^\pi(X) - B^\pi(Y))(s)| = \gamma \cdot \max_{s \in \mathcal{N}} |(\mathcal{P}^\pi \cdot (X - Y))(s)| \leq \gamma \cdot \max_{s \in \mathcal{N}} |(X - Y)(s)|$$

So invoking Banach Fixed-Point Theorem proves the following Theorem:

Theorem 5.4.1 (Policy Evaluation Convergence Theorem). *For a Finite MDP with $|\mathcal{N}| = m$ and $\gamma < 1$, if $V^\pi \in \mathbb{R}^m$ is the Value Function of the MDP when evaluated with a fixed policy $\pi : \mathcal{N} \times \mathcal{A} \to [0, 1]$, then V^π is the unique Fixed-Point of the Bellman Policy Operator $B^\pi : \mathbb{R}^m \to \mathbb{R}^m$, and*

$$\lim_{i \to \infty} (B^\pi)^i(V_0) \to V^\pi \text{ for all starting Value Functions } V_0 \in \mathbb{R}^m$$

This gives us the following iterative algorithm (known as the *Policy Evaluation* algorithm for fixed policy $\pi : \mathcal{N} \times \mathcal{A} \to [0, 1]$):

- Start with any Value Function $V_0 \in \mathbb{R}^m$

- Iterating over $i = 0, 1, 2, \ldots$, calculate in each iteration:

$$V_{i+1} = B^\pi(V_i) = \mathcal{R}^\pi + \gamma \mathcal{P}^\pi \cdot V_i$$

- Stop the algorithm when $d(V_i, V_{i+1}) = \max_{s \in \mathcal{N}} |(V_i - V_{i+1})(s)|$ is adequately small.

It pays to emphasize that Banach Fixed-Point Theorem not only assures convergence to the unique solution V^π (no matter what Value Function V_0 we start the algorithm with), it also assures a reasonable speed of convergence (dependent on the choice of starting Value Function V_0 and the choice of γ). Now let's write the code for Policy Evaluation.

```
DEFAULT_TOLERANCE = 1e-5
V = Mapping[NonTerminal[S], float]

def evaluate_mrp(
    mrp: FiniteMarkovRewardProcess[S],
    gamma: float
) -> Iterator[np.ndarray]:
    def update(v: np.ndarray) -> np.ndarray:
        return mrp.reward_function_vec + gamma * \
            mrp.get_transition_matrix().dot(v)

    v_0: np.ndarray = np.zeros(len(mrp.non_terminal_states))

    return iterate(update, v_0)

def almost_equal_np_arrays(
    v1: np.ndarray,
    v2: np.ndarray,
    tolerance: float = DEFAULT_TOLERANCE
) -> bool:
    return max(abs(v1 - v2)) < tolerance

def evaluate_mrp_result(
    mrp: FiniteMarkovRewardProcess[S],
    gamma: float
) -> V[S]:
    v_star: np.ndarray = converged(
        evaluate_mrp(mrp, gamma=gamma),
        done=almost_equal_np_arrays
    )
    return {s: v_star[i] for i, s in enumerate(mrp.non_terminal_states)}
```

The code should be fairly self-explanatory. Since the Policy Evaluation problem applies to Finite MRPs, the function `evaluate_mrp` above takes as input `mrp: FiniteMarkovDecisionProcess[S]` and a `gamma: float` to produce an `Iterator` on Value Functions represented as `np.ndarray` (for fast vector/matrix calculations). The function `update` in `evaluate_mrp` represents the application of the Bellman Policy Operator B^π. The function `evaluate_mrp_result` produces the Value Function for the given `mrp` and the given `gamma`, returning the last value function on the `Iterator` (which terminates based on the `almost_equal_np_arrays` function, considering the maximum of the absolute value differences across all states). Note that the return type of `evaluate_mrp_result` is `V[S]` which is an alias for `Mapping[NonTerminal[S], float]`, capturing the semantic of $\mathcal{N} \to \mathbb{R}$. Note that `evaluate_mrp` is useful for debugging (by looking at the trace of value functions in the execution of the Policy Evaluation algorithm) while `evaluate_mrp_result` produces the desired output Value Function.

Note that although we defined the Bellman Policy Operator B^π as operating on Value Functions of the π-implied MRP, we can also view the Bellman Policy Operator B^π as operating on Value Functions of an MDP. To support this MDP view, we express Equation (5.1) in terms of the MDP transitions/rewards specification, as follows:

$$B^\pi(V)(s) = \sum_{a \in \mathcal{A}} \pi(s, a) \cdot \mathcal{R}(s, a) + \gamma \sum_{a \in \mathcal{A}} \pi(s, a) \sum_{s' \in \mathcal{N}} \mathcal{P}(s, a, s') \cdot V(s') \text{ for all } s \in \mathcal{N} \quad (5.2)$$

If the number of non-terminal states of a given MRP is m, then the running time of each iteration is $O(m^2)$. Note though that to construct an MRP from a given MDP and a given policy, we have to perform $O(m^2 \cdot k)$ operations, where $k = |\mathcal{A}|$.

5.5 GREEDY POLICY

We had said earlier that we will be presenting 3 Dynamic Programming Algorithms. The first (Policy Evaluation), as we saw in the previous section, solves the MDP Prediction problem. The other two (that will present in the next two sections) solve the MDP Control problem. This section is a stepping stone from *Prediction* to *Control*. In this section, we define a function that is motivated by the idea of *improving a value function/improving a policy* with a "greedy" technique. Formally, the *Greedy Policy Function*

$$G : \mathbb{R}^m \to (\mathcal{N} \to \mathcal{A})$$

interpreted as a function mapping a Value Function V (represented as a vector) to a deterministic policy $\pi'_D : \mathcal{N} \to \mathcal{A}$, is defined as:

$$G(V)(s) = \pi'_D(s) = \arg\max_{a \in \mathcal{A}} \{\mathcal{R}(s,a) + \gamma \cdot \sum_{s' \in \mathcal{N}} \mathcal{P}(s,a,s') \cdot V(s')\} \text{ for all } s \in \mathcal{N} \quad (5.3)$$

Note that for any specific s, if two or more actions a achieve the maximization of $\mathcal{R}(s,a) + \gamma \cdot \sum_{s' \in \mathcal{N}} \mathcal{P}(s,a,s') \cdot V(s')$, then we use an arbitrary rule in breaking ties and assigning a single action a as the output of the above $\arg\max$ operation. We shall use Equation (5.3) in our mathematical exposition but we require a different (but equivalent) expression for $G(V)(s)$ to guide us with our code since the interface for `FiniteMarkovDecisionProcess` operates on $\mathcal{P}_R$, rather than $\mathcal{R}$ and $\mathcal{P}$. The equivalent expression for $G(V)(s)$ is as follows:

$$G(V)(s) = \arg\max_{a \in \mathcal{A}} \{\sum_{s' \in \mathcal{S}} \sum_{r \in \mathcal{D}} \mathcal{P}_R(s,a,r,s') \cdot (r + \gamma \cdot W(s'))\} \text{ for all } s \in \mathcal{N} \quad (5.4)$$

where $W \in \mathbb{R}^n$ is defined as:

$$W(s') = \begin{cases} V(s') & \text{if } s' \in \mathcal{N} \\ 0 & \text{if } s' \in \mathcal{T} = \mathcal{S} - \mathcal{N} \end{cases}$$

Note that in Equation (5.4), because we have to work with $\mathcal{P}_R$, we need to consider transitions to all states $s' \in \mathcal{S}$ (versus transition to all states $s' \in \mathcal{N}$ in Equation (5.3)), and so, we need to handle the transitions to states $s' \in \mathcal{T}$ carefully (essentially by using the W function as described above).

Now let's write some code to create this "greedy policy" from a given value function, guided by Equation (5.4).

```
import operator
def extended_vf(v: V[S], s: State[S]) -> float:
    def non_terminal_vf(st: NonTerminal[S], v=v) -> float:
        return v[st]
    return s.on_non_terminal(non_terminal_vf, 0.0)

def greedy_policy_from_vf(
    mdp: FiniteMarkovDecisionProcess[S, A],
    vf: V[S],
    gamma: float
) -> FiniteDeterministicPolicy[S, A]:
    greedy_policy_dict: Dict[S, A] = {}
```

```
for s in mdp.non_terminal_states:
    q_values: Iterator[Tuple[A, float]] = \
        ((a, mdp.mapping[s][a].expectation(
            lambda s_r: s_r[1] + gamma * extended_vf(vf, s_r[0])
        )) for a in mdp.actions(s))
    greedy_policy_dict[s.state] = \
        max(q_values, key=operator.itemgetter(1))[0]
return FiniteDeterministicPolicy(greedy_policy_dict)
```

As you can see above, the function `greedy_policy_from_vf` loops through all the non-terminal states that serve as keys in `greedy_policy_dict: Dict[S, A]`. Within this loop, we go through all the actions in $\mathcal{A}(s)$ and compute Q-Value $Q(s, a)$ as the sum (over all (s', r) pairs) of $\mathcal{P}_R(s, a, r, s') \cdot (r + \gamma \cdot W(s'))$, written as $\mathbb{E}_{(s',r) \sim \mathcal{P}_R}[r + \gamma \cdot W(s')]$. Finally, we calculate $\arg\max_a Q(s, a)$ for all non-terminal states s, and return it as a `FinitePolicy` (which is our greedy policy).

Note that the `extended_vf` represents the $W : \mathcal{S} \to \mathbb{R}$ function used in the right-hand-side of Equation (5.4), which is the usual value function when its argument is a non-terminal state and is the default value of 0 when its argument is a terminal state. We shall use the `extended_vf` function in other Dynamic Programming algorithms later in this chapter as they also involve the $W : \mathcal{S} \to \mathbb{R}$ function in the right-hand-side of their corresponding governing equation.

The word "Greedy" is a reference to the term "Greedy Algorithm", which means an algorithm that takes heuristic steps guided by locally-optimal choices in the hope of moving towards a global optimum. Here, the reference to *Greedy Policy* means if we have a policy π and its corresponding Value Function V^π (obtained say using Policy Evaluation algorithm), then applying the Greedy Policy function G on V^π gives us a deterministic policy $\pi'_D : \mathcal{N} \to \mathcal{A}$ that is hopefully "better" than π in the sense that $V^{\pi'_D}$ is "greater" than V^π. We shall now make this statement precise and show how to use the *Greedy Policy Function* to perform *Policy Improvement*.

5.6 POLICY IMPROVEMENT

Terms such as "better" or "improvement" refer to either Value Functions or to Policies (in the latter case, to Value Functions of an MDP evaluated with the policies). So what does it mean to say a Value Function $X : \mathcal{N} \to \mathbb{R}$ is "better" than a Value Function $Y : \mathcal{N} \to \mathbb{R}$? Here's the answer:

Definition 5.6.1 (Value Function Comparison). We say $X \geq Y$ for Value Functions $X, Y : \mathcal{N} \to \mathbb{R}$ of an MDP if and only if:

$$X(s) \geq Y(s) \text{ for all } s \in \mathcal{N}$$

If we are dealing with finite MDPs (with m non-terminal states), we'd represent the Value Functions as vector $X, Y \in \mathbb{R}^m$, and say that $X \geq Y$ if and only if $X(s) \geq Y(s)$ for all $s \in \mathcal{N}$.

So whenever you hear terms like "Better Value Function" or "Improved Value Function", you should interpret it to mean that the Value Function is *no worse for each of the states* (versus the Value Function it's being compared to).

So then, what about the claim of $\pi'_D = G(V^\pi)$ being "better" than π? The following important theorem by Richard Bellman (Bellman 1957b) provides the clarification:

Theorem 5.6.1 (Policy Improvement Theorem). *For a finite MDP, for any policy π,*

$$V^{\pi'_D} = V^{G(V^\pi)} \geq V^\pi$$

Proof. This proof is based on application of the Bellman Policy Operator on Value Functions of the given MDP (note: this MDP view of the Bellman Policy Operator is expressed in Equation (5.2)). We start by noting that applying the Bellman Policy Operator $\boldsymbol{B}^{\pi'_D}$ repeatedly, starting with the Value Function $\boldsymbol{V}^\pi$, will converge to the Value Function $\boldsymbol{V}^{\pi'_D}$. Formally,

$$\lim_{i \to \infty} (\boldsymbol{B}^{\pi'_D})^i(\boldsymbol{V}^\pi) = \boldsymbol{V}^{\pi'_D}$$

So the proof is complete if we prove that:

$$(\boldsymbol{B}^{\pi'_D})^{i+1}(\boldsymbol{V}^\pi) \geq (\boldsymbol{B}^{\pi'_D})^i(\boldsymbol{V}^\pi) \text{ for all } i = 0, 1, 2, \ldots$$

which means we get a non-decreasing sequence of Value Functions $[(\boldsymbol{B}^{\pi'_D})^i(\boldsymbol{V}^\pi)|i = 0, 1, 2, \ldots]$ with repeated applications of $\boldsymbol{B}^{\pi'_D}$ starting with the Value Function $\boldsymbol{V}^\pi$.

Let us prove this by induction. The base case (for $i = 0$) of the induction is to prove that:

$$\boldsymbol{B}^{\pi'_D}(\boldsymbol{V}^\pi) \geq \boldsymbol{V}^\pi$$

Note that for the case of the deterministic policy π'_D and Value Function $\boldsymbol{V}^\pi$, Equation (5.2) simplifies to:

$$\boldsymbol{B}^{\pi'_D}(\boldsymbol{V}^\pi)(s) = \mathcal{R}(s, \pi'_D(s)) + \gamma \sum_{s' \in \mathcal{N}} \mathcal{P}(s, \pi'_D(s), s') \cdot \boldsymbol{V}^\pi(s') \text{ for all } s \in \mathcal{N}$$

From Equation (5.3), we know that for each $s \in \mathcal{N}$, $\pi'_D(s) = G(\boldsymbol{V}^\pi)(s)$ is the action that maximizes $\{\mathcal{R}(s, a) + \gamma \sum_{s' \in \mathcal{N}} \mathcal{P}(s, a, s') \cdot \boldsymbol{V}^\pi(s')\}$. Therefore,

$$\boldsymbol{B}^{\pi'_D}(\boldsymbol{V}^\pi)(s) = \max_{a \in \mathcal{A}}\{\mathcal{R}(s, a) + \gamma \sum_{s' \in \mathcal{N}} \mathcal{P}(s, a, s') \cdot \boldsymbol{V}^\pi(s')\} = \max_{a \in \mathcal{A}} Q^\pi(s, a) \text{ for all } s \in \mathcal{N}$$

Let's compare this equation against the Bellman Policy Equation for π (below):

$$V^\pi(s) = \sum_{a \in \mathcal{A}} \pi(s, a) \cdot Q^\pi(s, a) \text{ for all } s \in \mathcal{N}$$

We see that $\boldsymbol{V}^\pi(s)$ is a weighted average of $Q^\pi(s, a)$ (with weights equal to probabilities $\pi(s, a)$ over choices of a) while $\boldsymbol{B}^{\pi'_D}(\boldsymbol{V}^\pi)(s)$ is the maximum (over choices of a) of $Q^\pi(s, a)$. Therefore,

$$\boldsymbol{B}^{\pi'_D}(\boldsymbol{V}^\pi) \geq \boldsymbol{V}^\pi$$

This establishes the base case of the proof by induction. Now to complete the proof, all we have to do is to prove:

If $(\boldsymbol{B}^{\pi'_D})^{i+1}(\boldsymbol{V}^\pi) \geq (\boldsymbol{B}^{\pi'_D})^i(\boldsymbol{V}^\pi)$, then $(\boldsymbol{B}^{\pi'_D})^{i+2}(\boldsymbol{V}^\pi) \geq (\boldsymbol{B}^{\pi'_D})^{i+1}(\boldsymbol{V}^\pi)$ for all $i = 0, 1, 2, \ldots$

Since $(\boldsymbol{B}^{\pi'_D})^{i+1}(\boldsymbol{V}^\pi) = \boldsymbol{B}^{\pi'_D}((\boldsymbol{B}^{\pi'_D})^i(\boldsymbol{V}^\pi))$, from the definition of Bellman Policy Operator (Equation (5.1)), we can write the following two equations:

$$(\boldsymbol{B}^{\pi'_D})^{i+2}(\boldsymbol{V}^\pi)(s) = \mathcal{R}(s, \pi'_D(s)) + \gamma \sum_{s' \in \mathcal{N}} \mathcal{P}(s, \pi'_D(s), s') \cdot (\boldsymbol{B}^{\pi'_D})^{i+1}(\boldsymbol{V}^\pi)(s') \text{ for all } s \in \mathcal{N}$$

$$(\boldsymbol{B}^{\pi'_D})^{i+1}(\boldsymbol{V}^\pi)(s) = \mathcal{R}(s, \pi'_D(s)) + \gamma \sum_{s' \in \mathcal{N}} \mathcal{P}(s, \pi'_D(s), s') \cdot (\boldsymbol{B}^{\pi'_D})^i(\boldsymbol{V}^\pi)(s') \text{ for all } s \in \mathcal{N}$$

Subtracting each side of the second equation from the first equation yields:

$$(B^{\pi'_D})^{i+2}(V^\pi)(s) - (B^{\pi'_D})^{i+1}(s)$$

$$= \gamma \sum_{s' \in \mathcal{N}} \mathcal{P}(s, \pi'_D(s), s') \cdot ((B^{\pi'_D})^{i+1}(V^\pi)(s') - (B^{\pi'_D})^i(V^\pi)(s'))$$

for all $s \in \mathcal{N}$

Since $\gamma \mathcal{P}(s, \pi'_D(s), s')$ consists of all non-negative values and since the induction step assumes $(B^{\pi'_D})^{i+1}(V^\pi)(s') \geq (B^{\pi'_D})^i(V^\pi)(s')$ for all $s' \in \mathcal{N}$, the right-hand-side of this equation is non-negative, meaning the left-hand-side of this equation is non-negative, i.e.,

$$(B^{\pi'_D})^{i+2}(V^\pi)(s) \geq (B^{\pi'_D})^{i+1}(V^\pi)(s) \text{ for all } s \in \mathcal{N}$$

This completes the proof by induction. □

The way to understand the above proof is to think in terms of how each stage of further application of $B^{\pi'_D}$ improves the Value Function. Stage 0 is when you have the Value Function V^π where we execute the policy π throughout the MDP. Stage 1 is when you have the Value Function $B^{\pi'_D}(V^\pi)$ where from each state s, we execute the policy π'_D for the first time step following s and then execute the policy π for all further time steps. This has the effect of improving the Value Function from Stage 0 (V^π) to Stage 1 ($B^{\pi'_D}(V^\pi)$). Stage 2 is when you have the Value Function $(B^{\pi'_D})^2(V^\pi)$ where from each state s, we execute the policy π'_D for the first two time steps following s and then execute the policy π for all further time steps. This has the effect of improving the Value Function from Stage 1 ($B^{\pi'_D}(V^\pi)$) to Stage 2 ($(B^{\pi'_D})^2(V^\pi)$). And so on ... each stage applies policy π'_D instead of policy π for one extra time step, which has the effect of improving the Value Function. Note that "improve" means $\geq$ (really means that the Value Function doesn't get worse for *any* of the states). These stages are simply the iterations of the Policy Evaluation algorithm (using policy π'_D) with starting Value Function V^π, building a non-decreasing sequence of Value Functions $[(B^{\pi'_D})^i(V^\pi)|i = 0, 1, 2, \ldots]$ that get closer and closer until they converge to the Value Function $V^{\pi'_D}$ that is $\geq V^\pi$ (hence, the term *Policy Improvement*).

The Policy Improvement Theorem yields our first Dynamic Programming algorithm (called *Policy Iteration*) to solve the MDP Control problem. The Policy Iteration algorithm is due to Ronald Howard (Howard 1960).

5.7 POLICY ITERATION ALGORITHM

The proof of the Policy Improvement Theorem has shown us how to start with the Value Function V^π (for a policy π), perform a greedy policy improvement to create a policy $\pi'_D = G(V^\pi)$, and then perform Policy Evaluation (with policy π'_D) with starting Value Function V^π, resulting in the Value Function $V^{\pi'_D}$ that is an improvement over the Value Function V^π we started with. Now note that we can do the same process again to go from π'_D and $V^{\pi'_D}$ to an improved policy π''_D and associated improved Value Function $V^{\pi''_D}$. And we can keep going in this way to create further improved policies and associated Value Functions, until there is no further improvement. This methodology of performing Policy Improvement together with Policy Evaluation using the improved policy, in an iterative manner (depicted in Figure 5.1), is known as the Policy Iteration algorithm (shown below).

- Start with any Value Function $V_0 \in \mathbb{R}^m$

- Iterating over $j = 0, 1, 2, \ldots$, calculate in each iteration:

$$\text{Deterministic Policy } \pi_{j+1} = G(V_j)$$

$$\text{Value Function } V_{j+1} = \lim_{i \to \infty} (B^{\pi_{j+1}})^i(V_j)$$

- Stop the algorithm when $d(V_j, V_{j+1}) = \max_{s \in \mathcal{N}} |(V_j - V_{j+1})(s)|$ is adequately small.

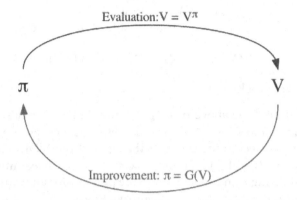

Figure 5.1 **Policy Iteration Loop**

So, the algorithm terminates when there is no further improvement to the Value Function. When this happens, the following should hold:

$$V_j = (B^{G(V_j)})^i(V_j) = V_{j+1} \text{ for all } i = 0, 1, 2, \ldots$$

In particular, this equation should hold for $i = 1$:

$$V_j(s) = B^{G(V_j)}(V_j)(s) = \mathcal{R}(s, G(V_j)(s)) + \gamma \sum_{s' \in \mathcal{N}} \mathcal{P}(s, G(V_j)(s), s') \cdot V_j(s') \text{ for all } s \in \mathcal{N}$$

From Equation (5.3), we know that for each $s \in \mathcal{N}$, $\pi_{j+1}(s) = G(V_j)(s)$ is the action that maximizes $\{\mathcal{R}(s, a) + \gamma \sum_{s' \in \mathcal{N}} \mathcal{P}(s, a, s') \cdot V_j(s')\}$. Therefore,

$$V_j(s) = \max_{a \in \mathcal{A}} \{\mathcal{R}(s, a) + \gamma \sum_{s' \in \mathcal{N}} \mathcal{P}(s, a, s') \cdot V_j(s')\} \text{ for all } s \in \mathcal{N}$$

But this, in fact, is the MDP State-Value Function Bellman Optimality Equation, which would mean that $V_j = V^*$, i.e., when V_{j+1} is identical to V_j, the Policy Iteration algorithm has converged to the Optimal Value Function. The associated deterministic policy at the convergence of the Policy Iteration algorithm ($\pi_j : \mathcal{N} \to \mathcal{A}$) is an Optimal Policy because $V^{\pi_j} = V_j \approx V^*$, meaning that evaluating the MDP with the deterministic policy π_j achieves the Optimal Value Function (depicted in Figure 5.2). This means the Policy Iteration algorithm solves the MDP Control problem. This proves the following Theorem:

Theorem 5.7.1 (Policy Iteration Convergence Theorem). *For a Finite MDP with $|\mathcal{N}| = m$ and $\gamma < 1$, Policy Iteration algorithm converges to the Optimal Value Function $V^* \in \mathbb{R}^m$ along with a Deterministic Optimal Policy $\pi_D^* : \mathcal{N} \to \mathcal{A}$, no matter which Value Function $V_0 \in \mathbb{R}^m$ we start the algorithm with.*

$$\pi^* \quad \xleftarrow{\qquad} \text{At Convergence} \xrightarrow{\qquad} \quad V^*$$

Figure 5.2 Policy Iteration Convergence

Now let's write some code for Policy Iteration Algorithm. Unlike Policy Evaluation which repeatedly operates on Value Functions (and returns a Value Function), Policy Iteration repeatedly operates on a pair of Value Function and Policy (and returns a pair of Value Function and Policy). In the code below, notice the type Tuple[V[S], FinitePolicy[S, A]] that represents a pair of Value Function and Policy. The function policy_iteration repeatedly applies the function update on a pair of Value Function and Policy. The update function, after splitting its input vf_policy into vf: V[S] and pi: FinitePolicy[S, A], creates an MRP (mrp: FiniteMarkovRewardProcess[S]) from the combination of the input mdp and pi. Then it performs a policy evaluation on mrp (using the evaluate_mrp_result function) to produce a Value Function policy_vf: V[S], and finally creates a greedy (improved) policy named improved_pi from policy_vf (using the previously-written function greedy_policy_from_vf). Thus the function update performs a Policy Evaluation followed by a Policy Improvement. Notice also that policy_iteration offers the option to perform the linear-algebra-solver-based computation of Value Function for a given policy (get_value_function_vec method of the mrp object), in case the state space is not too large. policy_iteration returns an Iterator on pairs of Value Function and Policy produced by this process of repeated Policy Evaluation and Policy Improvement. almost_equal_vf_pis is the function to decide termination based on the distance between two successive Value Functions produced by Policy Iteration. policy_iteration_result returns the final (optimal) pair of Value Function and Policy (from the Iterator produced by policy_iteration), based on the termination criterion of almost_equal_vf_pis.

```
DEFAULT_TOLERANCE = 1e-5
def policy_iteration(
    mdp: FiniteMarkovDecisionProcess[S, A],
    gamma: float,
    matrix_method_for_mrp_eval: bool = False
) -> Iterator[Tuple[V[S], FinitePolicy[S, A]]]:

    def update(vf_policy: Tuple[V[S], FinitePolicy[S, A]])\
            -> Tuple[V[S], FiniteDeterministicPolicy[S, A]]:

        vf, pi = vf_policy
        mrp: FiniteMarkovRewardProcess[S] = mdp.apply_finite_policy(pi)
        policy_vf: V[S] = {mrp.non_terminal_states[i]: v for i, v in
                        enumerate(mrp.get_value_function_vec(gamma))}\
            if matrix_method_for_mrp_eval else evaluate_mrp_result(mrp, gamma)
        improved_pi: FiniteDeterministicPolicy[S, A] = greedy_policy_from_vf(
            mdp,
            policy_vf,
            gamma
        )

        return policy_vf, improved_pi
    v_0: V[S] = {s: 0.0 for s in mdp.non_terminal_states}
    pi_0: FinitePolicy[S, A] = FinitePolicy(
        {s.state: Choose(mdp.actions(s)) for s in mdp.non_terminal_states}
    )
    return iterate(update, (v_0, pi_0))
def almost_equal_vf_pis(
    x1: Tuple[V[S], FinitePolicy[S, A]],
```

```
    x2: Tuple[V[S], FinitePolicy[S, A]]
) -> bool:
    return max(
        abs(x1[0][s] - x2[0][s]) for s in x1[0]
    ) < DEFAULT_TOLERANCE
def policy_iteration_result(
    mdp: FiniteMarkovDecisionProcess[S, A],
    gamma: float,
) -> Tuple[V[S], FiniteDeterministicPolicy[S, A]]:
    return converged(policy_iteration(mdp, gamma), done=almost_equal_vf_pis)
```

If the number of non-terminal states of a given MDP is m and the number of actions ($|\mathcal{A}|$) is k, then the running time of Policy Improvement is $O(m^2 \cdot k)$ and we've already seen before that each iteration of Policy Evaluation is $O(m^2 \cdot k)$.

5.8 BELLMAN OPTIMALITY OPERATOR AND VALUE ITERATION ALGORITHM

By making a small tweak to the definition of Greedy Policy Function in Equation (5.3) (changing the arg max to max), we define the *Bellman Optimality Operator*

$$\boldsymbol{B}^* : \mathbb{R}^m \to \mathbb{R}^m$$

as the following (non-linear) transformation of a vector (representing a Value Function) in the vector space $\mathbb{R}^m$

$$\boldsymbol{B}^*(\boldsymbol{V})(s) = \max_{a \in \mathcal{A}} \{\mathcal{R}(s, a) + \gamma \sum_{s' \in \mathcal{N}} \mathcal{P}(s, a, s') \cdot \boldsymbol{V}(s')\} \text{ for all } s \in \mathcal{N} \qquad (5.5)$$

We shall use Equation (5.5) in our mathematical exposition but we require a different (but equivalent) expression for $\boldsymbol{B}^*(\boldsymbol{V})(s)$ to guide us with our code since the interface for FiniteMarkovDecisionProcess operates on $\mathcal{P}_R$, rather than $\mathcal{R}$ and $\mathcal{P}$. The equivalent expression for $\boldsymbol{B}^*(\boldsymbol{V})(s)$ is as follows:

$$\boldsymbol{B}^*(\boldsymbol{V})(s) = \max_{a \in \mathcal{A}} \{\sum_{s' \in \mathcal{S}} \sum_{r \in \mathcal{D}} \mathcal{P}_R(s, a, r, s') \cdot (r + \gamma \cdot \boldsymbol{W}(s'))\} \text{ for all } s \in \mathcal{N} \qquad (5.6)$$

where $\boldsymbol{W} \in \mathbb{R}^n$ is defined (same as in the case of Equation (5.4)) as:

$$\boldsymbol{W}(s') = \begin{cases} \boldsymbol{V}(s') & \text{if } s' \in \mathcal{N} \\ 0 & \text{if } s' \in \mathcal{T} = \mathcal{S} - \mathcal{N} \end{cases}$$

Note that in Equation (5.6), because we have to work with $\mathcal{P}_R$, we need to consider transitions to all states $s' \in \mathcal{S}$ (versus transition to all states $s' \in \mathcal{N}$ in Equation (5.5)), and so, we need to handle the transitions to states $s' \in \mathcal{T}$ carefully (essentially by using the $\boldsymbol{W}$ function as described above).

For each $s \in \mathcal{N}$, the action $a \in \mathcal{A}$ that produces the maximization in (5.5) is the action prescribed by the deterministic policy π_D in (5.3). Therefore, if we apply the Bellman Policy Operator on any Value Function $\boldsymbol{V} \in \mathbb{R}^m$ using the Greedy Policy $G(\boldsymbol{V})$, it should be identical to applying the Bellman Optimality Operator. Therefore,

$$\boldsymbol{B}^{G(\boldsymbol{V})}(\boldsymbol{V}) = \boldsymbol{B}^*(\boldsymbol{V}) \text{ for all } \boldsymbol{V} \in \mathbb{R}^m \qquad (5.7)$$

In particular, it's interesting to observe that by specializing $\boldsymbol{V}$ to be the Value Function $\boldsymbol{V}^\pi$ for a policy π, we get:

$$\boldsymbol{B}^{G(\boldsymbol{V}^\pi)}(\boldsymbol{V}^\pi) = \boldsymbol{B}^*(\boldsymbol{V}^\pi)$$

which is a succinct representation of the first stage of Policy Evaluation with an improved policy $G(V^\pi)$ (note how all three of Bellman Policy Operator, Bellman Optimality Operator and Greedy Policy Function come together in this equation).

Much like how the Bellman Policy Operator B^π was motivated by the MDP Bellman Policy Equation (equivalently, the MRP Bellman Equation), Bellman Optimality Operator B^* is motivated by the MDP State-Value Function Bellman Optimality Equation (re-stated below):

$$V^*(s) = \max_{a \in \mathcal{A}}\{\mathcal{R}(s,a) + \gamma \sum_{s' \in \mathcal{N}} \mathcal{P}(s,a,s') \cdot V^*(s')\} \text{ for all } s \in \mathcal{N}$$

Therefore, we can express the MDP State-Value Function Bellman Optimality Equation succinctly as:

$$V^* = B^*(V^*)$$

which means $V^* \in \mathbb{R}^m$ is a Fixed-Point of the Bellman Optimality Operator $B^* : \mathbb{R}^m \to \mathbb{R}^m$.

Note that the definitions of the Greedy Policy Function and of the Bellman Optimality Operator that we have provided can be generalized to non-finite MDPs, and consequently we can generalize Equation (5.7) and the statement that V^* is a Fixed-Point of the Bellman Optimality Operator would still hold. However, in this chapter, since we are focused on developing algorithms for finite MDPs, we shall stick to the definitions we've provided for the case of finite MDPs.

Much like how we proved that B^π is a contraction function, we want to prove that B^* is a contraction function (under L^∞ norm) so we can take advantage of Banach Fixed-Point Theorem and solve the Control problem by iterative applications of the Bellman Optimality Operator B^*. So we need to prove that for all $X, Y \in \mathbb{R}^m$,

$$\max_{s \in \mathcal{N}} |(B^*(X) - B^*(Y))(s)| \leq \gamma \cdot \max_{s \in \mathcal{N}} |(X - Y)(s)|$$

This proof is a bit harder than the proof we did for B^π. Here we need to utilize two key properties of B^*.

1. Monotonicity Property, i.e, for all $X, Y \in \mathbb{R}^m$,

    If $X(s) \geq Y(s)$ for all $s \in \mathcal{N}$, then $B^*(X)(s) \geq B^*(Y)(s)$ for all $s \in \mathcal{N}$

Observe that for each state $s \in \mathcal{N}$ and each action $a \in \mathcal{A}$,

$$\{\mathcal{R}(s,a) + \gamma \sum_{s' \in \mathcal{N}} \mathcal{P}(s,a,s') \cdot X(s')\} - \{\mathcal{R}(s,a) + \gamma \sum_{s' \in \mathcal{N}} \mathcal{P}(s,a,s') \cdot Y(s')\}$$

$$= \gamma \sum_{s' \in \mathcal{N}} \mathcal{P}(s,a,s') \cdot (X(s') - Y(s')) \geq 0$$

Therefore for each state $s \in \mathcal{N}$,

$$B^*(X)(s) - B^*(Y)(s)$$

$$= \max_{a \in \mathcal{A}}\{\mathcal{R}(s,a) + \gamma \sum_{s' \in \mathcal{N}} \mathcal{P}(s,a,s') \cdot X(s')\} - \max_{a \in \mathcal{A}}\{\mathcal{R}(s,a) + \gamma \sum_{s' \in \mathcal{N}} \mathcal{P}(s,a,s') \cdot Y(s')\} \geq 0$$

2. Constant Shift Property, i.e., for all $X \in \mathbb{R}^m, c \in \mathbb{R}$,

$$B^*(X + c)(s) = B^*(X)(s) + \gamma c \text{ for all } s \in \mathcal{N}$$

In the above statement, adding a constant ($\in \mathbb{R}$) to a Value Function ($\in \mathbb{R}^m$) adds the constant point-wise to all states of the Value Function (to all dimensions of the vector representing the Value Function). In other words, a constant $\in \mathbb{R}$ might as well be treated as a Value Function with the same (constant) value for all states. Therefore,

$$\boldsymbol{B}^*(\boldsymbol{X} + c)(s) = \max_{a \in \mathcal{A}} \{ \mathcal{R}(s, a) + \gamma \sum_{s' \in \mathcal{N}} \mathcal{P}(s, a, s') \cdot (\boldsymbol{X}(s') + c) \}$$

$$= \max_{a \in \mathcal{A}} \{ \mathcal{R}(s, a) + \gamma \sum_{s' \in \mathcal{N}} \mathcal{P}(s, a, s') \cdot \boldsymbol{X}(s') \} + \gamma c = \boldsymbol{B}^*(\boldsymbol{X})(s) + \gamma c$$

With these two properties of $\boldsymbol{B}^*$ in place, let's prove that $\boldsymbol{B}^*$ is a contraction function. For given $\boldsymbol{X}, \boldsymbol{Y} \in \mathbb{R}^m$, assume:

$$\max_{s \in \mathcal{N}} |(\boldsymbol{X} - \boldsymbol{Y})(s)| = c$$

We can rewrite this as:

$$\boldsymbol{X}(s) - c \le \boldsymbol{Y}(s) \le \boldsymbol{X}(s) + c \text{ for all } s \in \mathcal{N}$$

Since $\boldsymbol{B}^*$ has the monotonicity property, we can apply $\boldsymbol{B}^*$ throughout the above double-inequality.

$$\boldsymbol{B}^*(\boldsymbol{X} - c)(s) \le \boldsymbol{B}^*(\boldsymbol{Y})(s) \le \boldsymbol{B}^*(\boldsymbol{X} + c)(s) \text{ for all } s \in \mathcal{N}$$

Since $\boldsymbol{B}^*$ has the constant shift property,

$$\boldsymbol{B}^*(\boldsymbol{X})(s) - \gamma c \le \boldsymbol{B}^*(\boldsymbol{Y})(s) \le \boldsymbol{B}^*(\boldsymbol{X})(s) + \gamma c \text{ for all } s \in \mathcal{N}$$

In other words,

$$\max_{s \in \mathcal{N}} |(\boldsymbol{B}^*(\boldsymbol{X}) - \boldsymbol{B}^*(\boldsymbol{Y}))(s)| \le \gamma c = \gamma \cdot \max_{s \in \mathcal{N}} |(\boldsymbol{X} - \boldsymbol{Y})(s)|$$

So, invoking Banach Fixed-Point Theorem proves the following Theorem:

Theorem 5.8.1 (Value Iteration Convergence Theorem). *For a Finite MDP with $|\mathcal{N}| = m$ and $\gamma < 1$, if $\boldsymbol{V}^* \in \mathbb{R}^m$ is the Optimal Value Function, then $\boldsymbol{V}^*$ is the unique Fixed-Point of the Bellman Optimality Operator $\boldsymbol{B}^* : \mathbb{R}^m \to \mathbb{R}^m$, and*

$$\lim_{i \to \infty} (\boldsymbol{B}^*)^i (\boldsymbol{V}_0) \to \boldsymbol{V}^* \text{ for all starting Value Functions } \boldsymbol{V}_0 \in \mathbb{R}^m$$

This gives us the following iterative algorithm, known as the *Value Iteration* algorithm, due to Richard Bellman (Bellman 1957a):

- Start with any Value Function $\boldsymbol{V}_0 \in \mathbb{R}^m$
- Iterating over $i = 0, 1, 2, \ldots$, calculate in each iteration:

$$\boldsymbol{V}_{i+1}(s) = \boldsymbol{B}^*(\boldsymbol{V}_i)(s) \text{ for all } s \in \mathcal{N}$$

- Stop the algorithm when $d(\boldsymbol{V}_i, \boldsymbol{V}_{i+1}) = \max_{s \in \mathcal{N}} |(\boldsymbol{V}_i - \boldsymbol{V}_{i+1})(s)|$ is adequately small.

It pays to emphasize that Banach Fixed-Point Theorem not only assures convergence to the unique solution $\boldsymbol{V}^*$ (no matter what Value Function $\boldsymbol{V}_0$ we start the algorithm with), it also assures a reasonable speed of convergence (dependent on the choice of starting Value Function $\boldsymbol{V}_0$ and the choice of γ).

5.9   OPTIMAL POLICY FROM OPTIMAL VALUE FUNCTION

Note that the Policy Iteration algorithm produces a policy together with a Value Function in each iteration. So, in the end, when we converge to the Optimal Value Function $V_j = V^*$ in iteration j, the Policy Iteration algorithm has a deterministic policy π_j associated with V_j such that:

$$V_j = V^{\pi_j} = V^*$$

and we refer to π_j as the Optimal Policy π^*, one that yields the Optimal Value Function V^*, i.e.,

$$V^{\pi^*} = V^*$$

But Value Iteration has no such policy associated with it since the entire algorithm is devoid of a policy representation and operates only with Value Functions. So now the question is: when Value Iteration converges to the Optimal Value Function $V_i = V^*$ in iteration i, how do we get hold of an Optimal Policy π^* such that:

$$V^{\pi^*} = V_i = V^*$$

The answer lies in the Greedy Policy function G. Equation (5.7) told us that:

$$B^{G(V)}(V) = B^*(V) \text{ for all } V \in \mathbb{R}^m$$

Specializing V to be V^*, we get:

$$B^{G(V^*)}(V^*) = B^*(V^*)$$

But we know that V^* is the Fixed-Point of the Bellman Optimality Operator B^*, i.e., $B^*(V^*) = V^*$. Therefore,

$$B^{G(V^*)}(V^*) = V^*$$

The above equation says V^* is the Fixed-Point of the Bellman Policy Operator $B^{G(V^*)}$. However, we know that $B^{G(V^*)}$ has a unique Fixed-Point equal to $V^{G(V^*)}$. Therefore,

$$V^{G(V^*)} = V^*$$

This says that evaluating the MDP with the deterministic greedy policy $G(V^*)$ (policy created from the Optimal Value Function V^* using the Greedy Policy Function G), in fact, achieves the Optimal Value Function V^*. In other words, $G(V^*)$ is the (Deterministic) Optimal Policy π^* we've been seeking.

Now let's write the code for Value Iteration. The function `value_iteration` returns an `Iterator` on Value Functions (of type `V[S]`) produced by the Value Iteration algorithm. It uses the function `update` for application of the Bellman Optimality Operator. `update` prepares the Q-Values for a state by looping through all the allowable actions for the state, and then calculates the maximum of those Q-Values (over the actions). The Q-Value calculation is same as what we saw in `greedy_policy_from_vf`: $\mathbb{E}_{(s',r) \sim \mathcal{P}_R}[r + \gamma \cdot W(s')]$, using the $\mathcal{P}_R$ probabilities represented in the `mapping` attribute of the `mdp` object (essentially Equation (5.6)). Note the use of the previously-written function `extended_vf` to handle the function $W : \mathcal{S} \to \mathbb{R}$ that appears in the definition of Bellman Optimality Operator in Equation (5.6). The function `value_iteration_result` returns the final (optimal) Value Function, together with its associated Optimal Policy. It simply returns the last Value Function of the `Iterator[V[S]]` returned by `value_iteration`, using the termination condition specified in `almost_equal_vfs`.

```
DEFAULT_TOLERANCE = 1e-5
def value_iteration(
    mdp: FiniteMarkovDecisionProcess[S, A],
    gamma: float
) -> Iterator[V[S]]:
    def update(v: V[S]) -> V[S]:
        return {s: max(mdp.mapping[s][a].expectation(
            lambda s_r: s_r[1] + gamma * extended_vf(v, s_r[0])
        ) for a in mdp.actions(s)) for s in v}

    v_0: V[S] = {s: 0.0 for s in mdp.non_terminal_states}
    return iterate(update, v_0)
def almost_equal_vfs(
    v1: V[S],
    v2: V[S],
    tolerance: float = DEFAULT_TOLERANCE
) -> bool:
    return max(abs(v1[s] - v2[s]) for s in v1) < tolerance
def value_iteration_result(
    mdp: FiniteMarkovDecisionProcess[S, A],
    gamma: float
) -> Tuple[V[S], FiniteDeterministicPolicy[S, A]]:
    opt_vf: V[S] = converged(
        value_iteration(mdp, gamma),
        done=almost_equal_vfs
    )
    opt_policy: FiniteDeterministicPolicy[S, A] = greedy_policy_from_vf(
        mdp,
        opt_vf,
        gamma
    )
    return opt_vf, opt_policy
```

If the number of non-terminal states of a given MDP is m and the number of actions ($|\mathcal{A}|$) is k, then the running time of each iteration of Value Iteration is $O(m^2 \cdot k)$.

We encourage you to play with the above implementations of Policy Evaluation, Policy Iteration and Value Iteration (code in the file rl/dynamic_programming.py) by running it on MDPs/Policies of your choice, and observing the traces of the algorithms.

5.10   REVISITING THE SIMPLE INVENTORY EXAMPLE

Let's revisit the simple inventory example. We shall consider the version with a space capacity since we want an example of a FiniteMarkovDecisionProcess. It will help us test our code for Policy Evaluation, Policy Iteration and Value Iteration. More importantly, it will help us identify the mathematical structure of the optimal policy of ordering for this store inventory problem. So let's take another look at the code we wrote in Chapter 4 to set up an instance of a SimpleInventoryMDPCap and a FiniteDeterministicPolicy (that we can use for Policy Evaluation).

```
user_capacity = 2
user_poisson_lambda = 1.0
user_holding_cost = 1.0
user_stockout_cost = 10.0

si_mdp: FiniteMarkovDecisionProcess[InventoryState, int] =\
    SimpleInventoryMDPCap(
        capacity=user_capacity,
        poisson_lambda=user_poisson_lambda,
        holding_cost=user_holding_cost,
        stockout_cost=user_stockout_cost
    )

fdp: FiniteDeterministicPolicy[InventoryState, int] = \
```

```
        FiniteDeterministicPolicy(
            {InventoryState(alpha, beta): user_capacity - (alpha + beta)
              for alpha in range(user_capacity + 1)
              for beta in range(user_capacity + 1 - alpha)}
)
```

Now let's write some code to evaluate si_mdp with the policy fdp.

```
from pprint import pprint
implied_mrp: FiniteMarkovRewardProcess[InventoryState] =\
    si_mdp.apply_finite_policy(fdp)
user_gamma = 0.9
pprint(evaluate_mrp_result(implied_mrp, gamma=user_gamma))
```

This prints the following Value Function.

```
{NonTerminal(state=InventoryState(on_hand=0, on_order=0)): -35.510518165628724,
 NonTerminal(state=InventoryState(on_hand=0, on_order=1)): -27.93217421014731,
 NonTerminal(state=InventoryState(on_hand=0, on_order=2)): -28.345029758390766,
 NonTerminal(state=InventoryState(on_hand=1, on_order=0)): -28.93217421014731,
 NonTerminal(state=InventoryState(on_hand=1, on_order=1)): -29.345029758390766,
 NonTerminal(state=InventoryState(on_hand=2, on_order=0)): -30.345029758390766}
```

Next, let's run Policy Iteration.

```
opt_vf_pi, opt_policy_pi = policy_iteration_result(
    si_mdp,
    gamma=user_gamma
)
pprint(opt_vf_pi)
print(opt_policy_pi)
```

This prints the following Optimal Value Function and Optimal Policy.

```
{NonTerminal(state=InventoryState(on_hand=1, on_order=0)): -28.660960231637507,
 NonTerminal(state=InventoryState(on_hand=0, on_order=2)): -27.991900091403533,
 NonTerminal(state=InventoryState(on_hand=0, on_order=1)): -27.660960231637507,
 NonTerminal(state=InventoryState(on_hand=0, on_order=0)): -34.894855781630035,
 NonTerminal(state=InventoryState(on_hand=1, on_order=1)): -28.991900091403533,
 NonTerminal(state=InventoryState(on_hand=2, on_order=0)): -29.991900091403533}
```

```
For State InventoryState(on_hand=0, on_order=0): Do Action 1
For State InventoryState(on_hand=0, on_order=1): Do Action 1
For State InventoryState(on_hand=0, on_order=2): Do Action 0
For State InventoryState(on_hand=1, on_order=0): Do Action 1
For State InventoryState(on_hand=1, on_order=1): Do Action 0
For State InventoryState(on_hand=2, on_order=0): Do Action 0
```

As we can see, the Optimal Policy is to not order if the Inventory Position (sum of On-Hand and On-Order) is greater than 1 unit and to order 1 unit if the Inventory Position is 0 or 1. Finally, let's run Value Iteration.

```
opt_vf_vi, opt_policy_vi = value_iteration_result(si_mdp, gamma=user_gamma
)
pprint(opt_vf_vi)
print(opt_policy_vi)
```

You'll see the output from Value Iteration matches the output produced from Policy Iteration—this is a good validation of our code correctness. We encourage you to play around with `user_capacity`, `user_poisson_lambda`, `user_holding_cost`, `user_stockout_cost` and `user_gamma`(code in `__main__` in rl/chapter3/simple_inventory_mdp_cap.py). As a valuable exercise, using this code, discover the mathematical structure of the Optimal Policy as a function of the above inputs.

5.11   GENERALIZED POLICY ITERATION

In this section, we dig into the structure of the Policy Iteration algorithm and show how this structure can be generalized. Let us start by looking at a 2-dimensional layout of how the Value Functions progress in Policy Iteration from the starting Value Function V_0 to the final Value Function V^*.

$$\pi_1 = G(V_0), V_0 \to B^{\pi_1}(V_0) \to (B^{\pi_1})^2(V_0) \to \ldots (B^{\pi_1})^i(V_0) \to \ldots V^{\pi_1} = V_1$$

$$\pi_2 = G(V_1), V_1 \to B^{\pi_2}(V_1) \to (B^{\pi_2})^2(V_1) \to \ldots (B^{\pi_2})^i(V_1) \to \ldots V^{\pi_2} = V_2$$

$$\ldots$$

$$\ldots$$

$$\pi_{j+1} = G(V_j), V_j \to B^{\pi_{j+1}}(V_j) \to (B^{\pi_{j+1}})^2(V_j) \to \ldots (B^{\pi_{j+1}})^i(V_j) \to \ldots V^{\pi_{j+1}} = V^*$$

Each row in the layout above represents the progression of the Value Function for a specific policy. Each row starts with the creation of the policy (for that row) using the Greedy Policy Function G, and the remainder of the row consists of successive applications of the Bellman Policy Operator (using that row's policy) until convergence to the Value Function for that row's policy. So each row starts with a Policy Improvement and the rest of the row is a Policy Evaluation. Notice how the end of one row dovetails into the start of the next row with application of the Greedy Policy Function G. It's also important to recognize that Greedy Policy Function as well as Bellman Policy Operator apply to *all states* in $\mathcal{N}$. So, in fact, the entire Policy Iteration algorithm has 3 nested loops. The outermost loop is over the rows in this 2-dimensional layout (each iteration in this outermost loop creates an improved policy). The loop within this outermost loop is over the columns in each row (each iteration in this loop applies the Bellman Policy Operator, i.e. the iterations of Policy Evaluation). The innermost loop is over each state in $\mathcal{N}$ since we need to sweep through all states in updating the Value Function when the Bellman Policy Operator is applied on a Value Function (we also need to sweep through all states in applying the Greedy Policy Function to improve the policy).

A higher-level view of Policy Iteration is to think of Policy Evaluation and Policy Improvement going back and forth iteratively—Policy Evaluation takes a policy and creates the Value Function for that policy, while Policy Improvement takes a Value Function and creates a Greedy Policy from it (that is improved relative to the previous policy). This was depicted in Figure 5.1. It is important to recognize that this loop of Policy Evaluation and Policy Improvement works to make the Value Function and the Policy increasingly consistent with each other, until we reach convergence when the Value Function and Policy become completely consistent with each other (as was illustrated in Figure 5.2).

We'd also like to share a visual of Policy Iteration that is quite popular in much of the literature on Dynamic Programming, originally appearing in Sutton and Barto's RL book (Richard S. Sutton and Barto 2018). It is the visual of Figure 5.3. It's a somewhat fuzzy sort of visual, but it has its benefits in terms of pedagogy of Policy Iteration. The idea behind this image is that the lower line represents the "policy line" indicating the progression of the policies as Policy Iteration algorithm moves along and the upper line represents the "value function line" indicating the progression of the Value Functions as Policy Iteration algorithm moves along. The arrows pointing towards the upper line ("value function line") represent a Policy Evaluation for a given policy π, yielding the point (Value Function) V^π on the upper line. The arrows pointing towards the lower line ("policy line") represent a Greedy Policy Improvement from a Value Function V^π, yielding the point (policy) $\pi' = G(V^\pi)$ on the lower line. The key concept here is that Policy Evaluation (arrows pointing to upper line) and Policy Improvement (arrows pointing to lower line) are "competing"—they "push in different directions" even as they aim to get the Value Function and Policy to be consistent with each other. This concept of simultaneously trying to compete and trying to be consistent might seem confusing and contradictory, so it deserves a proper explanation. Things become clear by noting that there are actually two notions of consistency between a Value Function V and Policy π.

1. The notion of the Value Function V being consistent with/close to the Value Function V^π of the policy π.
2. The notion of the Policy π being consistent with/close to the Greedy Policy $G(V)$ of the Value Function V.

Policy Evaluation aims for the first notion of consistency, but in the process, makes it worse in terms of the second notion of consistency. Policy Improvement aims for the second notion of consistency, but in the process, makes it worse in terms of the first notion of consistency. This also helps us understand the rationale for alternating between Policy Evaluation and Policy Improvement so that neither of the above two notions of consistency slip up too much (thanks to the alternating propping up of the two notions of consistency). Also, note that as Policy Iteration progresses, the upper line and lower line get closer and closer and the "pushing in different directions" looks more and more collaborative rather than competing (the gaps in consistency become lesser and lesser). In the end, the two lines intersect, when there is no more pushing to do for either of Policy Evaluation or Policy Improvement since at convergence, π^* and V^* have become completely consistent.

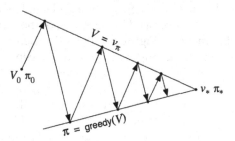

Figure 5.3 Progression Lines of Value Function and Policy in Policy Iteration (Image Credit: Sutton-Barto's RL Book)

Now we are ready to talk about a very important idea known as *Generalized Policy Iteration* that is emphasized throughout Sutton and Barto's RL book (Richard S. Sutton and Barto 2018) as the perspective that unifies all variants of DP as well as RL algorithms. Generalized Policy Iteration is the idea that we can evaluate the Value Function for a policy with *any* Policy Evaluation method, and we can improve a policy with *any* Policy Improvement method (not necessarily the methods used in the classical Policy Iteration DP algorithm). In particular, we'd like to emphasize the idea that neither of Policy Evaluation and Policy Improvement need to go fully towards the notion of consistency they are respectively striving for. As a simple example, think of modifying Policy Evaluation (say for a policy π) to not go all the way to V^π, but instead just perform say 3 Bellman Policy Evaluations. This means it would partially bridge the gap on the first notion of consistency (getting closer to V^π but not go all the way to V^π), but it would also mean not slipping up too much on the second notion of consistency. As another example, think of updating just 5 of the states (say in a large state space) with the Greedy Policy Improvement function (rather than the normal Greedy Policy Improvement function that operates on all the states). This means it would partially bridge the gap on the second notion of consistency (getting closer to $G(V^\pi)$ but not go all the way to $G(V^\pi)$), but it would also mean not slipping up too much on the first notion of consistency. A concrete example of Generalized Policy Iteration is, in fact, Value Iteration. In Value Iteration, we apply the Bellman Policy Operator just once before moving on to Policy Improvement. In a 2-dimensional layout, this is what Value Iteration looks like:

$$\pi_1 = G(V_0), V_0 \to B^{\pi_1}(V_0) = V_1$$
$$\pi_2 = G(V_1), V_1 \to B^{\pi_2}(V_1) = V_2$$
$$\bullet$$
$$\ldots$$

$$\ldots$$
$$\pi_{j+1} = G(V_j), V_j \to B^{\pi_{j+1}}(V_j) = V^*$$

So the greedy policy improvement step is unchanged, but Policy Evaluation is reduced to just a single Bellman Policy Operator application. In fact, pretty much all control algorithms in Reinforcement Learning can be viewed as special cases of Generalized Policy Iteration. In some of the simple versions of Reinforcement Learning Control algorithms, the Policy Evaluation step is done for just a single state (versus for all states in usual Policy Iteration, or even in Value Iteration) and the Policy Improvement step is also done for just a single state. So essentially these Reinforcement Learning Control algorithms are an alternating sequence of single-state policy evaluation and single-state policy improvement (where the single-state is the state produced by sampling or the state that is encountered in a real-world environment interaction). Figure 5.4 illustrates Generalized Policy Iteration as the shorter-length arrows (versus the longer-length arrows seen in Figure 5.3 for the usual Policy Iteration algorithm). Note how these shorter-length arrows don't go all the way to either the "value function line" or the "policy line" but they do go some part of the way towards the line they are meant to go towards at that stage in the algorithm.

We would go so far as to say that the Bellman Equations and the concept of Generalized Policy Iteration are the two most important concepts to internalize in the study of Reinforcement Learning, and we highly encourage you to think along the lines of these two ideas when we present several algorithms later in this book. The importance of the concept of Generalized Policy Iteration (GPI) might not be fully visible to you yet, but we hope that GPI will be your mantra by the time you finish this book. For now, let's just note the key takeaway regarding GPI—it is any algorithm to solve MDP control that alternates between *some form of* Policy Evaluation and *some form of* Policy Improvement. We will bring up GPI several times later in this book.

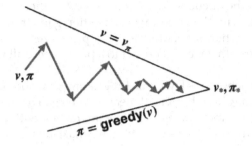

Figure 5.4 **Progression Lines of Value Function and Policy in Generalized Policy Iteration (Image Credit: Coursera Course on Fundamentals of RL)**

5.12 ASYNCHRONOUS DYNAMIC PROGRAMMING

The classical Dynamic Programming algorithms we have described in this chapter are qualified as *Synchronous* Dynamic Programming algorithms. The word *synchronous* refers to two things:

1. All states' values are updated in each iteration
2. The mathematical description of the algorithms corresponds to all the states' value updates to occur simultaneously. However, when implementing in code (in Python, where computation is serial and not parallel), this "simultaneous update" would be done by creating a new copy of the Value Function vector and sweeping through all states to assign values to the new copy from the values in the old copy.

In practice, Dynamic Programming algorithms are typically implemented as *Asynchronous* algorithms, where the above two constraints (all states updated simultaneously) are relaxed. The term *asynchronous* affords a lot of flexibility—we can update a subset of states in each iteration, and we can update states in any order we like. A natural outcome of this relaxation of the synchronous constraint is that we can maintain just one vector for the value function and update the values *in-place*. This has considerable benefits—an updated value for a state is immediately available for updates of other states (note: in synchronous, with the old and new value function vectors, one has to wait for the entire states sweep to be over until an updated state value is available for another state's update). In fact, in-place updates of value function is the norm in practical implementations of algorithms to solve the MDP Control problem.

Another feature of practical asynchronous algorithms is that we can prioritize the order in which state values are updated. There are many ways in which algorithms assign priorities, and we'll just highlight a simple but effective way of prioritizing state value updates. It's known as *prioritized sweeping*. We maintain a queue of the states, sorted by their "value function gaps" $g : \mathcal{N} \to \mathbb{R}$ (illustrated below as an example for Value Iteration):

$$g(s) = |V(s) - \max_{a \in \mathcal{A}} \{\mathcal{R}(s, a) + \gamma \cdot \sum_{s' \in \mathcal{N}} \mathcal{P}(s, a, s') \cdot V(s')\}| \text{ for all } s \in \mathcal{N}$$

After each state's value is updated with the Bellman Optimality Operator, we update the Value Function Gap for all the states whose Value Function Gap does get changed as a result of this state value update. These are exactly the states from which we have a probabilistic transition to the state whose value just got updated. What this also means is that we need to maintain the reverse transition dynamics in our data structure representation. So, after each state value update, the queue of states is resorted (by their value function gaps). We always pull out the state with the largest value function gap (from the top of the queue), and update the value function for that state. This prioritizes updates of states with the largest gaps, and it ensures that we quickly get to a point where all value function gaps are low enough.

Another form of Asynchronous Dynamic Programming worth mentioning here is *Real-Time Dynamic Programming* (RTDP). RTDP means we run a Dynamic Programming algorithm *while* the AI agent is experiencing real-time interaction with the environment. When a state is visited during the real-time interaction, we make an update for that state's value. Then, as we transition to another state as a result of the real-time interaction, we update that new state's value, and so on. Note also that in RTDP, the choice of action is the real-time action executed by the AI agent, which the environment responds to. This action choice is governed by the policy implied by the value function for the encountered state at that point in time in the real-time interaction.

Finally, we need to highlight that often special types of structures of MDPs can benefit from specific customizations of Dynamic Programming algorithms (typically, Asynchronous). One such specialization is when each state is encountered not more than once in each random sequence of state occurrences when an AI agent plays out an MDP, and when all such random sequences of the MDP terminate. This structure can be conceptualized as a Directed Acylic Graph wherein each non-terminal node in the Directed Acyclic Graph (DAG) represents a pair of non-terminal state and action, and each terminal node in the DAG represents a terminal state (the graph edges represent probabilistic transitions of the MDP). In this specialization, the MDP Prediction and Control problems can be solved in a fairly simple manner—by walking backwards on the DAG from the terminal nodes and setting the Value Function of visited states (in the backward DAG walk) using the Bellman Optimality Equation (for Control) or Bellman Policy Equation (for Prediction). Here we don't need the "iterate to convergence" approach of Policy Evaluation or Policy Iteration or Value Iteration. Rather, all these Dynamic Programming algorithms essentially reduce to a simple back-propagation of the Value Function on the DAG. This means, states are visited (and their Value Functions set) in the order determined by the reverse sequence of a Topological Sort on the DAG. We shall make this DAG back-propagation Dynamic Programming algorithm clear for a special DAG structure—Finite-Horizon MDPs—where all random sequences of the MDP terminate within a fixed number of time steps and each time step has a separate (from other time steps) set of states. This special case of Finite-Horizon MDPs is fairly common in Financial Applications and so, we cover it in detail in the next section.

5.13 FINITE-HORIZON DYNAMIC PROGRAMMING: BACKWARD INDUCTION

In this section, we consider a specialization of the DAG-structured MDPs described at the end of the previous section—one that we shall refer to as *Finite-Horizon MDPs*, where each sequence terminates within a fixed finite number of time steps T and each time step has a separate (from other time steps) set of countable states. So, all states at time-step T are terminal states and some states before time-step T could be terminal states. For all $t = 0, 1, \ldots, T$, denote the set of states for time step t as $\mathcal{S}_t$, the set of terminal states for time step t as $\mathcal{T}_t$ and the set of non-terminal states for time step t as $\mathcal{N}_t = \mathcal{S}_t - \mathcal{T}_t$ (note: $\mathcal{N}_T = \emptyset$). As mentioned previously, when the MDP is not time-homogeneous, we augment each state to include the index of the time step so that the augmented state at time step t is (t, s_t) for $s_t \in \mathcal{S}_t$. The entire MDP's (augmented) state space $\mathcal{S}$ is:

$$\{(t, s_t)|t = 0, 1, \ldots, T, s_t \in \mathcal{S}_t\}$$

We need a Python class to represent this augmented state space.

```python
@dataclass(frozen=True)
class WithTime(Generic[S]):
    state: S
    time: int = 0
```

The set of terminal states $\mathcal{T}$ is:

$$\{(t, s_t)|t = 0, 1, \ldots, T, s_t \in \mathcal{T}_t\}$$

As usual, the set of non-terminal states is denoted as $\mathcal{N} = \mathcal{S} - \mathcal{T}$.

We denote the set of rewards receivable by the AI agent at time t as $\mathcal{D}_t$ (countable subset of $\mathbb{R}$), and we denote the allowable actions for states in $\mathcal{N}_t$ as $\mathcal{A}_t$. In a more generic setting, as we shall represent in our code, each non-terminal state (t, s_t) has its own set of allowable actions, denoted $\mathcal{A}(s_t)$, However, for ease of exposition, here we shall treat all non-terminal states at a particular time step to have the same set of allowable actions $\mathcal{A}_t$. Let us denote the entire action space $\mathcal{A}$ of the MDP as the union of all the $\mathcal{A}_t$ over all $t = 0, 1, \ldots, T - 1$.

The state-reward transition probability function

$$\mathcal{P}_R : \mathcal{N} \times \mathcal{A} \times \mathcal{D} \times \mathcal{S} \to [0, 1]$$

is given by:

$$\mathcal{P}_R((t, s_t), a_{t'}, r_{t'}, (t', s_{t'})) = \begin{cases} (\mathcal{P}_R)_t(s_t, a_t, r_{t'}, s_{t'}) & \text{if } t' = t + 1 \text{ and } s_{t'} \in \mathcal{S}_{t'} \text{ and } r_{t'} \in \mathcal{D}_{t'} \\ 0 & \text{otherwise} \end{cases}$$

for all $t = 0, 1, \ldots T - 1, s_t \in \mathcal{N}_t, a_t \in \mathcal{A}_t, t' = 0, 1, \ldots, T$ where

$$(\mathcal{P}_R)_t : \mathcal{N}_t \times \mathcal{A}_t \times \mathcal{D}_{t+1} \times \mathcal{S}_{t+1} \to [0, 1]$$

are the separate state-reward transition probability functions for each of the time steps $t = 0, 1, \ldots, T - 1$ such that

$$\sum_{s_{t+1} \in \mathcal{S}_{t+1}} \sum_{r_{t+1} \in \mathcal{D}_{t+1}} (\mathcal{P}_R)_t(s_t, a_t, r_{t+1}, s_{t+1}) = 1$$

for all $t = 0, 1, \ldots, T - 1, s_t \in \mathcal{N}_t, a_t \in \mathcal{A}_t$.

So it is convenient to represent a finite-horizon MDP with separate state-reward transition probability functions $(\mathcal{P}_R)_t$ for each time step. Likewise, it is convenient to represent any policy of the MDP

$$\pi : \mathcal{N} \times \mathcal{A} \to [0, 1]$$

as:

$$\pi((t, s_t), a_t) = \pi_t(s_t, a_t)$$

where

$$\pi_t : \mathcal{N}_t \times \mathcal{A}_t \to [0, 1]$$

are the separate policies for each of the time steps $t = 0, 1, \ldots, T - 1$

So essentially we interpret π as being composed of the sequence $(\pi_0, \pi_1, \ldots, \pi_{T-1})$.

Consequently, the Value Function for a given policy π (equivalently, the Value Function for the π-implied MRP)

$$V^\pi : \mathcal{N} \to \mathbb{R}$$

can be conveniently represented in terms of a sequence of Value Functions

$$V_t^\pi : \mathcal{N}_t \to \mathbb{R}$$

for each of time steps $t = 0, 1, \ldots, T-1$, defined as:

$$V^\pi((t, s_t)) = V_t^\pi(s_t) \text{ for all } t = 0, 1, \ldots, T-1, s_t \in \mathcal{N}_t$$

Then, the Bellman Policy Equation can be written as:

$$V_t^\pi(s_t) = \sum_{s_{t+1} \in \mathcal{S}_{t+1}} \sum_{r_{t+1} \in \mathcal{D}_{t+1}} (\mathcal{P}_R^{\pi_t})_t (s_t, r_{t+1}, s_{t+1}) \cdot (r_{t+1} + \gamma \cdot W_{t+1}^\pi(s_{t+1})) \tag{5.8}$$

$$\text{for all } t = 0, 1, \ldots, T-1, s_t \in \mathcal{N}_t$$

where

$$W_t^\pi(s_t) = \begin{cases} V_t^\pi(s_t) & \text{if } s_t \in \mathcal{N}_t \\ 0 & \text{if } s_t \in \mathcal{T}_t \end{cases} \text{ for all } t = 1, 2, \ldots, T$$

and where $(\mathcal{P}_R^{\pi_t})_t : \mathcal{N}_t \times \mathcal{D}_{t+1} \times \mathcal{S}_{t+1}$ for all $t = 0, 1, \ldots, T-1$ represent the π-implied MRP's state-reward transition probability functions for the time steps, defined as:

$$(\mathcal{P}_R^{\pi_t})_t (s_t, r_{t+1}, s_{t+1}) = \sum_{a_t \in \mathcal{A}_t} \pi_t(s_t, a_t) \cdot (\mathcal{P}_R)_t(s_t, a_t, r_{t+1}, s_{t+1}) \text{ for all } t = 0, 1, \ldots, T-1$$

So for a Finite MDP, this yields a simple algorithm to calculate V_t^π for all t by simply decrementing down from $t = T-1$ to $t = 0$ and using Equation (5.8) to calculate V_t^π for all $t = 0, 1, \ldots, T-1$ from the known values of W_{t+1}^π (since we are decrementing in time index t).

This algorithm is the adaptation of Policy Evaluation to the finite horizon case with this simple technique of "stepping back in time" (known as *Backward Induction*). Let's write some code to implement this algorithm. We are given an MDP over the augmented (finite) state space WithTime[S], and a policy π (also over the augmented state space WithTime[S]). So, we can use the method apply_finite_policy in FiniteMarkovDecisionProcess[WithTime[S], A] to obtain the π-implied MRP of type FiniteMarkovRewardProcess[WithTime[S]].

Our first task to to "unwrap" the state-reward probability transition function $\mathcal{P}_R^\pi$ of this π-implied MRP into a time-indexed sequenced of state-reward probability transition functions $(\mathcal{P}_R^{\pi_t})_t, t = 0, 1, \ldots, T-1$. This is accomplished by the following function unwrap_finite_horizon_MRP (itertools.groupby groups the augmented states by their time step, and the function without_time strips the time step from the augmented states when placing the states in $(\mathcal{P}_R^\pi)_t$, i.e., Sequence[RewardTransition[S]]).

```
from itertools import groupby
StateReward = FiniteDistribution[Tuple[State[S], float]]
RewardTransition = Mapping[NonTerminal[S], StateReward[S]]
def unwrap_finite_horizon_MRP(
    process: FiniteMarkovRewardProcess[WithTime[S]]
) -> Sequence[RewardTransition[S]]:
    def time(x: WithTime[S]) -> int:
        return x.time

    def single_without_time(
        s_r: Tuple[State[WithTime[S]], float]
    ) -> Tuple[State[S], float]:
```

```
    if isinstance(s_r[0], NonTerminal):
        ret: Tuple[State[S], float] = (
            NonTerminal(s_r[0].state.state),
            s_r[1]
        )
    else:
        ret = (Terminal(s_r[0].state.state), s_r[1])
    return ret
def without_time(arg: StateReward[WithTime[S]]) -> StateReward[S]:
    return arg.map(single_without_time)

return [{NonTerminal(s.state): without_time(
    process.transition_reward(NonTerminal(s))
) for s in states} for _, states in groupby(
    sorted(
        (nt.state for nt in process.non_terminal_states),
        key=time
    ),
    key=time
)]
```

Now that we have the state-reward transition functions $(\mathcal{P}_R^{\pi_t})_t$ arranged in the form of a Sequence[RewardTransition[S]], we are ready to perform backward induction to calculate V_t^π. The following function evaluate accomplishes it with a straightforward use of Equation (5.8), as described above. Note the use of the previously-written extended_vf function, that represents the $W_t^\pi : \mathcal{S}_t \to \mathbb{R}$ function appearing on the right-hand-side of Equation (5.8).

```
def evaluate(
    steps: Sequence[RewardTransition[S]],
    gamma: float
) -> Iterator[V[S]]:
    v: List[V[S]] = []

    for step in reversed(steps):
        v.append({s: res.expectation(
            lambda s_r: s_r[1] + gamma * (
                extended_vf(v[-1], s_r[0]) if len(v) > 0 else 0.
            )
        ) for s, res in step.items()})

    return reversed(v)
```

If $|\mathcal{N}_t|$ is $O(m)$, then the running time of this algorithm is $O(m^2 \cdot T)$. However, note that it takes $O(m^2 \cdot k \cdot T)$ to convert the MDP to the π-implied MRP (where $|\mathcal{A}_t|$ is $O(k)$).

Now we move on to the Control problem—to calculate the Optimal Value Function and the Optimal Policy. Similar to the pattern seen so far, the Optimal Value Function

$$V^* : \mathcal{N} \to \mathbb{R}$$

can be conveniently represented in terms of a sequence of Value Functions

$$V_t^* : \mathcal{N}_t \to \mathbb{R}$$

for each of time steps $t = 0, 1, \ldots, T - 1$, defined as:

$$V^*((t, s_t)) = V_t^*(s_t) \text{ for all } t = 0, 1, \ldots, T - 1, s_t \in \mathcal{N}_t$$

Thus, the MDP State-Value Function Bellman Optimality Equation can be written as:

$$V_t^*(s_t) = \max_{a_t \in \mathcal{A}_t} \{ \sum_{s_{t+1} \in \mathcal{S}_{t+1}} \sum_{r_{t+1} \in \mathcal{D}_{t+1}} (\mathcal{P}_R)_t(s_t, a_t, r_{t+1}, s_{t+1}) \cdot (r_{t+1} + \gamma \cdot W_{t+1}^*(s_{t+1})) \}$$

$$\text{for all } t = 0, 1, \ldots, T - 1, s_t \in \mathcal{N}_t$$

(5.9)

where

$$W_t^*(s_t) = \begin{cases} V_t^*(s_t) & \text{if } s_t \in \mathcal{N}_t \\ 0 & \text{if } s_t \in \mathcal{T}_t \end{cases} \text{ for all } t = 1, 2, \ldots, T$$

The associated Optimal (Deterministic) Policy

$$(\pi_D^*)_t : \mathcal{N}_t \to \mathcal{A}_t$$

is defined as:

$$(\pi_D^*)_t(s_t) = \arg\max_{a_t \in \mathcal{A}_t} \{ \sum_{s_{t+1} \in \mathcal{S}_{t+1}} \sum_{r_{t+1} \in \mathcal{D}_{t+1}} (\mathcal{P}_R)_t(s_t, a_t, r_{t+1}, s_{t+1}) \cdot (r_{t+1} + \gamma \cdot W_{t+1}^*(s_{t+1})) \}$$

$$\text{for all } t = 0, 1, \ldots, T - 1, s_t \in \mathcal{N}_t$$

$$(5.10)$$

So for a Finite MDP, this yields a simple algorithm to calculate V_t^* for all t, by simply decrementing down from $t = T - 1$ to $t = 0$, using Equation (5.9) to calculate V_t^*, and Equation (5.10) to calculate $(\pi_D^*)_t$ for all $t = 0, 1, \ldots, T - 1$ from the known values of W_{t+1}^* (since we are decrementing in time index t).

This algorithm is the adaptation of Value Iteration to the finite horizon case with this simple technique of "stepping back in time" (known as *Backward Induction*). Let's write some code to implement this algorithm. We are given a MDP over the augmented (finite) state space WithTime[S]. So this MDP is of type FiniteMarkovDecisionProcess[WithTime[S], A]. Our first task to to "unwrap" the state-reward probability transition function $\mathcal{P}_R$ of this MDP into a time-indexed sequenced of state-reward probability transition functions $(\mathcal{P}_R)_t, t = 0, 1, \ldots, T - 1$. This is accomplished by the following function unwrap_finite_horizon_MDP (itertools.groupby groups the augmented states by their time step, and the function without_time strips the time step from the augmented states when placing the states in $(\mathcal{P}_R)_t$, i.e., Sequence[StateActionMapping[S, A]]).

```
from itertools import groupby
ActionMapping = Mapping[A, StateReward[S]]
StateActionMapping = Mapping[NonTerminal[S], ActionMapping[A, S]]
def unwrap_finite_horizon_MDP(
    process: FiniteMarkovDecisionProcess[WithTime[S], A]
) -> Sequence[StateActionMapping[S, A]]:
    def time(x: WithTime[S]) -> int:
        return x.time

    def single_without_time(
        s_r: Tuple[State[WithTime[S]], float]
    ) -> Tuple[State[S], float]:
        if isinstance(s_r[0], NonTerminal):
            ret: Tuple[State[S], float] = (
                NonTerminal(s_r[0].state.state),
                s_r[1]
            )
        else:
            ret = (Terminal(s_r[0].state.state), s_r[1])
        return ret

    def without_time(arg: ActionMapping[A, WithTime[S]]) -> \
            ActionMapping[A, S]:
        return {a: sr_distr.map(single_without_time)
                for a, sr_distr in arg.items()}

    return [{NonTerminal(s.state): without_time(
        process.mapping[NonTerminal(s)]
    ) for s in states} for _, states in groupby(
        sorted(
```

```
            (nt.state for nt in process.non_terminal_states),
            key=time
        ),
        key=time
    )]
```

Now that we have the state-reward transition functions $(\mathcal{P}_R)_t$ arranged in the form of a Sequence[StateActionMapping[S, A]], we are ready to perform backward induction to calculate V_t^*. The following function optimal_vf_and_policy accomplishes it with a straightforward use of Equations (5.9) and (5.10), as described above.

```
from operator import itemgetter
def optimal_vf_and_policy(
        steps: Sequence[StateActionMapping[S, A]],
        gamma: float
) -> Iterator[Tuple[V[S], FiniteDeterministicPolicy[S, A]]]:
    v_p: List[Tuple[V[S], FiniteDeterministicPolicy[S, A]]] = []

    for step in reversed(steps):
        this_v: Dict[NonTerminal[S], float] = {}
        this_a: Dict[S, A] = {}
        for s, actions_map in step.items():
            action_values = ((res.expectation(
                lambda s_r: s_r[1] + gamma * (
                    extended_vf(v_p[-1][0], s_r[0]) if len(v_p) > 0 else 0.
                )
            ), a) for a, res in actions_map.items())
            v_star, a_star = max(action_values, key=itemgetter(0))
            this_v[s] = v_star
            this_a[s.state] = a_star
        v_p.append((this_v, FiniteDeterministicPolicy(this_a)))

    return reversed(v_p)
```

If $|\mathcal{N}_t|$ is $O(m)$ for all t and $|\mathcal{A}_t|$ is $O(k)$, then the running time of this algorithm is $O(m^2 \cdot k \cdot T)$.

Note that these algorithms for finite-horizon finite MDPs do not require any "iterations to convergence" like we had for regular Policy Evaluation and Value Iteration. Rather, in these algorithms we simply walk back in time and immediately obtain the Value Function for each time step from the next time step's Value Function (which is already known since we walk back in time). This technique of "backpropagation of Value Function" goes by the name of *Backward Induction* algorithms, and is quite commonplace in many Financial applications (as we shall see later in this book). The above Backward Induction code is in the file rl/finite_horizon.py.

5.14   DYNAMIC PRICING FOR END-OF-LIFE/END-OF-SEASON OF A PRODUCT

Now we consider a rather important business application—Dynamic Pricing. We consider the problem of Dynamic Pricing for the case of products that reach their end of life or at the end of a season after which we don't want to carry the product anymore. We need to adjust the prices up and down dynamically depending on how much inventory of the product you have, how many days remain for end-of-life/end-of-season, and your expectations of customer demand as a function of price adjustments. To make things concrete, assume you own a super-market and you are T days away from Halloween. You have just received M Halloween masks from your supplier and you won't be receiving any more inventory during these final T days. You want to dynamically set the selling price of the Halloween masks at the start of each day in a manner that maximizes your *Expected Total Sales Revenue* for

Halloween masks from today until Halloween (assume no one will buy Halloween masks after Halloween).

Assume that for each of the T days, at the start of the day, you are required to select a price for that day from one of N prices $P_1, P_2, \ldots, P_N \in \mathbb{R}$, such that your selected price will be the selling price for all masks on that day. Assume that the customer demand for the number of Halloween masks on any day is governed by a Poisson probability distribution with mean $\lambda_i \in \mathbb{R}$ if you select that day's price to be P_i (where i is a choice among $1, 2, \ldots, N$). Note that on any given day, the demand could exceed the number of Halloween masks you have in the store, in which case the number of masks sold on that day will be equal to the number of masks you had at the start of that day.

A state for this MDP is given by a pair (t, I_t) where $t \in \{0, 1, \ldots, T\}$ denotes the time index and $I_t \in \{0, 1, \ldots, M\}$ denotes the inventory at time t. Using our notation from the previous section, $\mathcal{S}_t = \{0, 1, \ldots, M\}$ for all $t = 0, 1, \ldots, T$ so that $I_t \in \mathcal{S}_t$. $\mathcal{N}_t = \mathcal{S}_t$ for all $t = 0, 1, \ldots, T-1$ and $\mathcal{N}_T = \emptyset$. The action choices at time t can be represented by the choice of integers from 1 to N. Therefore, $\mathcal{A}_t = \{1, 2, \ldots, N\}$.

Note that:

$$I_0 = M, I_{t+1} = \max(0, I_t - d_t) \text{ for } 0 \leq t < T$$

where d_t is the random demand on day t governed by a Poisson distribution with mean λ_i if the action (index of the price choice) on day t is $i \in \mathcal{A}_t$. Also, note that the sales revenue on day t is equal to $\min(I_t, d_t) \cdot P_i$. Therefore, the state-reward probability transition function for time index t

$$(\mathcal{P}_R)_t : \mathcal{N}_t \times \mathcal{A}_t \times \mathcal{D}_{t+1} \times \mathcal{S}_{t+1} \to [0, 1]$$

is defined as:

$$(\mathcal{P}_R)_t(I_t, i, r_{t+1}, I_t - k) = \begin{cases} \frac{e^{-\lambda_i}\lambda_i^k}{k!} & \text{if } k < I_t \text{ and } r_{t+1} = k \cdot P_i \\ \sum_{j=I_t}^{\infty} \frac{e^{-\lambda_i}\lambda_i^j}{j!} & \text{if } k = I_t \text{ and } r_{t+1} = k \cdot P_i \\ 0 & \text{otherwise} \end{cases}$$

for all $0 \leq t < T$

Using the definition of $(\mathcal{P}_R)_t$ and using the boundary condition $W_T^*(I_T) = 0$ for all $I_T \in \{0, 1, \ldots, M\}$, we can perform the backward induction algorithm to calculate V_t^* and associated optimal (deterministic) policy $(\pi_D^*)_t$ for all $0 \leq t < T$.

Now let's write some code to represent this Dynamic Programming problem as a `FiniteMarkovDecisionProcess` and determine its optimal policy, i.e., the Optimal (Dynamic) Price at time step t for any available level of inventory I_t. The type $\mathcal{N}_t$ is int and the type $\mathcal{A}_t$ is also int. So we create an MDP of type `FiniteMarkovDecisionProcess[WithTime[int]`, `int]` (since the augmented state space is `WithTime[int]`). Our first task is to construct $\mathcal{P}_R$ of type:

```
Mapping[WithTime[int],
  Mapping[int, FiniteDistribution[Tuple[WithTime[int], float]]]]
```

In the class `ClearancePricingMDP` below, $\mathcal{P}_R$ is manufactured in `__init__` and is used to create the attribute `mdp: FiniteMarkovDecisionProces[WithTime[int], int]`. Since $\mathcal{P}_R$ is independent of time, we first create a single-step (time-invariant) MDP `single_step_mdp: FiniteMarkovDecisionProcess[int, int]` (think of this as the building-block MDP), and then use the function `finite_horizon_MDP` (from file rl/finite_horizon.py) to create `self.mdp` from `self.single_step_mdp`. The constructor argument `initial_inventory: int` represents the initial inventory M. The constructor argument `time_steps` represents the number of time steps T. The constructor argument `price_lambda_pairs` represents $[(P_i, \lambda_i)|1 \leq i \leq N]$.

```python
from scipy.stats import poisson
from rl.finite_horizon import finite_horizon_MDP
class ClearancePricingMDP:

    initial_inventory: int
    time_steps: int
    price_lambda_pairs: Sequence[Tuple[float, float]]
    single_step_mdp: FiniteMarkovDecisionProcess[int, int]
    mdp: FiniteMarkovDecisionProcess[WithTime[int], int]

    def __init__(
        self,
        initial_inventory: int,
        time_steps: int,
        price_lambda_pairs: Sequence[Tuple[float, float]]
    ):
        self.initial_inventory = initial_inventory
        self.time_steps = time_steps
        self.price_lambda_pairs = price_lambda_pairs
        distrs = [poisson(l) for _, l in price_lambda_pairs]
        prices = [p for p, _ in price_lambda_pairs]
        self.single_step_mdp: FiniteMarkovDecisionProcess[int, int] =\
            FiniteMarkovDecisionProcess({
                s: {i: Categorical(
                    {(s - k, prices[i] * k):
                     (distrs[i].pmf(k) if k < s else 1 - distrs[i].cdf(s - 1))
                     for k in range(s + 1)})
                    for i in range(len(prices))}
                for s in range(initial_inventory + 1)
            })
        self.mdp = finite_horizon_MDP(self.single_step_mdp, time_steps)
```

Now let's write two methods for this class:

- get_vf_for_policy that produces the Value Function for a given policy π, by first creating the π-implied MRP from mdp, then unwrapping the MRP into a sequence of state-reward transition probability functions $(\mathcal{P}_R^{\pi_t})_t$, and then performing backward induction using the previously-written function evaluate to calculate the Value Function.
- get_optimal_vf_and_policy that produces the Optimal Value Function and Optimal Policy, by first unwrapping self.mdp into a sequence of state-reward transition probability functions $(\mathcal{P}_R)_t$, and then performing backward induction using the previously-written function optimal_vf_and_policy to calculate the Optimal Value Function and Optimal Policy.

```python
from rl.finite_horizon import evaluate, optimal_vf_and_policy
    def get_vf_for_policy(
        self,
        policy: FinitePolicy[WithTime[int], int]
    ) -> Iterator[V[int]]:
        mrp: FiniteMarkovRewardProcess[WithTime[int]] \
            = self.mdp.apply_finite_policy(policy)
        return evaluate(unwrap_finite_horizon_MRP(mrp), 1.)

    def get_optimal_vf_and_policy(self)\
            -> Iterator[Tuple[V[int], FiniteDeterministicPolicy[int, int]]]:
        return optimal_vf_and_policy(unwrap_finite_horizon_MDP(self.mdp), 1.)
```

Now let's create a simple instance of ClearancePricingMDP for $M = 12, T = 8$ and 4 price choices: "Full Price", "30% Off", "50% Off", "70% Off" with respective mean daily demand of $0.5, 1.0, 1.5, 2.5$.

```
ii = 12
steps = 8
pairs = [(1.0, 0.5), (0.7, 1.0), (0.5, 1.5), (0.3, 2.5)]
cp: ClearancePricingMDP = ClearancePricingMDP(
    initial_inventory=ii,
    time_steps=steps,
    price_lambda_pairs=pairs
)
```

Now let us calculate its Value Function for a stationary policy that chooses "Full Price" if inventory is less than 2, otherwise "30% Off" if inventory is less than 5, otherwise "50% Off" if inventory is less than 8, otherwise "70% Off". Since we have a stationary policy, we can represent it as a single-step policy and combine it with the single-step MDP we had created above (attribute single_step_mdp) to create a single_step_mrp: FiniteMarkovRewardProcess[int]. Then we use the function finite_horizon_mrp (from file rl/finite_horizon.py) to create the entire (augmented state) MRP of type FiniteMarkovRewardProcess[WithTime[int]]. Finally, we unwrap this MRP into a sequence of state-reward transition probability functions and perform backward induction to calculate the Value Function for this stationary policy. Running the following code tells us that $V_0^{\pi}(12)$ is about 4.91 (assuming full price is 1), which is the Expected Revenue one would obtain over 8 days, starting with an inventory of 12, and executing this stationary policy (under the assumed demand distributions as a function of the price choices).

```
def policy_func(x: int) -> int:
    return 0 if x < 2 else (1 if x < 5 else (2 if x < 8 else 3))

stationary_policy: FiniteDeterministicPolicy[int, int] = \
    FiniteDeterministicPolicy({s: policy_func(s) for s in range(ii + 1)})

single_step_mrp: FiniteMarkovRewardProcess[int] = \
    cp.single_step_mdp.apply_finite_policy(stationary_policy)

vf_for_policy: Iterator[V[int]] = evaluate(
    unwrap_finite_horizon_MRP(finite_horizon_MRP(single_step_mrp, steps)),
    1.
)
```

Now let us determine what is the Optimal Policy and Optimal Value Function for this instance of ClearancePricingMDP. Running cp.get_optimal_vf_and_policy() and evaluating the Optimal Value Function for time step 0 and inventory of 12, i.e. $V_0^*(12)$, gives us a value of 5.64, which is the Expected Revenue we'd obtain over the 8 days if we executed the Optimal Policy.

Now let us plot the Optimal Price as a function of time steps and inventory levels.

```
import matplotlib.pyplot as plt
from matplotlib import cm
import numpy as np

prices = [[pairs[policy.act(s).value][0] for s in range(ii + 1)]
          for _, policy in cp.get_optimal_vf_and_policy()]

heatmap = plt.imshow(np.array(prices).T, origin='lower')
plt.colorbar(heatmap, shrink=0.5, aspect=5)
plt.xlabel("Time Steps")
plt.ylabel("Inventory")
plt.show()
```

Figure 5.5 shows us the image produced by the above code. The light shade is "Full Price", the medium shade is "30% Off" and the dark shade is "50% Off". This tells us that on day 0, the Optimal Price is "30% Off" (corresponding to State 12, i.e., for starting inventory $M = I_0 = 12$). However, if the starting inventory I_0 were less than 7, then the

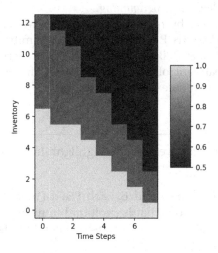

Figure 5.5 Optimal Policy Heatmap

Optimal Price is "Full Price". This makes intuitive sense because the lower the inventory, the less inclination we'd have to cut prices. We see that the thresholds for price cuts shift as time progresses (as we move horizontally in the figure). For instance, on Day 5, we set "Full Price" only if inventory has dropped below 3 (this would happen if we had a good degree of sales on the first 5 days), we set "30% Off" if inventory is 3 or 4 or 5, and we set "50% Off" if inventory is greater than 5. So even if we sold 6 units in the first 5 days, we'd offer "50% Off" because we have only 3 days remaining now and 6 units of inventory left. This makes intuitive sense. We see that the thresholds shift even further as we move to Days 6 and 7. We encourage you to play with this simple application of Dynamic Pricing by changing $M, T, N, [(P_i, \lambda_i)|1 \le i \le N]$ and studying how the Optimal Value Function changes and more importantly, studying the thresholds of inventory (under optimality) for various choices of prices and how these thresholds vary as time progresses.

5.15   GENERALIZATION TO NON-TABULAR ALGORITHMS

The Finite MDP algorithms covered in this chapter are called "tabular" algorithms. The word "tabular" (for "table") refers to the fact that the MDP is specified in the form of a finite data structure and the Value Function is also represented as a finite "table" of non-terminal states and values. These tabular algorithms typically make a sweep through all non-terminal states in each iteration to update the Value Function. This is not possible for large state spaces or infinite state spaces where we need some function approximation for the Value Function. The good news is that we can modify each of these tabular algorithms such that instead of sweeping through all the non-terminal states at each step, we simply sample an appropriate subset of non-terminal states, calculate the values for these sampled states with the appropriate Bellman calculations (just like in the tabular algorithms), and then create/update a function approximation (for the Value Function) with the sampled states' calculated values. The important point is that the fundamental structure of the algorithms and the fundamental principles (Fixed-Point and Bellman Operators) are still the same when we generalize from these tabular algorithms to function approximation-based algorithms. In Chapter 6, we cover generalization of these Dynamic Programming algorithms from tabular methods to function approximation methods. We call these algorithms *Approximate Dynamic Programming*.

We finish this chapter by referring you to the various excellent papers and books by Dimitri Bertsekas—(Dimitri P. Bertsekas 1981), (Dimitri P. Bertsekas 1983), (Dimitri P. Bertsekas 2005), (Dimitri P. Bertsekas 2012), (D. P. Bertsekas and Tsitsiklis 1996)—for a comprehensive treatment of the variants of DP, including Asynchronous DP, Finite-Horizon DP and Approximate DP.

5.16 SUMMARY OF KEY LEARNINGS FROM THIS CHAPTER

Before we end this chapter, we'd like to highlight the three highly important concepts we learnt in this chapter:

- Fixed-Point of Functions and Banach Fixed-Point Theorem: The simple concept of Fixed-Point of Functions that is profound in its applications, and the Banach Fixed-Point Theorem that enables us to construct iterative algorithms to solve problems with fixed-point formulations.
- Generalized Policy Iteration: The powerful idea of alternating between *any* method for Policy Evaluation and *any* method for Policy Improvement, including methods that are partial applications of Policy Evaluation or Policy Improvement. This generalized perspective unifies almost all of the algorithms that solve MDP Control problems.
- Backward Induction: A straightforward method to solve finite-horizon MDPs by simply backpropagating the Value Function from the horizon-end to the start.

CHAPTER 6

Function Approximation and Approximate Dynamic Programming

In Chapter 5, we covered Dynamic Programming algorithms where the MDP is specified in the form of a finite data structure and the Value Function is represented as a finite "table" of states and values. These Dynamic Programming algorithms swept through all states in each iteration to update the value function. But when the state space is large (as is the case in real-world applications), these Dynamic Programming algorithms won't work because:

1. A "tabular" representation of the MDP (or of the Value Function) won't fit within storage limits.
2. Sweeping through all states and their transition probabilities would be time-prohibitive (or simply impossible, in the case of infinite state spaces).

Hence, when the state space is very large, we need to resort to approximation of the Value Function. The Dynamic Programming algorithms would need to be suitably modified to their Approximate Dynamic Programming (abbreviated as ADP) versions. The good news is that it's not hard to modify each of the (tabular) Dynamic Programming algorithms such that instead of sweeping through all the states in each iteration, we simply sample an appropriate subset of the states, calculate the values for those states (with the same Bellman Operator calculations as for the case of tabular), and then create/update a function approximation (for the Value Function) using the sampled states' calculated values. Furthermore, if the set of transitions from a given state is large (or infinite), instead of using the explicit probabilities of those transitions, we can sample from the transitions probability distribution. The fundamental structure of the algorithms and the fundamental principles (Fixed-Point and Bellman Operators) would still be the same.

So, in this chapter, we do a quick review of function approximation, write some code for a couple of standard function approximation methods, and then utilize these function approximation methods to develop Approximate Dynamic Programming algorithms (in particular, Approximate Policy Evaluation, Approximate Value Iteration and Approximate Backward Induction). Since you are reading this book, it's highly likely that you are already familiar with the simple and standard function approximation methods such as linear function approximation and function approximation using neural networks supervised learning. So we shall go through the background on linear function approximation and neural networks supervised learning in a quick and terse manner, with the goal of developing some code for these methods that we can use not just for the ADP algorithms for this chapter, but also for RL algorithms later in the book. Note also that apart from approximation of State-Value Functions $\mathcal{N} \to \mathbb{R}$ and Action-Value Functions $\mathcal{N} \times \mathcal{A} \to \mathbb{R}$, these function approximation methods can also be used for approximation of Stochastic Policies $\mathcal{N} \times \mathcal{A} \to [0, 1]$ in Policy-based RL algorithms.

6.1   FUNCTION APPROXIMATION

In this section, we describe function approximation in a fairly generic setting (not specific to approximation of Value Functions or Policies). We denote the predictor variable as x, belonging to an arbitrary domain denoted $\mathcal{X}$ and the response variable as $y \in \mathbb{R}$. We treat x and y as unknown random variables and our goal is to estimate the probability distribution function f of the conditional random variable $y|x$ from data provided in the form of a sequence of (x, y) pairs. We shall consider parameterized functions f with the parameters denoted as w. The exact data type of w will depend on the specific form of function approximation. We denote the estimated probability of y conditional on x as $f(x; w)(y)$. Assume we are given data in the form of a sequence of n (x, y) pairs, as follows:

$$[(x_i, y_i)|1 \leq i \leq n]$$

The notion of estimating the conditional probability $\mathbb{P}[y|x]$ is formalized by solving for $w = w^*$ such that:

$$w^* = \arg\max_{w}\{\prod_{i=1}^{n} f(x_i; w)(y_i)\} = \arg\max_{w}\{\sum_{i=1}^{n} \log f(x_i; w)(y_i)\}$$

In other words, we shall be operating in the framework of Maximum Likelihood Estimation. We say that the data $[(x_i, y_i)|1 \leq i \leq n]$ specifies the *empirical probability distribution D* of $y|x$ and the function f (parameterized by w) specifies the *model probability distribution M* of $y|x$. With maximum likelihood estimation, we are essentially trying to reconcile the model probability distribution M with the empirical probability distribution D. Hence, maximum likelihood estimation is essentially minimization of a loss function defined as the cross-entropy $\mathcal{H}(D, M) = -\mathbb{E}_D[\log M]$ between the probability distributions D and M. The term *function approximation* refers to the fact that the model probability distribution M serves as an approximation to some true probability distribution that is only partially observed through the empirical probability distribution D.

Our framework will allow for incremental estimation wherein at each iteration t of the incremental estimation (for $t = 1, 2, \ldots$), data of the form

$$[(x_{t,i}, y_{t,i})|1 \leq i \leq n_t]$$

is used to update the parameters from w_{t-1} to w_t (parameters initialized at iteration $t = 0$ to w_0). This framework can be used to update the parameters incrementally with a gradient descent algorithm, either stochastic gradient descent (where a single (x, y) pair is used for each iteration's gradient calculation) or mini-batch gradient descent (where an appropriate subset of the available data is used for each iteration's gradient calculation) or simply re-using the entire data available for each iteration's gradient calculation (and consequent, parameters update). Moreover, the flexibility of our framework, allowing for incremental estimation, is particularly important for Reinforcement Learning algorithms wherein we update the parameters of the function approximation from the new data that is generated from each state transition as a result of interaction with either the real environment or a simulated environment.

Among other things, the estimate f (parameterized by w) gives us the model-expected value of y conditional on x, i.e.

$$\mathbb{E}_M[y|x] = \mathbb{E}_{f(x;w)}[y] = \int_{-\infty}^{+\infty} y \cdot f(x;w)(y) \cdot dy$$

We refer to $\mathbb{E}_M[y|x]$ as the function approximation's prediction for a given predictor variable x.

For the purposes of Approximate Dynamic Programming and Reinforcement Learning, the function approximation's prediction $\mathbb{E}[y|x]$ provides an estimate of the Value Function for any state (x takes the role of the *State*, and y takes the role of the *Return* following that State). In the case of function approximation for (stochastic) policies, x takes the role of the *State*, y takes the role of the *Action* for that policy, and $f(x;w)$ provides the probability distribution of actions for state x (corresponding to that policy). It's also worthwhile pointing out that the broader theory of function approximations covers the case of multi-dimensional y (where y is a real-valued vector, rather than scalar)—this allows us to solve classification problems as well as regression problems. However, for ease of exposition and for sufficient coverage of function approximation applications in this book, we only cover the case of scalar y.

Now let us write some code that captures this framework. We write an abstract base class `FunctionApprox` type-parameterized by `X` (to permit arbitrary data types $\mathcal{X}$), representing $f(x;w)$, with the following 3 key methods, each of which will work with inputs of generic `Iterable` type (`Iterable` is any data type that we can iterate over, such as `Sequence` types or `Iterator` type):

1. `@abstractmethod solve`: takes as input an `Iterable` of (x, y) pairs and solves for the optimal internal parameters w^* that minimizes the cross-entropy between the empirical probability distribution of the input data of (x, y) pairs and the model probability distribution $f(x;w)$. Some implementations of `solve` are iterative numerical methods and would require an additional input of `error_tolerance` that specifies the required precision of the best-fit parameters w^*. When an implementation of `solve` is an analytical solution not requiring an error tolerance, we specify the input `error_tolerance` as `None`. The output of `solve` is the `FunctionApprox` $f(x;w^*)$ (i.e., corresponding to the solved parameters w^*).

2. `update`: takes as input an `Iterable` of (x, y) pairs and updates the parameters w defining $f(x;w)$. The purpose of `update` is to perform an incremental (iterative) improvement to the parameters w, given the input data of (x, y) pairs in the current iteration. The output of `update` is the `FunctionApprox` corresponding to the updated parameters. Note that we should be able to `solve` based on an appropriate series of incremental updates (upto a specified `error_tolerance`).

3. `@abstractmethod evaluate`: takes as input an `Iterable` of x values and calculates $\mathbb{E}_M[y|x] = \mathbb{E}_{f(x;w)}[y]$ for each of the input x values, and outputs these expected values in the form of a `numpy.ndarray`.

As we've explained, an incremental update to the parameters w involves calculating a gradient and then using the gradient to adjust the parameters w. Hence the method `update` is supported by the following two abstract methods.

- `@abstractmethod objective_gradient`: computes the gradient of an objective function (call it $Obj(x, y)$) of the `FunctionApprox` with respect to the parameters w in the internal representation of the `FunctionApprox`. The gradient is output in the form of a `Gradient` type. The second argument `obj_deriv_out_fun` of the `objective_gradient` method represents the partial derivative of Obj with respect to an appropriate model-computed value (call it $Out(x)$), i.e., $\frac{\partial Obj(x,y)}{\partial Out(x)}$, when evaluated at a Sequence of x values and a Sequence of y values (to be obtained from the first argument xy_vals_seq of the objective_gradient method).

- @abstractmethod update_with_gradient: takes as input a Gradient and updates the internal parameters using the gradient values (e.g., gradient descent update to the parameters), returning the updated FunctionApprox.

The update method is written with $\frac{\partial Obj(x_i,y_i)}{\partial Out(x_i)}$ defined as follows, for each training data point (x_i, y_i):

$$\frac{\partial Obj(x_i, y_i)}{\partial Out(x_i)} = \mathbb{E}_M[y|x_i] - y_i$$

It turns out that for each concrete function approximation that we'd want to implement, if the Objective $Obj(x_i, y_i)$ is the cross-entropy loss function, we can identify a model-computed value $Out(x_i)$ (either the output of the model or an intermediate computation of the model) such that $\frac{\partial Obj(x_i,y_i)}{\partial Out(x_i)}$ is equal to the prediction error $\mathbb{E}_M[y|x_i] - y_i$ (for each training data point (x_i, y_i)), and we can come up with a numerical algorithm to compute $\nabla_w Out(x_i)$, so that by chain-rule, we have the required gradient:

$$\nabla_w Obj(x_i, y_i) = \frac{\partial Obj(x_i, y_i)}{\partial Out(x_i)} \cdot \nabla_w Out(x_i) = (\mathbb{E}_M[y|x_i] - y_i) \cdot \nabla_w Out(x_i)$$

The update method implements this chain-rule calculation, by setting obj_deriv_out_fun to be the prediction error $\mathbb{E}_M[y|x_i] - y_i$, delegating the calculation of $\nabla_w Out(x_i)$ to the concrete implementation of the abstract method objective_gradient

Note that the Gradient class contains a single attribute of type FunctionApprox so that a Gradient object can represent the gradient values in the form of the internal parameters of the FunctionApprox attribute (since each gradient value is simply a partial derivative with respect to an internal parameter).

We use the TypeVar F to refer to a concrete class that would implement the abstract interface of FunctionApprox.

```python
from abc import ABC, abstractmethod
import numpy as np

X = TypeVar('X')
F = TypeVar('F', bound='FunctionApprox')

class FunctionApprox(ABC, Generic[X]):

    @abstractmethod
    def objective_gradient(
        self: F,
        xy_vals_seq: Iterable[Tuple[X, float]],
        obj_deriv_out_fun: Callable[[Sequence[X], Sequence[float]], np.ndarray]
    ) -> Gradient[F]:
        pass

    @abstractmethod
    def evaluate(self, x_values_seq: Iterable[X]) -> np.ndarray:
        pass

    @abstractmethod
    def update_with_gradient(
        self: F,
        gradient: Gradient[F]
    ) -> F:
        pass
```

```
    def update(
        self: F,
        xy_vals_seq: Iterable[Tuple[X, float]]
    ) -> F:
        def deriv_func(x: Sequence[X], y: Sequence[float]) -> np.ndarray:
            return self.evaluate(x) - np.array(y)

        return self.update_with_gradient(
            self.objective_gradient(xy_vals_seq, deriv_func)
        )
    @abstractmethod
    def solve(
        self: F,
        xy_vals_seq: Iterable[Tuple[X, float]],
        error_tolerance: Optional[float] = None
    ) -> F:
        pass

@dataclass(frozen=True)
class Gradient(Generic[F]):
    function_approx: F
```

When concrete classes implementing FunctionApprox write the solve method in terms of the update method, they will need to check if a newly updated FunctionApprox is "close enough" to the previous FunctionApprox. So each of them will need to implement their own version of "Are two FunctionApprox instances within a certain error_tolerance of each other?". Hence, we need the following abstract method within:

```
    @abstractmethod
    def within(self: F, other: F, tolerance: float) -> bool:
        pass
```

Any concrete class that implement this abstract class FunctionApprox will need to implement these five abstract methods of FunctionApprox, based on the specific assumptions that the concrete class makes for f.

Next, we write some methods useful for classes that inherit from FunctionApprox. Firstly, we write a method called iterate_updates that takes as input a stream (Iterator) of Iterable of (x, y) pairs, and performs a series of incremental updates to the parameters w (each using the update method), with each update done for each Iterable of (x, y) pairs in the input stream xy_seq: Iterator[Iterable[Tuple[X, float]]]. iterate_updates returns an Iterator of FunctionApprox representing the successively updated FunctionApprox instances as a consequence of the repeated invocations to update. Note the use of the rl.iterate.accumulate function (a wrapped version of itertools.accumulate) that calculates accumulated results (including intermediate results) on an Iterable, based on a provided function to govern the accumulation. In the code below, the Iterable is the input stream xy_seq_stream and the function governing the accumulation is the update method of FunctionApprox.

```
import rl.iterate as iterate
    def iterate_updates(
        self: F,
        xy_seq_stream: Iterator[Iterable[Tuple[X, float]]]
    ) -> Iterator[F]:
        return iterate.accumulate(
            xy_seq_stream,
            lambda fa, xy: fa.update(xy),
            initial=self
        )
```

Next, we write a method called `rmse` to calculate the Root-Mean-Squared-Error of the predictions for x (using `evaluate`) relative to associated (supervisory) y, given as input an `Iterable` of (x, y) pairs. This method will be useful in testing the goodness of a `FunctionApprox` estimate.

```
def rmse(
    self,
    xy_vals_seq: Iterable[Tuple[X, float]]
) -> float:
    x_seq, y_seq = zip(*xy_vals_seq)
    errors: np.ndarray = self.evaluate(x_seq) - np.array(y_seq)
    return np.sqrt(np.mean(errors * errors))
```

Finally, we write a method `argmax` that takes as input an `Iterable` of x values and returns the x value that maximizes $\mathbb{E}_{f(x;w)}[y]$.

```
def argmax(self, xs: Iterable[X]) -> X:
    return list(xs)[np.argmax(self.evaluate(xs))]
```

The above code for `FunctionApprox` and `Gradient` is in the file rl/function_approx.py. rl/function_approx.py also contains the convenience methods `__add__` (to add two `FunctionApprox`), `__mul__` (to multiply a `FunctionApprox` with a real-valued scalar), and `__call__` (to treat a `FunctionApprox` object syntactically as a function taking an `x: X` as input, essentially a shorthand for `evaluate` on a single x value). `__add__` and `__mul__` are meant to perform element-wise addition and scalar-multiplication on the internal parameters w of the Function Approximation (see Appendix F on viewing Function Approximations as Vector Spaces). Likewise, it contains the methods `__add__` and `__mul__` for the `Gradient` class that simply delegates to the `__add__` and `__mul__` methods of the `FunctionApprox` within `Gradient`, and it also contains the method `zero` that returns a `Gradient` which is uniformly zero for each of the parameter values.

Now we are ready to cover a concrete but simple function approximation—the case of linear function approximation.

6.2 LINEAR FUNCTION APPROXIMATION

We define a sequence of feature functions

$$\phi_j : \mathcal{X} \to \mathbb{R} \text{ for each } j = 1, 2, \ldots, m$$

and we define $\phi : \mathcal{X} \to \mathbb{R}^m$ as:

$$\phi(x) = (\phi_1(x), \phi_2(x), \ldots, \phi_m(x)) \text{ for all } x \in \mathcal{X}$$

We treat $\phi(x)$ as a column vector for all $x \in \mathcal{X}$.

For linear function approximation, the internal parameters w are represented as a weights column-vector $\boldsymbol{w} = (w_1, w_2, \ldots, w_m) \in \mathbb{R}^m$. Linear function approximation is based on the assumption of a Gaussian distribution for $y|x$ with mean

$$\mathbb{E}_M[y|x] = \sum_{j=1}^{m} \phi_j(x) \cdot w_j = \phi(x)^T \cdot \boldsymbol{w}$$

and constant variance σ^2, i.e.,

$$\mathbb{P}[y|x] = f(x; \boldsymbol{w})(y) = \frac{1}{\sqrt{2\pi\sigma^2}} \cdot e^{-\frac{(y - \phi(x)^T \cdot \boldsymbol{w})^2}{2\sigma^2}}$$

So, the cross-entropy loss function (ignoring constant terms associated with σ^2) for a given set of data points $[x_i, y_i | 1 \le i \le n]$ is defined as:

$$\mathcal{L}(\boldsymbol{w}) = \frac{1}{2n} \cdot \sum_{i=1}^{n} (\boldsymbol{\phi}(x_i)^T \cdot \boldsymbol{w} - y_i)^2$$

Note that this loss function is identical to the mean-squared-error of the linear (in $\boldsymbol{w}$) predictions $\boldsymbol{\phi}(x_i)^T \cdot \boldsymbol{w}$ relative to the response values y_i associated with the predictor values x_i, over all $1 \le i \le n$.

If we include L^2 regularization (with λ as the regularization coefficient), then the regularized loss function is:

$$\mathcal{L}(\boldsymbol{w}) = \frac{1}{2n} (\sum_{i=1}^{n} (\boldsymbol{\phi}(x_i)^T \cdot \boldsymbol{w} - y_i)^2) + \frac{1}{2} \cdot \lambda \cdot |\boldsymbol{w}|^2$$

The gradient of $\mathcal{L}(\boldsymbol{w})$ with respect to $\boldsymbol{w}$ works out to:

$$\nabla_{\boldsymbol{w}} \mathcal{L}(\boldsymbol{w}) = \frac{1}{n} \cdot (\sum_{i=1}^{n} \boldsymbol{\phi}(x_i) \cdot (\boldsymbol{\phi}(x_i)^T \cdot \boldsymbol{w} - y_i)) + \lambda \cdot \boldsymbol{w}$$

We had said previously that for each concrete function approximation that we'd want to implement, if the Objective $Obj(x_i, y_i)$ is the cross-entropy loss function, we can identify a model-computed value $Out(x_i)$ (either the output of the model or an intermediate computation of the model) such that $\frac{\partial Obj(x_i, y_i)}{\partial Out(x_i)}$ is equal to the prediction error $\mathbb{E}_M[y|x_i] - y_i$ (for each training data point (x_i, y_i)), and we can come up with a numerical algorithm to compute $\nabla_{\boldsymbol{w}} Out(x_i)$, so that by chain-rule, we have the required gradient $\nabla_{\boldsymbol{w}} Obj(x_i, y_i)$ (without regularization). In the case of this linear function approximation, the model-computed value $Out(x_i)$ is simply the model prediction for predictor variable x_i, i.e.,

$$Out(x_i) = \mathbb{E}_M[y|x_i] = \boldsymbol{\phi}(x_i)^T \cdot \boldsymbol{w}$$

This is confirmed by noting that with $Obj(x_i, y_i)$ set to be the cross-entropy loss function $\mathcal{L}(\boldsymbol{w})$ and $Out(x_i)$ set to be the model prediction $\boldsymbol{\phi}(x_i)^T \cdot \boldsymbol{w}$ (for training data point (x_i, y_i)),

$$\frac{\partial Obj(x_i, y_i)}{\partial Out(x_i)} = \boldsymbol{\phi}(x_i)^T \cdot \boldsymbol{w} - y_i = \mathbb{E}_M[y|x_i] - y_i$$

$$\nabla_{\boldsymbol{w}} Out(x_i) = \nabla_{\boldsymbol{w}} (\boldsymbol{\phi}(x_i)^T \cdot \boldsymbol{w}) = \boldsymbol{\phi}(x_i)$$

We can solve for $\boldsymbol{w}^*$ by incremental estimation using gradient descent (change in $\boldsymbol{w}$ proportional to the gradient estimate of $\mathcal{L}(\boldsymbol{w})$ with respect to $\boldsymbol{w}$). If the (x_t, y_t) data at time t is:

$$[(x_{t,i}, y_{t,i}) | 1 \le i \le n_t],$$

then the gradient estimate $\mathcal{G}_{(x_t, y_t)}(\boldsymbol{w}_t)$ at time t is given by:

$$\mathcal{G}_{(x_t, y_t)}(\boldsymbol{w}_t) = \frac{1}{n} \cdot (\sum_{i=1}^{n_t} \boldsymbol{\phi}(x_{t,i}) \cdot (\boldsymbol{\phi}(x_{t,i})^T \cdot \boldsymbol{w}_t - y_{t,i})) + \lambda \cdot \boldsymbol{w}_t$$

which can be interpreted as the mean (over the data in iteration t) of the feature vectors $\boldsymbol{\phi}(x_{t,i})$ weighted by the (scalar) linear prediction errors $\boldsymbol{\phi}(x_{t,i})^T \cdot \boldsymbol{w}_t - y_{t,i}$ (plus regularization term $\lambda \cdot \boldsymbol{w}_t$).

Then, the update to the weights vector w is given by:

$$w_{t+1} = w_t - \alpha_t \cdot \mathcal{G}_{(x_t, y_t)}(w_t)$$

where α_t is the learning rate for the gradient descent at time t. To facilitate numerical convergence, we require α_t to be an appropriate function of time t. There are a number of numerical algorithms to achieve the appropriate time-trajectory of α_t. We shall go with one such numerical algorithm—ADAM (Kingma and Ba 2015), which we shall use not just for linear function approximation but later also for the deep neural network function approximation. Before we write code for linear function approximation, we need to write some helper code to implement the ADAM gradient descent algorithm.

We create an `@dataclass` `Weights` to represent and update the weights (i.e., internal parameters) of a function approximation. The `Weights` dataclass has 5 attributes: `adam_gradient` that captures the ADAM parameters, including the base learning rate and the decay parameters, `time` that represents how many times the weights have been updated, `weights` that represents the weight parameters of the function approximation as a numpy array (1-D array for linear function approximation and 2-D array for each layer of deep neural network function approximation), and the two ADAM cache parameters. The `@staticmethod` `create` serves as a factory method to create a new instance of the `Weights` dataclass. The `update` method of this `Weights` dataclass produces an updated instance of the `Weights` dataclass that represents the updated weight parameters together with the incremented `time` and the updated ADAM cache parameters. We will follow a programming design pattern wherein we don't update anything in-place—rather, we create a new object with updated values (using the `dataclasses.replace` function). This ensures we don't get unexpected/undesirable updates in-place, which are typically the cause of bugs in numerical code. Finally, we write the `within` method which will be required to implement the `within` method in the linear function approximation class as well as in the deep neural network function approximation class.

```
SMALL_NUM = 1e-6
from dataclasses import replace

@dataclass(frozen=True)
class AdamGradient:
    learning_rate: float
    decay1: float
    decay2: float

    @staticmethod
    def default_settings() -> AdamGradient:
        return AdamGradient(
            learning_rate=0.001,
            decay1=0.9,
            decay2=0.999
        )

@dataclass(frozen=True)
class Weights:
    adam_gradient: AdamGradient
    time: int
    weights: np.ndarray
    adam_cache1: np.ndarray
    adam_cache2: np.ndarray

    @staticmethod
    def create(
        adam_gradient: AdamGradient = AdamGradient.default_settings(),
        weights: np.ndarray,
        adam_cache1: Optional[np.ndarray] = None,
        adam_cache2: Optional[np.ndarray] = None
    ) -> Weights:
```

```
        return Weights(
            adam_gradient=adam_gradient,
            time=0,
            weights=weights,
            adam_cache1=np.zeros_like(
                weights
            ) if adam_cache1 is None else adam_cache1,
            adam_cache2=np.zeros_like(
                weights
            ) if adam_cache2 is None else adam_cache2
        )

    def update(self, gradient: np.ndarray) -> Weights:
        time: int = self.time + 1
        new_adam_cache1: np.ndarray = self.adam_gradient.decay1 * \
            self.adam_cache1 + (1 - self.adam_gradient.decay1) * gradient
        new_adam_cache2: np.ndarray = self.adam_gradient.decay2 * \
            self.adam_cache2 + (1 - self.adam_gradient.decay2) * gradient ** 2
        corrected_m: np.ndarray = new_adam_cache1 / \
            (1 - self.adam_gradient.decay1 ** time)
        corrected_v: np.ndarray = new_adam_cache2 / \
            (1 - self.adam_gradient.decay2 ** time)

        new_weights: np.ndarray = self.weights - \
            self.adam_gradient.learning_rate * corrected_m / \
            (np.sqrt(corrected_v) + SMALL_NUM)

        return replace(
            self,
            time=time,
            weights=new_weights,
            adam_cache1=new_adam_cache1,
            adam_cache2=new_adam_cache2,
        )

    def within(self, other: Weights[X], tolerance: float) -> bool:
        return np.all(np.abs(self.weights - other.weights) <= tolerance).item()
```

With this `Weights` class, we are ready to write the dataclass `LinearFunctionApprox` for linear function approximation, inheriting from the abstract base class `FunctionApprox`. It has attributes `feature_functions` that represents $\phi_j : \mathcal{X} \to \mathbb{R}$ for all $j = 1, 2, \ldots, m$, `regularization_coeff` that represents the regularization coefficient λ, `weights` which is an instance of the `Weights` class we wrote above, and `direct_solve` (which we will explain shortly). The static method `create` serves as a factory method to create a new instance of `LinearFunctionApprox`. The method `get_feature_values` takes as input an `x_values_seq`: `Iterable[X]` (representing a sequence or stream of $x \in \mathcal{X}$), and produces as output the corresponding feature vectors $\phi(x) \in \mathbb{R}^m$ for each x in the input. The feature vectors are output in the form of a 2-D numpy array, with each feature vector $\phi(x)$ (for each x in the input sequence) appearing as a row in the output 2-D numpy array (the number of rows in this numpy array is the length of the input `x_values_seq` and the number of columns is the number of feature functions). Note that often we want to include a bias term in our linear function approximation, in which case we need to prepend the sequence of feature functions we provide as input with an artificial feature function `lambda _: 1.` to represent the constant feature with value 1. This will ensure we have a bias weight in addition to each of the weights that serve as coefficients to the (non-artificial) feature functions.

The method `evaluate` (an abstract method in `FunctionApprox`) calculates the prediction $\mathbb{E}_M[y|x]$ for each input x as: $\phi(x)^T \cdot w = \sum_{j=1}^{m} \phi_j(x) \cdot w_i$. The method `objective_gradient` (from `FunctionApprox`) performs the calculation $\mathcal{G}_{(x_t,y_t)}(w_t)$ shown above: the mean of the feature vectors $\phi(x_{t,i})$ weighted by the (scalar) linear prediction errors $\phi(x_{t,i})^T \cdot w_t - y_{t,i}$ (plus regularization term $\lambda \cdot w_t$). The variable `obj_deriv_out` takes the role of the linear prediction errors, when `objective_gradient` is invoked by the `update` method through the method `update_with_gradient`. The method `update_with_gradient` (from `FunctionApprox`) updates the weights using the calculated gradient along with the ADAM cache updates (invoking the update method of the Weights class to ensure there are no in-place updates), and returns a new LinearFunctionApprox object containing the updated weights.

```python
from dataclasses import replace
@dataclass(frozen=True)
class LinearFunctionApprox(FunctionApprox[X]):

    feature_functions: Sequence[Callable[[X], float]]
    regularization_coeff: float
    weights: Weights
    direct_solve: bool

    @staticmethod
    def create(
        feature_functions: Sequence[Callable[[X], float]],
        adam_gradient: AdamGradient = AdamGradient.default_settings(),
        regularization_coeff: float = 0.,
        weights: Optional[Weights] = None,
        direct_solve: bool = True
    ) -> LinearFunctionApprox[X]:
        return LinearFunctionApprox(
            feature_functions=feature_functions,
            regularization_coeff=regularization_coeff,
            weights=Weights.create(
                adam_gradient=adam_gradient,
                weights=np.zeros(len(feature_functions))
            ) if weights is None else weights,
            direct_solve=direct_solve
        )

    def get_feature_values(self, x_values_seq: Iterable[X]) -> np.ndarray:
        return np.array(
            [[f(x) for f in self.feature_functions] for x in x_values_seq]
        )

    def objective_gradient(
        self,
        xy_vals_seq: Iterable[Tuple[X, float]],
        obj_deriv_out_fun: Callable[[Sequence[X], Sequence[float]], float]
    ) -> Gradient[LinearFunctionApprox[X]]:
        x_vals, y_vals = zip(*xy_vals_seq)
        obj_deriv_out: np.ndarray = obj_deriv_out_fun(x_vals, y_vals)
        features: np.ndarray = self.get_feature_values(x_vals)
        gradient: np.ndarray = \
            features.T.dot(obj_deriv_out) / len(obj_deriv_out) \
            + self.regularization_coeff * self.weights.weights
        return Gradient(replace(
            self,
            weights=replace(
                self.weights,
                weights=gradient
            )
        ))

    def evaluate(self, x_values_seq: Iterable[X]) -> np.ndarray:
        return np.dot(
            self.get_feature_values(x_values_seq),
            self.weights.weights
        )

    def update_with_gradient(
        self,
        gradient: Gradient[LinearFunctionApprox[X]]
    ) -> LinearFunctionApprox[X]:
        return replace(
            self,
            weights=self.weights.update(
                gradient.function_approx.weights.weights
```

```
    )
  )
```

We also require the within method, that simply delegates to the within method of the Weights class.

```
def within(self, other: FunctionApprox[X], tolerance: float) -> bool:
    if isinstance(other, LinearFunctionApprox):
        return self.weights.within(other.weights, tolerance)
    else:
        return False
```

The only method that remains to be written now is the solve method. Note that for linear function approximation, we can directly solve for w^* if the number of feature functions m is not too large. If the entire provided data is $[(x_i, y_i)|1 \leq i \leq n]$, then the gradient estimate based on this data can be set to 0 to solve for w^*, i.e.,

$$\frac{1}{n} \cdot (\sum_{i=1}^{n} \phi(x_i) \cdot (\phi(x_i)^T \cdot w^* - y_i)) + \lambda \cdot w^* = 0$$

We denote Φ as the n rows $\times$ m columns matrix defined as $\Phi_{i,j} = \phi_j(x_i)$ and the column vector $Y \in \mathbb{R}^n$ defined as $Y_i = y_i$. Then we can write the above equation as:

$$\frac{1}{n} \cdot \Phi^T \cdot (\Phi \cdot w^* - Y) + \lambda \cdot w^* = 0$$

$$\Rightarrow (\Phi^T \cdot \Phi + n\lambda \cdot I_m) \cdot w^* = \Phi^T \cdot Y$$

$$\Rightarrow w^* = (\Phi^T \cdot \Phi + n\lambda \cdot I_m)^{-1} \cdot \Phi^T \cdot Y$$

where I_m is the $m \times m$ identity matrix. Note that this direct linear-algebraic solution for solving a square linear system of equations of size m is computationally feasible only if m is not too large.

On the other hand, if the number of feature functions m is large, then we solve for w^* by repeatedly calling update. The attribute direct_solve: bool in LinearFunctionApprox specifies whether to perform a direct solution (linear algebra calculations shown above) or to perform a sequence of iterative (incremental) updates to w using gradient descent. The code below for the method solve does exactly this:

```
import itertools
import rl.iterate import iterate
    def solve(
        self,
        xy_vals_seq: Iterable[Tuple[X, float]],
        error_tolerance: Optional[float] = None
    ) -> LinearFunctionApprox[X]:
        if self.direct_solve:
            x_vals, y_vals = zip(*xy_vals_seq)
            feature_vals: np.ndarray = self.get_feature_values(x_vals)
            feature_vals_T: np.ndarray = feature_vals.T
            left: np.ndarray = np.dot(feature_vals_T, feature_vals) \
                + feature_vals.shape[0] * self.regularization_coeff * \
                np.eye(len(self.weights.weights))
            right: np.ndarray = np.dot(feature_vals_T, y_vals)
            ret = replace(
                self,
                weights=Weights.create(
                    adam_gradient=self.weights.adam_gradient,
                    weights=np.linalg.solve(left, right)
                )
            )
```

```
        )
    else:
        tol: float = 1e-6 if error_tolerance is None else error_tolerance
        def done(
            a: LinearFunctionApprox[X],
            b: LinearFunctionApprox[X],
            tol: float = tol
        ) -> bool:
            return a.within(b, tol)
        ret = iterate.converged(
            self.iterate_updates(itertools.repeat(list(xy_vals_seq))),
            done=done
        )
    return ret
```

The above code is in the file rl/function_approx.py.

6.3 NEURAL NETWORK FUNCTION APPROXIMATION

Now we generalize the linear function approximation to accommodate non-linear functions with a simple deep neural network, specifically a feed-forward fully-connected neural network. We work with the same notation $\phi(\cdot) = (\phi_1(\cdot), \phi_2(\cdot), \ldots, \phi_m(\cdot))$ for feature functions that we covered for the case of linear function approximation. Assume we have L hidden layers in the neural network. Layers numbered $l = 0, 1, \ldots, L - 1$ carry the hidden layer neurons and layer $l = L$ carries the output layer neurons.

A couple of things to note about our notation for vectors and matrices when performing linear algebra operations: Vectors will be treated as column vectors (including gradient of a scalar with respect to a vector). The gradient of a vector of dimension m with respect to a vector of dimension n is expressed as a Jacobian matrix with m rows and n columns. We use the notation $dim(v)$ to refer to the dimension of a vector v.

We denote the input to layer l as vector i_l and the output to layer l as vector o_l, for all $l = 0, 1, \ldots, L$. Denoting the predictor variable as $x \in \mathcal{X}$, the response variable as $y \in \mathbb{R}$, and the neural network as model M to predict the expected value of y conditional on x, we have:

$$i_0 = \phi(x) \in \mathbb{R}^m \text{ and } o_L = \mathbb{E}_M[y|x] \text{ and } i_{l+1} = o_l \text{ for all } l = 0, 1, \ldots, L - 1 \qquad (6.1)$$

We denote the parameters for layer l as the matrix w_l with $dim(o_l)$ rows and $dim(i_l)$ columns. Note that the number of neurons in layer l is equal to $dim(o_l)$. Since we are restricting ourselves to scalar y, $dim(o_L) = 1$ and so, the number of neurons in the output layer is 1.

The neurons in layer l define a linear transformation from layer input i_l to a variable we denote as s_l. Therefore,

$$s_l = w_l \cdot i_l \text{ for all } l = 0, 1, \ldots, L \qquad (6.2)$$

We denote the activation function of layer l as $g_l : \mathbb{R} \to \mathbb{R}$ for all $l = 0, 1, \ldots, L$. The activation function g_l applies point-wise on each dimension of vector s_l, so we take notational liberty with g_l by writing:

$$o_l = g_l(s_l) \text{ for all } l = 0, 1, \ldots, L \qquad (6.3)$$

Equations (6.1), (6.2) and (6.3) together define the calculation of the neural network prediction o_L (associated with the response variable y), given the predictor variable x.

This calculation is known as *forward-propagation* and will define the evaluate method of the deep neural network function approximation class we shall soon write.

Our goal is to derive an expression for the cross-entropy loss gradient $\nabla_{w_l}\mathcal{L}$ for all $l = 0, 1, \ldots, L$. For ease of understanding, our following exposition will be expressed in terms of the cross-entropy loss function for a single predictor variable input $x \in \mathcal{X}$ and its associated single response variable $y \in \mathbb{R}$ (the code will generalize appropriately to the cross-entropy loss function for a given set of data points $[x_i, y_i | 1 \leq i \leq n]$).

We can reduce this problem of calculating the cross-entropy loss gradient to the problem of calculating $P_l = \nabla_{s_l}\mathcal{L}$ for all $l = 0, 1, \ldots, L$, as revealed by the following chain-rule calculation:

$$\nabla_{w_l}\mathcal{L} = (\nabla_{s_l}\mathcal{L})^T \cdot \nabla_{w_l} s_l = P_l^T \cdot \nabla_{w_l} s_l = P_l \cdot i_l^T \text{ for all } l = 0, 1, \ldots L$$

Note that $P_l \cdot i_l^T$ represents the outer-product of the $dim(o_l)$-size vector P_l and the $dim(i_l)$-size vector i_l giving a matrix of size $dim(o_l) \times dim(i_l)$.

If we include L^2 regularization (with λ_l as the regularization coefficient for layer l), then:

$$\nabla_{w_l}\mathcal{L} = P_l \cdot i_l^T + \lambda_l \cdot w_l \text{ for all } l = 0, 1, \ldots, L \qquad (6.4)$$

Here's the summary of our notation:

Notation	Description
i_l	Vector Input to layer l for all $l = 0, 1, \ldots, L$
o_l	Vector Output of layer l for all $l = 0, 1, \ldots, L$
$\phi(x)$	Feature Vector for predictor variable x
y	Response variable associated with predictor variable x
w_l	Matrix of Parameters for layer l for all $l = 0, 1, \ldots, L$
$g_l(\cdot)$	Activation function for layer l for $l = 0, 1, \ldots, L$
s_l	$s_l = w_l \cdot i_l, o_l = g_l(s_l)$ for all $l = 0, 1, \ldots L$
P_l	$P_l = \nabla_{s_l}\mathcal{L}$ for all $l = 0, 1, \ldots, L$
λ_l	Regularization coefficient for layer l for all $l = 0, 1, \ldots, L$

Now that we have reduced the loss gradient calculation to calculation of P_l, we spend the rest of this section deriving the analytical calculation of P_l. The following theorem tells us that P_l has a recursive formulation that forms the core of the *back-propagation* algorithm for a feed-forward fully-connected deep neural network.

Theorem 6.3.1. *For all* $l = 0, 1, \ldots, L - 1$,

$$P_l = (w_{l+1}^T \cdot P_{l+1}) \circ g_l'(s_l)$$

where the symbol ∘ *represents the Hadamard Product, i.e., point-wise multiplication of two vectors of the same dimension.*

Proof. We start by applying the chain rule on P_l.

$$P_l = \nabla_{s_l}\mathcal{L} = (\nabla_{s_l} s_{l+1})^T \cdot \nabla_{s_{l+1}}\mathcal{L} = (\nabla_{s_l} s_{l+1})^T \cdot P_{l+1} \qquad (6.5)$$

Next, note that:

$$s_{l+1} = w_{l+1} \cdot g_l(s_l)$$

Therefore,

$$\nabla_{s_l} s_{l+1} = w_{l+1} \cdot Diagonal(g_l'(s_l))$$

where the notation $Diagonal(v)$ for an m-dimensional vector v represents an $m \times m$ diagonal matrix whose elements are the same (also in same order) as the elements of v.

Substituting this in Equation (6.5) yields:

$$P_l = (w_{l+1} \cdot Diagonal(g_l'(s_l)))^T \cdot P_{l+1} = Diagonal(g_l'(s_l)) \cdot w_{l+1}^T \cdot P_{l+1}$$

$$= g_l'(s_l) \circ (w_{l+1}^T \cdot P_{l+1}) = (w_{l+1}^T \cdot P_{l+1}) \circ g_l'(s_l)$$

□

Now all we need to do is to calculate $P_L = \nabla_{s_L} \mathcal{L}$ so that we can run this recursive formulation for P_l, estimate the loss gradient $\nabla_{w_l} \mathcal{L}$ for any given data (using Equation (6.4)), and perform gradient descent to arrive at w_l^* for all $l = 0, 1, \ldots L$.

Firstly, note that s_L, o_L, P_L are all scalars, so let's just write them as s_L, o_L, P_L respectively (without the bold-facing) to make it explicit in the derivation that they are scalars. Specifically, the gradient

$$\nabla_{s_L} \mathcal{L} = \frac{\partial \mathcal{L}}{\partial s_L}$$

To calculate $\frac{\partial \mathcal{L}}{\partial s_L}$, we need to assume a functional form for $\mathbb{P}[y|s_L]$. We work with a fairly generic exponential functional form for the probability distribution function:

$$p(y|\theta, \tau) = h(y, \tau) \cdot e^{\frac{\theta \cdot y - A(\theta)}{d(\tau)}}$$

where θ should be thought of as the "center" parameter (related to the mean) of the probability distribution and τ should be thought of as the "dispersion" parameter (related to the variance) of the distribution. $h(\cdot, \cdot), A(\cdot), d(\cdot)$ are general functions whose specializations define the family of distributions that can be modeled with this fairly generic exponential functional form (note that this structure is adopted from the framework of Generalized Linear Models).

For our neural network function approximation, we assume that τ is a constant, and we set θ to be s_L. So,

$$\mathbb{P}[y|s_L] = p(y|s_L, \tau) = h(y, \tau) \cdot e^{\frac{s_L \cdot y - A(s_L)}{d(\tau)}}$$

We require the scalar prediction of the neural network $o_L = g_L(s_L)$ to be equal to $\mathbb{E}_p[y|s_L]$. So the question is: What function $g_L : \mathbb{R} \to \mathbb{R}$ (in terms of the functional form of $p(y|s_L, \tau)$) would satisfy the requirement of $o_L = g_L(s_L) = \mathbb{E}_p[y|s_L]$? To answer this question, we first establish the following Lemma:

Lemma 6.3.2.

$$\mathbb{E}_p[y|s_L] = A'(s_L)$$

Proof. Since

$$\int_{-\infty}^{\infty} p(y|s_L, \tau) \cdot dy = 1,$$

the partial derivative of the left-hand-side of the above equation with respect to s_L is zero. In other words,

$$\frac{\partial \{\int_{-\infty}^{\infty} p(y|s_L, \tau) \cdot dy\}}{\partial s_L} = 0$$

Hence,

$$\frac{\partial\{\int_{-\infty}^{\infty} h(y,\tau) \cdot e^{\frac{s_L \cdot y - A(s_L)}{d(\tau)}} \cdot dy\}}{\partial s_L} = 0$$

Taking the partial derivative inside the integral, we get:

$$\int_{-\infty}^{\infty} h(y,\tau) \cdot e^{\frac{s_L \cdot y - A(s_L)}{d(\tau)}} \cdot \frac{y - A'(s_L)}{d(\tau)} \cdot dy = 0$$

$$\Rightarrow \int_{-\infty}^{\infty} p(y|s_L, \tau) \cdot (y - A'(s_L)) \cdot dy = 0$$

$$\Rightarrow \mathbb{E}_p[y|s_L] = A'(s_L)$$

$\square$

So to satisfy $o_L = g_L(s_L) = \mathbb{E}_p[y|s_L]$, we require that

$$o_L = g_L(s_L) = A'(s_L) \tag{6.6}$$

The above equation is important since it tells us that the output layer activation function $g_L(\cdot)$ must be set to be the derivative of the $A(\cdot)$ function. In the theory of generalized linear models, the derivative of the $A(\cdot)$ function serves as the *canonical link function* for a given probability distribution of the response variable conditional on the predictor variable.

Now we are equipped to derive a simple expression for P_L.

Theorem 6.3.3.
$$P_L = \frac{\partial \mathcal{L}}{\partial s_L} = \frac{o_L - y}{d(\tau)}$$

Proof. The Cross-Entropy Loss (Negative Log-Likelihood) for a single training data point (x, y) is given by:

$$\mathcal{L} = -\log(h(y,\tau)) + \frac{A(s_L) - s_L \cdot y}{d(\tau)}$$

Therefore,

$$P_L = \frac{\partial \mathcal{L}}{\partial s_L} = \frac{A'(s_L) - y}{d(\tau)}$$

But from Equation (6.6), we know that $A'(s_L) = o_L$. Therefore,

$$P_L = \frac{\partial \mathcal{L}}{\partial s_L} = \frac{o_L - y}{d(\tau)}$$

$\square$

At each iteration of gradient descent, we require an estimate of the loss gradient up to a constant factor. So we can ignore the constant $d(\tau)$ and simply say that $P_L = o_L - y$ (up to a constant factor). This is a rather convenient estimate of P_L for a given data point (x, y) since it represents the neural network prediction error for that data point. When presented with a sequence of data points $[(x_{t,i}, y_{t,i})|1 \leq i \leq n_t]$ in iteration t, we simply average the prediction errors across these presented data points. Then, beginning with this estimate of P_L, we can use the recursive formulation of P_l (Theorem 6.3.1) to calculate the gradient of the loss function (Equation (6.4)) with respect to all the parameters of the neural network (this is known as the back-propagation algorithm for a fully-connected feed-forward deep neural network).

Here are some common specializations of the functional form for the conditional probability distribution $\mathbb{P}[y|s_L]$, along with the corresponding canonical link function that serves as the activation function g_L of the output layer:

- Normal distribution $y \sim \mathcal{N}(\mu, \sigma^2)$:

$$s_L = \mu, \tau = \sigma, h(y, \tau) = \frac{e^{\frac{-y^2}{2\tau^2}}}{\sqrt{2\pi\tau}}, d(\tau) = \tau, A(s_L) = \frac{s_L^2}{2}$$

$$\Rightarrow o_L = g_L(s_L) = \mathbb{E}[y|s_L] = A'(s_L) = s_L$$

Hence, the output layer activation function g_L is the identity function. This means that the linear function approximation of the previous section is exactly the same as a neural network with 0 hidden layers (just the output layer) and with the output layer activation function equal to the identity function.

- Bernoulli distribution for binary-valued y, parameterized by p:

$$s_L = \log\left(\frac{p}{1-p}\right), \tau = 1, h(y, \tau) = 1, d(\tau) = 1, A(s_L) = \log\left(1 + e^{s_L}\right)$$

$$\Rightarrow o_L = g_L(s_L) = \mathbb{E}[y|s_L] = A'(s_L) = \frac{1}{1 + e^{-s_L}}$$

Hence, the output layer activation function g_L is the logistic function. This generalizes to softmax g_L when we generalize this framework to multivariate y, which in turn enables us to classify inputs x into a finite set of categories represented by y as one-hot-encodings.

- Poisson distribution for y parameterized by λ:

$$s_L = \log\lambda, \tau = 1, h(y, \tau) = \frac{1}{y!}, d(\tau) = 1, A(s_L) = e^{s_L}$$

$$\Rightarrow o_L = g_L(s_L) = \mathbb{E}[y|s_L] = A'(s_L) = e^{s_L}$$

Hence, the output layer activation function g_L is the exponential function.

Now we are ready to write a class for function approximation with the deep neural network framework described above. We assume that the activation functions $g_l(\cdot)$ are identical for all $l = 0, 1, \ldots, L-1$ (known as the hidden layers activation function) and the activation function $g_L(\cdot)$ is known as the output layer activation function. Note that often we want to include a bias term in the linear transformations of the layers. To include a bias term in layer 0, just like in the case of `LinearFuncApprox`, we prepend the sequence of feature functions we want to provide as input with an artificial feature function `lambda _: 1.` to represent the constant feature with value 1. This ensures we have a bias weight in layer 0 in addition to each of the weights (in layer 0) that serve as coefficients to the (non-artificial) feature functions. Moreover, we allow the specification of a `bias` boolean variable to enable a bias term in each if the layers $l = 1, 2, \ldots L$.

Before we develop the code for forward-propagation and back-propagation, we write a `@dataclass` to hold the configuration of a deep neural network (number of neurons in the layers, the `bias` boolean variable, hidden layers activation function and output layer activation function).

```
@dataclass(frozen=True)
class DNNSpec:
    neurons: Sequence[int]
    bias: bool
    hidden_activation: Callable[[np.ndarray], np.ndarray]
    hidden_activation_deriv: Callable[[np.ndarray], np.ndarray]
    output_activation: Callable[[np.ndarray], np.ndarray]
    output_activation_deriv: Callable[[np.ndarray], np.ndarray]
```

neurons is a sequence of length L specifying $dim(O_0), dim(O_1), \ldots, dim(O_{L-1})$ (note $dim(o_L)$ doesn't need to be specified since we know $dim(o_L) = 1$). If bias is set to be True, then $dim(I_l) = dim(O_{l-1}) + 1$ for all $l = 1, 2, \ldots L$ and so in the code below, when bias is True, we'll need to prepend the matrix representing I_l with a vector consisting of all 1s (to incorporate the bias term). Note that along with specifying the hidden and output layers activation functions $g_l(\cdot)$ defined as $g_l(s_l) = o_l$, we also specify the hidden layers activation function derivative (hidden_activation_deriv) and the output layer activation function derivative (output_activation_deriv) in the form of functions $h_l(\cdot)$ defined as $h_l(g(s_l)) = h_l(o_l) = g'_l(s_l)$ (as we know, this derivative is required in the back-propagation calculation). We shall soon see that in the code, $h_l(\cdot)$ is a more convenient specification than the direct specification of $g'_l(\cdot)$.

Now we write the @dataclass DNNApprox that implements the abstract base class FunctionApprox. It has attributes:

- feature_functions that represents $\phi_j : \mathcal{X} \to \mathbb{R}$ for all $j = 1, 2, \ldots, m$.
- dnn_spec that specifies the neural network configuration (instance of DNNSpec).
- regularization_coeff that represents the common regularization coefficient λ for the weights across all layers.
- weights which is a sequence of Weights objects (to represent and update the weights of all layers).

The method get_feature_values is identical to the case of LinearFunctionApprox producing a matrix with number of rows equal to the number of x values in its input x_values_seq: Iterable[X] and number of columns equal to the number of specified feature_functions.

The method forward_propagation implements the forward-propagation calculation that was covered earlier (combining Equations (6.1) (potentially adjusted for the bias term, as mentioned above), (6.2) and (6.3)). forward_propagation takes as input the same data type as the input of get_feature_values (x_values_seq: Iterable[X]) and returns a list with $L + 2$ numpy arrays. The last element of the returned list is a 1-D numpy array representing the final output of the neural network: $o_L = \mathbb{E}_M[y|x]$ for each of the x values in the input x_values_seq. The remaining $L + 1$ elements in the returned list are each 2-D numpy arrays, consisting of i_l for all $l = 0, 1, \ldots L$ (for each of the x values provided as input in x_values_seq).

The method evaluate (from FunctionApprox) returns the last element ($o_L = \mathbb{E}_M[y|x]$) of the list returned by forward_propagation.

The method backward_propagation is the most important method of DNNApprox, calculating $\nabla_{w_l} Obj$ for all $l = 0, 1, \ldots, L$, for some objective function Obj. We had said previously that for each concrete function approximation that we'd want to implement, if the Objective $Obj(x_i, y_i)$ is the cross-entropy loss function, we can identify a model-computed value $Out(x_i)$ (either the output of the model or an intermediate computation of the model) such that $\frac{\partial Obj(x_i, y_i)}{\partial Out(x_i)}$ is equal to the prediction error $\mathbb{E}_M[y|x_i] - y_i$ (for each training data point (x_i, y_i)), and we can come up with a numerical algorithm to compute $\nabla_w Out(x_i)$, so that by chain-rule, we have the required gradient $\nabla_w Obj(x_i, y_i)$ (without regularization). In the case of this DNN function approximation, the model-computed value $Out(x_i)$ is s_L. Thus,

$$\frac{\partial Obj(x_i, y_i)}{\partial Out(x_i)} = \frac{\partial \mathcal{L}}{\partial s_L} = P_L = o_L - y_i = \mathbb{E}_M[y|x_i] - y_i$$

`backward_propagation` takes two inputs:

1. `fwd_prop`: `Sequence[np.ndarray]` which represents the output of `forward_propagation` except for the last element (which is the final output of the neural network), i.e., a sequence of $L + 1$ 2-D numpy arrays representing the inputs to layers $l = 0, 1, \ldots L$ (for each of the `Iterable` of x-values provided as input to the neural network).
2. `obj_deriv_out`: `np.ndarray`, which represents the partial derivative of an arbitrary objective function Obj with respect to an arbitrary model-produced value Out, evaluated at each of the `Iterable` of (x, y) pairs that are provided as training data.

If we generalize the objective function from the cross-entropy loss function $\mathcal{L}$ to an arbitrary objective function Obj and define P_l to be $\nabla_{s_l} Obj$ (generalized from $\nabla_{s_l} \mathcal{L}$), then the output of `backward_propagation` would be equal to $P_l \cdot i_l^T$ (i.e., without the regularization term) for all $l = 0, 1, \ldots L$.

The first step in `backward_propagation` is to set P_L (variable `deriv` in the code) equal to `obj_deriv_out` (which in the case of cross-entropy loss as Obj and s_L as Out, reduces to the prediction error $\mathbb{E}_M[y|x_i] - y_i$). As we walk back through the layers of the DNN, the variable `deriv` represents $P_l = \nabla_{s_l} Obj$, evaluated for each of the values made available by `fwd_prop` (note that `deriv` is updated in each iteration of the loop reflecting Theorem 6.3.1: $P_l = (w_{l+1}^T \cdot P_{l+1}) \circ g_l'(s_l)$). Note also that the returned list `back_prop` is populated with the result of Equation (6.4): $\nabla_{w_l} Obj = P_l \cdot i_l^T$.

The method `objective_gradient` (from `FunctionApprox`) takes as input an `Iterable` of (x, y) pairs and the $\frac{\partial Obj}{\partial Out}$ function, invokes the `forward_propagation` method (to be passed as input to `backward_propagation`), then invokes `backward_propagation`, and finally adds on the regularization term $\lambda \cdot w_l$ to the output of `backward_propagation` to return the gradient $\nabla_{w_l} Obj$ for all $l = 0, 1, \ldots L$.

The method `update_with_gradient` (from `FunctionApprox`) takes as input a gradient (e.g., $\nabla_{w_l} Obj$), updates the weights w_l for all $l = 0, 1, \ldots, L$ along with the ADAM cache updates (invoking the `update` method of the `Weights` class to ensure there are no in-place updates), and returns a new instance of `DNNApprox` that contains the updated weights.

Finally, the method `solve` (from `FunctionApprox`) utilizes the method `iterate_updates` (inherited from `FunctionApprox`) along with the method `within` to perform a best-fit of the weights that minimizes the cross-entropy loss function (basically, a series of incremental updates based on gradient descent).

```
from dataclasses import replace
import itertools
import rl.iterate import iterate

@dataclass(frozen=True)
class DNNApprox(FunctionApprox[X]):

    feature_functions: Sequence[Callable[[X], float]]
    dnn_spec: DNNSpec
    regularization_coeff: float
    weights: Sequence[Weights]

    @staticmethod
    def create(
        feature_functions: Sequence[Callable[[X], float]],
        dnn_spec: DNNSpec,
        adam_gradient: AdamGradient = AdamGradient.default_settings(),
        regularization_coeff: float = 0.,
        weights: Optional[Sequence[Weights]] = None
```

```
    ) -> DNNApprox[X]:
        if weights is None:
            inputs: Sequence[int] = [len(feature_functions)] + \
                [n + (1 if dnn_spec.bias else 0)
                    for i, n in enumerate(dnn_spec.neurons)]
            outputs: Sequence[int] = list(dnn_spec.neurons) + [1]
            wts = [Weights.create(
                weights=np.random.randn(output, inp) / np.sqrt(inp),
                adam_gradient=adam_gradient
            ) for inp, output in zip(inputs, outputs)]
        else:
            wts = weights

        return DNNApprox(
            feature_functions=feature_functions,
            dnn_spec=dnn_spec,
            regularization_coeff=regularization_coeff,
            weights=wts
        )

    def get_feature_values(self, x_values_seq: Iterable[X]) -> np.ndarray:
        return np.array(
            [[f(x) for f in self.feature_functions] for x in x_values_seq]
        )

    def forward_propagation(
        self,
        x_values_seq: Iterable[X]
    ) -> Sequence[np.ndarray]:
        """
        :param x_values_seq: a n-length iterable of input points
        :return: list of length (L+2) where the first (L+1) values
                 each represent the 2-D input arrays (of size n x |i_l|),
                 for each of the (L+1) layers (L of which are hidden layers),
                 and the last value represents the output of the DNN (as a
                 1-D array of length n)
        """
        inp: np.ndarray = self.get_feature_values(x_values_seq)
        ret: List[np.ndarray] = [inp]
        for w in self.weights[:-1]:
            out: np.ndarray = self.dnn_spec.hidden_activation(
                np.dot(inp, w.weights.T)
            )
            if self.dnn_spec.bias:
                inp = np.insert(out, 0, 1., axis=1)
            else:
                inp = out
            ret.append(inp)
        ret.append(
            self.dnn_spec.output_activation(
                np.dot(inp, self.weights[-1].weights.T)
            )[:, 0]
        )
        return ret

    def evaluate(self, x_values_seq: Iterable[X]) -> np.ndarray:
        return self.forward_propagation(x_values_seq)[-1]

    def backward_propagation(
        self,
        fwd_prop: Sequence[np.ndarray],
        obj_deriv_out: np.ndarray
    ) -> Sequence[np.ndarray]:
        """
        :param fwd_prop represents the result of forward propagation (without
        the final output), a sequence of L 2-D np.ndarrays of the DNN.
        : param obj_deriv_out represents the derivative of the objective
        function with respect to the linear predictor of the final layer.

        :return: list (of length L+1) of |o_l| x |i_l| 2-D arrays,
```

```
                    i.e., same as the type of self.weights.weights    •
        This function computes the gradient (with respect to weights) of
        the objective where the output layer activation function
        is the canonical link function of the conditional distribution of y|x
        """
        deriv: np.ndarray = obj_deriv_out.reshape(1, -1)
        back_prop: List[np.ndarray] = [np.dot(deriv, fwd_prop[-1]) /
                                       deriv.shape[1]]
        # L is the number of hidden layers, n is the number of points
        # layer l deriv represents dObj/ds_l where s_l = i_l . weights_l
        # (s_l is the result of applying layer l without the activation func)
        for i in reversed(range(len(self.weights) - 1)):
            # deriv_l is a 2-D array of dimension |o_l| x n
            # The recursive formulation of deriv is as follows:
            # deriv_{l-1} = (weights_l^T inner deriv_l) haddamard g'(s_{l-1}),
            # which is ((|i_l| x |o_l|) inner (|o_l| x n)) haddamard
            # (|i_l| x n), which is (|i_l| x n) = (|o_{l-1}| x n)
            # Note: g'(s_{l-1}) is expressed as hidden layer activation
            # derivative as a function of o_{l-1} (=i_l).
            deriv = np.dot(self.weights[i + 1].weights.T, deriv) * \
                self.dnn_spec.hidden_activation_deriv(fwd_prop[i + 1].T)
            # If self.dnn_spec.bias is True, then i_l = o_{l-1} + 1, in which
            # case # the first row of the calculated deriv is removed to yield
            # a 2-D array of dimension |o_{l-1}| x n.
            if self.dnn_spec.bias:
                deriv = deriv[1:]
            # layer l gradient is deriv_l inner fwd_prop[l], which is
            # of dimension (|o_l| x n) inner (n x (|i_l|)) = |o_l| x |i_l|
            back_prop.append(np.dot(deriv, fwd_prop[i]) / deriv.shape[1])
        return back_prop[::-1]

    def objective_gradient(
        self,
        xy_vals_seq: Iterable[Tuple[X, float]],
        obj_deriv_out_fun: Callable[[[Sequence[X], Sequence[float]], float]
    ) -> Gradient[DNNApprox[X]]:
        x_vals, y_vals = zip(*xy_vals_seq)
        obj_deriv_out: np.ndarray = obj_deriv_out_fun(x_vals, y_vals)
        fwd_prop: Sequence[np.ndarray] = self.forward_propagation(x_vals)[:-1]
        gradient: Sequence[np.ndarray] = \
            [x + self.regularization_coeff * self.weights[i].weights
             for i, x in enumerate(self.backward_propagation(
                 fwd_prop=fwd_prop,
                 obj_deriv_out=obj_deriv_out
             ))]
        return Gradient(replace(
            self,
            weights=[replace(w, weights=g) for
                     w, g in zip(self.weights, gradient)]
        ))

    def solve(
        self,
        xy_vals_seq: Iterable[Tuple[X, float]],
        error_tolerance: Optional[float] = None
    ) -> DNNApprox[X]:
        tol: float = 1e-6 if error_tolerance is None else error_tolerance

        def done(
            a: DNNApprox[X],
            b: DNNApprox[X],
            tol: float = tol
        ) -> bool:
            return a.within(b, tol)

        return iterate.converged(
            self.iterate_updates(itertools.repeat(list(xy_vals_seq))),
            done=done
        )
```

```
def within(self, other: FunctionApprox[X], tolerance: float) -> bool:
    if isinstance(other, DNNApprox):
        return all(w1.within(w2, tolerance)
                   for w1, w2 in zip(self.weights, other.weights))
    else:
        return False
```

All of the above code is in the file rl/function_approx.py.

A comprehensive treatment of function approximations using Deep Neural Networks can be found in the Deep Learning book by Goodfellow, Bengio, Courville (Goodfellow, Bengio, and Courville 2016).

Let us now write some code to create function approximations with LinearFunctionApprox and DNNApprox, given a stream of data from a simple data model—one that has some noise around a linear function. Here's some code to create an Iterator of (x, y) pairs (where $x = (x_1, x_2, x_3)$) for the data model:

$$y = 2 + 10x_1 + 4x_2 - 6x_3 + \mathcal{N}(0, 0.3)$$

```
def example_model_data_generator() -> Iterator[Tuple[Triple, float]]:
    coeffs: Aug_Triple = (2., 10., 4., -6.)
    d = norm(loc=0., scale=0.3)

    while True:
        pt: np.ndarray = np.random.randn(3)
        x_val: Triple = (pt[0], pt[1], pt[2])
        y_val: float = coeffs[0] + np.dot(coeffs[1:], pt) + \
            d.rvs(size=1)[0]
        yield (x_val, y_val)
```

Next, we wrap this in an Iterator that returns a certain number of (x, y) pairs upon each request for data points.

```
def data_seq_generator(
    data_generator: Iterator[Tuple[Triple, float]],
    num_pts: int
) -> Iterator[DataSeq]:
    while True:
        pts: DataSeq = list(islice(data_generator, num_pts))
        yield pts
```

Now let's write a function to create a LinearFunctionApprox.

```
def feature_functions():
    return [lambda _: 1., lambda x: x[0], lambda x: x[1], lambda x: x[2]]

def adam_gradient():
    return AdamGradient(
        learning_rate=0.1,
        decay1=0.9,
        decay2=0.999
    )

def get_linear_model() -> LinearFunctionApprox[Triple]:
    ffs = feature_functions()
    ag = adam_gradient()
    return LinearFunctionApprox.create(
        feature_functions=ffs,
        adam_gradient=ag,
        regularization_coeff=0.,
        direct_solve=True
    )
```

Likewise, let's write a function to create a DNNApprox with 1 hidden layer with 2 neurons and a little bit of regularization since this deep neural network is somewhat over-parameterized to fit the data generated from the linear data model with noise.

```
def get_dnn_model() -> DNNApprox[Triple]:
    ffs = feature_functions()
    ag = adam_gradient()

    def relu(arg: np.ndarray) -> np.ndarray:
        return np.vectorize(lambda x: x if x > 0. else 0.)(arg)

    def relu_deriv(res: np.ndarray) -> np.ndarray:
        return np.vectorize(lambda x: 1. if x > 0. else 0.)(res)

    def identity(arg: np.ndarray) -> np.ndarray:
        return arg

    def identity_deriv(res: np.ndarray) -> np.ndarray:
        return np.ones_like(res)

    ds = DNNSpec(
        neurons=[2],
        bias=True,
        hidden_activation=relu,
        hidden_activation_deriv=relu_deriv,
        output_activation=identity,
        output_activation_deriv=identity_deriv
    )

    return DNNApprox.create(
        feature_functions=ffs,
        dnn_spec=ds,
        adam_gradient=ag,
        regularization_coeff=0.05
    )
```

Now let's write some code to do a `direct_solve` with the `LinearFunctionApprox` based on the data from the data model we have set up.

```
training_num_pts: int = 1000
test_num_pts: int = 10000
training_iterations: int = 200
data_gen: Iterator[Tuple[Triple, float]] = example_model_data_generator()
training_data_gen: Iterator[DataSeq] = data_seq_generator(
    data_gen,
    training_num_pts
)
test_data: DataSeq = list(islice(data_gen, test_num_pts))

direct_solve_lfa: LinearFunctionApprox[Triple] = \
    get_linear_model().solve(next(training_data_gen))
direct_solve_rmse: float = direct_solve_lfa.rmse(test_data)
```

Running the above code, we see that the Root-Mean-Squared-Error (`direct_solve_rmse`) is indeed 0.3, matching the standard deviation of the noise in the linear data model (which is used above to generate the training data as well as the test data).

Now let us perform stochastic gradient descent with instances of `LinearFunctionApprox` and `DNNApprox` and examine the Root-Mean-Squared-Errors on the two function approximations as a function of number of iterations in the gradient descent.

```
linear_model_rmse_seq: Sequence[float] = \
    [lfa.rmse(test_data) for lfa in islice(
        get_linear_model().iterate_updates(training_data_gen),
        training_iterations
    )]
dnn_model_rmse_seq: Sequence[float] = \
    [dfa.rmse(test_data) for dfa in islice(
        get_dnn_model().iterate_updates(training_data_gen),
        training_iterations
    )]
```

The plot of `linear_model_rmse_seq` and `dnn_model_rmse_seq` is shown in Figure 6.1.

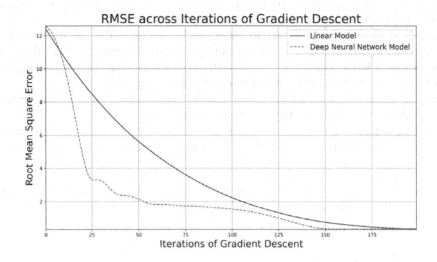

Figure 6.1 **SGD Convergence**

6.4 TABULAR AS A FORM OF FUNCTIONAPPROX

Now we consider a simple case where we have a fixed and finite set of x-values $\mathcal{X} = \{x_1, x_2, \ldots, x_n\}$, and any data set of (x, y) pairs made available to us needs to have its x-values from within this finite set $\mathcal{X}$. The prediction $\mathbb{E}[y|x]$ for each $x \in \mathcal{X}$ needs to be calculated only from the y-values associated with this x within the data set of (x, y) pairs. In other words, the y-values in the data associated with other x should not influence the prediction for x. Since we'd like the prediction for x to be $\mathbb{E}[y|x]$, it would make sense for the prediction for a given x to be *some sort of average* of all the y-values associated with x within the data set of (x, y) pairs seen so far. This simple case is referred to as *Tabular* because we can store all $x \in \mathcal{X}$ together with their corresponding predictions $\mathbb{E}[y|x]$ in a finite data structure (loosely referred to as a "table").

So the calculations for Tabular prediction of $\mathbb{E}[y|x]$ is particularly straightforward. What is interesting though is the fact that Tabular prediction actually fits the interface of FunctionApprox in terms of the following three methods that we have emphasized as the essence of FunctionApprox:

- the solve method, that would simply take the average of all the y-values associated with each x in the given data set, and store those averages in a dictionary data structure.
- the update method, that would update the current averages in the dictionary data structure, based on the new data set of (x, y) pairs that is provided.
- the evaluate method, that would simply look up the dictionary data structure for the y-value averages associated with each x-value provided as input.

This view of Tabular prediction as a special case of FunctionApprox also permits us to cast the tabular algorithms of Dynamic Programming and Reinforcement Learning as special cases of the function approximation versions of the algorithms (using the Tabular class we develop below).

So now let us write the code for @dataclass Tabular as an implementation of the abstract base class FunctionApprox. The attributes of @dataclass Tabular are:

- values_map which is a dictionary mapping each x-value to the average of the y-values associated with x that have been seen so far in the data.
- counts_map which is a dictionary mapping each x-value to the count of y-values associated with x that have been seen so far in the data. We need to track the count of y-values associated with each x because this enables us to update values_map appropriately upon seeing a new y-value associated with a given x.
- count_to_weight_func which defines a function from number of y-values seen so far (associated with a given x) to the weight assigned to the most recent y. This enables us to do a weighted average of the y-values seen so far, controlling the emphasis to be placed on more recent y-values relative to previously seen y-values (associated with a given x).

The evaluate, objective_gradient, update_with_gradient, solve and within methods should now be self-explanatory.

```
from dataclasses import field, replace

@dataclass(frozen=True)
class Tabular(FunctionApprox[X]):
    values_map: Mapping[X, float] = field(default_factory=lambda: {})
    counts_map: Mapping[X, int] = field(default_factory=lambda: {})
    count_to_weight_func: Callable[[int], float] = \
        field(default_factory=lambda: lambda n: 1.0 / n)

    def objective_gradient(
        self,
        xy_vals_seq: Iterable[Tuple[X, float]],
        obj_deriv_out_fun: Callable[[Sequence[X], Sequence[float]], float]
    ) -> Gradient[Tabular[X]]:
        x_vals, y_vals = zip(*xy_vals_seq)
        obj_deriv_out: np.ndarray = obj_deriv_out_fun(x_vals, y_vals)
        sums_map: Dict[X, float] = defaultdict(float)
        counts_map: Dict[X, int] = defaultdict(int)
        for x, o in zip(x_vals, obj_deriv_out):
            sums_map[x] += o
            counts_map[x] += 1
        return Gradient(replace(
            self,
            values_map={x: sums_map[x] / counts_map[x] for x in sums_map},
            counts_map=counts_map
        ))

    def evaluate(self, x_values_seq: Iterable[X]) -> np.ndarray:
        return np.array([self.values_map.get(x, 0.) for x in x_values_seq])

    def update_with_gradient(
        self,
        gradient: Gradient[Tabular[X]]
    ) -> Tabular[X]:
        values_map: Dict[X, float] = dict(self.values_map)
        counts_map: Dict[X, int] = dict(self.counts_map)
        for key in gradient.function_approx.values_map:
            counts_map[key] = counts_map.get(key, 0) + \
                gradient.function_approx.counts_map[key]
            weight: float = self.count_to_weight_func(counts_map[key])
            values_map[key] = values_map.get(key, 0.) - \
                weight * gradient.function_approx.values_map[key]
        return replace(
            self,
            values_map=values_map,
            counts_map=counts_map
        )

    def solve(
        self,
        xy_vals_seq: Iterable[Tuple[X, float]],
```

```
        error_tolerance: Optional[float] = None
    ) -> Tabular[X]:
        values_map: Dict[X, float] = {}
        counts_map: Dict[X, int] = {}
        for x, y in xy_vals_seq:
            counts_map[x] = counts_map.get(x, 0) + 1
            weight: float = self.count_to_weight_func(counts_map[x])
            values_map[x] = weight * y + (1 - weight) * values_map.get(x, 0.)
        return replace(
            self,
            values_map=values_map,
            counts_map=counts_map
        )

    def within(self, other: FunctionApprox[X], tolerance: float) -> bool:
        if isinstance(other, Tabular):
            return all(abs(self.values_map[s] - other.values_map.get(s, 0.))
                       <= tolerance for s in self.values_map)
        return False
```

Here's a valuable insight: This *Tabular* setting is actually a special case of linear function approximation by setting a feature function $\phi_i(\cdot)$ for each x_i as: $\phi_i(x_i) = 1$ and $\phi_i(x) = 0$ for each $x \neq x_i$ (i.e., $\phi_i(\cdot)$ is the indicator function for x_i, and the Φ matrix is the identity matrix), and the corresponding weights w_i equal to the average of the y-values associated with x_i in the given data. This also means that the count_to_weights_func plays the role of the learning rate function (as a function of the number of iterations in stochastic gradient descent).

When we implement Approximate Dynamic Programming (ADP) algorithms with the @abstractclass FunctionApprox (later in this chapter), using the Tabular class (for FunctionApprox) enables us to specialize the ADP algorithm implementation to the Tabular DP algorithms (that we covered in Chapter 5). Note that in the tabular DP algorithms, the set of finite states take the role of $\mathcal{X}$ and the Value Function for a given state $x = s$ takes the role of the "predicted" y-value associated with x. We also note that in the tabular DP algorithms, in each iteration of sweeping through all the states, the Value Function for a state $x = s$ is set to the current y value (not the average of all y-values seen so far). The current y-value is simply the right-hand-side of the Bellman Equation corresponding to the tabular DP algorithm. Consequently, when using Tabular class for tabular DP, we'd need to set count_to_weight_func to be the function lambda _: 1 (this is because a weight of 1 for the current y-value sets values_map[x] equal to the current y-value).

Likewise, when we implement RL algorithms (using @abstractclass FunctionApprox) later in this book, using the Tabular class (for FunctionApprox) specializes the RL algorithm implementation to Tabular RL. In Tabular RL, we average all the Returns seen so far for a given state. If we choose to do a plain average (equal importance for all y-values seen so far, associated with a given x), then in the Tabular class, we'd need to set count_to_weights_func to be the function lambda n: 1. / n.

We want to emphasize that although tabular algorithms are just a special case of algorithms with function approximation, we give special coverage in this book to tabular algorithms because they help us conceptualize the core concepts in a simple (tabular) setting without the distraction of some of the details and complications in the apparatus of function approximation.

Now we are ready to write algorithms for Approximate Dynamic Programming (ADP). Before we go there, it pays to emphasize that we have described and implemented a fairly generic framework for gradient-based estimation of function approximations, given arbitrary training data. It can be used for arbitrary objective functions and arbitrary functional forms/neural networks (beyond the concrete classes we implemented). We encourage you to explore implementing and using this function approximation code for other types of objectives and other types of functional forms/neural networks.

6.5 APPROXIMATE POLICY EVALUATION

The first Approximate Dynamic Programming (ADP) algorithm we cover is Approximate Policy Evaluation, i.e., evaluating the Value Function for a Markov Reward Process (MRP). Approximate Policy Evaluation is fundamentally the same as Tabular Policy Evaluation in terms of repeatedly applying the Bellman Policy Operator B^π on the Value Function $V : \mathcal{N} \to \mathbb{R}$. However, unlike Tabular Policy Evaluation algorithm, here the Value Function $V(\cdot)$ is set up and updated as an instance of FunctionApprox rather than as a table of values for the non-terminal states. This is because unlike Tabular Policy Evaluation which operates on an instance of a FiniteMarkovRewardProcess, Approximate Policy Evaluation algorithm operates on an instance of MarkovRewardProcess. So we do not have an enumeration of states of the MRP, and we do not have the transition probabilities of the MRP. This is typical in many real-world problems where the state space is either very large or is continuous-valued, and the transitions could be too many or could be continuous-valued transitions. So, here's what we do to overcome these challenges:

- We specify a sampling probability distribution of non-terminal states (argument non_terminal_states_distribution in the code below) from which we shall sample a specified number (num_state_samples in the code below) of non-terminal states, and construct a list of those sampled non-terminal states (nt_states in the code below) in each iteration. The type of this probability distribution of non-terminal states is aliased as follows (this type will be used not just for Approximate Dynamic Programming algorithms, but also for Reinforcement Learning algorithms):

```
NTStateDistribution = Distribution[NonTerminal[S]]
```

- We sample pairs of (next state s', reward r) from a given non-terminal state s, and calculate the expectation $\mathbb{E}[r + \gamma \cdot V(s')]$ by averaging $r + \gamma \cdot V(s')$ across the sampled pairs. Note that the method expectation of a Distribution object performs a sampled expectation. $V(s')$ is obtained from the function approximation instance of FunctionApprox that is being updated in each iteration. The type of the function approximation of the Value Function is aliased as follows (this type will be used not just for Approximate Dynamic Programming algorithms, but also for Reinforcement Learning Algorithms).

```
ValueFunctionApprox = FunctionApprox[NonTerminal[S]]
```

- The sampled list of non-terminal states s comprise our x-values and the associated sampled expectations described above comprise our y-values. This list of (x, y) pairs are used to update the approximation of the Value Function in each iteration (producing a new instance of ValueFunctionApprox using its update method).

The entire code is shown below. The evaluate_mrp method produces an Iterator on ValueFunctionApprox instances, and the code that calls evaluate_mrp can decide when/how to terminate the iterations of Approximate Policy Evaluation.

```
from rl.iterate import iterate
def evaluate_mrp(
    mrp: MarkovRewardProcess[S],
    gamma: float,
    approx_0: ValueFunctionApprox[S],
```

```
        non_terminal_states_distribution: NTStateDistribution[S],
        num_state_samples: int
) -> Iterator[ValueFunctionApprox[S]]:

    def update(v: ValueFunctionApprox[S]) -> ValueFunctionApprox[S]:
        nt_states: Sequence[NonTerminal[S]] = \
            non_terminal_states_distribution.sample_n(num_state_samples)

        def return_(s_r: Tuple[State[S], float]) -> float:
            s1, r = s_r
            return r + gamma * extended_vf(v, s1)

        return v.update(
            [(s, mrp.transition_reward(s).expectation(return_))
             for s in nt_states]
        )

    return iterate(update, approx_0)
```

Notice the function `extended_vf` used to evaluate the Value Function for the next state transitioned to. However, the next state could be terminal or non-terminal, and the Value Function is only defined for non-terminal states. `extended_vf` utilizes the method `on_non_terminal` we had written in Chapter 3 when designing the `State` class—it evaluates to the default value of 0 for a terminal state (and evaluates the given `ValueFunctionApprox` for a non-terminal state).

```
def extended_vf(vf: ValueFunctionApprox[S], s: State[S]) -> float:
    return s.on_non_terminal(vf, 0.0)
```

`extended_vf` will be useful not just for Approximate Dynamic Programming algorithms, but also for Reinforcement Learning algorithms.

6.6   APPROXIMATE VALUE ITERATION

Now that we've understood and coded Approximate Policy Evaluation (to solve the Prediction problem), we can extend the same concepts to Approximate Value Iteration (to solve the Control problem). The code below in `value_iteration` is almost the same as the code above in `evaluate_mrp`, except that instead of a `MarkovRewardProcess`, here we have a `MarkovDecisionProcess`, and instead of the Bellman Policy Operator update, here we have the Bellman Optimality Operator update. Therefore, in the Value Function update, we maximize the Q-value function (over all actions a) for each non-terminal state s. Also, similar to `evaluate_mrp`, `value_iteration` produces an `Iterator` on `ValueFunctionApprox` instances, and the code that calls `value_iteration` can decide when/how to terminate the iterations of Approximate Value Iteration.

```
from rl.iterate import iterate
def value_iteration(
    mdp: MarkovDecisionProcess[S, A],
    gamma: float,
    approx_0: ValueFunctionApprox[S],
    non_terminal_states_distribution: NTStateDistribution[S],
    num_state_samples: int
) -> Iterator[ValueFunctionApprox[S]]:

    def update(v: ValueFunctionApprox[S]) -> ValueFunctionApprox[S]:
        nt_states: Sequence[NonTerminal[S]] = \
            non_terminal_states_distribution.sample_n(num_state_samples)

        def return_(s_r: Tuple[State[S], float]) -> float:
            s1, r = s_r
            return r + gamma * extended_vf(v, s1)

        return v.update(
```

```
            [(s, max(mdp.step(s, a).expectation(return_)
                     for a in mdp.actions(s)))
             for s in nt_states]
    )
    return iterate(update, approx_0)
```

6.7 FINITE-HORIZON APPROXIMATE POLICY EVALUATION

Next, we move on to Approximate Policy Evaluation in a finite-horizon setting, meaning we perform Approximate Policy Evaluation with a backward induction algorithm, much like how we did backward induction for finite-horizon Tabular Policy Evaluation. We will of course make the same types of adaptations from Tabular to Approximate as we did in the functions evaluate_mrp and value_iteration above.

In the backward_evaluate code below, the input argument mrp_f0_mu_triples is a list of triples, with each triple corresponding to each non-terminal time step in the finite horizon. Each triple consists of:

- An instance of MarkovRewardProcess—note that each time step has its own instance of MarkovRewardProcess representation of transitions from non-terminal states s in a time step t to the (state s', reward r) pairs in the next time step $t + 1$ (variable mrp in the code below).

- An instance of ValueFunctionApprox to capture the approximate Value Function for the time step (variable approx0 in the code below represents the initial ValueFunctionApprox instances).

- A sampling probability distribution of non-terminal states in the time step (variable mu in the code below).

The backward induction code below should be pretty self-explanatory. Note that in backward induction, we don't invoke the update method of FunctionApprox like we did in the non-finite-horizon cases—here we invoke the solve method which internally performs a series of updates on the FunctionApprox for a given time step (until we converge to within a specified level of error_tolerance). In the non-finite-horizon cases, it was okay to simply do a single update in each iteration because we revisit the same set of states in further iterations. Here, once we converge to an acceptable ValueFunctionApprox (using solve) for a specific time step, we won't be performing any more updates to the Value Function for that time step (since we move on to the next time step, in reverse). backward_evaluate returns an Iterator over ValueFunctionApprox objects, from time step 0 to the horizon time step. We should point out that in the code below, we've taken special care to handle terminal states (that occur either at the end of the horizon or can even occur before the end of the horizon)—this is done using the extended_vf function we'd written earlier.

```
MRP_FuncApprox_Distribution = Tuple[MarkovRewardProcess[S],
                                    ValueFunctionApprox[S],
                                    NTStateDistribution[S]]

def backward_evaluate(
    mrp_f0_mu_triples: Sequence[MRP_FuncApprox_Distribution[S]],
    gamma: float,
    num_state_samples: int,
    error_tolerance: float
) -> Iterator[ValueFunctionApprox[S]]:
    v: List[ValueFunctionApprox[S]] = []

    for i, (mrp, approx0, mu) in enumerate(reversed(mrp_f0_mu_triples)):
```

```
        def return_(s_r: Tuple[State[S], float], i=i) -> float:
            s1, r = s_r
            return r + gamma * (extended_vf(v[i-1], s1) if i > 0 else 0.)
        v.append(
            approx0.solve(
                [(s, mrp.transition_reward(s).expectation(return_))
                    for s in mu.sample_n(num_state_samples)],
                error_tolerance
            )
        )
    return reversed(v)
```

6.8   FINITE-HORIZON APPROXIMATE VALUE ITERATION

Now that we've understood and coded finite-horizon Approximate Policy Evaluation
(to solve the finite-horizon Prediction problem), we can extend the same concepts to
finite-horizon Approximate Value Iteration (to solve the finite-horizon Control problem). The code below in `back_opt_vf_and_policy` is almost the same as the code
above in `backward_evaluate`, except that instead of `MarkovRewardProcess`, here we have
`MarkovDecisionProcess`. For each non-terminal time step, we maximize the Q-Value function (over all actions a) for each non-terminal state s. `back_opt_vf_and_policy` returns an
Iterator over pairs of `ValueFunctionApprox` and `DeterministicPolicy` objects (representing
the Optimal Value Function and the Optimal Policy respectively), from time step 0 to the
horizon time step.

```
from rl.distribution import Constant
from operator import itemgetter
MDP_FuncApproxV_Distribution = Tuple[
    MarkovDecisionProcess[S, A],
    ValueFunctionApprox[S],
    NTStateDistribution[S]
]
def back_opt_vf_and_policy(
    mdp_f0_mu_triples: Sequence[MDP_FuncApproxV_Distribution[S, A]],
    gamma: float,
    num_state_samples: int,
    error_tolerance: float
) -> Iterator[Tuple[ValueFunctionApprox[S], DeterministicPolicy[S, A]]]:
    vp: List[Tuple[ValueFunctionApprox[S], DeterministicPolicy[S, A]]] = []

    for i, (mdp, approx0, mu) in enumerate(reversed(mdp_f0_mu_triples)):
        def return_(s_r: Tuple[State[S], float], i=i) -> float:
            s1, r = s_r
            return r + gamma * (extended_vf(vp[i-1][0], s1) if i > 0 else 0.)

        this_v = approx0.solve(
            [(s, max(mdp.step(s, a).expectation(return_)
                    for a in mdp.actions(s)))
                for s in mu.sample_n(num_state_samples)],
            error_tolerance
        )

        def deter_policy(state: S) -> A:
            return max(
                ((mdp.step(NonTerminal(state), a).expectation(return_), a)
                    for a in mdp.actions(NonTerminal(state))),
                key=itemgetter(0)
            )[1]

        vp.append((this_v, DeterministicPolicy(deter_policy)))
    return reversed(vp)
```

6.9 FINITE-HORIZON APPROXIMATE Q-VALUE ITERATION

The above code for Finite-Horizon Approximate Value Iteration extends the Finite-Horizon Backward Induction Value Iteration algorithm of Chapter 5 by treating the Value Function as a function approximation instead of an exact tabular representation. However, there is an alternative (and arguably simpler and more effective) way to solve the Finite-Horizon Control problem—we can perform backward induction on the optimal Action-Value (Q-value) Function instead of the optimal (State-)Value Function. In general, the key advantage of working with the optimal Action Value function is that it has all the information necessary to extract the optimal State-Value function and the optimal Policy (since we just need to perform a max / arg max over all the actions for any non-terminal state). This contrasts with the case of working with the optimal State-Value function which requires us to also avail of the transition probabilities, rewards and discount factor in order to extract the optimal policy. We shall see later that Reinforcement Learning algorithms for Control work with Action-Value (Q-Value) Functions for this very reason.

Performing backward induction on the optimal Q-value function means that knowledge of the optimal Q-value function for a given time step t immediately gives us the optimal State-Value function and the optimal policy for the same time step t. This contrasts with performing backward induction on the optimal State-Value function—knowledge of the optimal State-Value function for a given time step t cannot give us the optimal policy for the same time step t (for that, we need the optimal State-Value function for time step $t + 1$ and furthermore, we also need the t to $t + 1$ state/reward transition probabilities).

So now we develop an algorithm that works with a function approximation for the Q-Value function and steps back in time similar to the backward induction we had performed earlier for the (State-)Value function. Just like we defined an alias type ValueFunctionApprox for the State-Value function, we define an alias type QValueFunctionApprox for the Action-Value function, as follows:

```
QValueFunctionApprox = FunctionApprox[Tuple[NonTerminal[S], A]]
```

The code below in back_opt_qvf is quite similar to back_opt_vf_and_policy above. The key difference is that we have QValueFunctionApprox in the input to the function rather than ValueFunctionApprox to reflect the fact that we are approximating $Q_t^* : \mathcal{N}_t \times \mathcal{A}_t \to \mathbb{R}$ for all time steps t in the finite horizon. For each non-terminal time step, we express the Q-value function (for a set of sample non-terminal states s and for all actions a) in terms of the Q-value function approximation of the next time step. This is essentially the MDP Action-Value Function Bellman Optimality Equation for the finite-horizon case (adapted to function approximation). back_opt_qvf returns an Iterator over QValueFunctionApprox (representing the Optimal Q-Value Function), from time step 0 to the horizon time step. We can then obtain V_t^* (Optimal State-Value Function) and π_t^* for each t by simply performing a max / arg max over all actions $a \in \mathcal{A}_t$ of $Q_t^*(s, a)$ for any $s \in \mathcal{N}_t$.

```
MDP_FuncApproxQ_Distribution = Tuple[
    MarkovDecisionProcess[S, A],
    QValueFunctionApprox[S, A],
    NTStateDistribution[S]
]

def back_opt_qvf(
    mdp_f0_mu_triples: Sequence[MDP_FuncApproxQ_Distribution[S, A]],
    gamma: float,
    num_state_samples: int,
    error_tolerance: float
) -> Iterator[QValueFunctionApprox[S, A]]:
    horizon: int = len(mdp_f0_mu_triples)
    qvf: List[QValueFunctionApprox[S, A]] = []
```

```
for i, (mdp, approx0, mu) in enumerate(reversed(mdp_f0_mu_triples)):
    def return_(s_r: Tuple[State[S], float], i=i) -> float:
        s1, r = s_r
        next_return: float = max(
            qvf[i-1]((s1, a)) for a in
            mdp_f0_mu_triples[horizon - i][0].actions(s1)
        ) if i > 0 and isinstance(s1, NonTerminal) else 0.
        return r + gamma * next_return

    this_qvf = approx0.solve(
        [((s, a), mdp.step(s, a).expectation(return_))
         for s in mu.sample_n(num_state_samples) for a in mdp.actions(s)],
        error_tolerance
    )

    qvf.append(this_qvf)
return reversed(qvf)
```

We should also point out here that working with the optimal Q-value function (rather than the optimal State-Value function) in the context of ADP prepares us nicely for RL because RL algorithms typically work with the optimal Q-value function instead of the optimal State-Value function.

All of the above code for Approximate Dynamic Programming (ADP) algorithms is in the file rl/approximate_dynamic_programming.py. We encourage you to create instances of MarkovRewardProcess and MarkovDecisionProcess (including finite-horizon instances) and play with the above ADP code with different choices of function approximations, non-terminal state sampling distributions, and number of samples. A simple but valuable exercise is to reproduce the tabular versions of these algorithms by using the Tabular implementation of FunctionApprox (note: the count_to_weights_func would then need to be lambda _: 1.) in the above ADP functions.

6.10 HOW TO CONSTRUCT THE NON-TERMINAL STATES DISTRIBUTION

Each of the above ADP algorithms takes as input probability distribution(s) of non-terminal states. You may be wondering how one constructs the probability distribution of non-terminal states so you can feed it as input to any of these ADP algorithm. There is no simple, crisp answer to this. But we will provide some general pointers in this section on how to construct the probability distribution of non-terminal states.

Let us start with Approximate Policy Evaluation and Approximate Value Iteration algorithms. They require as input the probability distribution of non-terminal states. For Approximate Value Iteration algorithm, a natural choice would be evaluate the Markov Decision Process (MDP) with a uniform policy (equal probability for each action, from any state) to construct the implied Markov Reward Process (MRP), and then infer the stationary distribution of its Markov Process, using some special property of the Markov Process (for instance, if it's a finite-states Markov Process, we might be able to perform the matrix calculations we covered in Chapter 3 to calculate the stationary distribution). The stationary distribution would serve as the probability distribution of non-terminal states to be used by the Approximate Value Iteration algorithm. For Approximate Policy Evaluation algorithm, we do the same stationary distribution calculation with the given MRP. If we cannot take advantage of any special properties of the given MDP/MRP, then we can run a simulation with the simulate method in MarkovRewardProcess (inherited from MarkovProcess) and create a SampledDistribution of non-terminal states based on the non-terminal states reached by the sampling traces after a sufficiently large (but fixed) number of time steps (this is essentially an estimate of the stationary distribution). If the above choices are infeasible or computationally expensive, then a simple and neutral choice is to use a uniform distribution over the states.

Next, we consider the backward induction ADP algorithms for finite-horizon MDPs/MRPs. Our job here is to infer the distribution of non-terminal states for each time step in the finite horizon. Sometimes you can take advantage of the mathematical structure of the underlying Markov Process to come up with an analytical expression (exact or approximate) for the probability distribution of non-terminal states at any time step for the underlying Markov Process of the MRP/implied-MRP. For instance, if the Markov Process is described by a stochastic differential equation (SDE) and if we are able to solve the SDE, we would know the analytical expression for the probability distribution of non-terminal states. If we cannot take advantage of any such special properties, then we can generate sampling traces by time-incrementally sampling from the state-transition probability distributions of each of the Markov Reward Processes at each time step (if we are solving a Control problem, then we create implied-MRPs by evaluating the given MDPs with a uniform policy). The states reached by these sampling traces at any fixed time step provide a SampledDistribution of non-terminal states for that time step. If the above choices are infeasible or computationally expensive, then a simple and neutral choice is to use a uniform distribution over the non-terminal states for each time step.

We will write some code in Chapter 8 to create a SampledDistribution of non-terminal states for each time step of a finite-horizon problem by stitching together samples of state transitions at each time step. If you are curious about this now, you can take a peek at the code in rl/chapter7/asset_alloc_discrete.py.

6.11   KEY TAKEAWAYS FROM THIS CHAPTER

- The Function Approximation interface involves two key methods—A) updating the parameters of the Function Approximation based on training data available from each iteration of a data stream and B) evaluating the expectation of the response variable whose conditional probability distribution is modeled by the Function Approximation. Linear Function Approximation and Deep Neural Network Function Approximation are the two main Function Approximations we've implemented and will be using in the rest of the book.
- Tabular satisfies the interface of Function Approximation, and can be viewed as a special case of linear function approximation with feature functions set to be indicator functions for each of the x values.
- All the Tabular DP algorithms can be generalized to ADP algorithms replacing tabular Value Function updates with updates to Function Approximation parameters (where the Function Approximation represents the Value Function). Sweep over all states in the tabular case is replaced by sampling states in the ADP case. Expectation calculations in Bellman Operators are handled in ADP as averages of the corresponding calculations over transition samples (versus calculations using explicit transition probabilities in the tabular algorithms).

II

Modeling Financial Applications

CHAPTER 7

Utility Theory

7.1 INTRODUCTION TO THE CONCEPT OF UTILITY

This chapter marks the beginning of Module II, where we cover a set of financial applications that can be solved with Dynamic Programming or Reinforcement Learning Algorithms. A fundamental feature of many financial applications cast as Stochastic Control problems is that the *Rewards* of the modeled MDP are Utility functions in order to capture the tradeoff between financial returns and risk. So this chapter is dedicated to the topic of *Financial Utility*. We begin with developing an understanding of what *Utility* means from a broad Economic perspective, then zoom into the concept of Utility from a financial/monetary perspective and finally show how Utility functions can be designed to capture individual preferences of "risk-taking-inclination" when it comes to specific financial applications.

Utility Theory is a vast and important topic in Economics, and we won't cover it in detail in this book—rather, we will focus on the aspects of Utility Theory that are relevant for the Financial Applications we cover in this book. But it pays to have some familiarity with the general concept of Utility in Economics. The term *Utility* (in Economics) refers to the abstract concept of an individual's preferences over choices of products or services or activities (or more generally, over choices of certain abstract entities analyzed in Economics). Let's say you are offered 3 options to spend your Saturday afternoon: A) lie down on your couch and listen to music, B) baby-sit your neighbor's kid and earn some money or C) play a game of tennis with your friend. We really don't know how to compare these 3 options in a formal/analytical manner. But we tend to be fairly decisive (instinctively) in picking among disparate options of this type. Utility Theory aims to formalize making choices by assigning a real number to each presented choice, and then picking the choice with the highest assigned number. The assigned real number for each choice represents the "value"/"worth" of the choice, noting that the "value"/"worth" is often an implicit/instinctive value that needs to be made explicit. In our example, the numerical value for each choice is not something concrete or precise like number of dollars earned on a choice or the amount of time spent on a choice—rather it is a more abstract notion of an individual's "happiness" or "satisfaction" associated with a choice. In this example, you might say you prefer option A) because you feel lazy today (so, no tennis) and you care more about enjoying some soothing music after a long work-week than earning a few extra bucks through baby-sitting. Thus, you are comparing different attributes like money, relaxation and pleasure. This can get more complicated if your friend is offered these options, and say your friend chooses option C). If you see your friend's choice, you might then instead choose option C) because you perceive the "collective value" (for you and your friend together) to be highest if you both choose option C).

DOI: 10.1201/9781003229193-7

We won't go any further on this topic of abstract Utility, but we hope the above example provides the basic intuition for the broad notion of Utility in Economics as preferences over choices by assigning a numerical value for each choice. For a deeper study of Utility Theory, we refer you to The Handbook of Utility Theory (Barbera, Hammond, and Seidl 1998). In this book, we focus on the narrower notion of *Utility of Money* because money is what we care about when it comes to financial applications. However, Utility of Money is not so straightforward because different people respond to different levels of money in different ways. Moreover, in many financial applications, utility functions help us determine the tradeoff between financial return and risk, and this involves (challenging) assessments of the likelihood of various outcomes. The next section develops the intuition on these concepts.

7.2 A SIMPLE FINANCIAL EXAMPLE

To warm up to the concepts associated with Financial Utility Theory, let's start with a simple financial example. Consider a casino game where your financial gain/loss is based on the outcome of tossing a fair coin (HEAD or TAIL outcomes). Let's say you will be paid $1000 if the coin shows HEAD on the toss, and let's say you would be required to pay $500 if the coin shows TAIL on the toss. Now the question is: How much would you be willing to pay upfront to play this game? Your first instinct might be to say: "I'd pay $250 upfront to play this game because that's my expected payoff, based on the probability of the outcomes" $(250 = 0.5(1000) + 0.5(-500))$. But after you think about it carefully, you might alter your answer to be: "I'd pay a little less than $250". When pressed for why the fair upfront cost for playing the game should be less than $250, you might say: "I need to be compensated for taking the risk".

What does the word "risk" mean? It refers to the degree of variation in the outcomes ($1000 versus -$500). But then why would you say you need to be compensated for being exposed to this variation in outcomes? If −$500 makes you unhappy, $1000 should make you happy, and so, shouldn't we average out the happiness to the tune of $250? Well, not quite. Let's understand the word "happiness" (or call it "satisfaction")—this is the notion of utility of outcomes. Let's say you did pay $250 upfront to play the game. Then the coin toss outcome of HEAD is a net gain of $1000 − $250 = $750 and the coin toss outcome of TAIL is a net gain of −$500 − $250 = −$750 (i.e., net loss of $750). Now let's say the HEAD outcome gain of $750 gives you "happiness" of say 100 units. If the TAIL outcome loss of $750 gives you "unhappiness" of 100 units, then "happiness" and "unhappiness" levels cancel out, and in that case, it would be fair to pay $250 upfront to play the game. But it turns out that for most people, the "happiness"/"satisfaction" levels are asymmetric. If the "happiness" for $750 gain is 100 units, then the "unhappiness" for $750 loss is typically more than 100 units (let's say for you it's 120 units). This means you will pay an upfront amount X (less than $250) such that the difference in Utilities of $1000 and X is exactly the difference in the Utilities of X and −$500. Let's say this X amounts of $180. The gap of $70 ($250 − $180) represents your compensation for taking the risk, and it really comes down to the asymmetry in your assignment of utility to the outcomes.

Note that the degree of asymmetry of utility ("happiness" versus "unhappiness" for equal gains versus losses) is fairly individualized. Your utility assignment to outcomes might be different from your friend's. Your friend might be more asymmetric in assessing utility of the two outcomes and might assign 100 units of "happiness" for the gain outcome and 150 units of "unhappiness" for the loss outcome. So then your friend would pay an upfront amount X lower than the amount of $180 you paid upfront to play this game. Let's say the X for your friend works out to $100, so his compensation for taking the risk is $250 − $100 = $150, significantly more than your $70 of compensation for taking the same risk.

Thus we see that each individual's asymmetry in utility assignment to different outcomes results in this psychology of "I need to be compensated for taking this risk". We refer to this individualized demand of "compensation for risk" as the attitude of *Risk-Aversion*. It means that individuals have differing degrees of discomfort with taking risk, and they want to be compensated commensurately for taking risk. The amount of compensation they seek is called *Risk-Premium*. The more Risk-Averse an individual is, the more Risk-Premium the individual seeks. In the example above, your friend was more risk-averse than you. Your risk-premium was $70 and your friend's risk-premium was $150. But the most important concept that you are learning here is that the root-cause of Risk-Aversion is the asymmetry in the assignment of utility to outcomes of opposite sign and same magnitude. We have introduced this notion of "asymmetry of utility" in a simple, intuitive manner with this example, but we will soon embark on developing the formal theory for this notion, and introduce a simple and elegant mathematical framework for Utility Functions, Risk-Aversion and Risk-Premium.

A quick note before we get into the mathematical framework—you might be thinking that a typical casino would actually charge you a bit more than $250 upfront for playing the above game (because the casino needs to make a profit, on an expected basis), and people are indeed willing to pay this amount at a typical casino. So what about the risk-aversion we talked about earlier? The crucial point here is that people who play at casinos are looking for entertainment and excitement emanating purely from the psychological aspects of experiencing risk. They are willing to pay money for this entertainment and excitement, and this payment is separate from the cost of pure financial utility that we described above. So if people knew the true odds of pure-chance games of the type we described above and if people did not care for entertainment and excitement value of risk-taking in these games, focusing purely on financial utility, then what they'd be willing to pay upfront to play such a game will be based on the type of calculations we outlined above (meaning for the example we described, they'd typically pay less than $250 upfront to play the game).

7.3 THE SHAPE OF THE UTILITY FUNCTION

We seek a "valuation formula" for the amount we'd pay upfront to sign-up for situations like the simple example above, where we have uncertain outcomes with varying payoffs for the outcomes. Intuitively, we see that the amount we'd pay:

- Increases as the Mean of the outcome increases
- Decreases as the Variance of the outcome (i.e., Risk) increases
- Decreases as our Personal Risk-Aversion increases

The last two properties above enable us to establish the Risk-Premium. Now let us understand the nature of Utility as a function of financial outcomes. The key is to note that Utility is a non-linear function of financial outcomes. We call this non-linear function as the Utility function—it represents the "happiness"/"satisfaction" as a function of money. You should think of the concept of Utility in terms of *Utility of Consumption* of money, i.e., what exactly do the financial gains fetch you in your life or business. This is the idea of "value" (utility) derived from consuming the financial gains (or the negative utility of requisite financial recovery from monetary losses). So now let us look at another simple example to illustrate the concept of Utility of Consumption, this time not of consumption of money, but of consumption of cookies (to make the concept vivid and intuitive). Figure 7.1 shows two curves—we refer to the lower curve as the marginal satisfaction (utility) curve and the upper curve as the accumulated satisfaction (utility) curve. Marginal Utility refers to the *incremental satisfaction* we gain from an additional unit of consumption and Accumulated Utility refers to the *aggregate satisfaction* obtained from a certain number of units of consumption (in continuous-space, you can think of accumulated utility function as the integral, over consumption, of marginal utility function). In this example, we are consuming (i.e., eating) cookies. The marginal satisfaction curve tells us that the first cookie we eat provides us with 100 units of satisfaction (i.e., utility). The second cookie provides us 80 units of satisfaction, which is intuitive because you are not as hungry after eating the first cookie compared to before eating the first cookie. Also, the emotions of biting into the first cookie are extra positive because of the novelty of the experience. When you get to your 5th cookie, although you are still enjoying the cookie, you don't enjoy it as nearly as much as the first couple of cookies. The marginal satisfaction curve shows this—the 5th cookie provides us 30 units of satisfaction, and the 10th cookie provides us only 10 units of satisfaction. If we'd keep going, we might even find that the marginal satisfaction turns negative (as in, one might feel too full or maybe even feel like throwing up).

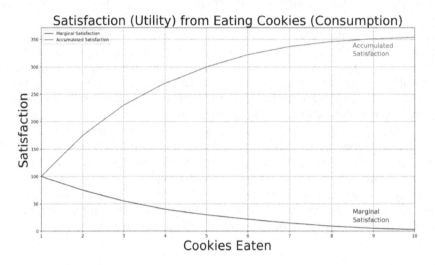

Figure 7.1 Utility Curve

So, we see that the marginal utility function is a decreasing function. Hence, accumulated utility function is a concave function. The accumulated utility function is the Utility of Consumption function (call it U) that we've been discussing so far. Let us denote the number of cookies eaten as x, and so the total "satisfaction" (utility) after eating x cookies is referred to as $U(x)$. In our financial examples, x would be amount of money one has at one's disposal and is typically an uncertain outcome, i.e., x is a random variable with an associated probability distribution. The extent of asymmetry in utility assignments for gains versus losses that we saw earlier manifests as extent of concavity of the $U(\cdot)$ function (which as we've discussed earlier, determines the extent of Risk-Aversion).

Now let's examine the concave nature of the Utility function for financial outcomes with another illustrative example. Let's say you have to pick between two situations:

- In Situation 1, you have a 10% probability of winning a million dollars (and 90% probability of winning 0).
- In Situation 2, you have a 0.1% probability of winning a billion dollars (and 99.9% probability of winning 0).

The expected winning in Situation 1 is $100,000 and the expected winning in Situation 2 is $1,000,000 (i.e., 10 times more than Situation 1). If you analyzed this naively as winning expectation maximization, you'd choose Situation 2. But most people would choose Situation 1. The reason for this is that the Utility of a billion dollars is nowhere close to 1000 times the utility of a million dollars (except for some very wealth people perhaps). In fact, the ratio of Utility of a billion dollars to Utility of a million dollars might be more like 10. So, the choice of Situation 1 over Situation 2 is usually quite clear—it's about Utility expectation maximization. So if the Utility of 0 dollars is 0 units, the Utility of a million dollars is say 1000 units, and the Utility of a billion dollars is say 10,000 units (i.e., 10 times that of a million dollars), then we see that the Utility of financial gains is a fairly concave function.

7.4 CALCULATING THE RISK-PREMIUM

Note that the concave nature of the $U(\cdot)$ function implies that:

$$\mathbb{E}[U(x)] < U(\mathbb{E}[x])$$

We define *Certainty-Equivalent Value* x_{CE} as:

$$x_{CE} = U^{-1}(\mathbb{E}[U(x)])$$

Certainty-Equivalent Value represents the certain amount we'd pay to consume an uncertain outcome. This is the amount of $180 you were willing to pay to play the casino game of the previous section.

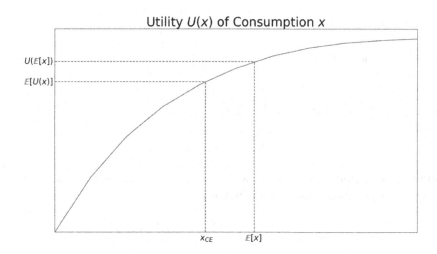

Figure 7.2 **Certainty-Equivalent Value**

Figure 7.2 illustrates this concept of Certainty-Equivalent Value in graphical terms. Next, we define Risk-Premium in two different conventions:

- **Absolute Risk-Premium** π_A:
$$\pi_A = \mathbb{E}[x] - x_{CE}$$

- **Relative Risk-Premium** π_R:
$$\pi_R = \frac{\pi_A}{\mathbb{E}[x]} = \frac{\mathbb{E}[x] - x_{CE}}{\mathbb{E}[x]} = 1 - \frac{x_{CE}}{\mathbb{E}[x]}$$

Now we develop mathematical formalism to derive formulas for Risk-Premia π_A and π_R in terms of the extent of Risk-Aversion and the extent of Risk itself. To lighten notation, we refer to $\mathbb{E}[x]$ as $\bar{x}$ and Variance of x as σ_x^2. We write $U(x)$ as the Taylor series expansion around $\bar{x}$ and ignore terms beyond quadratic in the expansion, as follows:

$$U(x) \approx U(\bar{x}) + U'(\bar{x}) \cdot (x - \bar{x}) + \frac{1}{2} U''(\bar{x}) \cdot (x - \bar{x})^2$$

Taking the expectation of $U(x)$ in the above formula, we get:

$$\mathbb{E}[U(x)] \approx U(\bar{x}) + \frac{1}{2} \cdot U''(\bar{x}) \cdot \sigma_x^2$$

Next, we write the Taylor-series expansion for $U(x_{CE})$ around $\bar{x}$ and ignore terms beyond linear in the expansion, as follows:

$$U(x_{CE}) \approx U(\bar{x}) + U'(\bar{x}) \cdot (x_{CE} - \bar{x})$$

Since $\mathbb{E}[U(x)] = U(x_{CE})$ (by definition of x_{CE}), the above two expressions are approximately the same. Hence,

$$U'(\bar{x}) \cdot (x_{CE} - \bar{x}) \approx \frac{1}{2} \cdot U''(\bar{x}) \cdot \sigma_x^2 \qquad (7.1)$$

From Equation (7.1), Absolute Risk-Premium

$$\pi_A = \bar{x} - x_{CE} \approx -\frac{1}{2} \cdot \frac{U''(\bar{x})}{U'(\bar{x})} \cdot \sigma_x^2$$

We refer to the function:

$$A(x) = -\frac{U''(x)}{U'(x)}$$

as the *Absolute Risk-Aversion* function. Therefore,

$$\pi_A \approx \frac{1}{2} \cdot A(\bar{x}) \cdot \sigma_x^2$$

In multiplicative uncertainty settings, we focus on the variance $\sigma_{\frac{x}{\bar{x}}}^2$ of $\frac{x}{\bar{x}}$. So in multiplicative settings, we focus on the Relative Risk-Premium:

$$\pi_R = \frac{\pi_A}{\bar{x}} \approx -\frac{1}{2} \cdot \frac{U''(\bar{x}) \cdot \bar{x}}{U'(\bar{x})} \cdot \frac{\sigma_x^2}{\bar{x}^2} = -\frac{1}{2} \cdot \frac{U''(\bar{x}) \cdot \bar{x}}{U'(\bar{x})} \cdot \sigma_{\frac{x}{\bar{x}}}^2$$

We refer to the function

$$R(x) = -\frac{U''(x) \cdot x}{U'(x)}$$

as the *Relative Risk-Aversion* function. Therefore,

$$\pi_R \approx \frac{1}{2} \cdot R(\bar{x}) \cdot \sigma_{\frac{x}{\bar{x}}}^2$$

Now let's take stock of what we've learning here. We've shown that Risk-Premium is proportional to the product of:

- Extent of Risk-Aversion: either $A(\bar{x})$ or $R(\bar{x})$
- Extent of uncertainty of outcome (i.e., Risk): either σ_x^2 or $\sigma_{\frac{x}{\bar{x}}}^2$

We've expressed the extent of Risk-Aversion to be proportional to the negative ratio of:

- Concavity of the Utility function (at $\bar{x}$): $-U''(\bar{x})$
- Slope of the Utility function (at $\bar{x}$): $U'(\bar{x})$

So for typical optimization problems in financial applications, we maximize $\mathbb{E}[U(x)]$ (not $\mathbb{E}[x]$), which in turn amounts to maximization of $x_{CE} = \mathbb{E}[x] - \pi_A$. If we refer to $\mathbb{E}[x]$ as our "Expected Return on Investment" (or simply "Return" for short) and π_A as the "risk-adjustment" due to risk-aversion and uncertainty of outcomes, then x_{CE} can be conceptualized as "risk-adjusted-return". Thus, in financial applications, we seek to maximize risk-adjusted-return x_{CE} rather than just the return $\mathbb{E}[x]$. It pays to emphasize here that the idea of maximizing risk-adjusted-return is essentially the idea of maximizing expected utility, and that the utility function is a representation of an individual's risk-aversion.

Note that Linear Utility function $U(x) = a + bx$ implies *Risk-Neutrality* (i.e., when one doesn't demand any compensation for taking risk). Next, we look at typically-used Utility functions $U(\cdot)$ with:

- Constant Absolute Risk-Aversion (CARA)
- Constant Relative Risk-Aversion (CRRA)

7.5   CONSTANT ABSOLUTE RISK-AVERSION (CARA)

Consider the Utility function $U : \mathbb{R} \to \mathbb{R}$, parameterized by $a \in \mathbb{R}$, defined as:

$$U(x) = \begin{cases} \frac{1-e^{-ax}}{a} & \text{for } a \neq 0 \\ x & \text{for } a = 0 \end{cases}$$

Firstly, note that $U(x)$ is continuous with respect to a for all $x \in \mathbb{R}$ since:

$$\lim_{a \to 0} \frac{1 - e^{-ax}}{a} = x$$

Now let us analyze the function $U(\cdot)$ for any fixed a. We note that for all $a \in \mathbb{R}$:

- $U(0) = 0$
- $U'(x) = e^{-ax} > 0$ for all $x \in \mathbb{R}$
- $U''(x) = -a \cdot e^{-ax}$

This means $U(\cdot)$ is a monotonically increasing function passing through the origin, and its curvature has the opposite sign as that of a (note: no curvature when $a = 0$).

So now we can calculate the Absolute Risk-Aversion function:

$$A(x) = \frac{-U''(x)}{U'(x)} = a$$

So we see that the Absolute Risk-Aversion function is the constant value a. Consequently, we say that this Utility function corresponds to *Constant Absolute Risk-Aversion (CARA)*. The parameter a is referred to as the Coefficient of CARA. The magnitude of positive a signifies the degree of risk-aversion. $a = 0$ is the case of being Risk-Neutral. Negative values of a mean one is "risk-seeking", i.e., one will pay to take risk (the opposite of risk-aversion) and the magnitude of negative a signifies the degree of risk-seeking.

If the random outcome $x \sim \mathcal{N}(\mu, \sigma^2)$, then using Equation (A.5) from Appendix A, we get:

$$\mathbb{E}[U(x)] = \begin{cases} \frac{1-e^{-a\mu + \frac{a^2\sigma^2}{2}}}{a} & \text{for } a \neq 0 \\ \mu & \text{for } a = 0 \end{cases}$$

$$x_{CE} = \mu - \frac{a\sigma^2}{2}$$

$$\text{Absolute Risk Premium } \pi_A = \mu - x_{CE} = \frac{a\sigma^2}{2}$$

For optimization problems where we need to choose across probability distributions where σ^2 is a function of μ, we seek the distribution that maximizes $x_{CE} = \mu - \frac{a\sigma^2}{2}$. This clearly illustrates the concept of "risk-adjusted-return" because μ serves as the "return" and the risk-adjustment $\frac{a\sigma^2}{2}$ is proportional to the product of risk-aversion a and risk (i.e., variance in outcomes) σ^2.

7.6   A PORTFOLIO APPLICATION OF CARA

Let's say we are given \$1 to invest and hold for a horizon of 1 year. Let's say our portfolio investment choices are:

- A risky asset with Annual Return $\sim \mathcal{N}(\mu, \sigma^2)$, $\mu \in \mathbb{R}, \sigma \in \mathbb{R}^+$.
- A riskless asset with Annual Return $= r \in \mathbb{R}$.

Our task is to determine the allocation π (out of the given \$1) to invest in the risky asset (so, $1 - \pi$ is invested in the riskless asset) so as to maximize the Expected Utility of Consumption of Portfolio Wealth in 1 year. Note that we allow π to be unconstrained, i.e., π can be any real number from $-\infty$ to $+\infty$. So, if $\pi > 0$, we buy the risky asset and if $\pi < 0$, we "short-sell" the risky asset. Investing π in the risky asset means in 1 year, the risky asset's value will be a normal distribution $\mathcal{N}(\pi(1 + \mu), \pi^2\sigma^2)$. Likewise, if $1 - \pi > 0$, we lend $1 - \pi$ (and will be paid back $(1 - \pi)(1 + r)$ in 1 year), and if $1 - \pi < 0$, we borrow $1 - \pi$ (and need to pay back $(1 - \pi)(1 + r)$ in 1 year).

Portfolio Wealth W in 1 year is given by:

$$W \sim \mathcal{N}(1 + r + \pi(\mu - r), \pi^2\sigma^2)$$

We assume CARA Utility with $a \neq 0$, so:

$$U(W) = \frac{1 - e^{-aW}}{a}$$

We know that maximizing $\mathbb{E}[U(W)]$ is equivalent to maximizing the Certainty-Equivalent Value of Wealth W, which in this case (using the formula for x_{CE} in the section on CARA) is given by:

$$1 + r + \pi(\mu - r) - \frac{a\pi^2\sigma^2}{2}$$

This is a quadratic concave function of π for $a > 0$, and so, taking its derivative with respect to π and setting it to 0 gives us the optimal investment fraction in the risky asset (π^*) as follows:

$$\pi^* = \frac{\mu - r}{a\sigma^2}$$

7.7 CONSTANT RELATIVE RISK-AVERSION (CRRA)

Consider the Utility function $U : \mathbb{R}^+ \to \mathbb{R}$, parameterized by $\gamma \in \mathbb{R}$, defined as:

$$U(x) = \begin{cases} \frac{x^{1-\gamma}-1}{1-\gamma} & \text{for } \gamma \neq 1 \\ \log(x) & \text{for } \gamma = 1 \end{cases}$$

Firstly, note that $U(x)$ is continuous with respect to γ for all $x \in \mathbb{R}^+$ since:

$$\lim_{\gamma \to 1} \frac{x^{1-\gamma}-1}{1-\gamma} = \log(x)$$

Now let us analyze the function $U(\cdot)$ for any fixed γ. We note that for all $\gamma \in \mathbb{R}$:

- $U(1) = 0$
- $U'(x) = x^{-\gamma} > 0$ for all $x \in \mathbb{R}^+$
- $U''(x) = -\gamma \cdot x^{-1-\gamma}$

This means $U(\cdot)$ is a monotonically increasing function passing through $(1, 0)$, and its curvature has the opposite sign as that of γ (note: no curvature when $\gamma = 0$).

So now we can calculate the Relative Risk-Aversion function:

$$R(x) = \frac{-U''(x) \cdot x}{U'(x)} = \gamma$$

So we see that the Relative Risk-Aversion function is the constant value γ. Consequently, we say that this Utility function corresponds to *Constant Relative Risk-Aversion (CRRA)*. The parameter γ is referred to as the Coefficient of CRRA. The magnitude of positive γ signifies the degree of risk-aversion. $\gamma = 0$ yields the Utility function $U(x) = x - 1$ and is the case of being Risk-Neutral. Negative values of γ mean one is "risk-seeking", i.e., one will pay to take risk (the opposite of risk-aversion) and the magnitude of negative γ signifies the degree of risk-seeking.

If the random outcome x is lognormal, with $\log(x) \sim \mathcal{N}(\mu, \sigma^2)$, then making a substitution $y = \log(x)$, expressing $\mathbb{E}[U(x)]$ as $\mathbb{E}[U(e^y)]$, and using Equation (A.5) in Appendix A, we get:

$$\mathbb{E}[U(x)] = \begin{cases} \frac{e^{\mu(1-\gamma) + \frac{\sigma^2}{2}(1-\gamma)^2} - 1}{1-\gamma} & \text{for } \gamma \neq 1 \\ \mu & \text{for } \gamma = 1 \end{cases}$$

$$x_{CE} = e^{\mu + \frac{\sigma^2}{2}(1-\gamma)}$$

$$\text{Relative Risk Premium } \pi_R = 1 - \frac{x_{CE}}{\bar{x}} = 1 - e^{-\frac{\sigma^2 \gamma}{2}}$$

For optimization problems where we need to choose across probability distributions where σ^2 is a function of μ, we seek the distribution that maximizes $\log(x_{CE}) = \mu + \frac{\sigma^2}{2}(1 - \gamma)$. Just like in the case of CARA, this clearly illustrates the concept of "risk-adjusted-return" because $\mu + \frac{\sigma^2}{2}$ serves as the "return" and the risk-adjustment $\frac{\gamma \sigma^2}{2}$ is proportional to the product of risk-aversion γ and risk (i.e., variance in outcomes) σ^2.

7.8 A PORTFOLIO APPLICATION OF CRRA

This application of CRRA is a special case of Merton's Portfolio Problem (Merton 1969) that we shall cover in its full generality in Chapter 8. This section requires us to have some basic familiarity with Stochastic Calculus (covered in Appendix C), specifically Ito Processes and Ito's Lemma. Here we consider the single-decision version of Merton's Portfolio Problem where our portfolio investment choices are:

- A risky asset, evolving in continuous time, with value denoted S_t at time t, whose movements are defined by the Ito process:

$$dS_t = \mu \cdot S_t \cdot dt + \sigma \cdot S_t \cdot dz_t$$

  where $\mu \in \mathbb{R}, \sigma \in \mathbb{R}^+$ are given constants, and z_t is 1-dimensional standard Brownian Motion.
- A riskless asset, growing continuously in time, with value denoted R_t at time t, whose growth is defined by the ordinary differential equation:

$$dR_t = r \cdot R_t \cdot dt$$

where $r \in \mathbb{R}$ is a given constant.

We are given \$1 to invest over a period of 1 year. We are asked to maintain a constant fraction of investment of wealth (denoted $\pi \in \mathbb{R}$) in the risky asset at each time t (with $1 - \pi$ as the fraction of investment in the riskless asset at each time t). Note that to maintain a constant fraction of investment in the risky asset, we need to continuously rebalance the portfolio of the risky asset and riskless asset. Our task is to determine the constant π that maximizes the Expected Utility of Consumption of Wealth at the end of 1 year. We allow π to be unconstrained, i.e., π can take any value from $-\infty$ to $+\infty$. Positive π means we have a "long" position in the risky asset and negative π means we have a "short" position in the risky asset. Likewise, positive $1 - \pi$ means we are lending money at the riskless interest rate of r and negative $1 - \pi$ means we are borrowing money at the riskless interest rate of r.

We denote the Wealth at time t as W_t. Without loss of generality, assume $W_0 = 1$. Since W_t is the portfolio wealth at time t, the value of the investment in the risky asset at time t would need to be $\pi \cdot W_t$ and the value of the investment in the riskless asset at time t would need to be $(1 - \pi) \cdot W_t$. Therefore, the change in the value of the risky asset investment from time t to time $t + dt$ is:

$$\mu \cdot \pi \cdot W_t \cdot dt + \sigma \cdot \pi \cdot W_t \cdot dz_t$$

Likewise, the change in the value of the riskless asset investment from time t to time $t + dt$ is:

$$r \cdot (1 - \pi) \cdot W_t \cdot dt$$

Therefore, the infinitesimal change in portfolio wealth dW_t from time t to time $t + dt$ is given by:

$$dW_t = (r + \pi(\mu - r)) \cdot W_t \cdot dt + \pi \cdot \sigma \cdot W_t \cdot dz_t$$

Note that this is an Ito process defining the stochastic evolution of portfolio wealth. Applying Ito's Lemma (see Appendix C) on $\log W_t$ gives us:

$$d(\log W_t) = ((r + \pi(\mu - r)) \cdot W_t \cdot \frac{1}{W_t} - \frac{\pi^2 \cdot \sigma^2 \cdot W_t^2}{2} \cdot \frac{1}{W_t^2}) \cdot dt + \pi \cdot \sigma \cdot W_t \cdot \frac{1}{W_t} \cdot dz_t$$

$$= (r + \pi(\mu - r) - \frac{\pi^2 \sigma^2}{2}) \cdot dt + \pi \cdot \sigma \cdot dz_t$$

Therefore,

$$\log W_t = \int_0^t (r + \pi(\mu - r) - \frac{\pi^2 \sigma^2}{2}) \cdot du + \int_0^t \pi \cdot \sigma \cdot dz_u$$

Using the martingale property and Ito Isometry for the Ito integral $\int_0^t \pi \cdot \sigma \cdot dz_u$ (see Appendix C), we get:

$$\log W_1 \sim \mathcal{N}(r + \pi(\mu - r) - \frac{\pi^2 \sigma^2}{2}, \pi^2 \sigma^2)$$

We assume CRRA Utility with $\gamma \neq 0$, so:

$$U(W_1) = \begin{cases} \frac{W_1^{1-\gamma}-1}{1-\gamma} & \text{for } \gamma \neq 1 \\ \log(W_1) & \text{for } \gamma = 1 \end{cases}$$

We know that maximizing $\mathbb{E}[U(W_1)]$ is equivalent to maximizing the Certainty-Equivalent Value of W_1, hence also equivalent to maximizing the log of the Certainty-Equivalent Value of W_1, which in this case (using the formula for x_{CE} from the section on CRRA) is given by:

$$r + \pi(\mu - r) - \frac{\pi^2 \sigma^2}{2} + \frac{\pi^2 \sigma^2 (1 - \gamma)}{2}$$

$$= r + \pi(\mu - r) - \frac{\pi^2 \sigma^2 \gamma}{2}$$

This is a quadratic concave function of π for $\gamma > 0$, and so, taking its derivative with respect to π and setting it to 0 gives us the optimal investment fraction in the risky asset (π^*) as follows:

$$\pi^* = \frac{\mu - r}{\gamma \sigma^2}$$

7.9   KEY TAKEAWAYS FROM THIS CHAPTER

- An individual's financial risk-aversion is represented by the concave nature of the individual's Utility as a function of financial outcomes.
- Risk-Premium (compensation an individual seeks for taking financial risk) is roughly proportional to the individual's financial risk-aversion and the measure of uncertainty in financial outcomes.
- Risk-Adjusted-Return in finance should be thought of as the Certainty-Equivalent-Value, whose Utility is the Expected Utility across uncertain (risky) financial outcomes.

CHAPTER 8

Dynamic Asset-Allocation and Consumption

This chapter covers the first of five financial applications of Stochastic Control covered in this book. This financial application deals with the topic of investment management for not just a financial company, but more broadly for any corporation or for any individual. The nuances for specific companies and individuals can vary considerably but what is common across these entities is the need to:

- Periodically decide how one's investment portfolio should be split across various choices of investment assets—the key being how much money to invest in more risky assets (which have potential for high returns on investment) versus less risky assets (that tend to yield modest returns on investment). This problem of optimally allocating capital across investment assets of varying risk-return profiles relates to the topic of Utility Theory we covered in Chapter 7. However, in this chapter, we deal with the further challenge of adjusting one's allocation of capital across assets, as time progresses. We refer to this feature as *Dynamic Asset Allocation* (the word dynamic refers to the adjustment of capital allocation to adapt to changing circumstances)

- Periodically decide how much capital to leave in one's investment portfolio versus how much money to consume for one's personal needs/pleasures (or for a corporation's operational requirements) by extracting money from one's investment portfolio. Extracting money from one's investment portfolio can mean potentially losing out on investment growth opportunities, but the flip side of this is the Utility of Consumption that a corporation/individual desires. Noting that ultimately our goal is to maximize total utility of consumption over a certain time horizon, this decision of investing versus consuming really amounts to the timing of consumption of one's money over the given time horizon.

Thus, this problem constitutes the dual and dynamic decisioning of asset-allocation and consumption. To gain an intuitive understanding of the challenge of this dual dynamic decisioning problem, let us consider this problem from the perspective of personal finance in a simplified setting.

8.1 OPTIMIZATION OF PERSONAL FINANCE

Personal Finances can be very simple for some people (earn a monthly salary, spend the entire salary) and can be very complicated for some other people (e.g., those who own multiple businesses in multiple countries and have complex assets and liabilities). Here we shall consider a situation that is relatively simple but includes sufficient nuances to provide you with the essential elements of the general problem of dynamic asset-allocation and consumption. Let's say your personal finances consist of the following aspects:

- *Receiving money*: This could include your periodic salary, which typically remains constant for a period of time, but can change if you get a promotion or if you get a new job. This also includes money you liquidate from your investment portfolio, e.g., if you sell some stock, and decide not to re-invest in other investment assets. This also includes interest you earn from your savings account or from some bonds you might own. There are many other ways one can *receive money*, some fixed regular payments and some uncertain in terms of payment quantity and timing, and we won't enumerate all the different ways of *receiving money*. We just want to highlight here that *receiving money* at various points in time is one of the key financial aspects in one's life.

- *Consuming money*: The word "consume" refers to "spending". Note that one needs to *consume money* periodically to satisfy basic needs like shelter, food and clothing. The rent or mortgage you pay on your house is one example—it may be a fixed amount every month, but if your mortgage rate is a floating rate, it is subject to variation. Moreover, if you move to a new house, the rent or mortgage can be different. The money you spend on food and clothing also constitutes *consuming money*. This can often be fairly stable from one month to the next, but if you have a newborn baby, it might require additional expenses of the baby's food, clothing and perhaps also toys. Then there is *consumption of money* that are beyond the "necessities"—things like eating out at a fancy restaurant on the weekend, taking a summer vacation, buying a luxury car or an expensive watch. One gains "satisfaction"/"happiness" (i.e., *Utility*) from this *consumption of money*. The key point here is that we need to periodically make a decision on how much to spend (*consume money*) on a weekly or monthly basis. One faces a tension in the dynamic decision between *consuming money* (that gives us *Consumption Utility*) and *saving money* (which is the money we put in our investment portfolio in the hope of the money growing, so we can consume potentially larger amounts of money in the future).

- *Investing Money*: Let us suppose there are a variety of investment assets you can invest in—simple savings account giving small interest, exchange-traded stocks (ranging from value stocks to growth stocks, with their respective risk-return tradeoffs), real-estate (the house you bought and live in is indeed considered an investment asset), commodities such as gold, paintings etc. We call the composition of money invested in these assets as one's investment portfolio (see Appendix B for a quick introduction to Portfolio Theory). Periodically, we need to decide if one should play safe by putting most of one's money in a savings account, or if we should allocate investment capital mostly in stocks, or if we should be more speculative and invest in an early-stage startup or in a rare painting. Reviewing the composition and potentially re-allocating capital (referred to as re-balancing one's portfolio) is the problem of dynamic asset-allocation. Note also that we can put some of our *received money* into our investment portfolio (meaning we choose to not consume that money right away). Likewise, we can extract some money out of our investment portfolio so we can *consume money*. The decisions of insertion and extraction of money into/from our investment portfolio is essentially the dynamic money-consumption decision we make, which goes together with the dynamic asset-allocation decision.

The above description has hopefully given you a flavor of the dual and dynamic decisioning of asset-allocation and consumption. Ultimately, our personal goal is to maximize the Expected Aggregated Utility of Consumption of Money over our lifetime (and perhaps, also include the Utility of Consumption of Money for one's spouse and children, after one dies). Since investment portfolios are stochastic in nature and since we have to periodically make decisions on asset-allocation and consumption, you can see that this has all the ingredients of a Stochastic Control problem, and hence can be modeled as a Markov Decision Process (albeit typically fairly complicated, since real-life finances have plenty of nuances). Here's a rough and informal sketch of what that MDP might look like (bear in mind that we will formalize the MDP for simplified cases later in this chapter):

- States: The *State* can be quite complex in general, but mainly it consists of one's age (to keep track of the time to reach the MDP horizon), the quantities of money invested in each investment asset, the valuation of the assets invested in, and potentially also other aspects like one's job/career situation (required to make predictions of future salary possibilities).

- Actions: The *Action* is two-fold. Firstly, it's the vector of investment amounts one chooses to make at each time step (the time steps are at the periodicity at which we review our investment portfolio for potential re-allocation of capital across assets). Secondly, it's the quantity of money one chooses to consume that is *flexible/optional* (i.e., beyond the fixed payments like rent that we are committed to make).

- Rewards: The *Reward* is the Utility of Consumption of Money that we deemed as flexible/optional—it corresponds to the second part of the *Action*.

- Model: The *Model* (probabilities of next state and reward, given current state and action) can be fairly complex in most real-life situations. The hardest aspect is the prediction of what might happen tomorrow in our life and career (we need this prediction since it determines our future likelihood to receive money, consume money and invest money). Moreover, the uncertain movements of investment assets would need to be captured by our model.

Since our goal here was to simply do a rough and informal sketch, the above coverage of the MDP is very hazy but we hope you get a sense for what the MDP might look like. Now we are ready to take a simple special case of this MDP which does away with many of the real-world frictions and complexities, yet retains the key features (in particular, the dual dynamic decisioning aspect). This simple special case was the subject of Merton's Portfolio Problem (Merton 1969) which he formulated and solved in 1969 in a landmark paper. A key feature of his formulation was that time is continuous and so, *state* (based on asset prices) evolves as a continuous-time stochastic process, and actions (asset-allocation and consumption) are made continuously. We cover the important parts of his paper in the next section. Note that our coverage below requires some familiarity with Stochastic Calculus (covered in Appendix C) and with the Hamilton-Jacobi-Bellman Equation (covered in Appendix D), which is the continuous-time analog of Bellman's Optimality Equation.

8.2 MERTON'S PORTFOLIO PROBLEM AND SOLUTION

Now we describe Merton's Portfolio problem and derive its analytical solution, which is one of the most elegant solutions in Mathematical Economics. The solution structure will provide tremendous intuition for how the asset-allocation and consumption decisions depend on not just the state variables but also on the problem inputs.

We denote time as t and say that current time is $t = 0$. Assume that you have just retired (meaning you won't be earning any money for the rest of your life) and that you are going to live for T more years (T is a fixed real number). So, in the language of the previous section, you will not be *receiving money* for the rest of your life, other than the option of extracting money from your investment portfolio. Also assume that you have no fixed payments to make like mortgage, subscriptions etc. (assume that you have already paid for a retirement service that provides you with your essential food, clothing and other services). This means all of your *money consumption* is flexible/optional, i.e., you have a choice of consuming any real non-negative number at any point in time. All of the above are big (and honestly, unreasonable) assumptions but they help keep the problem simple enough for analytical tractability. In spite of these over-simplified assumptions, the problem formulation still captures the salient aspects of dual dynamic decisioning of asset-allocation and consumption while eliminating the clutter of A) *receiving money* from external sources and B) *consuming money* that is of a non-optional nature.

We define wealth at any time t (denoted W_t) as the aggregate market value of your investment assets. Note that since no external money is received and since all consumption is optional, W_t is your "net-worth". Assume there are a fixed number n of risky assets and a single riskless asset. Assume that each risky asset has a known normal distribution of returns. Now we make a couple of big assumptions for analytical tractability:

- You are allowed to buy or sell any fractional quantities of assets at any point in time (i.e., in continuous time).
- There are no transaction costs with any of the buy or sell transactions in any of the assets.

You start with wealth W_0 at time $t = 0$. As mentioned earlier, the goal is to maximize your expected lifetime-aggregated Utility of Consumption of money with the actions at any point in time being two-fold: Asset-Allocation and Consumption (Consumption being equal to the capital extracted from the investment portfolio at any point in time). Note that since there is no external source of money and since all capital extracted from the investment portfolio at any point in time is immediately consumed, you are never adding capital to your investment portfolio. The growth of the investment portfolio can happen only from growth in the market value of assets in your investment portfolio. Lastly, we assume that the Consumption Utility function is Constant Relative Risk-Aversion (CRRA), which we covered in Chapter 7.

For ease of exposition, we formalize the problem setting and derive Merton's beautiful analytical solution for the case of $n = 1$ (i.e., only 1 risky asset). The solution generalizes in a straightforward manner to the case of $n > 1$ risky assets, so it pays to keep the notation and explanations simple, emphasizing intuition rather than heavy technical details.

Since we are operating in continuous-time, the risky asset follows a stochastic process (denoted S)—specifically an Ito Process (introductory background on Ito Processes and Ito's Lemma covered in Appendix C), as follows:

$$dS_t = \mu \cdot S_t \cdot dt + \sigma \cdot S_t \cdot dz_t$$

where $\mu \in \mathbb{R}, \sigma \in \mathbb{R}^+$ are fixed constants (note that for n assets, we would instead work with a vector for μ and a matrix for σ).

The riskless asset has no uncertainty associated with it and has a fixed rate of growth in continuous-time, so the valuation of the riskless asset R_t at time t is given by:

$$dR_t = r \cdot R_t \cdot dt$$

Assume $r \in \mathbb{R}$ is a fixed constant, representing the instantaneous riskless growth of money. We denote the consumption of wealth (equal to extraction of money from the investment portfolio) per unit time (at time t) as $c(t, W_t) \geq 0$ to make it clear that the consumption (our decision at any time t) will in general depend on both time t and wealth W_t. Note that we talk about "rate of consumption in time" because consumption is assumed to be continuous in time. As mentioned earlier, we denote wealth at time t as W_t (note that W is a stochastic process too). We assume that $W_t > 0$ for all $t \geq 0$. This is a reasonable assumption to make as it manifests in constraining the consumption (extraction from investment portfolio) to ensure wealth remains positive. We denote the fraction of wealth allocated to the risky asset at time t as $\pi(t, W_t)$. Just like consumption c, risky-asset allocation fraction π is a function of time t and wealth W_t. Since there is only one risky asset, the fraction of wealth allocated to the riskless asset at time t is $1 - \pi(t, W_t)$. Unlike the constraint $c(t, W_t) \geq 0$, $\pi(t, W_t)$ is assumed to be unconstrained. Note that $c(t, W_t)$ and $\pi(t, W_t)$ together constitute the decision (MDP action) at time t. To keep our notation light, we shall write c_t for $c(t, W_t)$ and π_t for $\pi(t, W_t)$, but please do recognize throughout the derivation that both are functions of wealth W_t at time t as well as of time t itself. Finally, we assume that the Utility of Consumption function is defined as:

$$U(x) = \frac{x^{1-\gamma}}{1 - \gamma}$$

for a risk-aversion parameter $\gamma \neq 1$. This Utility function is essentially the CRRA Utility function (ignoring the constant term $\frac{-1}{1-\gamma}$) that we covered in Chapter 7 for $\gamma \neq 1$. γ is the Coefficient of CRRA equal to $\frac{-x \cdot U''(x)}{U'(x)}$. We will not cover the case of CRRA Utility function for $\gamma = 1$ (i.e., $U(x) = \log(x)$), but we encourage you to work out the derivation for $U(x) = \log(x)$ as an exercise.

Due to our assumption of no addition of money to our investment portfolio of the risky asset S_t and riskless asset R_t and due to our assumption of no transaction costs of buying/selling any fractional quantities of risky as well as riskless assets, the time-evolution for wealth should be conceptualized as a continuous adjustment of the allocation π_t and continuous extraction from the portfolio (equal to continuous consumption c_t).

Since the value of the risky asset investment at time t is $\pi_t \cdot W_t$, the change in the value of the risky asset investment from time t to time $t + dt$ is:

$$\mu \cdot \pi_t \cdot W_t \cdot dt + \sigma \cdot \pi_t \cdot W_t \cdot dz_t$$

Likewise, since the value of the riskless asset investment at time t is $(1 - \pi_t) \cdot W_t$, the change in the value of the riskless asset investment from time t to time $t + dt$ is:

$$r \cdot (1 - \pi_t) \cdot W_t \cdot dt$$

Therefore, the infinitesimal change in wealth dW_t from time t to time $t + dt$ is given by:

$$dW_t = ((r + \pi_t \cdot (\mu - r)) \cdot W_t - c_t) \cdot dt + \pi_t \cdot \sigma \cdot W_t \cdot dz_t \tag{8.1}$$

Note that this is an Ito process defining the stochastic evolution of wealth.

Our goal is to determine optimal $(\pi(t, W_t), c(t, W_t))$ at any time t to maximize:

$$\mathbb{E}[\int_t^T \frac{e^{-\rho(s-t)} \cdot c_s^{1-\gamma}}{1 - \gamma} \cdot ds + \frac{e^{-\rho(T-t)} \cdot B(T) \cdot W_T^{1-\gamma}}{1 - \gamma} \mid W_t]$$

where $\rho \geq 0$ is the utility discount rate to account for the fact that future utility of consumption might be less than current utility of consumption, and $B(\cdot)$ is known as the "bequest" function (think of this as the money you will leave for your family when you die at time T). We can solve this problem for arbitrary bequest $B(T)$ but for simplicity, we shall consider

$B(T) = \epsilon^\gamma$ where $0 < \epsilon \ll 1$, meaning "no bequest". We require the bequest to be ϵ^γ rather than 0 for technical reasons, that will become apparent later.

We should think of this problem as a continuous-time Stochastic Control problem where the MDP is defined as below:

- The *State* at time t is (t, W_t)
- The *Action* at time t is (π_t, c_t)
- The *Reward* per unit time at time $t < T$ is:

$$U(c_t) = \frac{c_t^{1-\gamma}}{1-\gamma}$$

and the *Reward* at time T is:

$$B(T) \cdot U(W_T) = \epsilon^\gamma \cdot \frac{W_T^{1-\gamma}}{1-\gamma}$$

The *Return* at time t is the accumulated discounted *Reward*:

$$\int_t^T e^{-\rho(s-t)} \cdot \frac{c_s^{1-\gamma}}{1-\gamma} \cdot ds + \frac{e^{-\rho(T-t)} \cdot \epsilon^\gamma \cdot W_T^{1-\gamma}}{1-\gamma}$$

Our goal is to find the *Policy* : $(t, W_t) \to (\pi_t, c_t)$ that maximizes the *Expected Return*. Note the important constraint that $c_t \geq 0$, but π_t is unconstrained.

Our first step is to write out the Hamilton-Jacobi-Bellman (HJB) Equation (the analog of the Bellman Optimality Equation in continuous-time). We denote the Optimal Value Function as V^* such that the Optimal Value for wealth W_t at time t is $V^*(t, W_t)$. Note that unlike Section 5.13 in Chapter 5 where we denoted the Optimal Value Function as a time-indexed sequence $V_t^*(\cdot)$, here we make t an explicit functional argument of V^*. This is because in the continuous-time setting, we are interested in the time-differential of the Optimal Value Function. Appendix D provides the derivation of the general HJB formulation (Equation (D.1) in Appendix D)—this general HJB Equation specializes here to the following:

$$\max_{\pi_t, c_t}\{\mathbb{E}_t[dV^*(t, W_t) + \frac{c_t^{1-\gamma}}{1-\gamma} \cdot dt\} = \rho \cdot V^*(t, W_t) \cdot dt \tag{8.2}$$

Now use Ito's Lemma on dV^*, remove the dz_t term since it's a martingale, and divide throughout by dt to produce the HJB Equation in partial-differential form for any $0 \leq t < T$, as follows (the general form of this transformation appears as Equation (D.2) in Appendix D):

$$\max_{\pi_t, c_t}\{\frac{\partial V^*}{\partial t} + \frac{\partial V^*}{\partial W_t} \cdot ((\pi_t(\mu-r)+r)W_t - c_t) + \frac{\partial^2 V^*}{\partial W_t^2} \cdot \frac{\pi_t^2 \cdot \sigma^2 \cdot W_t^2}{2} + \frac{c_t^{1-\gamma}}{1-\gamma}\} = \rho \cdot V^*(t, W_t) \tag{8.3}$$

This HJB Equation is subject to the terminal condition:

$$V^*(T, W_T) = \epsilon^\gamma \cdot \frac{W_T^{1-\gamma}}{1-\gamma}$$

Let us write Equation (8.3) more succinctly as:

$$\max_{\pi_t, c_t} \Phi(t, W_t; \pi_t, c_t) = \rho \cdot V^*(t, W_t) \tag{8.4}$$

It pays to emphasize again that we are working with the constraints $W_t > 0, c_t \geq 0$ for $0 \leq t < T$

To find optimal π_t^*, c_t^*, we take the partial derivatives of $\Phi(t, W_t; \pi_t, c_t)$ with respect to π_t and c_t, and equate to 0 (first-order conditions for Φ). The partial derivative of Φ with respect to π_t is:

$$(\mu - r) \cdot \frac{\partial V^*}{\partial W_t} + \frac{\partial^2 V^*}{\partial W_t^2} \cdot \pi_t \cdot \sigma^2 \cdot W_t = 0$$

$$\Rightarrow \pi_t^* = \frac{-\frac{\partial V^*}{\partial W_t} \cdot (\mu - r)}{\frac{\partial^2 V^*}{\partial W_t^2} \cdot \sigma^2 \cdot W_t} \tag{8.5}$$

The partial derivative of Φ with respect to c_t is:

$$-\frac{\partial V^*}{\partial W_t} + (c_t^*)^{-\gamma} = 0$$

$$\Rightarrow c_t^* = \left(\frac{\partial V^*}{\partial W_t}\right)^{\frac{-1}{\gamma}} \tag{8.6}$$

Now substitute π_t^* (from Equation (8.5))and c_t^* (from Equation (8.6)) in $\Phi(t, W_t; \pi_t, c_t)$ (in Equation (8.3)) and equate to $\rho \cdot V^*(t, W_t)$. This gives us the Optimal Value Function Partial Differential Equation (PDE):

$$\frac{\partial V^*}{\partial t} - \frac{(\mu - r)^2}{2\sigma^2} \cdot \frac{\left(\frac{\partial V^*}{\partial W_t}\right)^2}{\frac{\partial^2 V^*}{\partial W_t^2}} + \frac{\partial V^*}{\partial W_t} \cdot r \cdot W_t + \frac{\gamma}{1 - \gamma} \cdot \left(\frac{\partial V^*}{\partial W_t}\right)^{\frac{\gamma-1}{\gamma}} = \rho \cdot V^*(t, W_t) \tag{8.7}$$

The boundary condition for this PDE is:

$$V^*(T, W_T) = \epsilon^\gamma \cdot \frac{W_T^{1-\gamma}}{1 - \gamma}$$

The second-order conditions for Φ are satisfied under the assumptions: $c_t^* > 0, W_t > 0, \frac{\partial^2 V^*}{\partial W_t^2} < 0$ for all $0 \leq t < T$ (we will later show that these are all satisfied in the solution we derive), and for concave $U(\cdot)$, i.e., $\gamma > 0$

Next, we want to reduce the PDE (8.7) to an Ordinary Differential Equation (ODE) so we can solve the (simpler) ODE. Towards this goal, we surmise with a guess solution in terms of a deterministic function (f) of time:

$$V^*(t, W_t) = f(t)^\gamma \cdot \frac{W_t^{1-\gamma}}{1 - \gamma} \tag{8.8}$$

Then,

$$\frac{\partial V^*}{\partial t} = \gamma \cdot f(t)^{\gamma-1} \cdot f'(t) \cdot \frac{W_t^{1-\gamma}}{1 - \gamma} \tag{8.9}$$

$$\frac{\partial V^*}{\partial W_t} = f(t)^\gamma \cdot W_t^{-\gamma} \tag{8.10}$$

$$\frac{\partial^2 V^*}{\partial W_t^2} = -f(t)^\gamma \cdot \gamma \cdot W_t^{-\gamma-1} \tag{8.11}$$

Substituting the guess solution in the PDE, we get the simple ODE:

$$f'(t) = \nu \cdot f(t) - 1 \tag{8.12}$$

where

$$\nu = \frac{\rho - (1 - \gamma) \cdot \left(\frac{(\mu-r)^2}{2\sigma^2\gamma} + r\right)}{\gamma}$$

We note that the bequest function $B(T) = \epsilon^\gamma$ proves to be convenient in order to fit the guess solution for $t = T$. This means the boundary condition for this ODE is: $f(T) = \epsilon$. Consequently, this ODE together with this boundary condition has a simple enough solution, as follows:

$$f(t) = \begin{cases} \frac{1+(\nu\epsilon-1)\cdot e^{-\nu(T-t)}}{\nu} & \text{for } \nu \neq 0 \\ T - t + \epsilon & \text{for } \nu = 0 \end{cases} \tag{8.13}$$

Substituting V^* (from Equation (8.8)) and its partial derivatives (from Equations (8.9), (8.10) and (8.11)) in Equations (8.5) and (8.6), we get:

$$\pi^*(t, W_t) = \frac{\mu - r}{\sigma^2\gamma} \tag{8.14}$$

$$c^*(t, W_t) = \frac{W_t}{f(t)} = \begin{cases} \frac{\nu \cdot W_t}{1+(\nu\epsilon-1)\cdot e^{-\nu(T-t)}} & \text{for } \nu \neq 0 \\ \frac{W_t}{T-t+\epsilon} & \text{for } \nu = 0 \end{cases} \tag{8.15}$$

Finally, substituting the solution for $f(t)$ (Equation (8.13)) in Equation (8.8), we get:

$$V^*(t, W_t) = \begin{cases} \frac{(1+(\nu\epsilon-1)\cdot e^{-\nu(T-t)})^\gamma}{\nu^\gamma} \cdot \frac{W_t^{1-\gamma}}{1-\gamma} & \text{for } \nu \neq 0 \\ \frac{(T-t+\epsilon)^\gamma \cdot W_t^{1-\gamma}}{1-\gamma} & \text{for } \nu = 0 \end{cases} \tag{8.16}$$

Note that $f(t) > 0$ for all $0 \leq t < T$ (for all ν) ensures $W_t > 0$, $c_t^* > 0$, $\frac{\partial^2 V^*}{\partial W_t^2} < 0$. This ensures the constraints $W_t > 0$ and $c_t \geq 0$ are satisfied and the second-order conditions for Φ are also satisfied. A very important lesson in solving Merton's Portfolio problem is the fact that the HJB Formulation is key and that this solution approach provides a template for similar continuous-time stochastic control problems.

8.3 DEVELOPING INTUITION FOR THE SOLUTION TO MERTON'S PORTFOLIO PROBLEM

The solution for $\pi^*(t, W_t)$ and $c^*(t, W_t)$ are surprisingly simple. $\pi^*(t, W_t)$ is a constant, i.e., it is independent of both of the state variables t and W_t. This means that no matter what wealth we carry and no matter how close we are to the end of the horizon (i.e., no matter what our age is), we should invest the same fraction of our wealth in the risky asset (likewise for the case of n risky assets). The simplifying assumptions in Merton's Portfolio problem statement did play a part in the simplicity of the solution, but the fact that $\pi^*(t, W_t)$ is a constant is still rather surprising. The simplicity of the solution means that asset allocation is straightforward—we just need to keep re-balancing to maintain this constant fraction of our wealth in the risky asset. We expect our wealth to grow over time and so, the capital in the risky asset would also grow proportionately.

The form of the solution for $c^*(t, W_t)$ is extremely intuitive—the excess return of the risky asset ($\mu - r$) shows up in the numerator, which makes sense, since one would expect to invest a higher fraction of one's wealth in the risky asset if it gives us a higher excess return. It also makes sense that the volatility σ of the risky asset (squared) shows up in the denominator (the greater the volatility, the less we'd allocate to the risky asset, since we are typically risk-averse, i.e., $\gamma > 0$). Likewise, it makes since that the coefficient of CRRA γ

shows up in the denominator since a more risk-averse individual (greater value of γ) will want to invest less in the risky asset.

The Optimal Consumption Rate $c^*(t, W_t)$ should be conceptualized in terms of the *Optimal Fractional Consumption Rate*, i.e., the Optimal Consumption Rate $c^*(t, W_t)$ as a fraction of the Wealth W_t. Note that the Optimal Fractional Consumption Rate depends only on t (it is equal to $\frac{1}{f(t)}$). This means no matter what our wealth is, we should be extracting a fraction of our wealth on a daily/monthly/yearly basis that is only dependent on our age. Note also that if $\epsilon < \frac{1}{\nu}$, the Optimal Fractional Consumption Rate increases as time progresses. This makes intuitive sense because when we have many more years to live, we'd want to consume less and invest more to give the portfolio more ability to grow, and when we get close to our death, we increase our consumption (since the optimal is "to die broke", assuming no bequest).

Now let us understand how the Wealth process evolves. Let us substitute for $\pi^*(t, W_t)$ (from Equation (8.14)) and $c^*(t, W_t)$ (from Equation (8.15)) in the Wealth process defined in Equation (8.1). This yields the following Wealth process W^* when we asset-allocate optimally and consume optimally:

$$dW_t^* = (r + \frac{(\mu - r)^2}{\sigma^2 \gamma} - \frac{1}{f(t)}) \cdot W_t^* \cdot dt + \frac{\mu - r}{\sigma \gamma} \cdot W_t^* \cdot dz_t \qquad (8.17)$$

The first thing to note about this Wealth process is that it is a lognormal process of the form covered in Section C.7 of Appendix C. The lognormal volatility (fractional dispersion) of this wealth process is constant $(= \frac{\mu - r}{\sigma \gamma})$. The lognormal drift (fractional drift) is independent of the wealth but is dependent on time $(= r + \frac{(\mu - r)^2}{\sigma^2 \gamma} - \frac{1}{f(t)})$. From the solution of the general lognormal process derived in Section C.7 of Appendix C, we conclude that:

$$\mathbb{E}[W_t^*] = W_0 \cdot e^{(r + \frac{(\mu - r)^2}{\sigma^2 \gamma})t} \cdot e^{-\int_0^t \frac{du}{f(u)}} = \begin{cases} W_0 \cdot e^{(r + \frac{(\mu - r)^2}{\sigma^2 \gamma})t} \cdot (1 - \frac{1 - e^{-\nu t}}{1 + (\nu \epsilon - 1) \cdot e^{-\nu T}}) & \text{if } \nu \neq 0 \\ W_0 \cdot e^{(r + \frac{(\mu - r)^2}{\sigma^2 \gamma})t} \cdot (1 - \frac{t}{T + \epsilon}) & \text{if } \nu = 0 \end{cases}$$
$$(8.18)$$

Since we assume no bequest, we should expect the Wealth process to keep growing up to some point in time and then fall all the way down to 0 when time runs out (i.e., when $t = T$). We shall soon write the code for Equation (8.18) and plot the graph for this rise and fall. An important point to note is that although the wealth process growth varies in time (expected wealth growth rate = $r + \frac{(\mu - r)^2}{\sigma^2 \gamma} - \frac{1}{f(t)}$ as seen from Equation (8.17)), the variation (in time) of the wealth process growth is only due to the fractional consumption rate varying in time. If we ignore the fractional consumption rate $(= \frac{1}{f(t)})$, then what we get is the Expected Portfolio Annual Return of $r + \frac{(\mu - r)^2}{\sigma^2 \gamma}$ which is a constant (does not depend on either time t or on Wealth W_t^*). Now let us write some code to calculate the time-trajectories of Expected Wealth, Fractional Consumption Rate, Expected Wealth Growth Rate and Expected Portfolio Annual Return.

The code should be pretty self-explanatory. We will just provide a few explanations of variables in the code that may not be entirely obvious: `portfolio_return` calculates the Expected Portfolio Annual Return, `nu` calculates the value of ν, `f` represents the function $f(t)$, `wealth_growth_rate` calculates the Expected Wealth Growth Rate as a function of time t. The `expected_wealth` method assumes $W_0 = 1$.

```
@dataclass(frozen=True)
class MertonPortfolio:
    mu: float
    sigma: float
    r: float
```

```
rho: float
horizon: float
gamma: float
epsilon: float = 1e-6

def excess(self) -> float:
    return self.mu - self.r

def variance(self) -> float:
    return self.sigma * self.sigma

def allocation(self) -> float:
    return self.excess() / (self.gamma * self.variance())

def portfolio_return(self) -> float:
    return self.r + self.allocation() * self.excess()

def nu(self) -> float:
    return (self.rho - (1 - self.gamma) * self.portfolio_return()) / \
        self.gamma

def f(self, time: float) -> float:
    remaining: float = self.horizon - time
    nu = self.nu()
    if nu == 0:
        ret = remaining + self.epsilon
    else:
        ret = (1 + (nu * self.epsilon - 1) * exp(-nu * remaining)) / nu
    return ret

def fractional_consumption_rate(self, time: float) -> float:
    return 1 / self.f(time)

def wealth_growth_rate(self, time: float) -> float:
    return self.portfolio_return() - self.fractional_consumption_rate(time)

def expected_wealth(self, time: float) -> float:
    base: float = exp(self.portfolio_return() * time)
    nu = self.nu()
    if nu == 0:
        ret = base * (1 - (1 - exp(-nu * time)) /
                      (1 + (nu * self.epsilon - 1) *
                       exp(-nu * self.horizon)))
    else:
        ret = base * (1 - time / (self.horizon + self.epsilon))
    return ret
```

The above code is in the file rl/chapter7/merton_solution_graph.py. We highly encourage you to experiment by changing the various inputs in this code $(T, \mu, \sigma, r, \rho, \gamma)$ and visualize how the results change. Doing this will help build tremendous intuition.

A rather interesting observation is that if $r + \frac{(\mu-r)^2}{\sigma^2\gamma} > \frac{1}{f(0)}$ and $\epsilon < \frac{1}{\nu}$, then the Fractional Consumption Rate is initially less than the Expected Portfolio Annual Return and over time, the Fractional Consumption Rate becomes greater than the Expected Portfolio Annual Return. This illustrates how the optimal behavior is to consume modestly and invest more when one is younger, then to gradually increase the consumption as one ages, and finally to ramp up the consumption sharply when one is close to the end of one's life. Figure 8.1 shows the visual for this (along with the Expected Wealth Growth Rate) using the above code for input values of: $T = 20, \mu = 10\%, \sigma = 10\%, r = 2\%, \rho = 1\%, \gamma = 2.0$.

Figure 8.2 shows the time-trajectory of the expected wealth based on Equation (8.18) for the same input values as listed above. Notice how the Expected Wealth rises in a convex shape for several years since the consumption during all these years is quite modest, and then the shape of the Expected Wealth curve turns concave at about 12 years, peaks at about 16 years (when Fractional Consumption Rate rises to equal Expected Portfolio Annual Return), and then falls precipitously in the last couple of years (as the Consumption increasingly drains the Wealth down to 0).

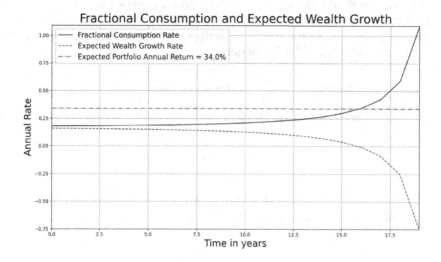

Figure 8.1 Portfolio Return and Consumption Rate

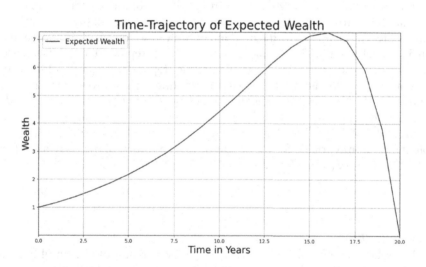

Figure 8.2 Expected Wealth Time-Trajectory

8.4 A DISCRETE-TIME ASSET-ALLOCATION EXAMPLE

In this section, we cover a discrete-time version of the problem that lends itself to analytical tractability, much like Merton's Portfolio Problem in continuous-time. We are given wealth W_0 at time 0. At each of discrete time steps labeled $t = 0, 1, \ldots, T - 1$, we are allowed to allocate the wealth W_t at time t to a portfolio of a risky asset and a riskless asset in an unconstrained manner with no transaction costs. The risky asset yields a random return $\sim \mathcal{N}(\mu, \sigma^2)$ over each single time step (for a given $\mu \in \mathbb{R}$ and a given $\sigma \in \mathbb{R}^+$). The riskless asset yields a constant return denoted by r over each single time step (for a given $r \in \mathbb{R}$). We assume that there is no consumption of wealth at any time $t < T$, and that we liquidate and consume the wealth W_T at time T. So our goal is simply to maximize the Expected

Utility of Wealth at the final time step $t = T$ by dynamically allocating $x_t \in \mathbb{R}$ in the risky asset and the remaining $W_t - x_t$ in the riskless asset for each $t = 0, 1, \ldots, T - 1$. Assume the single-time-step discount factor is γ and that the Utility of Wealth at the final time step $t = T$ is given by the following CARA function:

$$U(W_T) = \frac{1 - e^{-aW_T}}{a} \text{ for some fixed } a \neq 0$$

Thus, the problem is to maximize, for each $t = 0, 1, \ldots, T - 1$, over choices of $x_t \in \mathbb{R}$, the value:

$$\mathbb{E}[\gamma^{T-t} \cdot \frac{1 - e^{-aW_T}}{a} | (t, W_t)]$$

Since γ^{T-t} and a are constants, this is equivalent to maximizing, for each $t = 0, 1, \ldots, T - 1$, over choices of $x_t \in \mathbb{R}$, the value:

$$\mathbb{E}[\frac{-e^{-aW_T}}{a} | (t, W_t)] \tag{8.19}$$

We formulate this problem as a *Continuous States* and *Continuous Actions* discrete-time finite-horizon MDP by specifying its *State Transitions*, *Rewards* and *Discount Factor* precisely. The problem then is to solve the MDP's Control problem to find the Optimal Policy.

The terminal time for the finite-horizon MDP is T and hence, all the states at time $t = T$ are terminal states. We shall follow the notation of finite-horizon MDPs that we had covered in Section 5.13 of Chapter 5. The *State* $s_t \in \mathcal{S}_t$ at any time step $t = 0, 1, \ldots, T$ consists of the wealth W_t. The decision (*Action*) $a_t \in \mathcal{A}_t$ at any time step $t = 0, 1, \ldots, T - 1$ is the quantity of investment in the risky asset ($= x_t$). Hence, the quantity of investment in the riskless asset at time t will be $W_t - x_t$. A deterministic policy at time t (for all $t = 0, 1, \ldots T - 1$) is denoted as π_t, and hence, we write: $\pi_t(W_t) = x_t$. Likewise, an optimal deterministic policy at time t (for all $t = 0, 1, \ldots, T - 1$) is denoted as π_t^*, and hence, we write: $\pi_t^*(W_t) = x_t^*$.

Denote the random variable for the single-time-step return of the risky asset from time t to time $t + 1$ as $Y_t \sim \mathcal{N}(\mu, \sigma^2)$ for all $t = 0, 1, \ldots T - 1$. So,

$$W_{t+1} = x_t \cdot (1 + Y_t) + (W_t - x_t) \cdot (1 + r) = x_t \cdot (Y_t - r) + W_t \cdot (1 + r) \tag{8.20}$$

for all $t = 0, 1, \ldots, T - 1$.

The MDP *Reward* is 0 for all $t = 0, 1, \ldots, T - 1$. As a result of the simplified objective (8.19) above, the MDP *Reward* for $t = T$ is the following random quantity:

$$\frac{-e^{-aW_T}}{a}$$

We set the MDP discount factor to be $\gamma = 1$ (again, because of the simplified objective (8.19) above).

We denote the Value Function at time t (for all $t = 0, 1, \ldots, T - 1$) for a given policy $\pi = (\pi_0, \pi_1, \ldots, \pi_{T-1})$ as:

$$V_t^\pi(W_t) = \mathbb{E}_\pi[\frac{-e^{-aW_T}}{a} | (t, W_t)]$$

We denote the Optimal Value Function at time t (for all $t = 0, 1, \ldots, T - 1$) as:

$$V_t^*(W_t) = \max_\pi V_t^\pi(W_t) = \max_\pi \{\mathbb{E}_\pi[\frac{-e^{-aW_T}}{a} | (t, W_t)]\}$$

The Bellman Optimality Equation is:

$$V_t^*(W_t) = \max_{x_t} Q_t^*(W_t, x_t) = \max_{x_t}\{\mathbb{E}_{Y_t \sim \mathcal{N}(\mu, \sigma^2)}[V_{t+1}^*(W_{t+1})]\}$$

for all $t = 0, 1, \ldots, T - 2$, and

$$V_{T-1}^*(W_{T-1}) = \max_{x_{T-1}} Q_{T-1}^*(W_{T-1}, x_{T-1}) = \max_{x_{T-1}}\{\mathbb{E}_{Y_{T-1} \sim \mathcal{N}(\mu, \sigma^2)}[\frac{-e^{-aW_T}}{a}]\}$$

where Q_t^* is the Optimal Action-Value Function at time t for all $t = 0, 1, \ldots, T - 1$.

We make an educated guess for the functional form of the Optimal Value Function as:

$$V_t^*(W_t) = -b_t \cdot e^{-c_t \cdot W_t} \tag{8.21}$$

where b_t, c_t are independent of the wealth W_t for all $t = 0, 1, \ldots, T - 1$. Next, we express the Bellman Optimality Equation using this functional form for the Optimal Value Function:

$$V_t^*(W_t) = \max_{x_t}\{\mathbb{E}_{Y_t \sim \mathcal{N}(\mu, \sigma^2)}[-b_{t+1} \cdot e^{-c_{t+1} \cdot W_{t+1}}]\}$$

Using Equation (8.20), we can write this as:

$$V_t^*(W_t) = \max_{x_t}\{\mathbb{E}_{Y_t \sim \mathcal{N}(\mu, \sigma^2)}[-b_{t+1} \cdot e^{-c_{t+1} \cdot (x_t \cdot (Y_t - r) + W_t \cdot (1+r))}]\}$$

The expectation of this exponential form (under the normal distribution) evaluates to:

$$V_t^*(W_t) = \max_{x_t}\{-b_{t+1} \cdot e^{-c_{t+1} \cdot (1+r) \cdot W_t - c_{t+1} \cdot (\mu - r) \cdot x_t + c_{t+1}^2 \cdot \frac{\sigma^2}{2} \cdot x_t^2}\} \tag{8.22}$$

Since $V_t^*(W_t) = \max_{x_t} Q_t^*(W_t, x_t)$, from Equation (8.22), we can infer the functional form for $Q_t^*(W_t, x_t)$ in terms of b_{t+1} and c_{t+1}:

$$Q_t^*(W_t, x_t) = -b_{t+1} \cdot e^{-c_{t+1} \cdot (1+r) \cdot W_t - c_{t+1} \cdot (\mu - r) \cdot x_t + c_{t+1}^2 \cdot \frac{\sigma^2}{2} \cdot x_t^2} \tag{8.23}$$

Since the right-hand-side of the Bellman Optimality Equation (8.22) involves a max over x_t, we can say that the partial derivative of the term inside the max with respect to x_t is 0. This enables us to write the Optimal Allocation x_t^* in terms of c_{t+1}, as follows:

$$-c_{t+1} \cdot (\mu - r) + \sigma^2 \cdot c_{t+1}^2 \cdot x_t^* = 0$$

$$\Rightarrow x_t^* = \frac{\mu - r}{\sigma^2 \cdot c_{t+1}} \tag{8.24}$$

Next, we substitute this maximizing x_t^* in the Bellman Optimality Equation (Equation (8.22)):

$$V_t^*(W_t) = -b_{t+1} \cdot e^{-c_{t+1} \cdot (1+r) \cdot W_t - \frac{(\mu - r)^2}{2\sigma^2}}$$

But since

$$V_t^*(W_t) = -b_t \cdot e^{-c_t \cdot W_t}$$

we can write the following recursive equations for b_t and c_t:

$$b_t = b_{t+1} \cdot e^{-\frac{(\mu - r)^2}{2\sigma^2}}$$

$$c_t = c_{t+1} \cdot (1 + r)$$

We can calculate b_{T-1} and c_{T-1} from the knowledge of the MDP *Reward* $\frac{-e^{-aW_T}}{a}$ (Utility of Terminal Wealth) at time $t = T$, which will enable us to unroll the above recursions for b_t and c_t for all $t = 0, 1, \ldots, T - 2$.

$$V_{T-1}^*(W_{T-1}) = \max_{x_{T-1}}\{\mathbb{E}_{Y_{T-1}\sim\mathcal{N}(\mu,\sigma^2)}[\frac{-e^{-aW_T}}{a}]\}$$

From Equation (8.20), we can write this as:

$$V_{T-1}^*(W_{T-1}) = \max_{x_{T-1}}\{\mathbb{E}_{Y_{T-1}\sim\mathcal{N}(\mu,\sigma^2)}[\frac{-e^{-a(x_{T-1}\cdot(Y_{T-1}-r)+W_{T-1}\cdot(1+r))}}{a}]\}$$

Using the result in Equation (A.9) in Appendix A, we can write this as:

$$V_{T-1}^*(W_{T-1}) = \frac{-e^{-\frac{(\mu-r)^2}{2\sigma^2}-a\cdot(1+r)\cdot W_{T-1}}}{a}$$

Therefore,

$$b_{T-1} = \frac{e^{-\frac{(\mu-r)^2}{2\sigma^2}}}{a}$$

$$c_{T-1} = a \cdot (1+r)$$

Now we can unroll the above recursions for b_t and c_t for all $t = 0, 1, \ldots T - 2$ as:

$$b_t = \frac{e^{-\frac{(\mu-r)^2\cdot(T-t)}{2\sigma^2}}}{a}$$

$$c_t = a \cdot (1+r)^{T-t}$$

Substituting the solution for c_{t+1} in Equation (8.24) gives us the solution for the Optimal Policy:

$$\pi_t^*(W_t) = x_t^* = \frac{\mu - r}{\sigma^2 \cdot a \cdot (1+r)^{T-t-1}} \tag{8.25}$$

for all $t = 0, 1, \ldots, T-1$. Note that the optimal action at time step t (for all $t = 0, 1, \ldots, T-1$) does not depend on the state W_t at time t (it only depends on the time t). Hence, the optimal policy $\pi_t^*(\cdot)$ for a fixed time t is a constant deterministic policy function.

Substituting the solutions for b_t and c_t in Equation (8.21) gives us the solution for the Optimal Value Function:

$$V_t^*(W_t) = \frac{-e^{-\frac{(\mu-r)^2(T-t)}{2\sigma^2}}}{a} \cdot e^{-a(1+r)^{T-t}\cdot W_t} \tag{8.26}$$

for all $t = 0, 1, \ldots, T - 1$.

Substituting the solutions for b_{t+1} and c_{t+1} in Equation (8.23) gives us the solution for the Optimal Action-Value Function:

$$Q_t^*(W_t, x_t) = \frac{-e^{-\frac{(\mu-r)^2(T-t-1)}{2\sigma^2}}}{a} \cdot e^{-a(1+r)^{T-t}\cdot W_t - a(\mu-r)(1+r)^{T-t-1}\cdot x_t + \frac{(a\sigma(1+r)^{T-t-1})^2}{2}\cdot x_t^2} \tag{8.27}$$

for all $t = 0, 1, \ldots, T - 1$.

8.5 PORTING TO REAL-WORLD

We have covered a continuous-time setting and a discrete-time setting with simplifying assumptions that provide analytical tractability. The specific simplifying assumptions that enabled analytical tractability were:

- Normal distribution of asset returns
- CRRA/CARA assumptions
- Frictionless markets/trading (no transaction costs, unconstrained and continuous prices/allocation amounts/consumption)

But real-world problems involving dynamic asset-allocation and consumption are not so simple and clean. We have arbitrary, more complex asset price movements. Utility functions don't fit into simple CRRA/CARA formulas. In practice, trading often occurs in discrete space—asset prices, allocation amounts and consumption are often discrete quantities. Moreover, when we change our asset allocations or liquidate a portion of our portfolio to consume, we incur transaction costs. Furthermore, trading doesn't always happen in continuous-time—there are typically specific windows of time where one is locked-out from trading or there are trading restrictions. Lastly, many investments are illiquid (e.g., real-estate) or simply not allowed to be liquidated until a certain horizon (e.g., retirement funds), which poses major constraints on extracting money from one's portfolio for consumption. So even though prices/allocation amounts/consumption might be close to being continuous-variables, the other above-mentioned frictions mean that we don't get the benefits of calculus that we obtained in the simple examples we covered.

With the above real-world considerations, we need to tap into Dynamic Programming—more specifically, Approximate Dynamic Programming since real-world problems have large state spaces and large action spaces (even if these spaces are not continuous, they tend to be close to continuous). Appropriate function approximation of the Value Function is key to solving these problems. Implementing a full-blown real-world investment and consumption management system is beyond the scope of this book, but let us implement an illustrative example that provides sufficient understanding of how a full-blown real-world example would be implemented. We have to keep things simple enough and yet sufficiently general. So here is the setting we will implement for:

- One risky asset and one riskless asset.
- Finite number of time steps (discrete-time setting akin to Section 8.4).
- No consumption (i.e., no extraction from the investment portfolio) until the end of the finite horizon, and hence, without loss of generality, we set the discount factor equal to 1.
- Arbitrary distribution of return for the risky asset, allowing the distribution of returns to vary over time (`risky_return_distributions: Sequence[Distribution[float]]` in the code below).
- The return on the riskless asset, varying in time (`riskless_returns: Sequence[float]` in the code below).
- Arbitrary Utility Function (`utility_func: Callable[[float], float]` in the code below).
- Finite number of choices of investment amounts in the risky asset at each time step (`risky_alloc_choices: Sequence[float]` in the code below).
- Arbitrary distribution of initial wealth W_0 (`initial_wealth_distribution: Distribution[float]` in the code below).

The code in the class `AssetAllocDiscrete` below is fairly self-explanatory. We use the function `back_opt_qvf` covered in Section 6.9 of Chapter 6 to perform backward induction on the optimal Q-Value Function. Since the state space is continuous, the optimal Q-Value Function is represented as a `QValueFunctionApprox` (specifically, as a `DNNApprox`). Moreover, since we are working with a generic distribution of returns that govern the state transitions of this MDP, we need to work with the methods of the abstract class `MarkovDecisionProcess` (and not the class `FiniteMarkovDecisionProcess`). The method `backward_induction_qvf` below makes the call to `back_opt_qvf`. Since the risky returns distribution is arbitrary and since the utility function is arbitrary, we don't have prior knowledge of the functional form of the Q-Value function. Hence, the user of the class `AssetAllocDiscrete` also needs to provide the set of feature functions (`feature_functions` in the code below) and the specification of a deep neural network to represent the Q-Value function (`dnn_spec` in the code below). The rest of the code below is mainly about preparing the input `mdp_f0_mu_triples` to be passed to `back_opt_qvf`. As was explained in Section 6.9 of Chapter 6, `mdp_f0_mu_triples` is a sequence (for each time step) of the following triples:

- A `MarkovDecisionProcess[float, float]` object, which in the code below is prepared by the method `get_mdp`. *State* is the portfolio wealth (`float` type) and *Action* is the quantity of investment in the risky asset (also of `float` type). `get_mdp` creates a class `AssetAllocMDP` that implements the abstract class `MarkovDecisionProcess`. To do so, we need to implement the `step` method and the `actions` method. The `step` method returns an instance of `SampledDistribution`, which is based on the `sr_sampler_func` that returns a sample of the pair of next state (next time step's wealth) and reward, given the current state (current wealth) and action (current time step's quantity of investment in the risky asset).
- A `QValueFunctionApprox[float], float]` object, prepared by `get_qvf_func_approx`. This method sets up a `DNNApprox[Tuple[NonTerminal[float], float]]` object that represents a neural-network function approximation for the optimal Q-Value Function. So the input to this neural network would be a `Tuple[NonTerminal[float], float]` representing a (state, action) pair.
- An `NTStateDistribution[float]` object prepared by `get_states_distribution`, which returns a `SampledDistribution[NonTerminal[float]]` representing the distribution of non-terminal states (distribution of portfolio wealth) at each time step.

The `SampledDistribution[NonTerminal[float]]` is prepared by `states_sampler_func` that generates a sampling trace by sampling the state-transitions (portfolio wealth transitions) from time 0 to the given time step in a time-incremental manner (invoking the `sample` method of the risky asset's return `Distributions` and the `sample` method of a uniform distribution over the action choices specified by `risky_alloc_choices`).

```
from rl.distribution import Distribution, SampledDistribution, Choose
from rl.function_approx import DNNSpec, AdamGradient, DNNApprox
from rl.approximate_dynamic_programming import back_opt_qvf, QValueFunctionApprox
from operator import itemgetter
import numpy as np

@dataclass(frozen=True)
class AssetAllocDiscrete:
    risky_return_distributions: Sequence[Distribution[float]]
    riskless_returns: Sequence[float]
    utility_func: Callable[[float], float]
    risky_alloc_choices: Sequence[float]
    feature_functions: Sequence[Callable[[Tuple[float, float]], float]]
    dnn_spec: DNNSpec
    initial_wealth_distribution: Distribution[float]

    def time_steps(self) -> int:
```

```python
            return len(self.risky_return_distributions)

    def uniform_actions(self) -> Choose[float]:
        return Choose(self.risky_alloc_choices)

    def get_mdp(self, t: int) -> MarkovDecisionProcess[float, float]:
        distr: Distribution[float] = self.risky_return_distributions[t]
        rate: float = self.riskless_returns[t]
        alloc_choices: Sequence[float] = self.risky_alloc_choices
        steps: int = self.time_steps()
        utility_f: Callable[[float], float] = self.utility_func

        class AssetAllocMDP(MarkovDecisionProcess[float, float]):

            def step(
                self,
                wealth: NonTerminal[float],
                alloc: float
            ) -> SampledDistribution[Tuple[State[float], float]]:

                def sr_sampler_func(
                    wealth=wealth,
                    alloc=alloc
                ) -> Tuple[State[float], float]:
                    next_wealth: float = alloc * (1 + distr.sample()) \
                        + (wealth.state - alloc) * (1 + rate)
                    reward: float = utility_f(next_wealth) \
                        if t == steps - 1 else 0.
                    next_state: State[float] = Terminal(next_wealth) \
                        if t == steps - 1 else NonTerminal(next_wealth)
                    return (next_state, reward)

                return SampledDistribution(
                    sampler=sr_sampler_func,
                    expectation_samples=1000
                )

            def actions(self, wealth: NonTerminal[float]) -> Sequence[float]:
                return alloc_choices

        return AssetAllocMDP()

    def get_qvf_func_approx(self) -> \
            DNNApprox[Tuple[NonTerminal[float], float]]:

        adam_gradient: AdamGradient = AdamGradient(
            learning_rate=0.1,
            decay1=0.9,
            decay2=0.999
        )
        ffs: List[Callable[[Tuple[NonTerminal[float], float]], float]] = []
        for f in self.feature_functions:
            def this_f(pair: Tuple[NonTerminal[float], float], f=f) -> float:
                return f((pair[0].state, pair[1]))
            ffs.append(this_f)

        return DNNApprox.create(
            feature_functions=ffs,
            dnn_spec=self.dnn_spec,
            adam_gradient=adam_gradient
        )

    def get_states_distribution(self, t: int) -> \
            SampledDistribution[NonTerminal[float]]:

        actions_distr: Choose[float] = self.uniform_actions()

        def states_sampler_func() -> NonTerminal[float]:
            wealth: float = self.initial_wealth_distribution.sample()
            for i in range(t):
                distr: Distribution[float] = self.risky_return_distributions[i]
                rate: float = self.riskless_returns[i]
                alloc: float = actions_distr.sample()
                wealth = alloc * (1 + distr.sample()) + \
```

```
                        (wealth - alloc) * (1 + rate)
                return NonTerminal(wealth)

        return SampledDistribution(states_sampler_func)

    def backward_induction_qvf(self) -> \
            Iterator[QValueFunctionApprox[float, float]]:

        init_fa: DNNApprox[Tuple[NonTerminal[float], float]] = \
            self.get_qvf_func_approx()

        mdp_f0_mu_triples: Sequence[Tuple[
            MarkovDecisionProcess[float, float],
            DNNApprox[Tuple[NonTerminal[float], float]],
            SampledDistribution[NonTerminal[float]]
        ]] = [(
            self.get_mdp(i),
            init_fa,
            self.get_states_distribution(i)
        ) for i in range(self.time_steps())]

        num_state_samples: int = 300
        error_tolerance: float = 1e-6

        return back_opt_qvf(
            mdp_f0_mu_triples=mdp_f0_mu_triples,
            gamma=1.0,
            num_state_samples=num_state_samples,
            error_tolerance=error_tolerance
        )
```

The above code is in the file rl/chapter7/asset_alloc_discrete.py. We encourage you to create a few different instances of AssetAllocDiscrete by varying its inputs (try different return distributions, different utility functions, different action spaces). But how do we know the code above is correct? We need a way to test it. A good test is to specialize the inputs to fit the setting of Section 8.4 for which we have a closed-form solution to compare against. So let us write some code to specialize the inputs to fit this setting. Since the above code has been written with an educational motivation rather than an efficient-computation motivation, the convergence of the backward induction ADP algorithm is going to be slow. So we shall test it on a small number of time steps and provide some assistance for fast convergence (using limited knowledge from the closed-form solution in specifying the function approximation). We write code below to create an instance of AssetAllocDiscrete with time steps $T = 4$, $\mu = 13\%$, $\sigma = 20\%$, $r = 7\%$, coefficient of CARA $a = 1.0$. We set up risky_return_distributions as a sequence of identical Gaussian distributions, riskless_returns as a sequence of identical riskless rate of returns, and utility_func as a lambda parameterized by the coefficient of CARA a. We know from the closed-form solution that the optimal allocation to the risky asset for each of time steps $t = 0, 1, 2, 3$ is given by:

$$x_t^* = \frac{1.5}{1.07^{4-t}}$$

Therefore, we set risky_alloc_choices (action choices) in the range $[1.0, 2.0]$ in increments of 0.1 to see if our code can hit the correct values within the 0.1 granularity of action choices.

To specify feature_functions and dnn_spec, we need to leverage the functional form of the closed-form solution for the Action-Value function (i.e., Equation (8.27)). We observe that we can write this as:

$$Q_t^*(W_t, x_t) = -sign(a) \cdot e^{-(\alpha_0 + \alpha_1 \cdot W_t + \alpha_2 \cdot x_t + \alpha_3 \cdot x_t^2)}$$

where

$$\alpha_0 = \frac{(\mu - r)^2(T - t - 1)}{2\sigma^2} + \log(|a|)$$

$$\alpha_1 = a(1 + r)^{T-t}$$

$$\alpha_2 = a(\mu - r)(1 + r)^{T-t-1}$$

$$\alpha_3 = -\frac{(a\sigma(1 + r)^{T-t-1})^2}{2}$$

This means, the function approximation for Q_t^* can be set up with a neural network with no hidden layers, with the output layer activation function as $g(S) = -sign(a) \cdot e^{-S}$, and with the feature functions as:

$$\phi_1((W_t, x_t)) = 1$$

$$\phi_2((W_t, x_t)) = W_t$$

$$\phi_3((W_t, x_t)) = x_t$$

$$\phi_4((W_t, x_t)) = x_t^2$$

We set `initial_wealth_distribution` to be a normal distribution with a mean of `init_wealth` (set equal to 1.0 below) and a standard distribution of `init_wealth_stdev` (set equal to a small value of 0.1 below).

```
from rl.distribution import Gaussian

steps: int = 4
mu: float = 0.13
sigma: float = 0.2
r: float = 0.07
a: float = 1.0
init_wealth: float = 1.0
init_wealth_stdev: float = 0.1

excess: float = mu - r
var: float = sigma * sigma
base_alloc: float = excess / (a * var)

risky_ret: Sequence[Gaussian] = [Gaussian(mu=mu, sigma=sigma)
                                 for _ in range(steps)]
riskless_ret: Sequence[float] = [r for _ in range(steps)]
utility_function: Callable[[float], float] = lambda x: - np.exp(-a * x) / a
alloc_choices: Sequence[float] = np.linspace(
    2 / 3 * base_alloc,
    4 / 3 * base_alloc,
    11
)
feature_funcs: Sequence[Callable[[Tuple[float, float]], float]] = \
    [
        lambda _: 1.,
        lambda w_x: w_x[0],
        lambda w_x: w_x[1],
        lambda w_x: w_x[1] * w_x[1]
    ]
dnn: DNNSpec = DNNSpec(
    neurons=[],
    bias=False,
    hidden_activation=lambda x: x,
    hidden_activation_deriv=lambda y: np.ones_like(y),
    output_activation=lambda x: - np.sign(a) * np.exp(-x),
    output_activation_deriv=lambda y: -y
)
init_wealth_distr: Gaussian = Gaussian(
    mu=init_wealth,
```

```
    sigma=init_wealth_stdev
)
aad: AssetAllocDiscrete = AssetAllocDiscrete(
    risky_return_distributions=risky_ret,
    riskless_returns=riskless_ret,
    utility_func=utility_function,
    risky_alloc_choices=alloc_choices,
    feature_functions=feature_funcs,
    dnn_spec=dnn,
    initial_wealth_distribution=init_wealth_distr
)
```

Next, we perform the Q-Value backward induction, step through the returned iterator (fetching the Q-Value function for each time step from $t = 0$ to $t = T - 1$), and evaluate the Q-values at the init_wealth (for each time step) for all alloc_choices. Performing a max and arg max over the alloc_choices at the init_wealth gives us the Optimal Value function and the Optimal Policy for each time step for wealth equal to init_wealth.

```
from pprint import pprint
it_qvf: Iterator[QValueFunctionApprox[float, float]] = \
    aad.backward_induction_qvf()
for t, q in enumerate(it_qvf):
    print(f"Time {t:d}")
    print()
    opt_alloc: float = max(
        ((q((NonTerminal(init_wealth), ac)), ac) for ac in alloc_choices),
        key=itemgetter(0)
    )[1]
    val: float = max(q((NonTerminal(init_wealth), ac))
                    for ac in alloc_choices)
    print(f"Opt Risky Allocation = {opt_alloc:.3f}, Opt Val = {val:.3f}")
    print("Optimal Weights below:")
    for wts in q.weights:
        pprint(wts.weights)
    print()
```

This prints the following:

```
Time 0

Opt Risky Allocation = 1.200, Opt Val = -0.225
Optimal Weights below:
array([[ 0.13318188,  1.31299678,  0.07327264, -0.03000281]])


Time 1

Opt Risky Allocation = 1.300, Opt Val = -0.257
Optimal Weights below:
array([[ 0.08912411,  1.22479503,  0.07002802, -0.02645654]])


Time 2

Opt Risky Allocation = 1.400, Opt Val = -0.291
Optimal Weights below:
array([[ 0.03772409,  1.144612  ,  0.07373166, -0.02566819]])


Time 3
```

```
Opt Risky Allocation = 1.500, Opt Val = -0.328
Optimal Weights below:
array([[ 0.00126822,  1.0700996 ,  0.05798272, -0.01924149]])
```

Now let's compare these results against the closed-form solution.

```
for t in range(steps):
    print(f"Time {t:d}")
    print()
    left: int = steps - t
    growth: float = (1 + r) ** (left - 1)
    alloc: float = base_alloc / growth
    val: float = - np.exp(- excess * excess * left / (2 * var)
                          - a * growth * (1 + r) * init_wealth) / a
    bias_wt: float = excess * excess * (left - 1) / (2 * var) + \
        np.log(np.abs(a))
    w_t_wt: float = a * growth * (1 + r)
    x_t_wt: float = a * excess * growth
    x_t2_wt: float = - var * (a * growth) ** 2 / 2

    print(f"Opt Risky Allocation = {alloc:.3f}, Opt Val = {val:.3f}")
    print(f"Bias Weight = {bias_wt:.3f}")
    print(f"W_t Weight = {w_t_wt:.3f}")
    print(f"x_t Weight = {x_t_wt:.3f}")
    print(f"x_t^2 Weight = {x_t2_wt:.3f}")
    print()
```

This prints the following:

```
Time 0

Opt Risky Allocation = 1.224, Opt Val = -0.225
Bias Weight = 0.135
W_t Weight = 1.311
x_t Weight = 0.074
x_t^2 Weight = -0.030

Time 1

Opt Risky Allocation = 1.310, Opt Val = -0.257
Bias Weight = 0.090
W_t Weight = 1.225
x_t Weight = 0.069
x_t^2 Weight = -0.026

Time 2

Opt Risky Allocation = 1.402, Opt Val = -0.291
Bias Weight = 0.045
W_t Weight = 1.145
x_t Weight = 0.064
x_t^2 Weight = -0.023

Time 3

Opt Risky Allocation = 1.500, Opt Val = -0.328
Bias Weight = 0.000
```

```
W_t Weight = 1.070
x_t Weight = 0.060
x_t^2 Weight = -0.020
```

As mentioned previously, this serves as a good test for the correctness of the implementation of `AssetAllocDiscrete`.

We need to point out here that the general case of dynamic asset allocation and consumption for a large number of risky assets will involve a continuous-valued action space of high dimension. This means ADP algorithms will have challenges in performing the max / arg max calculation across this large and continuous action space. Even many of the RL algorithms find it challenging to deal with very large action spaces. Sometimes we can take advantage of the specifics of the control problem to overcome this challenge. But in a general setting, these large/continuous action space require special types of RL algorithms that are well suited to tackle such action spaces. One such class of RL algorithms is Policy Gradient Algorithms that we shall learn in Chapter 14.

8.6 KEY TAKEAWAYS FROM THIS CHAPTER

- A fundamental problem in Mathematical Finance is that of jointly deciding on A) optimal investment allocation (among risky and riskless investment assets) and B) optimal consumption, over a finite horizon. Merton, in his landmark paper from 1969, provided an elegant closed-form solution under assumptions of continuous-time, normal distribution of returns on the assets, CRRA utility, and frictionless transactions.
- In a more general setting of the above problem, we need to model it as an MDP. If the MDP is not too large and if the asset return distributions are known, we can employ finite-horizon ADP algorithms to solve it. However, in typical real-world situations, the action space can be quite large and the asset return distributions are unknown. This points to RL, and specifically RL algorithms that are well suited to tackle large action spaces (such as Policy Gradient Algorithms).

CHAPTER 9

Derivatives Pricing and Hedging

In this chapter, we cover two applications of MDP Control regarding financial derivatives' pricing and hedging (the word *hedging* refers to reducing or eliminating market risks associated with a derivative). The first application is to identify the optimal time/state to exercise an American Option (a type of financial derivative) in an idealized market setting (akin to the "frictionless" market setting of Merton's Portfolio problem from Chapter 8). Optimal exercise of an American Option is the key to determining its fair price. The second application is to identify the optimal hedging strategy for derivatives in real-world situations (technically referred to as *incomplete markets*, a term we will define shortly). The optimal hedging strategy of a derivative is the key to determining its fair price in the real-world (incomplete market) setting. Both of these applications can be cast as Markov Decision Processes where the Optimal Policy gives the Optimal Hedging/Optimal Exercise in the respective applications, leading to the fair price of the derivatives under consideration. Casting these derivatives applications as MDPs means that we can tackle them with Dynamic Programming or Reinforcement Learning algorithms, providing an interesting and valuable alternative to the traditional methods of pricing derivatives.

In order to understand and appreciate the modeling of these derivatives applications as MDPs, one requires some background in the classical theory of derivatives pricing. Unfortunately, thorough coverage of this theory is beyond the scope of this book, and we refer you to Tomas Bjork's book on Arbitrage Theory in Continuous Time (Björk 2005) for a thorough understanding of this theory. We shall spend much of this chapter covering the very basics of this theory, and in particular explaining the key technical concepts (such as arbitrage, replication, risk-neutral measure, market-completeness etc.) in a simple and intuitive manner. In fact, we shall cover the theory for the very simple case of discrete-time with a single-period. While that is nowhere near enough to do justice to the rich continuous-time theory of derivatives pricing and hedging, this is the best we can do in a single chapter. The good news is that MDP-modeling of the two problems we want to solve—optimal exercise of american options and optimal hedging of derivatives in a real-world (incomplete market) setting—doesn't require one to have a thorough understanding of the classical theory. Rather, an intuitive understanding of the key technical and economic concepts should suffice, which we bring to life in the simple setting of discrete-time with a single-period. We start this chapter with a quick introduction to derivatives, next we describe the simple setting of a single-period with formal mathematical notation, covering the key concepts (arbitrage, replication, risk-neutral measure, market-completeness etc.), state and prove the all-important fundamental theorems of asset pricing (only for the single-period setting),

DOI: 10.1201/9781003229193-9

and finally show how these two derivatives applications can be cast as MDPs, along with the appropriate algorithms to solve the MDPs.

9.1 A BRIEF INTRODUCTION TO DERIVATIVES

If you are reading this book, you likely already have some familiarity with Financial Derivatives (or at least have heard of them, given that derivatives were at the center of the 2008 financial crisis). In this section, we sketch an overview of financial derivatives and refer you to the book by John Hull (Hull 2010) for a thorough coverage of Derivatives. The term "Derivative" is based on the word "derived"—it refers to the fact that a derivative is a financial instrument whose structure and hence, value is derived from the *performance* of an underlying entity or entities (which we shall simply refer to as "underlying"). The underlying can be pretty much any financial entity—it could be a stock, currency, bond, basket of stocks, or something more exotic like another derivative. The term *performance* also refers to something fairly generic—it could be the price of a stock or commodity, it could be the interest rate a bond yields, it could be the average price of a stock over a time interval, it could be a market-index, or it could be something more exotic like the implied volatility of an option (which itself is a type of derivative). Technically, a derivative is a legal contract between the derivative buyer and seller that either:

- Entitles the derivative buyer to cashflow (which we'll refer to as derivative *payoff*) at future point(s) in time, with the payoff being contingent on the underlying's performance (i.e., the payoff is a precise mathematical function of the underlying's performance, e.g., a function of the underlying's price at a future point in time). This type of derivative is known as a "lock-type" derivative.
- Provides the derivative buyer with choices at future points in time, upon making which, the derivative buyer can avail of cashflow (i.e., *payoff*) that is contingent on the underlying's performance. This type of derivative is known as an "option-type" derivative (the word "option" referring to the choice or choices the buyer can make to trigger the contingent payoff).

Although both "lock-type" and "option-type" derivatives can both get very complex (with contracts running over several pages of legal descriptions), we now illustrate both these types of derivatives by going over the most basic derivative structures. In the following descriptions, current time (when the derivative is bought/sold) is denoted as time $t = 0$.

9.1.1 Forwards

The most basic form of Forward Contract involves specification of:

- A future point in time $t = T$ (we refer to T as expiry of the forward contract).
- The fixed payment K to be made by the forward contract buyer to the seller at time $t = T$.

In addition, the contract establishes that at time $t = T$, the forward contract seller needs to deliver the underlying (say a stock with price S_t at time t) to the forward contract buyer. This means at time $t = T$, effectively the payoff for the buyer is $S_T - K$ (likewise, the payoff for the seller is $K - S_T$). This is because the buyer, upon receiving the underlying from the seller, can immediately sell the underlying in the market for the price of S_T and so, would have made a gain of $S_T - K$ (note $S_T - K$ can be negative, in which case the payoff for the buyer is negative).

The problem of forward contract "pricing" is to determine the fair value of K so that the price of this forward contract derivative at the time of contract creation is 0. As time t progresses, the underlying price might fluctuate, which would cause a movement away from the initial price of 0. If the underlying price increases, the price of the forward would naturally increase (and if the underlying price decreases, the price of the forward would naturally decrease). This is an example of a "lock-type" derivative since neither the buyer nor the seller of the forward contract need to make any choices at time $t = T$. Rather, the payoff for the buyer is determined directly by the formula $S_T - K$ and the payoff for the seller is determined directly by the formula $K - S_T$.

9.1.2 European Options

The most basic forms of European Options are European Call and Put Options. The most basic European Call Option contract involves specification of:

- A future point in time $t = T$ (we refer to T as the expiry of the Call Option).
- Underlying Price K known as *Strike Price*.

The contract gives the buyer (owner) of the European Call Option the right, but not the obligation, to buy the underlying at time $t = T$ at the price of K. Since the option owner doesn't have the obligation to buy, if the price S_T of the underlying at time $t = T$ ends up being equal to or below K, the rational decision for the option owner would be to not buy (at price K), which would result in a payoff of 0 (in this outcome, we say that the call option is *out-of-the-money*). However, if $S_T > K$, the option owner would make an instant profit of $S_T - K$ by *exercising* her right to buy the underlying at the price of K. Hence, the payoff in this case is $S_T - K$ (in this outcome, we say that the call option is *in-the-money*). We can combine the two cases and say that the payoff is $f(S_T) = \max(S_T - K, 0)$. Since the payoff is always non-negative, the call option owner would need to pay for this privilege. The amount the option owner would need to pay to own this call option is known as the fair price of the call option. Identifying the value of this fair price is the highly celebrated problem of *Option Pricing* (which you will learn more about as this chapter progresses).

A European Put Option is very similar to a European Call Option with the only difference being that the owner of the European Put Option has the right (but not the obligation) to *sell* the underlying at time $t = T$ at the price of K. This means that the payoff is $f(S_T) = \max(K - S_T, 0)$. Payoffs for these Call and Put Options are known as "hockey-stick" payoffs because if you plot the $f(\cdot)$ function, it is a flat line on the *out-of-the-money* side and a sloped line on the *in-the-money* side. Such European Call and Put Options are "Option-Type" (and not "Lock-Type") derivatives since they involve a choice to be made by the option owner (the choice of exercising the right to buy/sell at the Strike Price K). However, it is possible to construct derivatives with the same payoff as these European Call/Put Options by simply writing in the contract that the option owner will get paid $\max(S_T - K, 0)$ (in case of Call Option) or will get paid $\max(K - S_T, 0)$ (in case of Put Option) at time $t = T$. Such derivatives contracts do away with the option owner's exercise choice and hence, they are "Lock-Type" contracts. There is a subtle difference—setting these derivatives up as "Option-Type" means the option owner might act "irrationally"— the call option owner might mistakenly buy even if $S_T < K$, or the call option owner might for some reason forget/neglect to exercise her option even when $S_T > K$. Setting up such contracts as "Lock-Type" takes away the possibilities of these types of irrationalities from the option owner.

A more general European Derivative involves an arbitrary function $f(\cdot)$ (generalizing from the hockey-stick payoffs) and could be set up as "Option-Type" or "Lock-Type".

9.1.3 American Options

The term "European" above refers to the fact that the option to exercise is available only at a fixed point in time $t = T$. Even if it is set up as "Lock-Type", the term "European" typically means that the payoff can happen only at a fixed point in time $t = T$. This is in contrast to American Options. The most basic forms of American Options are American Call and Put Options. American Call and Put Options are essentially extensions of the corresponding European Call and Put Options by allowing the buyer (owner) of the American Option to exercise the option to buy (in the case of Call) or sell (in the case of Put) at any time $t \leq T$. The allowance of exercise at any time at or before the expiry time T can often be a tricky financial decision for the option owner. At each point in time when the American Option is *in-the-money* (i.e., positive payoff upon exercise), the option owner might be tempted to exercise and collect the payoff but might as well be thinking that if she waits, the option might become more *in-the-money* (i.e., prospect of a bigger payoff if she waits for a while). Hence, it's clear that an American Option is always of the "Option-Type" (and not "Lock-Type") since the timing of the decision (option) to exercise is very important in the case of an American Option. This also means that the problem of pricing an American Option (the fair price the buyer would need to pay to own an American Option) is much harder than the problem of pricing an European Option.

So what purpose do derivatives serve? There are actually many motivations for different market participants, but we'll just list two key motivations. The first reason is to protect against adverse market movements that might damage the value of one's portfolio (this is known as *hedging*). As an example, buying a put option can reduce or eliminate the risk associated with ownership of the underlying. The second reason is operational or financial convenience in trading to express a speculative view of market movements. For instance, if one thinks a stock will increase in value by 50% over the next two years, instead of paying say $100,000 to buy the stock (hoping to make $50,000 after two years), one can simply buy a call option on $100,000 of the stock (paying the option price of say $5,000). If the stock price indeed appreciates by 50% after 2 years, one makes $50,000 − $5,000 = $45,000. Although one made $5000 less than the alternative of simply buying the stock, the fact that one needs to pay $5000 (versus $50,000) to enter into the trade means the potential *return on investment* is much higher.

Next, we embark on the journey of learning how to value derivatives, i.e., how to figure out the fair price that one would be willing to buy or sell the derivative for at any point in time. As mentioned earlier, the general theory of derivatives pricing is quite rich and elaborate (based on continuous-time stochastic processes), and we don't cover it in this book. Instead, we provide intuition for the core concepts underlying derivatives pricing theory in the context of a simple, special case—that of discrete-time with a single-period. We formalize this simple setting in the next section.

9.2 NOTATION FOR THE SINGLE-PERIOD SIMPLE SETTING

Our simple setting involves discrete time with a single-period from $t = 0$ to $t = 1$. Time $t = 0$ has a single state which we shall refer to as the "Spot" state. Time $t = 1$ has n random outcomes formalized by the sample space $\Omega = \{\omega_1, \ldots, \omega_n\}$. The probability distribution of this finite sample space is given by the probability mass function

$$\mu : \Omega \to [0, 1]$$

such that

$$\sum_{i=1}^{n} \mu(\omega_i) = 1$$

This simple single-period setting involves $m + 1$ fundamental assets $A_0, A_1, \ldots, A_m$ where A_0 is a riskless asset (i.e., its price will evolve deterministically from $t = 0$ to $t = 1$) and $A_1, \ldots, A_m$ are risky assets. We denote the Spot Price (at $t = 0$) of A_j as $S_j^{(0)}$ for all $j = 0, 1, \ldots, m$. We denote the Price of A_j in ω_i as $S_j^{(i)}$ for all $j = 0, \ldots, m, i = 1, \ldots, n$. Assume that all asset prices are real numbers, i.e., in $\mathbb{R}$ (negative prices are typically unrealistic, but we still assume it for simplicity of exposition). For convenience, we normalize the Spot Price (at $t = 0$) of the riskless asset A_O to be 1. Therefore,

$$S_0^{(0)} = 1 \text{ and } S_0^{(i)} = 1 + r \text{ for all } i = 1, \ldots, n$$

where r represents the constant riskless rate of growth. We should interpret this riskless rate of growth as the "time value of money" and $\frac{1}{1+r}$ as the riskless discount factor corresponding to the "time value of money".

9.3 PORTFOLIOS, ARBITRAGE AND RISK-NEUTRAL PROBABILITY MEASURE

We define a portfolio as a vector $\theta = (\theta_0, \theta_1, \ldots, \theta_m) \in \mathbb{R}^{m+1}$, representing the number of units held in the assets $A_j, j = 0, 1, \ldots, m$. The Spot Value (at $t = 0$) of portfolio θ, denoted by $V_\theta^{(0)}$, is:

$$V_\theta^{(0)} = \sum_{j=0}^{m} \theta_j \cdot S_j^{(0)} \tag{9.1}$$

The Value of portfolio θ in random outcome ω_i (at $t = 1$), denoted by $V_\theta^{(i)}$, is:

$$V_\theta^{(i)} = \sum_{j=0}^{m} \theta_j \cdot S_j^{(i)} \text{ for all } i = 1, \ldots, n \tag{9.2}$$

Next, we cover an extremely important concept in Mathematical Economics/Finance—the concept of *Arbitrage*. An Arbitrage Portfolio θ is one that "makes money from nothing". Formally, an arbitrage portfolio is a portfolio θ such that:

- $V_\theta^{(0)} \leq 0$
- $V_\theta^{(i)} \geq 0$ for all $i = 1, \ldots, n$
- There exists an $i \in \{1, \ldots, n\}$ such that $\mu(\omega_i) > 0$ and $V_\theta^{(i)} > 0$

Thus, with an Arbitrage Portfolio, we never end up (at $t = 0$) with less value than what we start with (at $t = 1$), and we end up with expected value strictly greater than what we start with. This is the formalism of the notion of *arbitrage*, i.e., "making money from nothing". Arbitrage allows market participants to make infinite returns. In an efficient market, arbitrage would disappear as soon as it appears since market participants would immediately exploit it and through the process of exploiting the arbitrage, immediately eliminate the arbitrage. Hence, Finance Theory typically assumes "arbitrage-free" markets (i.e., financial markets with no arbitrage opportunities).

Next, we describe another important concept in Mathematical Economics/Finance—the concept of a *Risk-Neutral Probability Measure*. Consider a Probability Distribution $\pi : \Omega \to [0, 1]$ such that

$$\pi(\omega_i) = 0 \text{ if and only if } \mu(\omega_i) = 0 \text{ for all } i = 1, \ldots, n$$

Then, π is said to be a Risk-Neutral Probability Measure if:

$$S_j^{(0)} = \frac{1}{1+r} \cdot \sum_{i=1}^{n} \pi(\omega_i) \cdot S_j^{(i)} \text{ for all } j = 0, 1, \ldots, m \tag{9.3}$$

So for each of the $m+1$ assets, the asset spot price (at $t = 0$) is the riskless rate-discounted expectation (under π) of the asset price at $t = 1$. The term "risk-neutral" here is the same as the term "risk-neutral" we used in Chapter 7, meaning it's a situation where one doesn't need to be compensated for taking risk (the situation of a linear utility function). However, we are not saying that the market is risk-neutral—if that were the case, the market probability measure μ would be a risk-neutral probability measure. We are simply defining π as a *hypothetical construct* under which each asset's spot price is equal to the riskless rate-discounted expectation (under π) of the asset's price at $t = 1$. This means that under the hypothetical π, there's no return in excess of r for taking on the risk of probabilistic outcomes at $t = 1$ (note: outcome probabilities are governed by the hypothetical π). Hence, we refer to π as a risk-neutral probability measure. The purpose of this hypothetical construct π is that it helps in the development of Derivatives Pricing and Hedging Theory, as we shall soon see. The actual probabilities of outcomes in Ω are governed by μ, and not π.

Before we cover the two fundamental theorems of asset pricing, we need to cover an important lemma that we will utilize in the proofs of the two fundamental theorems of asset pricing.

Lemma 9.3.1. *For any portfolio* $\theta = (\theta_0, \theta_1, \ldots, \theta_m) \in \mathbb{R}^{m+1}$ *and any risk-neutral probability measure* $\pi : \Omega \to [0, 1]$,

$$V_\theta^{(0)} = \frac{1}{1+r} \cdot \sum_{i=1}^{n} \pi(\omega_i) \cdot V_\theta^{(i)}$$

Proof. Using Equations (9.1), (9.3) and (9.2), the proof is straightforward:

$$V_\theta^{(0)} = \sum_{j=0}^{m} \theta_j \cdot S_j^{(0)} = \sum_{j=0}^{m} \theta_j \cdot \frac{1}{1+r} \cdot \sum_{i=1}^{n} \pi(\omega_i) \cdot S_j^{(i)}$$

$$= \frac{1}{1+r} \cdot \sum_{i=1}^{n} \pi(\omega_i) \cdot \sum_{j=0}^{m} \theta_j \cdot S_j^{(i)} = \frac{1}{1+r} \cdot \sum_{i=1}^{n} \pi(\omega_i) \cdot V_\theta^{(i)}$$

□

Now we are ready to cover the two fundamental theorems of asset pricing (sometimes, also referred to as the fundamental theorems of arbitrage and the fundamental theorems of finance!). We start with the first fundamental theorem of asset pricing, which associates absence of arbitrage with existence of a risk-neutral probability measure.

9.4 FIRST FUNDAMENTAL THEOREM OF ASSET PRICING (1ST FTAP)

Theorem 9.4.1 (First Fundamental Theorem of Asset Pricing (1st FTAP)). *Our simple setting of discrete time with single-period will not admit arbitrage portfolios if and only if there exists a Risk-Neutral Probability Measure.*

Proof. First, we prove the easy implication—if there exists a Risk-Neutral Probability Measure π, then we cannot have any arbitrage portfolios. Let's review what it takes to have an arbitrage portfolio $\theta = (\theta_0, \theta_1, \ldots, \theta_m)$. The following are two of the three conditions to be satisfied to qualify as an arbitrage portfolio θ (according to the definition of arbitrage portfolio we gave above):

- $V_\theta^{(i)} \geq 0$ for all $i = 1, \dots, n$

- There exists an $i \in \{1, \dots, n\}$ such that $\mu(\omega_i) > 0 \; (\Rightarrow \pi(\omega_i) > 0)$ and $V_\theta^{(i)} > 0$

But if these two conditions are satisfied, the third condition $V_\theta^{(0)} \leq 0$ cannot be satisfied because from Lemma (9.3.1), we know that:

$$V_\theta^{(0)} = \frac{1}{1+r} \cdot \sum_{i=1}^{n} \pi(\omega_i) \cdot V_\theta^{(i)}$$

which is strictly greater than 0, given the two conditions stated above. Hence, all three conditions cannot be simultaneously satisfied which eliminates the possibility of arbitrage for any portfolio θ.

Next, we prove the reverse (harder to prove) implication—if a risk-neutral probability measure doesn't exist, then there exists an arbitrage portfolio θ. We define $\mathbb{V} \subset \mathbb{R}^m$ as the set of vectors $v = (v_1, \dots, v_m)$ such that

$$v_j = \frac{1}{1+r} \cdot \sum_{i=1}^{n} \mu(\omega_i) \cdot S_j^{(i)} \text{ for all } j = 1, \dots, m$$

with $\mathbb{V}$ defined as spanning over all possible probability distributions $\mu : \Omega \to [0, 1]$. $\mathbb{V}$ is a bounded, closed, convex polytope in $\mathbb{R}^m$. By the definition of a risk-neutral probability measure, we can say that if a risk-neutral probability measure doesn't exist, the vector $(S_1^{(0)}, \dots, S_m^{(0)}) \notin \mathbb{V}$. The Hyperplane Separation Theorem implies that there exists a non-zero vector $(\theta_1, \dots, \theta_m)$ such that for any $v = (v_1, \dots, v_m) \in \mathbb{V}$,

$$\sum_{j=1}^{m} \theta_j \cdot v_j > \sum_{j=1}^{m} \theta_j \cdot S_j^{(0)}$$

In particular, consider vectors v corresponding to the corners of $\mathbb{V}$, those for which the full probability mass is on a particular $\omega_i \in \Omega$, i.e.,

$$\sum_{j=1}^{m} \theta_j \cdot \left(\frac{1}{1+r} \cdot S_j^{(i)} \right) > \sum_{j=1}^{m} \theta_j \cdot S_j^{(0)} \text{ for all } i = 1, \dots, n$$

Since this is a strict inequality, we will be able to choose a $\theta_0 \in \mathbb{R}$ such that:

$$\sum_{j=1}^{m} \theta_j \cdot \left(\frac{1}{1+r} \cdot S_j^{(i)} \right) > -\theta_0 > \sum_{j=1}^{m} \theta_j \cdot S_j^{(0)} \text{ for all } i = 1, \dots, n$$

Therefore,

$$\frac{1}{1+r} \cdot \sum_{j=0}^{m} \theta_j \cdot S_j^{(i)} > 0 > \sum_{j=0}^{m} \theta_j \cdot S_j^{(0)} \text{ for all } i = 1, \dots, n$$

This can be rewritten in terms of the Values of portfolio $\theta = (\theta_0, \theta_1, \dots, \theta)$ at $t = 0$ and $t = 1$, as follows:

$$\frac{1}{1+r} \cdot V_\theta^{(i)} > 0 > V_\theta^{(0)} \text{ for all } i = 1, \dots, n$$

Thus, we can see that all three conditions in the definition of arbitrage portfolio are satisfied and hence, $\theta = (\theta_0, \theta_1, \dots, \theta_m)$ is an arbitrage portfolio.

□

Now we are ready to move on to the second fundamental theorem of asset pricing, which associates replication of derivatives with a unique risk-neutral probability measure.

9.5 SECOND FUNDAMENTAL THEOREM OF ASSET PRICING (2ND FTAP)

Before we state and prove the 2nd FTAP, we need some definitions.

Definition 9.5.1. A Derivative D (in our simple setting of discrete-time with a single-period) is specified as a vector payoff at time $t = 1$, denoted as:

$$(V_D^{(1)}, V_D^{(2)}, \ldots, V_D^{(n)})$$

where $V_D^{(i)}$ is the payoff of the derivative in random outcome ω_i for all $i = 1, \ldots, n$

Definition 9.5.2. A Portfolio $\theta = (\theta_0, \theta_1, \ldots, \theta_m) \in \mathbb{R}^{m+1}$ is a *Replicating Portfolio* for derivative D if:

$$V_D^{(i)} = V_\theta^{(i)} = \sum_{j=0}^{m} \theta_j \cdot S_j^{(i)} \text{ for all } i = 1, \ldots, n \qquad (9.4)$$

The negatives of the components $(\theta_0, \theta_1, \ldots, \theta_m)$ are known as the *hedges* for D since they can be used to offset the risk in the payoff of D at $t = 1$.

Definition 9.5.3. An arbitrage-free market (i.e., a market devoid of arbitrage) is said to be *Complete* if every derivative in the market has a replicating portfolio.

Theorem 9.5.1 (Second Fundamental Theorem of Asset Pricing (2nd FTAP)). *A market (in our simple setting of discrete-time with a single-period) is Complete if and only if there is a unique Risk-Neutral Probability Measure.*

Proof. We will first prove that in an arbitrage-free market, if every derivative has a replicating portfolio (i.e., the market is complete), then there is a unique risk-neutral probability measure. We define n special derivatives (known as *Arrow-Debreu securities*), one for each random outcome in Ω at $t = 1$. We define the time $t = 1$ payoff of *Arrow-Debreu security* D_k (for each of $k = 1, \ldots, n$) as follows:

$$V_{D_k}^{(i)} = \mathbb{I}_{i=k} \text{ for all } i = 1, \ldots, n$$

where $\mathbb{I}$ represents the indicator function. This means the payoff of derivative D_k is 1 for random outcome ω_k and 0 for all other random outcomes.

Since each derivative has a replicating portfolio, denote $\theta^{(k)} = (\theta_0^{(k)}, \theta_1^{(k)}, \ldots, \theta_m^{(k)})$ as the replicating portfolio for D_k for each $k = 1, \ldots, m$. Therefore, for each $k = 1, \ldots, m$:

$$V_{\theta^{(k)}}^{(i)} = \sum_{j=0}^{m} \theta_j^{(k)} \cdot S_j^{(i)} = V_{D_k}^{(i)} = \mathbb{I}_{i=k} \text{ for all } i = 1, \ldots, n$$

Using Lemma (9.3.1), we can write the following equation for any risk-neutral probability measure π, for each $k = 1, \ldots, m$:

$$\sum_{j=0}^{m} \theta_j^{(k)} \cdot S_j^{(0)} = V_{\theta^{(k)}}^{(0)} = \frac{1}{1+r} \cdot \sum_{i=1}^{n} \pi(\omega_i) \cdot V_{\theta^{(k)}}^{(i)} = \frac{1}{1+r} \cdot \sum_{i=1}^{n} \pi(\omega_i) \cdot \mathbb{I}_{i=k} = \frac{1}{1+r} \cdot \pi(\omega_k)$$

We note that the above equation is satisfied for a unique $\pi : \Omega \to [0, 1]$, defined as:

$$\pi(\omega_k) = (1 + r) \cdot \sum_{j=0}^{m} \theta_j^{(k)} \cdot S_j^{(0)} \text{ for all } k = 1, \ldots, n$$

which implies that we have a unique risk-neutral probability measure.

Next, we prove the other direction of the 2nd FTAP. We need to prove that if there exists a risk-neutral probability measure π and if there exists a derivative D with no replicating portfolio, then we can construct a risk-neutral probability measure different than π.

Consider the following vectors in the vector space $\mathbb{R}^n$

$$v = (V_D^{(1)}, \ldots, V_D^{(n)}) \text{ and } v_j = (S_j^{(1)}, \ldots, S_j^{(n)}) \text{ for all } j = 0, 1, \ldots, m$$

Since D does not have a replicating portfolio, v is not in the span of $\{v_0, v_1, \ldots, v_m\}$, which means $\{v_0, v_1, \ldots, v_m\}$ do not span $\mathbb{R}^n$. Hence, there exists a non-zero vector $u = (u_1, \ldots, u_n) \in \mathbb{R}^n$ orthogonal to each of $v_0, v_1, \ldots, v_m$, i.e.,

$$\sum_{i=1}^{n} u_i \cdot S_j^{(i)} = 0 \text{ for all } j = 0, 1, \ldots, n \tag{9.5}$$

Note that $S_0^{(i)} = 1 + r$ for all $i = 1, \ldots, n$ and so,

$$\sum_{i=1}^{n} u_i = 0 \tag{9.6}$$

Define $\pi' : \Omega \to \mathbb{R}$ as follows (for some $\epsilon \in \mathbb{R}^+$):

$$\pi'(\omega_i) = \pi(\omega_i) + \epsilon \cdot u_i \text{ for all } i = 1, \ldots, n \tag{9.7}$$

To establish π' as a risk-neutral probability measure different than π, note:

- Since $\sum_{i=1}^{n} \pi(\omega_i) = 1$ and since $\sum_{i=1}^{n} u_i = 0$, $\sum_{i=1}^{n} \pi'(\omega_i) = 1$

- Construct $\pi'(\omega_i) > 0$ for each i where $\pi(\omega_i) > 0$ by making $\epsilon > 0$ sufficiently small, and set $\pi'(\omega_i) = 0$ for each i where $\pi(\omega_i) = 0$

- From Equations (9.7), (9.3) and (9.5), we have for each $j = 0, 1, \ldots, m$:

$$\frac{1}{1+r} \cdot \sum_{i=1}^{n} \pi'(\omega_i) \cdot S_j^{(i)} = \frac{1}{1+r} \cdot \sum_{i=1}^{n} \pi(\omega_i) \cdot S_j^{(i)} + \frac{\epsilon}{1+r} \cdot \sum_{i=1}^{n} u_i \cdot S_j^{(i)} = S_j^{(0)}$$

$\square$

Together, the two FTAPs classify markets into:

- Market with arbitrage $\Leftrightarrow$ No risk-neutral probability measure
- Complete (arbitrage-free) market $\Leftrightarrow$ Unique risk-neutral probability measure
- Incomplete (arbitrage-free) market $\Leftrightarrow$ Multiple risk-neutral probability measures

The next topic is derivatives pricing that is based on the concepts of *replication of derivatives* and *risk-neutral probability measures*, and so is tied to the concepts of *arbitrage* and *completeness*.

9.6 DERIVATIVES PRICING IN SINGLE-PERIOD SETTING

In this section, we cover the theory of derivatives pricing for our simple setting of discrete-time with a single-period. To develop the theory of how to price a derivative, first we need to define the notion of a *Position*.

Definition 9.6.1. A *Position* involving a derivative D is the combination of holding some units in D and some units in the fundamental assets $A_0, A_1, \ldots, A_m$, which can be formally represented as a vector $\gamma_D = (\alpha, \theta_0, \theta_1, \ldots, \theta_m) \in \mathbb{R}^{m+2}$ where α denotes the units held in derivative D and α_j denotes the units held in A_j for all $j = 0, 1 \ldots, m$.

Therefore, a *Position* is an extension of the Portfolio concept that includes a derivative. Hence, we can naturally extend the definition of *Portfolio Value* to *Position Value*, and we can also extend the definition of *Arbitrage Portfolio* to *Arbitrage Position*.

We need to consider derivatives pricing in three market situations:

- When the market is complete
- When the market is incomplete
- When the market has arbitrage

9.6.1 Derivatives Pricing When Market Is Complete

Theorem 9.6.1. *For our simple setting of discrete-time with a single-period, if the market is complete, then any derivative D with replicating portfolio $\theta = (\theta_0, \theta_1, \ldots, \theta_m)$ has price at time $t = 0$ (denoted as value $V_D^{(0)}$):*

$$V_D^{(0)} = V_\theta^{(0)} = \sum_{j=0}^{n} \theta_j \cdot S_j^{(i)} \tag{9.8}$$

Furthermore, if the unique risk-neutral probability measure is $\pi : \Omega \to [0, 1]$, then:

$$V_D^{(0)} = \frac{1}{1+r} \cdot \sum_{i=1}^{n} \pi(\omega_i) \cdot V_D^{(i)} \tag{9.9}$$

Proof. It seems quite reasonable that since θ is the replicating portfolio for D, the value of the replicating portfolio at time $t = 0$ (equal to $V_\theta^{(0)} = \sum_{j=0}^{n} \theta_j \cdot S_j^{(i)}$) should be the price (at $t = 0$) of derivative D. However, we will formalize the proof by first arguing that any candidate derivative price for D other than $V_\theta^{(0)}$ leads to arbitrage, thus dismissing those other candidate derivative prices, and then argue that with $V_\theta^{(0)}$ as the price of derivative D, we eliminate the possibility of an arbitrage position involving D.

Consider candidate derivative prices $V_\theta^{(0)} - x$ for any positive real number x. Position $(1, -\theta_0 + x, -\theta_1, \ldots, -\theta_m)$ has value $x \cdot (1+r) > 0$ in each of the random outcomes at $t = 1$. But this position has spot ($t = 0$) value of 0, which means this is an Arbitrage Position, rendering these candidate derivative prices invalid. Next, consider candidate derivative prices $V_\theta^{(0)} + x$ for any positive real number x. Position $(-1, \theta_0 + x, \theta_1, \ldots, \theta_m)$ has value $x \cdot (1+r) > 0$ in each of the random outcomes at $t = 1$. But this position has spot ($t = 0$) value of 0, which means this is an Arbitrage Position, rendering these candidate derivative prices invalid as well. So every candidate derivative price other than $V_\theta^{(0)}$ is invalid. Now our goal is to *establish* $V_\theta^{(0)}$ as the derivative price of D by showing that we eliminate the possibility of an arbitrage position in the market involving D if $V_\theta^{(0)}$ is indeed the derivative price.

Firstly, note that $V_\theta^{(0)}$ can be expressed as the riskless rate-discounted expectation (under π) of the payoff of D at $t = 1$, i.e.,

$$V_\theta^{(0)} = \sum_{j=0}^{m} \theta_j \cdot S_j^{(0)} = \sum_{j=0}^{m} \theta_j \cdot \frac{1}{1+r} \cdot \sum_{i=1}^{n} \pi(\omega_i) \cdot S_j^{(i)} = \frac{1}{1+r} \cdot \sum_{i=1}^{n} \pi(\omega_i) \cdot \sum_{j=0}^{m} \theta_j \cdot S_j^{(i)}$$

$$= \frac{1}{1+r} \cdot \sum_{i=1}^{n} \pi(\omega_i) \cdot V_D^{(i)} \quad (9.10)$$

Now consider an *arbitrary portfolio* $\beta = (\beta_0, \beta_1, \dots, \beta_m)$. Define a position $\gamma_D = (\alpha, \beta_0, \beta_1, \dots, \beta_m)$. Assuming the derivative price $V_D^{(0)}$ is equal to $V_\theta^{(0)}$, the Spot Value (at $t = 0$) of position γ_D, denoted $V_{\gamma_D}^{(0)}$, is:

$$V_{\gamma_D}^{(0)} = \alpha \cdot V_\theta^{(0)} + \sum_{j=0}^{m} \beta_j \cdot S_j^{(0)} \quad (9.11)$$

Value of position γ_D in random outcome ω_i (at $t = 1$), denoted $V_{\gamma_D}^{(i)}$, is:

$$V_{\gamma_D}^{(i)} = \alpha \cdot V_D^{(i)} + \sum_{j=0}^{m} \beta_j \cdot S_j^{(i)} \quad \text{for all } i = 1, \dots, n \quad (9.12)$$

Combining the linearity in Equations (9.3), (9.10), (9.11) and (9.12), we get:

$$V_{\gamma_D}^{(0)} = \frac{1}{1+r} \cdot \sum_{i=1}^{n} \pi(\omega_i) \cdot V_{\gamma_D}^{(i)} \quad (9.13)$$

So the position spot value (at $t = 0$) is the riskless rate-discounted expectation (under π) of the position value at $t = 1$. For any γ_D (containing any arbitrary portfolio β), with derivative price $V_D^{(0)}$ equal to $V_\theta^{(0)}$, if the following two conditions are satisfied:

- $V_{\gamma_D}^{(i)} \geq 0$ for all $i = 1, \dots, n$

- There exists an $i \in \{1, \dots, n\}$ such that $\mu(\omega_i) > 0$ ($\Rightarrow \pi(\omega_i) > 0$) and $V_{\gamma_D}^{(i)} > 0$

then:

$$V_{\gamma_D}^{(0)} = \frac{1}{1+r} \cdot \sum_{i=1}^{n} \pi(\omega_i) \cdot V_{\gamma_D}^{(i)} > 0$$

This eliminates any arbitrage possibility if D is priced at $V_\theta^{(0)}$.

To summarize, we have eliminated all candidate derivative prices other than $V_\theta^{(0)}$, and we have established the price $V_\theta^{(0)}$ as the correct price of D in the sense that we eliminate the possibility of an arbitrage position involving D if the price of D is $V_\theta^{(0)}$.

Finally, we note that with the derivative price $V_D^{(0)} = V_\theta^{(0)}$, from Equation (9.10), we have:

$$V_D^{(0)} = \frac{1}{1+r} \cdot \sum_{i=1}^{n} \pi(\omega_i) \cdot V_D^{(i)}$$

$\square$

Now let us consider the special case of 1 risky asset ($m = 1$) and 2 random outcomes ($n = 2$), which we will show is a Complete Market. To lighten notation, we drop the subscript 1 on the risky asset price. Without loss of generality, we assume $S^{(1)} < S^{(2)}$. No-arbitrage requires:

$$S^{(1)} \leq (1+r) \cdot S^{(0)} \leq S^{(2)}$$

Assuming absence of arbitrage and invoking 1st FTAP, there exists a risk-neutral probability measure π such that:

$$S^{(0)} = \frac{1}{1+r} \cdot (\pi(\omega_1) \cdot S^{(1)} + \pi(\omega_2) \cdot S^{(2)})$$

$$\pi(\omega_1) + \pi(\omega_2) = 1$$

With 2 linear equations and 2 variables, this has a straightforward solution, as follows:

$$\pi(\omega_1) = \frac{S^{(2)} - (1+r) \cdot S^{(0)}}{S^{(2)} - S^{(1)}}$$

$$\pi(\omega_2) = \frac{(1+r) \cdot S^{(0)} - S^{(1)}}{S^{(2)} - S^{(1)}}$$

Conditions $S^{(1)} < S^{(2)}$ and $S^{(1)} \leq (1+r) \cdot S^{(0)} \leq S^{(2)}$ ensure that $0 \leq \pi(\omega_1), \pi(\omega_2) \leq 1$. Also note that this is a unique solution for $\pi(\omega_1), \pi(\omega_2)$, which means that the risk-neutral probability measure is unique, implying that this is a complete market.

We can use these probabilities to price a derivative D as:

$$V_D^{(0)} = \frac{1}{1+r} \cdot (\pi(\omega_1) \cdot V_D^{(1)} + \pi(\omega_2) \cdot V_D^{(2)})$$

Now let us try to form a replicating portfolio (θ_0, θ_1) for D

$$V_D^{(1)} = \theta_0 \cdot (1+r) + \theta_1 \cdot S^{(1)}$$

$$V_D^{(2)} = \theta_0 \cdot (1+r) + \theta_1 \cdot S^{(2)}$$

Solving this yields Replicating Portfolio (θ_0, θ_1) as follows:

$$\theta_0 = \frac{1}{1+r} \cdot \frac{V_D^{(1)} \cdot S^{(2)} - V_D^{(2)} \cdot S^{(1)}}{S^{(2)} - S^{(1)}} \text{ and } \theta_1 = \frac{V_D^{(2)} - V_D^{(1)}}{S^{(2)} - S^{(1)}} \tag{9.14}$$

Note that the derivative price can also be expressed as:

$$V_D^{(0)} = \theta_0 + \theta_1 \cdot S^{(0)}$$

9.6.2 Derivatives Pricing When Market Is Incomplete

Theorem (9.6.1) assumed a complete market, but what about an incomplete market? Recall that an incomplete market means some derivatives can't be replicated. Absence of a replicating portfolio for a derivative precludes usual no-arbitrage arguments. The 2nd FTAP says that in an incomplete market, there are multiple risk-neutral probability measures which means there are multiple derivative prices (each consistent with no-arbitrage).

To develop intuition for derivatives pricing when the market is incomplete, let us consider the special case of 1 risky asset ($m = 1$) and 3 random outcomes ($n = 3$), which we will show is an Incomplete Market. To lighten notation, we drop the subscript 1 on the risky asset price. Without loss of generality, we assume $S^{(1)} < S^{(2)} < S^{(3)}$. No-arbitrage requires:

$$S^{(1)} \leq S^{(0)} \cdot (1+r) \leq S^{(3)}$$

Assuming absence of arbitrage and invoking the 1st FTAP, there exists a risk-neutral probability measure π such that:

$$S^{(0)} = \frac{1}{1+r} \cdot (\pi(\omega_1) \cdot S^{(1)} + \pi(\omega_2) \cdot S^{(2)} + \pi(\omega_3) \cdot S^{(3)})$$

$$\pi(\omega_1) + \pi(\omega_2) + \pi(\omega_3) = 1$$

So we have 2 equations and 3 variables, which implies there are multiple solutions for π. Each of these solutions for π provides a valid price for a derivative D.

$$V_D^{(0)} = \frac{1}{1+r} \cdot (\pi(\omega_1) \cdot V_D^{(1)} + \pi(\omega_2) \cdot V_D^{(2)} + \pi(\omega_3) \cdot V_D^{(3)})$$

Now let us try to form a replicating portfolio (θ_0, θ_1) for D

$$V_D^{(1)} = \theta_0 \cdot (1+r) + \theta_1 \cdot S^{(1)}$$

$$V_D^{(2)} = \theta_0 \cdot (1+r) + \theta_1 \cdot S^{(2)}$$

$$V_D^{(3)} = \theta_0 \cdot (1+r) + \theta_1 \cdot S^{(3)}$$

3 equations & 2 variables implies there is no replicating portfolio for *some D*. This means this is an Incomplete Market.

So with multiple risk-neutral probability measures (and consequent, multiple derivative prices), how do we go about determining how much to buy/sell derivatives for? One approach to handle derivative pricing in an incomplete market is the technique called *Superhedging*, which provides upper and lower bounds for the derivative price. The idea of Superhedging is to create a portfolio of fundamental assets whose Value *dominates* the derivative payoff in *all* random outcomes at $t = 1$. Superhedging Price is the smallest possible Portfolio Spot ($t = 0$) Value among all such Derivative-Payoff-Dominating portfolios. Without getting into too many details of the Superhedging technique (out of scope for this book), we shall simply sketch the outline of this technique for our simple setting.

We note that for our simple setting of discrete-time with a single-period, this is a constrained linear optimization problem:

$$\min_{\theta} \sum_{j=0}^{m} \theta_j \cdot S_j^{(0)} \text{ such that } \sum_{j=0}^{m} \theta_j \cdot S_j^{(i)} \geq V_D^{(i)} \text{ for all } i = 1, \dots, n \tag{9.15}$$

Let $\theta^* = (\theta_0^*, \theta_1^*, \dots, \theta_m^*)$ be the solution to Equation (9.15). Let SP be the Superhedging Price $\sum_{j=0}^{m} \theta_j^* \cdot S_j^{(0)}$.

After establishing feasibility, we define the Lagrangian $J(\theta, \lambda)$ as follows:

$$J(\theta, \lambda) = \sum_{j=0}^{m} \theta_j \cdot S_j^{(0)} + \sum_{i=1}^{n} \lambda_i \cdot (V_D^{(i)} - \sum_{j=0}^{m} \theta_j \cdot S_j^{(i)})$$

So there exists $\lambda = (\lambda_1, \dots, \lambda_n)$ that satisfy the following KKT conditions:

$$\lambda_i \geq 0 \text{ for all } i = 1, \dots, n$$

$$\lambda_i \cdot (V_D^{(i)} - \sum_{j=0}^{m} \theta_j^* \cdot S_j^{(i)}) = 0 \text{ for all } i = 1, \ldots, n \text{ (Complementary Slackness)}$$

$$\nabla_\theta J(\theta^*, \lambda) = 0 \Rightarrow S_j^{(0)} = \sum_{i=1}^{n} \lambda_i \cdot S_j^{(i)} \text{ for all } j = 0, 1, \ldots, m$$

This implies $\lambda_i = \frac{\pi(\omega_i)}{1+r}$ for all $i = 1, \ldots, n$ for a risk-neutral probability measure $\pi : \Omega \rightarrow [0, 1]$ (λ can be thought of as "discounted probabilities").

Define Lagrangian Dual

$$L(\lambda) = \inf_\theta J(\theta, \lambda)$$

Then, Superhedging Price

$$SP = \sum_{j=0}^{m} \theta_j^* \cdot S_j^{(0)} = \sup_\lambda L(\lambda) = \sup_\lambda \inf_\theta J(\theta, \lambda)$$

Complementary Slackness and some linear algebra over the space of risk-neutral probability measures $\pi : \Omega \rightarrow [0, 1]$ enables us to argue that:

$$SP = \sup_\pi \sum_{i=1}^{n} \frac{\pi(\omega_i)}{1+r} \cdot V_D^{(i)}$$

This means the Superhedging Price is the least upper-bound of the riskless rate-discounted expectation of derivative payoff across each of the risk-neutral probability measures in the incomplete market, which is quite an intuitive thing to do amidst multiple risk-neutral probability measures.

Likewise, the *Subhedging* price SB is defined as:

$$\max_\theta \sum_{j=0}^{m} \theta_j \cdot S_j^{(0)} \text{ such that } \sum_{j=0}^{m} \theta_j \cdot S_j^{(i)} \leq V_D^{(i)} \text{ for all } i = 1, \ldots, n$$

Likewise arguments enable us to establish:

$$SB = \inf_\pi \sum_{i=1}^{n} \frac{\pi(\omega_i)}{1+r} \cdot V_D^{(i)}$$

This means the Subhedging Price is the highest lower-bound of the riskless rate-discounted expectation of derivative payoff across each of the risk-neutral probability measures in the incomplete market, which is quite an intuitive thing to do amidst multiple risk-neutral probability measures.

So this technique provides an lower bound (SB) and an upper bound (SP) for the derivative price, meaning:

- A price outside these bounds leads to an arbitrage
- Valid prices must be established within these bounds

But often these bounds are not tight and so, not useful in practice.

The alternative approach is to identify hedges that maximize Expected Utility of the combination of the derivative along with its hedges, for an appropriately chosen market/trader Utility Function (as covered in Chapter 7). The Utility function is a specification of reward-versus-risk preference that effectively chooses the risk-neutral probability measure (and hence, Price).

Consider a concave Utility function $U : \mathbb{R} \to \mathbb{R}$ applied to the Value in each random outcome $\omega_i, i = 1, \ldots n$, at $t = 1$ (e.g., $U(x) = \frac{1-e^{-ax}}{a}$ where $a \in \mathbb{R}$ is the degree of risk-aversion). Let the real-world probabilities be given by $\mu : \Omega \to [0, 1]$. Denote $V_D = (V_D^{(1)}, \ldots, V_D^{(n)})$ as the payoff of Derivative D at $t = 1$. Let us say that you buy the derivative D at $t = 0$ and will receive the random outcome-contingent payoff V_D at $t = 1$. Let x be the candidate derivative price for D, which means you will pay a cash quantity of x at $t = 0$ for the privilege of receiving the payoff V_D at $t = 1$. We refer to the candidate hedge as Portfolio $\theta = (\theta_0, \theta_1, \ldots, \theta_m)$, representing the units held in the fundamental assets.

Note that at $t = 0$, the cash quantity x you'd be paying to buy the derivative and the cash quantity you'd be paying to buy the Portfolio θ should sum to 0 (note: either of these cash quantities can be positive or negative, but they need to sum to 0 since "money can't just appear or disappear"). Formally,

$$x + \sum_{j=0}^{m} \theta_j \cdot S_j^{(0)} = 0 \tag{9.16}$$

Our goal is to solve for the appropriate values of x and θ based on an *Expected Utility* consideration (that we are about to explain). Consider the Utility of the position consisting of derivative D together with portfolio θ in random outcome ω_i at $t = 1$:

$$U(V_D^{(i)} + \sum_{j=0}^{m} \theta_j \cdot S_j^{(i)})$$

So, the Expected Utility of this position at $t = 1$ is given by:

$$\sum_{i=1}^{n} \mu(\omega_i) \cdot U(V_D^{(i)} + \sum_{j=0}^{m} \theta_j \cdot S_j^{(i)}) \tag{9.17}$$

Noting that $S_0^{(0)} = 1, S_0^{(i)} = 1 + r$ for all $i = 1, \ldots, n$, we can substitute for the value of $\theta_0 = -(x + \sum_{j=1}^{m} \theta_j \cdot S_j^{(0)})$ (obtained from Equation (9.16)) in the above Expected Utility expression (9.17), so as to rewrite this Expected Utility expression in terms of just $(\theta_1, \ldots, \theta_m)$ (call it $\theta_{1:m}$) as:

$$g(V_D, x, \theta_{1:m}) = \sum_{i=1}^{n} \mu(\omega_i) \cdot U(V_D^{(i)} - (1 + r) \cdot x + \sum_{j=1}^{m} \theta_j \cdot (S_j^{(i)} - (1 + r) \cdot S_j^{(0)}))$$

We define the *Price* of D as the "breakeven value" x^* such that:

$$\max_{\theta_{1:m}} g(V_D, x^*, \theta_{1:m}) = \max_{\theta_{1:m}} g(0, 0, \theta_{1:m})$$

The core principle here (known as *Expected-Utility-Indifference Pricing*) is that introducing a $t = 1$ payoff of V_D together with a derivative price payment of x^* at $t = 0$ keeps the Maximum Expected Utility unchanged.

The $(\theta_1^*, \ldots, \theta_m^*)$ that achieve $\max_{\theta_{1:m}} g(V_D, x^*, \theta_{1:m})$ and $\theta_0^* = -(x^* + \sum_{j=1}^{m} \theta_j^* \cdot S_j^{(0)})$ are the requisite hedges associated with the derivative price x^*. Note that the Price of V_D will NOT be the negative of the Price of $-V_D$, hence these prices simply serve as bid prices or ask prices, depending on whether one pays or receives the random outcomes-contingent payoff V_D.

To develop some intuition for what this solution looks like, let us now write some code for the case of 1 risky asset (i.e., $m = 1$). To make things interesting, we will write code for the case where the risky asset price at $t = 1$ (denoted S) follows a normal distribution $S \sim \mathcal{N}(\mu, \sigma^2)$. This means we have a continuous (rather than discrete) set of values for the risky asset price at $t = 1$. Since there are more than 2 random outcomes at time $t = 1$, this is the case of an Incomplete Market. Moreover, we assume the CARA utility function:

$$U(y) = \frac{1 - e^{-a \cdot y}}{a}$$

where a is the CARA coefficient of risk-aversion.

We refer to the units of investment in the risky asset as α and the units of investment in the riskless asset as β. Let S_0 be the spot ($t = 0$) value of the risky asset (riskless asset value at $t = 0$ is 1). Let $f(S)$ be the payoff of the derivative D at $t = 1$. So, the price of derivative D is the breakeven value x^* such that:

$$\max_{\alpha} \mathbb{E}_{S \sim \mathcal{N}(\mu, \sigma^2)} \left[\frac{1 - e^{-a \cdot (f(S) - (1+r) \cdot x^* + \alpha \cdot (S - (1+r) \cdot S_0))}}{a} \right]$$

$$= \max_{\alpha} \mathbb{E}_{S \sim \mathcal{N}(\mu, \sigma^2)} \left[\frac{1 - e^{-a \cdot (\alpha \cdot (S - (1+r) \cdot S_0))}}{a} \right] \quad (9.18)$$

The maximizing value of α (call it α^*) on the left-hand-side of Equation (9.18) along with $\beta^* = -(x^* + \alpha^* \cdot S_0)$ are the requisite hedges associated with the derivative price x^*.

We set up a @dataclass `MaxExpUtility` with attributes to represent the risky asset spot price S_0 (`risky_spot`), the riskless rate r (`riskless_rate`), mean μ of S (`risky_mean`), standard deviation σ of S (`risky_stdev`), and the payoff function $f(\cdot)$ of the derivative (`payoff_func`).

```
@dataclass(frozen=True)
class MaxExpUtility:
    risky_spot: float    # risky asset price at t=0
    riskless_rate: float    # riskless asset price grows from 1 to 1+r
    risky_mean: float    # mean of risky asset price at t=1
    risky_stdev: float    # std dev of risky asset price at t=1
    payoff_func: Callable[[float], float]    # derivative payoff at t=1
```

Before we write code to solve the derivatives pricing and hedging problem for an incomplete market, let us write code to solve the problem for a complete market (as this will serve as a good comparison against the incomplete market solution). For a complete market, the risky asset has two random prices at $t = 1$: prices $\mu + \sigma$ and $\mu - \sigma$, with probabilities of 0.5 each. As we've seen in Section 9.6.1, we can perfectly replicate a derivative payoff in this complete market situation as it amounts to solving 2 linear equations in 2 unknowns (solution shown in Equation (9.14)). The number of units of the requisite hedges are simply the negatives of the replicating portfolio units. The method `complete_mkt_price_and_hedges` (of the `MaxExpUtility` class) shown below implements this solution, producing a dictionary comprising of the derivative price (`price`) and the hedge units α (`alpha`) and β (`beta`).

```
def complete_mkt_price_and_hedges(self) -> Mapping[str, float]:
    x = self.risky_mean + self.risky_stdev
    z = self.risky_mean - self.risky_stdev
    v1 = self.payoff_func(x)
    v2 = self.payoff_func(z)
    alpha = (v1 - v2) / (z - x)
    beta = - 1 / (1 + self.riskless_rate) * (v1 + alpha * x)
    price = - (beta + alpha * self.risky_spot)
    return {"price": price, "alpha": alpha, "beta": beta}
```

Next, we write a helper method `max_exp_util_for_zero` (to handle the right-hand-side of Equation (9.18)) that calculates the maximum expected utility for the special case of a derivative with payoff equal to 0 in all random outcomes at $t = 1$, i.e., it calculates:

$$\max_{\alpha} \mathbb{E}_{S \sim \mathcal{N}(\mu, \sigma^2)} \left[\frac{1 - e^{-a \cdot (-(1+r) \cdot c + \alpha \cdot (S - (1+r) \cdot S_0))}}{a} \right]$$

where c is cash paid at $t = 0$ (so, $c = -(\alpha \cdot S_0 + \beta)$).

The method `max_exp_util_for_zero` accepts as input `c: float` (representing the cash c paid at $t = 0$) and `risk_aversion_param: float` (representing the CARA coefficient of risk aversion a). Refering to Section A.4.1 in Appendix A, we have a closed-form solution to this maximization problem:

$$\alpha^* = \frac{\mu - (1+r) \cdot S_0}{a \cdot \sigma^2}$$

$$\beta^* = -(c + \alpha^* \cdot S_0)$$

Substituting α^* in the Expected Utility expression above gives the following maximum value for the Expected Utility for this special case:

$$\frac{1 - e^{-a \cdot (-(1+r) \cdot c + \alpha^* \cdot (\mu - (1+r) \cdot S_0)) + \frac{(a \cdot \alpha^* \cdot \sigma)^2}{2}}}{a} = \frac{1 - e^{a \cdot (1+r) \cdot c - \frac{(\mu - (1+r) \cdot S_0)^2}{2\sigma^2}}}{a}$$

```python
def max_exp_util_for_zero(
    self,
    c: float,
    risk_aversion_param: float
) -> Mapping[str, float]:
    ra = risk_aversion_param
    er = 1 + self.riskless_rate
    mu = self.risky_mean
    sigma = self.risky_stdev
    s0 = self.risky_spot
    alpha = (mu - s0 * er) / (ra * sigma * sigma)
    beta = - (c + alpha * self.risky_spot)
    max_val = (1 - np.exp(-ra * (-er * c + alpha * (mu - s0 * er))
               + (ra * alpha * sigma) ** 2 / 2)) / ra
    return {"alpha": alpha, "beta": beta, "max_val": max_val}
```

Next, we write a method `max_exp_util` that calculates the maximum expected utility for the general case of a derivative with an arbitrary payoff $f(\cdot)$ at $t = 1$ (provided as input `pf: Callable[[float, float]]` below), i.e., it calculates:

$$\max_{\alpha} \mathbb{E}_{S \sim \mathcal{N}(\mu, \sigma^2)} \left[\frac{1 - e^{-a \cdot (f(S) - (1+r) \cdot c + \alpha \cdot (S - (1+r) \cdot S_0))}}{a} \right]$$

Clearly, this has no closed-form solution since $f(\cdot)$ is an arbitrary payoff. The method `max_exp_util` uses the `scipy.integrate.quad` function to calculate the expectation as an integral of the CARA utility function of $f(S) - (1+r) \cdot c + \alpha \cdot (S - (1+r) \cdot S_0)$ multiplied by the probability density of $\mathcal{N}(\mu, \sigma^2)$, and then uses the `scipy.optimize.minimize_scalar` function to perform the maximization over values of α.

```python
from scipy.integrate import quad
from scipy.optimize import minimize_scalar

def max_exp_util(
    self,
    c: float,
    pf: Callable[[float], float],
```

```
    risk_aversion_param: float
) -> Mapping[str, float]:
    sigma2 = self.risky_stdev * self.risky_stdev
    mu = self.risky_mean
    s0 = self.risky_spot
    er = 1 + self.riskless_rate
    factor = 1 / np.sqrt(2 * np.pi * sigma2)

    integral_lb = self.risky_mean - self.risky_stdev * 6
    integral_ub = self.risky_mean + self.risky_stdev * 6

    def eval_expectation(alpha: float, c=c) -> float:

        def integrand(rand: float, alpha=alpha, c=c) -> float:
            payoff = pf(rand) - er * c\
                     + alpha * (rand - er * s0)
            exponent = -(0.5 * (rand - mu) * (rand - mu) / sigma2
                         + risk_aversion_param * payoff)
            return (1 - factor * np.exp(exponent)) / risk_aversion_param

        return -quad(integrand, integral_lb, integral_ub)[0]

    res = minimize_scalar(eval_expectation)
    alpha_star = res["x"]
    max_val = - res["fun"]
    beta_star = - (c + alpha_star * s0)
    return {"alpha": alpha_star, "beta": beta_star, "max_val": max_val}
```

Finally, it's time to put it all together—the method max_exp_util_price_and_hedge below calculates the maximizing x^* in Equation (9.18). First, we call max_exp_util_for_zero (with c set to 0) to calculate the right-hand-side of Equation (9.18). Next, we create a wrapper function prep_func around max_exp_util, which is provided as input to scipt.optimize.root_scalar to solve for x^* in the right-hand-side of Equation (9.18). Plugging x^* (opt_price in the code below) in max_exp_util provides the hedges α^* and β^* (alpha and beta in the code below).

```
from scipy.optimize import root_scalar
def max_exp_util_price_and_hedge(
    self,
    risk_aversion_param: float
) -> Mapping[str, float]:
    meu_for_zero = self.max_exp_util_for_zero(
        0.,
        risk_aversion_param
    )["max_val"]

    def prep_func(pr: float) -> float:
        return self.max_exp_util(
            pr,
            self.payoff_func,
            risk_aversion_param
        )["max_val"] - meu_for_zero

    lb = self.risky_mean - self.risky_stdev * 10
    ub = self.risky_mean + self.risky_stdev * 10
    payoff_vals = [self.payoff_func(x) for x in np.linspace(lb, ub, 1001)]
    lb_payoff = min(payoff_vals)
    ub_payoff = max(payoff_vals)

    opt_price = root_scalar(
        prep_func,
        bracket=[lb_payoff, ub_payoff],
        method="brentq"
    ).root

    hedges = self.max_exp_util(
        opt_price,
        self.payoff_func,
        risk_aversion_param
```

```
)
alpha = hedges["alpha"]
beta = hedges["beta"]
return {"price": opt_price, "alpha": alpha, "beta": beta}
```

The above code for the class MaxExpUtility is in the file rl/chapter8/max_exp_utility.py. As ever, we encourage you to play with various choices of S_0, r, μ, σ, f to create instances of MaxExpUtility, analyze the obtained prices/hedges, and plot some graphs to develop intuition on how the results change as a function of the various inputs.

Running this code for $S_0 = 100, r = 5\%, \mu = 110, \sigma = 25$ when buying a call option (European since we have only one time period) with strike price $= 105$, the method complete_mkt_price_and hedges gives an option price of 11.43, risky asset hedge units of -0.6 (i.e., we hedge the risk of owning the call option by short-selling 60% of the risky asset) and riskless asset hedge units of 48.57 (i.e., we take the $60 proceeds of short-sale less the $11.43 option price payment $= \$48.57$ of cash and invest in a riskless bank account earning 5% interest). As mentioned earlier, this is the perfect hedge if we had a complete market (i.e., two random outcomes). Running this code for the same inputs for an incomplete market (calling the method max_exp_util_price_and_hedge for risk-aversion parameter values of $a = 0.3, 0.6, 0.9$ gives us the following results:

```
--- Risk Aversion Param = 0.30 ---
{'price': 23.279, 'alpha': -0.473, 'beta': 24.055}
--- Risk Aversion Param = 0.60 ---
{'price': 12.669, 'alpha': -0.487, 'beta': 35.998}
--- Risk Aversion Param = 0.90 ---
{'price': 8.865, 'alpha': -0.491, 'beta': 40.246}
```

We note that the call option price is quite high (23.28) when the risk-aversion is low at $a = 0.3$ (relative to the complete market price of 11.43) but the call option price drops to 12.67 and 8.87 for $a = 0.6$ and $a = 0.9$, respectively. This makes sense since if you are more risk-averse (high a), then you'd be less willing to take the risk of buying a call option and hence, would want to pay less to buy the call option. Note how the risky asset short-sale is significantly less (~47% – ~49%) compared the to the risky asset short-sale of 60% in the case of a complete market. The varying investments in the riskless asset (as a function of the risk-aversion a) essentially account for the variation in option prices (as a function of a). Figure 9.1 provides tremendous intuition on how the hedges work for the case of a complete market and for the cases of an incomplete market with the 3 choices of risk-aversion parameters. Note that we have plotted the negatives of the hedge portfolio values at $t = 1$ so as to visualize them appropriately relative to the payoff of the call option. Note that the hedge portfolio value is a linear function of the risky asset price at $t = 1$. Notice how the slope and intercept of the hedge portfolio value changes for the 3 risk-aversion scenarios and how they compare against the complete market hedge portfolio value.

Now let us consider the case of selling the same call option. In our code, the only change we make is to make the payoff function lambda x: - max(x - 105.0, 0) instead of lambda x: max(x - 105.0, 0) to reflect the fact that we are now selling the call option and so, our payoff will be the negative of that of an owner of the call option.

With the same inputs of $S_0 = 100, r = 5\%, \mu = 110, \sigma = 25$, and for the same risk-aversion parameter values of $a = 0.3, 0.6, 0.9$, we get the following results:

```
--- Risk Aversion Param = 0.30 ---
{'price': -6.307, 'alpha': 0.527, 'beta': -46.395}
--- Risk Aversion Param = 0.60 ---
{'price': -32.317, 'alpha': 0.518, 'beta': -19.516}
```

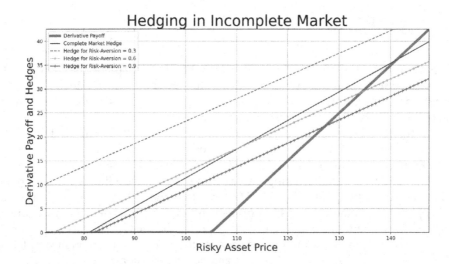

Figure 9.1 **Hedges When Buying a Call Option**

```
--- Risk Aversion Param = 0.90 ---
{'price': -44.236, 'alpha': 0.517, 'beta': -7.506}
```

We note that the sale price demand for the call option is quite low (6.31) when the risk-aversion is low at $a = 0.3$ (relative to the complete market price of 11.43) but the sale price demand for the call option rises sharply to 32.32 and 44.24 for $a = 0.6$ and $a = 0.9$, respectively. This makes sense since if you are more risk-averse (high a), then you'd be less willing to take the risk of selling a call option and hence, would want to charge more for the sale of the call option. Note how the risky asset hedge units are less ($\sim 52\% - 53\%$) compared to the risky asset hedge units (60%) in the case of a complete market. The varying riskless borrowing amounts (as a function of the risk-aversion a) essentially account for the variation in option prices (as a function of a). Figure 9.2 provides the visual intuition on how the hedges work for the 3 choices of risk-aversion parameters (along with the hedges for the complete market, for reference).

Note that each buyer and each seller might have a different level of risk-aversion, meaning each of them would have a different buy price bid/different sale price ask. A transaction can occur between a buyer and a seller (with potentially different risk-aversion levels) if the buyer's bid matches the seller's ask.

9.6.3 Derivatives Pricing When Market Has Arbitrage

Finally, we arrive at the case where the market has arbitrage. This is the case where there is no risk-neutral probability measure and there can be multiple replicating portfolios (which can lead to arbitrage). So this is the case where we are unable to price derivatives. To provide intuition for the case of a market with arbitrage, we consider the special case of 2 risky assets ($m = 2$) and 2 random outcomes ($n = 2$), which we will show is a Market with Arbitrage. Without loss of generality, we assume $S_1^{(1)} < S_1^{(2)}$ and $S_2^{(1)} < S_2^{(2)}$. Let us try to determine a risk-neutral probability measure π:

$$S_1^{(0)} = e^{-r} \cdot (\pi(\omega_1) \cdot S_1^{(1)} + \pi(\omega_2) \cdot S_1^{(2)})$$
$$S_2^{(0)} = e^{-r} \cdot (\pi(\omega_1) \cdot S_2^{(1)} + \pi(\omega_2) \cdot S_2^{(2)})$$

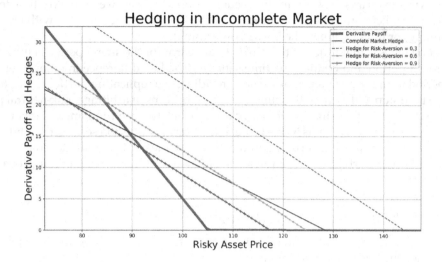

Figure 9.2 **Hedges When Selling a Call Option**

$$\pi(\omega_1) + \pi(\omega_2) = 1$$

3 equations and 2 variables implies that there is no risk-neutral probability measure π for various sets of values of $S_1^{(1)}, S_1^{(2)}, S_2^{(1)}, S_2^{(2)}$. Let's try to form a replicating portfolio $(\theta_0, \theta_1, \theta_2)$ for a derivative D:

$$V_D^{(1)} = \theta_0 \cdot e^r + \theta_1 \cdot S_1^{(1)} + \theta_2 \cdot S_2^{(1)}$$

$$V_D^{(2)} = \theta_0 \cdot e^r + \theta_1 \cdot S_1^{(2)} + \theta_2 \cdot S_2^{(2)}$$

2 equations and 3 variables implies that there are multiple replicating portfolios. Each such replicating portfolio yields a price for D as:

$$V_D^{(0)} = \theta_0 + \theta_1 \cdot S_1^{(0)} + \theta_2 \cdot S_2^{(0)}$$

Select two such replicating portfolios with different $V_D^{(0)}$. The combination of one of these replicating portfolios with the negative of the other replicating portfolio is an Arbitrage Portfolio because:

- They cancel off each other's portfolio value in each $t = 1$ states.
- The combined portfolio value can be made to be negative at $t = 0$ (by appropriately choosing the replicating portfolio to negate).

So this is a market that admits arbitrage (no risk-neutral probability measure).

9.7   DERIVATIVES PRICING IN MULTI-PERIOD/CONTINUOUS-TIME

Now that we have understood the key concepts of derivatives pricing/hedging for the simple setting of discrete-time with a single-period, it's time to do an overview of derivatives pricing/hedging theory in the full-blown setting of multiple time-periods and in continuous-time. While an adequate coverage of this theory is beyond the scope of this book, we will sketch an overview in this section. Along the way, we will cover two derivatives pricing applications that can be modeled as MDPs (and hence, tackled with Dynamic Programming or Reinforcement Learning Algorithms).

The good news is that much of the concepts we learnt for the single-period setting carry over to multi-period and continuous-time settings. The key difference in going over from single-period to multi-period is that we need to adjust the replicating portfolio (i.e., adjust θ) at each time step. Other than this difference, the concepts of arbitrage, risk-neutral probability measures, complete market etc. carry over. In fact, the two fundamental theorems of asset pricing also carry over. It is indeed true that in the multi-period setting, no-arbitrage is equivalent to the existence of a risk-neutral probability measure and market completeness (i.e., replication of derivatives) is equivalent to having a unique risk-neutral probability measure.

9.7.1 Multi-Period Complete-Market Setting

We learnt in the single-period setting that if the market is complete, there are two equivalent ways to conceptualize derivatives pricing:

- Solve for the replicating portfolio (i.e., solve for the units in the fundamental assets that would replicate the derivative payoff), and then calculate the derivative price as the value of this replicating portfolio at $t = 0$.
- Calculate the probabilities of random-outcomes for the unique risk-neutral probability measure, and then calculate the derivative price as the riskless rate-discounted expectation (under this risk-neutral probability measure) of the derivative payoff.

It turns out that even in the multi-period setting, when the market is complete, we can calculate the derivative price (not just at $t = 0$, but at any random outcome at any future time) with either of the above two (equivalent) methods, as long as we appropriately adjust the fundamental assets' units in the replicating portfolio (depending on the random outcome) as we move from one time step to the next. It is important to note that when we alter the fundamental assets' units in the replicating portfolio at each time step, we need to respect the constraint that money cannot enter or leave the replicating portfolio (i.e., it is a *self-financing replicating portfolio* with the replicating portfolio value remaining unchanged in the process of altering the units in the fundamental assets). It is also important to note that the alteration in units in the fundamental assets is dependent on the prices of the fundamental assets (which are random outcomes as we move forward from one time step to the next). Hence, the fundamental assets' units in the replicating portfolio evolve as random variables, while respecting the self-financing constraint. Therefore, the replicating portfolio in a multi-period setting in often referred to as a *Dynamic Self-Financing Replicating Portfolio* to reflect the fact that the replicating portfolio is adapting to the changing prices of the fundamental assets. The negatives of the fundamental assets' units in the replicating portfolio form the hedges for the derivative.

To ensure that the market is complete in a multi-period setting, we need to assume that the market is "frictionless"—that we can trade in real-number quantities in any fundamental asset and that there are no transaction costs for any trades at any time step. From a computational perspective, we walk back in time from the final time step (call it $t = T$) to $t = 0$, and calculate the fundamental assets' units in the replicating portfolio in a "backward recursive manner". As in the case of the single-period setting, each backward-recursive step from outcomes at time $t + 1$ to a specific outcome at time t simply involves solving a linear system of equations where each unknown is the replicating portfolio units in a specific fundamental asset and each equation corresponds to the value of the replicating portfolio at a specific outcome at time $t + 1$ (which is established recursively). The market is complete if there is a unique solution to each linear system of equations (for each time t and for each outcome at time t) in this backward-recursive computation. This gives us not just the replicating portfolio (and consequently, hedges) at each outcome at each time step, but also the price at each outcome at each time step (the price is equal to the value of the calculated replicating portfolio at that outcome at that time step).

Equivalently, we can do a backward-recursive calculation in terms of the risk-neutral probability measures, with each risk-neutral probability measure giving us the transition probabilities from an outcome at time step t to outcomes at time step $t + 1$. Again, in a complete market, it amounts to a unique solution of each of these linear system of equations. For each of these linear system of equations, an unknown is a transition probability to a time $t + 1$ outcome and an equation corresponds to a specific fundamental asset's prices at the time $t + 1$ outcomes. This calculation is popularized (and easily understood) in the simple context of a Binomial Options Pricing Model. We devote Section 9.8 to coverage of the original Binomial Options Pricing Model and model it as a Finite-State Finite-Horizon MDP (and utilize the Finite-Horizon DP code developed in Chapter 5 to solve the MDP).

9.7.2 Continuous-Time Complete-Market Setting

To move on from multi-period to continuous-time, we simply make the time-periods smaller and smaller, and take the limit of the time-period tending to zero. We need to preserve the complete-market property as we do this, which means that we can trade in real-number units without transaction costs in continuous-time. As we've seen before, operating in continuous-time allows us to tap into stochastic calculus, which forms the foundation of much of the rich theory of continuous-time derivatives pricing/hedging. With this very rough and high-level overview, we refer you to Tomas Bjork's book on Arbitrage Theory in Continuous Time (Björk 2005) for a thorough understanding of this theory.

To provide a sneak-peek into this rich continuous-time theory, we've sketched in Appendix E the derivation of the famous Black-Scholes equation and its solution for the case of European Call and Put Options.

So to summarize, we are in good shape to price/hedge in a multi-period and continuous-time setting if the market is complete. But what if the market is incomplete (which is typical in a real-world situation)? Founded on the Fundamental Theorems of Asset Pricing (which applies to multi-period and continuous-time settings as well), there is indeed considerable literature on how to price in incomplete markets for multi-period/continuous-time, which includes the superhedging approach as well as the *Expected-Utility-Indifference* approach, that we had covered in Subsection 9.6.2 for the simple setting of discrete-time with single-period. However, in practice, these approaches are not adopted as they fail to capture real-world nuances adequately. Besides, most of these approaches lead to fairly wide price bounds that are not particularly useful in practice. In Section 9.10, we extend the *Expected-Utility-Indifference* approach that we had covered for the single-period setting to the multi-period setting. It turns out that this approach can be modeled as an MDP, with the adjustments to the hedge quantities at each time step as the actions of the MDP—calculating the optimal policy gives us the optimal derivative hedging strategy and the associated optimal value function gives us the derivative price. This approach is applicable to real-world situations and one can even incorporate all the real-world frictions in one's MDP to build a practical solution for derivatives trading (covered in Section 9.10).

9.8 OPTIMAL EXERCISE OF AMERICAN OPTIONS CAST AS A FINITE MDP

In this section, we tackle the pricing of American Options in a discrete-time, multi-period setting, assuming the market is complete. To satisfy market completeness, we need to assume that the market is "frictionless"—that we can trade in real-number quantities in any fundamental asset and that there are no transaction costs for any trades at any time step. In particular, we employ the Binomial Options Pricing Model to solve for the price (and hedges) of American Options. The original Binomial Options Pricing Model was developed to price (and hedge) options (including American Options) on an underlying whose price evolves according to a lognormal stochastic process, with the stochastic process approximated in the form of a simple discrete-time, finite-horizon, finite-states process that enables enormous computational tractability. The lognormal stochastic process is basically of the same form as the stochastic process of the underlying price in the Black-Scholes model (covered in Appendix E). However, the underlying price process in the Black-Scholes model is specified in the real-world probability measure whereas here we specify the underlying price process in the risk-neutral probability measure. This is because here we will employ the pricing method of riskless rate-discounted expectation (under the risk-neutral probability measure) of the option payoff. Recall that in the single-period setting, the underlying asset price's expected rate of growth is calibrated to be equal to the riskless rate r, under the risk-neutral probability measure. This calibration applies even in the multi-period and continuous-time settings. For a continuous-time lognormal stochastic process, the lognormal drift will hence be equal to r in the risk-neutral probability measure (rather than μ in the real-world probability measure, as per the Black-Scholes model). Precisely, the stochastic process S for the underlying price in the risk-neutral probability measure is:

$$dS_t = r \cdot S_t \cdot dt + \sigma \cdot S_t \cdot dz_t \tag{9.19}$$

where σ is the lognormal dispersion (often referred to as "lognormal volatility"—we will simply call it volatility for the rest of this section). If you want to develop a thorough understanding of the broader topic of change of probability measures and how it affects the drift term (beyond the scope of this book, but an important topic in continuous-time financial pricing theory), we refer you to the technical material on Radon-Nikodym Derivative and Girsanov Theorem.

The Binomial Options Pricing Model serves as a discrete-time, finite-horizon, finite-states approximation to this continuous-time process, and is essentially an extension of the single-period model we had covered earlier for the case of a single fundamental risky asset. We've learnt previously that in the single-period case for a single fundamental risky asset, in order to be a complete market, we need to have exactly two random outcomes. We basically extend this "two random outcomes" pattern to each outcome at each time step, by essentially growing out a "binary tree". But there is a caveat—with a binary tree, we end up with an exponential (2^i) number of outcomes after i time steps. To contain the exponential growth, we construct a "recombining tree", meaning an "up move" followed by a "down move" ends up in the same underlying price outcome as a "down move" followed by an "up move" (as illustrated in Figure 9.3). Thus, we have $i+1$ price outcomes after i time steps in this "recombining tree". We conceptualize the ascending-sorted sequence of $i + 1$ price outcomes as the (time step = i) states $\mathcal{S}_i = \{0, 1, \ldots, i\}$ (since the price movements form a discrete-time, finite-states Markov Process). Since we are modeling a lognormal process, we model the discrete-time price moves as multiplicative to the price. We denote $S_{i,j}$ as the price after i time steps in state j (for any $i \in \mathbb{Z}_{\geq 0}$ and for any $0 \leq j \leq i$). So the two random prices resulting from $S_{i,j}$ are $S_{i+1,j+1} = S_{i,j} \cdot u$ and $S_{i+1,j} = S_{i,j} \cdot d$ for some constants u and d (that are calibrated). The important point is that u and d remain constant across time steps i and across states j at each time step i (as seen in Figure 9.3).

Let q be the probability of the "up move" (typically, we use p to denote real-world probability and q to denote the risk-neutral probability) so that $1 - q$ is the probability of the "down move". Just like u and d, the value of q is kept constant across time steps i and across

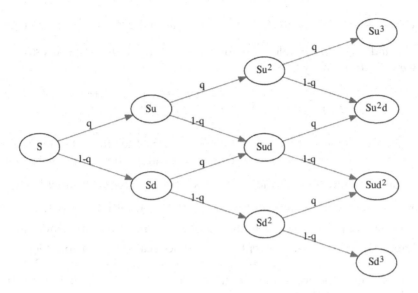

Figure 9.3 **Binomial Option Pricing Model (Binomial Tree)**

states j at each time step i (as seen in Figure 9.3). q, u and d need to be calibrated so that the probability distribution of log-price-ratios $\{\log\left(\frac{S_{i,0}}{S_{0,0}}\right), \log\left(\frac{S_{i,1}}{S_{0,0}}\right), \ldots, \log\left(\frac{S_{i,i}}{S_{0,0}}\right)\}$ after i time steps (with each time step of interval $\frac{T}{n}$ for a given expiry time $T \in \mathbb{R}^+$ and a fixed number of time steps $n \in \mathbb{Z}^+$) serves as a good approximation to $\mathcal{N}((r - \frac{\sigma^2}{2})\frac{iT}{n}, \frac{\sigma^2 iT}{n})$ (that we know to be the risk-neutral probability distribution of $\log\left(\frac{S_{\frac{iT}{n}}}{S_0}\right)$ in the continuous-time process defined by Equation (9.19), as derived in Section C.7 in Appendix C), for all $i = 0, 1, \ldots, n$. Note that the starting price $S_{0,0}$ of this discrete-time approximation process is equal to the starting price S_0 of the continuous-time process.

This calibration of q, u and d can be done in a variety of ways and there are indeed several variants of Binomial Options Pricing Models with different choices of how q, u and d are calibrated. We shall implement the choice made in the original Binomial Options Pricing Model that was proposed in a seminal paper by Cox, Ross, Rubinstein (Cox, Ross, and Rubinstein 1979). Their choice is best understood in two steps:

- As a first step, ignore the drift term $r \cdot S_t \cdot dt$ of the lognormal process, and assume the underlying price follows the martingale process $dS_t = \sigma \cdot S_t \cdot dz_t$. They chose d to be equal to $\frac{1}{u}$ and calibrated u such that for any $i \in \mathbb{Z}_{\geq 0}$, for any $0 \leq j \leq i$, the variance of two equal-probability random outcomes $\log\left(\frac{S_{i+1,j+1}}{S_{i,j}}\right) = \log(u)$ and $\log\left(\frac{S_{i+1,j}}{S_{i,j}}\right) = \log(d) = -\log(u)$ is equal to the variance $\frac{\sigma^2 T}{n}$ of the normally-distributed random variable $\log\left(\frac{S_{t+\frac{T}{n}}}{S_t}\right)$ for any $t \geq 0$ (assuming the process $dS_t = \sigma \cdot S_t \cdot dz_t$). This yields:

$$\log^2(u) = \frac{\sigma^2 T}{n} \Rightarrow u = e^{\sigma\sqrt{\frac{T}{n}}}$$

- As a second step, q needs to be calibrated to account for the drift term $r \cdot S_t \cdot dt$ in the lognormal process under the risk-neutral probability measure. Specifically, q is adjusted so that for any $i \in \mathbb{Z}_{\geq 0}$, for any $0 \leq j \leq i$, the mean of the two random outcomes $\frac{S_{i+1,j+1}}{S_{i,j}} = u$ and $\frac{S_{i+1,j}}{S_{i,j}} = \frac{1}{u}$ is equal to the mean $e^{\frac{rT}{n}}$ of the lognormally-distributed random variable $\frac{S_{t+\frac{T}{n}}}{S_t}$ for any $t \geq 0$ (assuming the process $dS_t = r \cdot S_t \cdot dt + \sigma \cdot S_t \cdot dz_t$). This yields:

$$qu + \frac{1-q}{u} = e^{\frac{rT}{n}} \Rightarrow q = \frac{u \cdot e^{\frac{rT}{n}} - 1}{u^2 - 1} = \frac{e^{\frac{rT}{n} + \sigma\sqrt{\frac{T}{n}}} - 1}{e^{2\sigma\sqrt{\frac{T}{n}}} - 1}$$

This calibration for u and q ensures that as $n \to \infty$ (i.e., time step interval $\frac{T}{n} \to 0$), the mean and variance of the binomial distribution after i time steps matches the mean $(r - \frac{\sigma^2}{2})\frac{iT}{n}$ and variance $\frac{\sigma^2 iT}{n}$ of the normally-distributed random variable $\log\left(\frac{S_{\frac{iT}{n}}}{S_0}\right)$ in the continuous-time process defined by Equation (9.19), for all $i = 0, 1, \ldots n$. Note that $\log\left(\frac{S_{i,j}}{S_{0,0}}\right)$ follows a random walk Markov Process (reminiscent of the random walk examples in Chapter 3) with each movement in state space scaled by a factor of $\log(u)$.

Thus, we have the parameters u and q that fully specify the Binomial Options Pricing Model. Now we get to the application of this model. We are interested in using this model for optimal exercise (and hence, pricing) of American Options. This is in contrast to the Black-Scholes Partial Differential Equation which only enabled us to price options with a fixed payoff at a fixed point in time (e.g., European Call and Put Options). Of course, a special case of American Options is indeed European Options. It's important to note that here we are tackling the much harder problem of the ideal timing of exercise of an American Option—the Binomial Options Pricing Model is well suited for this.

As mentioned earlier, we want to model the problem of Optimal Exercise of American Options as a discrete-time, finite-horizon, finite-states MDP. We set the terminal time to be $t = T + 1$, meaning all the states at time $T + 1$ are terminal states. Here we will utilize the states and state transitions (probabilistic price movements of the underlying) given by the Binomial Options Pricing Model as the states and state transitions in the MDP. The MDP actions in each state will be binary—either exercise the option (and immediately move to a terminal state) or don't exercise the option (i.e., continue on to the next time step's random state, as given by the Binomial Options Pricing Model). If the exercise action is chosen, the MDP reward is the option payoff. If the continue action is chosen, the reward is 0. The discount factor γ is $e^{-\frac{rT}{n}}$ since (as we've learnt in the single-period case), the price (which translates here to the Optimal Value Function) is defined as the riskless rate-discounted expectation (under the risk-neutral probability measure) of the option payoff. In the multi-period setting, the overall discounting amounts to composition (multiplication) of each time step's discounting (which is equal to γ) and the overall risk-neutral probability measure amounts to the composition of each time step's risk-neutral probability measure (which is specified by the calibrated value q).

Now let's write some code to determine the Optimal Exercise of American Options (and hence, the price of American Options) by modeling this problem as a discrete-time, finite-horizon, finite-states MDP. We create a dataclass OptimalExerciseBinTree whose attributes are spot_price (specifying the current, i.e., time=0 price of the underlying), payoff (specifying the option payoff, when exercised), expiry (specifying the time T to expiration of the American Option), rate (specifying the riskless rate r), vol (specifying the lognormal volatility σ), and num_steps (specifying the number n of time steps in the binomial tree). Note that each time step is of interval $\frac{T}{n}$ (which is implemented below in the method dt). Note also that the payoff function is fairly generic taking two arguments—the first argument is the time at which the option is exercised, and the second argument is the underlying price at the time the option is exercised. Note that for a typical American Call or Put Option, the payoff does not depend on time and the dependency on the underlying price is the standard "hockey-stick" payoff that we are now fairly familiar with (however, we designed the interface to allow for more general option payoff functions).

The set of states S_i at time step i (for all $0 \leq i \leq T+1$) is: $\{0, 1, \ldots, i\}$ and the method state_price below calculates the price in state j at time step i as:

$$S_{i,j} = S_{0,0} \cdot e^{(2j-i)\sigma\sqrt{\frac{T}{n}}}$$

Finally, the method get_opt_vf_and_policy calculates u (up_factor) and q (up_prob), prepares the requisite state-reward transitions (conditional on current state and action) to move from one time step to the next, and passes along the constructed time-sequenced transitions to rl.finite_horizon.get_opt_vf_and_policy (which we had written in Chapter 5) to perform the requisite backward induction and return an Iterator on pairs of V[int] and FiniteDeterministicPolicy[int, bool]. Note that the states at any time-step i are the integers from 0 to i and hence, represented as int, and the actions are represented as bool (True for exercise and False for continue). Note that we represent an early terminal state (in case of option exercise before expiration of the option) as -1.

```python
from rl.distribution import Constant, Categorical
from rl.finite_horizon import optimal_vf_and_policy
from rl.dynamic_programming import V
from rl.policy import FiniteDeterministicPolicy
@dataclass(frozen=True)
class OptimalExerciseBinTree:

    spot_price: float
    payoff: Callable[[float, float], float]
    expiry: float
    rate: float
    vol: float
    num_steps: int

    def dt(self) -> float:
        return self.expiry / self.num_steps

    def state_price(self, i: int, j: int) -> float:
        return self.spot_price * np.exp((2 * j - i) * self.vol *
                                        np.sqrt(self.dt()))

    def get_opt_vf_and_policy(self) -> \
            Iterator[Tuple[V[int], FiniteDeterministicPolicy[int, bool]]]:
        dt: float = self.dt()
        up_factor: float = np.exp(self.vol * np.sqrt(dt))
        up_prob: float = (np.exp(self.rate * dt) * up_factor - 1) / \
            (up_factor * up_factor - 1)
        return optimal_vf_and_policy(
            steps=[
                {NonTerminal(j): {
                    True: Constant(
                        (
                            Terminal(-1),
                            self.payoff(i * dt, self.state_price(i, j))
                        )
                    ),
                    False: Categorical(
                        {
                            (NonTerminal(j + 1), 0.): up_prob,
                            (NonTerminal(j), 0.): 1 - up_prob
                        }
                    )
                } for j in range(i + 1)}
                for i in range(self.num_steps + 1)
            ],
            gamma=np.exp(-self.rate * dt)
        )
```

Now we want to try out this code on an American Call Option and American Put Option. We know that it is never optimal to exercise an American Call Option before the option expiration. The reason for this is as follows: Upon early exercise (say at time $\tau < T$), we borrow cash K (to pay for the purchase of the underlying) and own the underlying (valued at S_τ). So, at option expiration T, we owe cash $K \cdot e^{r(T-\tau)}$ and own the underlying valued at S_T, which is an overall value at time T of $S_T - K \cdot e^{r(T-\tau)}$. We argue that this value is always less than the value $\max(S_T - K, 0)$ we'd obtain at option expiration T if we'd made the choice to not exercise early. If the call option ends up in-the-money at option expiration T (i.e., $S_T > K$), then $S_T - K \cdot e^{r(T-\tau)}$ is less than the value $S_T - K$ we'd get by exercising at option expiration T. If the call option ends up not being in-the-money at option expiration T (i.e., $S_T \le K$), then $S_T - K \cdot e^{r(T-\tau)} < 0$ which is less than the 0 payoff we'd obtain at option expiration T. Hence, we are always better off waiting until option expiration (i.e. it is never optimal to exercise a call option early, no matter how much in-the-money we get before option expiration). Hence, the price of an American Call Option should be equal to the price of an European Call Option with the same strike price and expiration time. However, for an American Put Option, it is indeed sometimes optimal to exercise early and hence, the price of an American Put Option is greater then the price of an European Put Option with the same strike price and expiration time. Thus, it is interesting to ask the question: For each time $t < T$, what is the threshold of underlying price S_t below which it is optimal to exercise an American Put Option? It is interesting to view this threshold as a function of time (we call this function as the optimal exercise boundary of an American Put Option). One would expect that this optimal exercise boundary rises as one gets closer to the option expiration T. But exactly what shape does this optimal exercise boundary have? We can answer this question by analyzing the optimal policy at each time step—we just need to find the state k at each time step i such that the Optimal Policy $\pi_i^*(\cdot)$ evaluates to True for all states $j \le k$ (and evaluates to False for all states $j > k$). We write the following method to calculate the Optimal Exercise Boundary:

```
def option_exercise_boundary(
    self,
    policy_seq: Sequence[FiniteDeterministicPolicy[int, bool]],
    is_call: bool
) -> Sequence[Tuple[float, float]]:
    dt: float = self.dt()
    ex_boundary: List[Tuple[float, float]] = []
    for i in range(self.num_steps + 1):
        ex_points = [j for j in range(i + 1)
                     if policy_seq[i].action_for[j] and
                     self.payoff(i * dt, self.state_price(i, j)) > 0]
        if len(ex_points) > 0:
            boundary_pt = min(ex_points) if is_call else max(ex_points)
            ex_boundary.append(
                (i * dt, opt_ex_bin_tree.state_price(i, boundary_pt))
            )
    return ex_boundary
```

option_exercise_boundary takes as input policy_seq which represents the sequence of optimal policies π_i^* for each time step $0 \le i \le T$, and produces as output the sequence of pairs $(\frac{iT}{n}, B_i)$ where

$$B_i = \max_{j : \pi_i^*(j) = True} S_{i,j}$$

with the little detail that we only consider those states j for which the option payoff is positive. For some time steps i, none of the states j qualify as $\pi_i^*(j) = True$, in which case we don't include that time step i in the output sequence.

To compare the results of American Call and Put Option Pricing on this Binomial Options Pricing Model against the corresponding European Options prices, we write the following method to implement the Black-Scholes closed-form solution (derived as Equations E.7 and E.8 in Appendix E):

```python
from scipy.stats import norm
    def european_price(self, is_call: bool, strike: float) -> float:
        sigma_sqrt: float = self.vol * np.sqrt(self.expiry)
        d1: float = (np.log(self.spot_price / strike) +
                    (self.rate + self.vol ** 2 / 2.) * self.expiry) \
            / sigma_sqrt
        d2: float = d1 - sigma_sqrt
        if is_call:
            ret = self.spot_price * norm.cdf(d1) - \
                strike * np.exp(-self.rate * self.expiry) * norm.cdf(d2)
        else:
            ret = strike * np.exp(-self.rate * self.expiry) * norm.cdf(-d2) - \
                self.spot_price * norm.cdf(-d1)
        return ret
```

Here's some code to price an American Put Option (changing is_call to True will price American Call Options):

```python
from rl.gen_utils.plot_funcs import plot_list_of_curves

spot_price_val: float = 100.0
strike: float = 100.0
is_call: bool = False
expiry_val: float = 1.0
rate_val: float = 0.05
vol_val: float = 0.25
num_steps_val: int = 300

if is_call:
    opt_payoff = lambda _, x: max(x - strike, 0)
else:
    opt_payoff = lambda _, x: max(strike - x, 0)

opt_ex_bin_tree: OptimalExerciseBinTree = OptimalExerciseBinTree(
    spot_price=spot_price_val,
    payoff=opt_payoff,
    expiry=expiry_val,
    rate=rate_val,
    vol=vol_val,
    num_steps=num_steps_val
)

vf_seq, policy_seq = zip(*opt_ex_bin_tree.get_opt_vf_and_policy())
ex_boundary: Sequence[Tuple[float, float]] = \
    opt_ex_bin_tree.option_exercise_boundary(policy_seq, is_call)
time_pts, ex_bound_pts = zip(*ex_boundary)
label = ("Call" if is_call else "Put") + " Option Exercise Boundary"
plot_list_of_curves(
    list_of_x_vals=[time_pts],
    list_of_y_vals=[ex_bound_pts],
    list_of_colors=["b"],
    list_of_curve_labels=[label],
    x_label="Time",
    y_label="Underlying Price",
    title=label
)

european: float = opt_ex_bin_tree.european_price(is_call, strike)
print(f"European Price = {european:.3f}")

am_price: float = vf_seq[0][NonTerminal(0)]
print(f"American Price = {am_price:.3f}")
```

This prints as output:

```
European Price = 7.459
American Price = 7.971
```

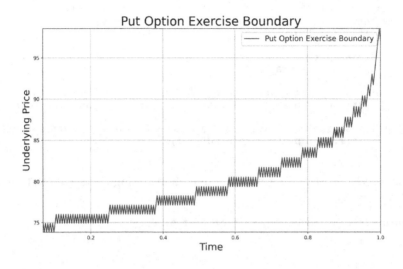

Figure 9.4 **Put Option Exercise Boundary**

So we can see that the price of this American Put Option is significantly higher than the price of the corresponding European Put Option. The exercise boundary produced by this code is shown in Figure 9.4. The locally-jagged nature of the exercise boundary curve is because of the "diamond-like" local-structure of the underlying prices at the nodes in the binomial tree. We can see that when the time to expiry is large, it is not optimal to exercise unless the underlying price drops significantly. It is only when the time to expiry becomes quite small that the optimal exercise boundary rises sharply towards the strike price value.

Changing is_call to True (and not changing any of the other inputs) prints as output:

```
European Price = 12.336
American Price = 12.328
```

This is a numerical validation of our proof above that it is never optimal to exercise an American Call Option before option expiration.

The above code is in the file rl/chapter8/optimal_exercise_bin_tree.py. As ever, we encourage you to play with various choices of inputs to develop intuition for how American Option Pricing changes as a function of the inputs (and how American Put Option Exercise Boundary changes). Note that you can specify the option payoff as any arbitrary function of time and the underlying price.

9.9 GENERALIZING TO OPTIMAL-STOPPING PROBLEMS

In this section, we generalize the problem of Optimal Exercise of American Options to the problem of Optimal Stopping in Stochastic Calculus, which has several applications in Mathematical Finance, including pricing of exotic derivatives. After defining the Optimal Stopping problem, we show how this problem can be modeled as an MDP (generalizing the MDP modeling of Optimal Exercise of American Options), which affords us the ability to solve them with Dynamic Programming or Reinforcement Learning algorithms.

First, we define the concept of *Stopping Time*. Informally, Stopping Time τ is a random time (time as a random variable) at which a given stochastic process exhibits certain behavior. Stopping time is defined by a *stopping policy* to decide whether to continue or stop a stochastic process based on the stochastic process' current and past values. Formally, it is a random variable τ such that the event $\{\tau \leq t\}$ is in the σ-algebra $\mathcal{F}_t$ of the stochastic process, for all t. This means the stopping decision (i.e., *stopping policy*) of whether $\tau \leq t$ only depends on information up to time t, i.e., we have all the information required to make the stopping decision at any time t.

A simple example of Stopping Time is *Hitting Time* of a set A for a process X. Informally, it is the first time when X takes a value within the set A. Formally, Hitting Time T_{X_A} is defined as:

$$T_{X,A} = \min\{t \in \mathbb{R} | X_t \in A\}$$

A simple and common example of Hitting Time is the first time a process exceeds a certain fixed threshold level. As an example, we might say we want to sell a stock when the stock price exceeds \$100. This \$100 threshold constitutes our stopping policy, which determines the stopping time (hitting time) in terms of when we want to sell the stock (i.e., exit owning the stock). Different people may have different criterion for exiting owning the stock (your friend's threshold might be \$90), and each person's criterion defines their own stopping policy and hence, their own stopping time random variable.

Now that we have defined Stopping Time, we are ready to define the Optimal Stopping problem. *Optimal Stopping* for a stochastic process X is a function $W(\cdot)$ whose domain is the set of potential initial values of the stochastic process and co-domain is the length of time for which the stochastic process runs, defined as:

$$W(x) = \max_{\tau} \mathbb{E}[H(X_\tau)|X_0 = x]$$

where τ is a set of stopping times of X and $H(\cdot)$ is a function from the domain of the stochastic process values to the set of real numbers.

Intuitively, you should think of Optimal Stopping as searching through many Stopping Times (i.e., many Stopping Policies), and picking out the best Stopping Policy—the one that maximizes the expected value of a function $H(\cdot)$ applied on the stochastic process at the stopping time.

Unsurprisingly (noting the connection to Optimal Control in an MDP), $W(\cdot)$ is called the Value function, and H is called the Reward function. Note that sometimes we can have several stopping times that maximize $\mathbb{E}[H(X_\tau)]$, and we say that the optimal stopping time is the smallest stopping time achieving the maximum value. We mentioned above that Optimal Exercise of American Options is a special case of Optimal Stopping. Let's understand this specialization better:

- X is the stochastic process for the underlying's price in the risk-neutral probability measure.
- x is the underlying security's current price.
- τ is a set of exercise times, each exercise time corresponding to a specific policy of option exercise (i.e., specific stopping policy).
- $W(\cdot)$ is the American Option price as a function of the underlying's current price x.
- $H(\cdot)$ is the option payoff function (with riskless-rate discounting built into $H(\cdot)$).

Now let us define Optimal Stopping problems as control problems in Markov Decision Processes (MDPs).

- The MDP *State* at time t is X_t.
- The MDP *Action* is Boolean: Stop the Process or Continue the Process.
- The MDP *Reward* is always 0, except upon Stopping, when it is equal to $H(X_\tau)$.
- The MDP *Discount Factor* γ is equal to 1.
- The MDP probabilistic-transitions are governed by the Stochastic Process X.

A specific policy corresponds to a specific stopping-time random variable τ, the Optimal Policy π^* corresponds to the stopping-time τ^* that yields the maximum (over τ) of $\mathbb{E}[H(X_\tau)|X_0 = x]$, and the Optimal Value Function V^* corresponds to the maximum value of $\mathbb{E}[H(X_\tau)|X_0 = x]$.

For discrete time steps, the Bellman Optimality Equation is:

$$V^*(X_t) = \max(H(X_t), \mathbb{E}[V^*(X_{t+1})|X_t])$$

Thus, we see that Optimal Stopping is the solution to the above Bellman Optimality Equation (solving the Control problem of the MDP described above). For a finite number of time steps, we can run a backward induction algorithm from the final time step back to time step 0 (essentially a generalization of the backward induction we did with the Binomial Options Pricing Model to determine Optimal Exercise of American Options).

Many derivatives pricing problems (and indeed many problems in the broader space of Mathematical Finance) can be cast as Optimal Stopping and hence can be modeled as MDPs (as described above). The important point here is that this enables us to employ Dynamic Programming or Reinforcement Learning algorithms to identify optimal stopping policy for exotic derivatives (which typically yields a pricing algorithm for exotic derivatives). When the state space is large (e.g., when the payoff depends on several underlying assets or when the payoff depends on the history of underlying's prices, such as Asian Options-payoff with American exercise feature), the classical algorithms used in the finance industry for exotic derivatives pricing are not computationally tractable. This points to the use of Reinforcement Learning algorithms which tend to be good at handling large state spaces by effectively leveraging sampling and function approximation methodologies in the context of solving the Bellman Optimality Equation. Hence, we propose Reinforcement Learning as a promising alternative technique to pricing of certain exotic derivatives that can be cast as Optimal Stopping problems. We will discuss this more after having covered Reinforcement Learning algorithms.

9.10   PRICING/HEDGING IN AN INCOMPLETE MARKET CAST AS AN MDP

In Subsection 9.6.2, we developed a pricing/hedging approach based on *Expected-Utility-Indifference* for the simple setting of discrete-time with single-period, when the market is incomplete. In this section, we extend this approach to the case of discrete-time with multi-period. In the single-period setting, the solution is rather straightforward as it amounts to an unconstrained multi-variate optimization together with a single-variable root-solver. Now when we extend this solution approach to the multi-period setting, it amounts to a sequential/dynamic optimal control problem. Although this is far more complex than the single-period setting, the good news is that we can model this solution approach for the multi-period setting as a Markov Decision Process. This section will be dedicated to modeling this solution approach as an MDP, which gives us enormous flexibility in capturing the real-world nuances. Besides, modeling this approach as an MDP permits us to tap into some of the recent advances in Deep Learning and Reinforcement Learning (i.e. Deep Reinforcement Learning). Since we haven't yet learnt about Reinforcement Learning algorithms, this section won't cover the algorithmic aspects (i.e., how to solve the MDP)—it will simply cover how to model the MDP for the *Expected-Utility-Indifference* approach to pricing/hedging derivatives in an incomplete market.

Before we get into the MDP modeling details, it pays to remind that in an incomplete market, we have multiple risk-neutral probability measures and hence, multiple valid derivative prices (each consistent with no-arbitrage). This means the market/traders need to "choose" a suitable risk-neutral probability measure (which amounts to choosing one out of the many valid derivative prices). In practice, this "choice" is typically made in ad-hoc and inconsistent ways. Hence, our proposal of making this "choice" in a mathematically-disciplined manner by noting that ultimately a trader is interested in maximizing the "risk-adjusted return" of a derivative together with its hedges (by sequential/dynamic adjustment of the hedge quantities). Once we take this view, it is reminiscent of the *Asset Allocation* problem we covered in Chapter 8 and the maximization objective is based on the specification of preference for trading risk versus return (which in turn, amounts to specification of a Utility function). Therefore, similar to the Asset Allocation problem, the decision at each time step is the set of adjustments one needs to make to the hedge quantities. With this rough overview, we are now ready to formalize the MDP model for this approach to multi-period pricing/hedging in an incomplete market. For ease of exposition, we simplify the problem setup a bit, although the approach and model we describe below essentially applies to more complex, more frictionful markets as well. Our exposition below is an adaptation of the treatment in the Deep Hedging paper by Buehler, Gonon, Teichmann, Wood, Mohan, Kochems (Bühler et al. 2018).

Assume we have a portfolio of m derivatives and we refer to our collective position across the portfolio of m derivatives as D. Assume each of these m derivatives expires by time T (i.e., all of their contingent cashflows will transpire by time T). We model the problem as a discrete-time finite-horizon MDP with the terminal time at $t = T+1$ (i.e., all states at time $t = T+1$ are terminal states). We require the following notation to model the MDP:

- Denote the derivatives portfolio-aggregated *Contingent Cashflows* at time t as $X_t \in \mathbb{R}$.
- Assume we have n assets trading in the market that would serve as potential hedges for our derivatives position D.
- Denote the number of units held in the hedge positions at time t as $\boldsymbol{\alpha}_t \in \mathbb{R}^n$.
- Denote the cashflows per unit of hedges at time t as $\boldsymbol{Y}_t \in \mathbb{R}^n$.
- Denote the prices per unit of hedges at time t as $\boldsymbol{P}_t \in \mathbb{R}^n$.
- Denote the trading account value at time t as $\beta_t \in \mathbb{R}$.

We will use the notation that we have previously used for discrete-time finite-horizon MDPs, i.e., we will use time-subscripts in our notation.

We denote the State Space at time t (for all $0 \le t \le T+1$) as $\mathcal{S}_t$ and a specific state at time t as $s_t \in \mathcal{S}_t$. Among other things, the key ingredients of s_t include: $\boldsymbol{\alpha}_t, \boldsymbol{P}_t, \beta_t, D$. In practice, s_t will include many other components (in general, any market information relevant to hedge trading decisions). However, for simplicity (motivated by ease of articulation), we assume s_t is simply the 4-tuple:

$$s_t := (\boldsymbol{\alpha}_t, \boldsymbol{P}_t, \beta_t, D)$$

We denote the Action Space at time t (for all $0 \le t \le T$) as $\mathcal{A}_t$ and a specific action at time t as $a_t \in \mathcal{A}_t$. a_t represents the number of units of hedges traded at time t (i.e., adjustments to be made to the hedges at each time step). Since there are n hedge positions (n assets to be traded), $a_t \in \mathbb{R}^n$, i.e., $\mathcal{A}_t \subseteq \mathbb{R}^n$. Note that for each of the n assets, its corresponding component in a_t is positive if we buy the asset at time t and negative if we sell the asset at time t. Any trading restrictions (e.g., constraints on short-selling) will essentially manifest themselves in terms of the exact definition of $\mathcal{A}_t$ as a function of s_t.

State transitions are essentially defined by the random movements of prices of the assets that make up the potential hedges, i.e., $\mathbb{P}[\boldsymbol{P}_{t+1}|\boldsymbol{P}_t]$. In practice, this is available either as an explicit transition-probabilities model, or more likely available in the form of a *simulator*, that produces an on-demand sample of the next time step's prices, given the current time step's prices. Either way, the internals of $\mathbb{P}[\boldsymbol{P}_{t+1}|\boldsymbol{P}_t]$ are estimated from actual market data and realistic trading/market assumptions. The practical details of how to estimate these internals are beyond the scope of this book—it suffices to say here that this estimation is a form of supervised learning, albeit fairly nuanced due to the requirement of capturing the complexities of market-price behavior. For the following description of the MDP, simply assume that we have access to $\mathbb{P}[\boldsymbol{P}_{t+1}|\boldsymbol{P}_t]$ in *some form*.

It is important to pay careful attention to the sequence of events at each time step $t = 0, \ldots, T$, described below:

1. Observe the state $s_t := (\boldsymbol{\alpha}_t, \boldsymbol{P}_t, \beta_t, D)$.
2. Perform action (trades) $\boldsymbol{a}_t$, which produces trading account value change $= -\boldsymbol{a}_t^T \cdot \boldsymbol{P}_t$ (note: this is an inner-product in $\mathbb{R}^n$).
3. These trades incur transaction costs, for example equal to $\gamma \cdot abs(\boldsymbol{a}_t^T) \cdot \boldsymbol{P}_t$ for some $\gamma \in \mathbb{R}^+$ (note: *abs*, denoting absolute value, applies point-wise on $\boldsymbol{a}_t^T \in \mathbb{R}^n$, and then we take its inner-product with $\boldsymbol{P}_t \in \mathbb{R}^n$).
4. Update α_t as:
$$\boldsymbol{\alpha}_{t+1} = \boldsymbol{\alpha}_t + \boldsymbol{a}_t$$

   At termination, we need to force-liquidate, which establishes the constraint: $\boldsymbol{a}_T = -\boldsymbol{\alpha}_T$.
5. Realize end-of-time-step cashflows from the derivatives position D as well as from the (updated) hedge positions. This is equal to $X_{t+1} + \boldsymbol{\alpha}_{t+1}^T \cdot \boldsymbol{Y}_{t+1}$ (note: $\boldsymbol{\alpha}_{t+1}^T \cdot \boldsymbol{Y}_{t+1}$ is an inner-product in $\mathbb{R}^n$).
6. Update trading account value β_t as:
$$\beta_{t+1} = \beta_t - \boldsymbol{a}_t^T \cdot \boldsymbol{P}_t - \gamma \cdot abs(\boldsymbol{a}_t^T) \cdot \boldsymbol{P}_t + X_{t+1} + \boldsymbol{\alpha}_{t+1}^T \cdot \boldsymbol{Y}_{t+1}$$
7. MDP Reward $r_{t+1} = 0$ for all $t = 0, \ldots, T-1$ and $r_{T+1} = U(\beta_{T+1})$ for an appropriate concave Utility function (based on the extent of risk-aversion).
8. Hedge prices evolve from $\boldsymbol{P}_t$ to $\boldsymbol{P}_{t+1}$, based on price-transition model of $\mathbb{P}[\boldsymbol{P}_{t+1}|\boldsymbol{P}_t]$.

Assume we now want to enter into an incremental position of derivatives-portfolio D' in m' derivatives. We denote the combined position as $D \cup D'$. We want to determine the *Price* of the incremental position D', as well as the hedging strategy for $D \cup D'$.

Denote the Optimal Value Function at time t (for all $0 \leq t \leq T$) as $V_t^* : \mathcal{S}_t \to \mathbb{R}$. Pricing of D' is based on the principle that introducing the incremental position of D' together with a calibrated cash payment/receipt (Price of D') at $t = 0$ should leave the Optimal Value (at $t = 0$) unchanged. Precisely, the Price of D' is the value x^* such that

$$V_0^*((\boldsymbol{\alpha}_0, \boldsymbol{P}_0, \beta_0 - x^*, D \cup D')) = V_0^*((\boldsymbol{\alpha}_0, \boldsymbol{P}_0, \beta_0, D))$$

This Pricing principle is known as the principle of *Indifference Pricing*. The hedging strategy for $D \cup D'$ at time t (for all $0 \leq t < T$) is given by the associated Optimal Deterministic Policy $\pi_t^* : \mathcal{S}_t \to \mathcal{A}_t$.

9.11 KEY TAKEAWAYS FROM THIS CHAPTER

- The concepts of Arbitrage, Completeness and Risk-Neutral Probability Measure.
- The two fundamental theorems of Asset Pricing.

- Pricing of derivatives in a complete market in two equivalent ways: A) Based on construction of a replicating portfolio and B) Based on riskless rate-discounted expectation in the risk-neutral probability measure.
- Optimal Exercise of American Options (and its generalization to Optimal Stopping problems) cast as an MDP Control problem.
- Pricing and Hedging of Derivatives in an Incomplete (real-world) Market cast as an MDP Control problem.

CHAPTER 10

Order-Book Trading Algorithms

In this chapter, we venture into the world of Algorithmic Trading and specifically, we cover a couple of problems involving a trading *Order Book* that can be cast as Markov Decision Processes, and hence tackled with Dynamic Programming or Reinforcement Learning. We start the chapter by covering the basics of how trade orders are submitted and executed on an *Order Book*, a structure that allows for efficient transactions between buyers and sellers of a financial asset. Without loss of generality, we refer to the financial asset being traded on the Order Book as a "stock" and the number of units of the asset as "shares". Next, we will explain how a large trade can significantly shift the Order Book, a phenomenon known as *Price Impact*. Finally, we will cover the two algorithmic trading problems that can be cast as MDPs. The first problem is Optimal Execution of the sale of a large number of shares of a stock so as to yield the maximum utility of sales proceeds over a finite horizon. This involves breaking up the sale of the shares into appropriate pieces and selling those pieces at the right times so as to achieve the goal of maximizing the utility of sales proceeds. Hence, it is an MDP Control problem where the actions are the number of shares sold at each time step. The second problem is Optimal Market-Making, i.e., the optimal *bids* (willingness to buy a certain number of shares at a certain price) and *asks* (willingness to sell a certain number of shares at a certain price) to be submitted on the Order Book. Again, by optimal, we mean maximization of the utility of revenues generated by the market-maker over a finite-horizon (market-makers generate revenue through the spread, i.e. the gap between the bid and ask prices they offer). This is also an MDP Control problem where the actions are the bid and ask prices along with the bid and ask shares at each time step.

For a deeper study on the topics of Order Book, Price Impact, Order Execution, Market-Making (and related topics), we refer you to the comprehensive treatment in Olivier Gueant's book (Gueant 2016).

10.1 BASICS OF ORDER BOOK AND PRICE IMPACT

Some of the financial literature refers to the Order Book as Limit Order Book (abbreviated as LOB) but we will stick with the lighter language—Order Book, abbreviated as OB. The Order Book is essentially a data structure that facilitates matching stock buyers with stock sellers (i.e., an electronic marketplace). Figure 10.1 depicts a simplified view of an order book. In this order book market, buyers and sellers express their intent to trade by submitting *bids* (intent to buy) and *asks* (intent to sell). These expressions of intent to buy or sell are known as Limit Orders (abbreviated as LO). The word "limit" in Limit Order refers to

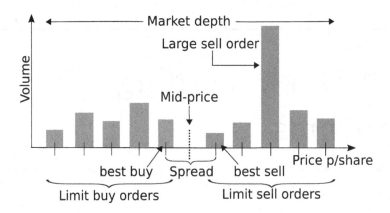

Figure 10.1 Trading Order Book (Image Credit: `https://nms.kcl.ac.uk/rll/enrique-miranda/index.html`)

the fact that one is interested in buying only below a certain price level (and likewise, one is interested in selling only above a certain price level). Each LO is comprised of a price P and number of shares N. A bid, i.e., Buy LO (P, N) states willingness to buy N shares at a price less than or equal to P. Likewise, an ask, i.e., a Sell LO (P, N) states willingness to sell N shares at a price greater than or equal to P.

Note that multiple traders might submit LOs with the same price. The order book aggregates the number of shares at each unique price, and the OB data structure is typically presented for trading in the form of this aggregated view. Thus, the OB data structure can be represented as two sorted lists of (Price, Size) pairs:

$$\text{Buy LOs (Bids): } [(P_i^{(b)}, N_i^{(b)}) \mid 0 \leq i < m], P_i^{(b)} > P_j^{(b)} \text{ for } i < j$$

$$\text{Sell LOs (Asks): } [(P_i^{(a)}, N_i^{(a)}) \mid 0 \leq i < n], P_i^{(a)} < P_j^{(a)} \text{ for } i < j$$

Note that the Buy LOs are arranged in descending order and the Sell LOs are arranged in ascending order to signify the fact that the beginning of each list consists of the most important (best-price) LOs.

Now let's learn about some of the standard terminology:

- We refer to $P_0^{(b)}$ as *The Best Bid Price* (lightened to *Best Bid*) to signify that it is the highest offer to buy and hence, the *best* price for a seller to transact with.
- Likewise, we refer to $P_0^{(a)}$ as *The Ask Price* (lightened to *Best Ask*) to signify that it is the lowest offer to sell and hence, the *best* price for a buyer to transact with.
- $\frac{P_0^{(a)} + P_0^{(b)}}{2}$ is referred to as the *The Mid Price* (lightened to *Mid*).
- $P_0^{(a)} - P_0^{(b)}$ is referred to as *The Best Bid-Ask Spread* (lightened to *Spread*).
- $P_{n-1}^{(a)} - P_{m-1}^{(b)}$ is referred to as *The Market Depth* (lightened to *Depth*).

Although an actual real-world trading order book has many other details, we believe this simplified coverage is adequate for the purposes of core understanding of order book trading and to navigate the problems of optimal order execution and optimal market-making. Apart from Limit Orders, traders can express their interest to buy/sell with another type of order—a *Market Order* (abbreviated as MO). A Market Order (MO) states one's intent to buy/sell N shares at the *best possible price(s)* available on the OB at the time of MO submission. So, an LO is keen on price and not so keen on time (willing to wait to get the price one wants) while an MO is keen on time (desire to trade right away) and not so keen on price (will take whatever the best LO price is on the OB). So now let us understand the actual transactions that occur between LOs and MOs (buy and sell interactions, and how the OB changes as a result of these interactions). Firstly, we note that in normal trading activity, a newly submitted sell LO's price is typically above the price of the best buy LO on the OB. But if a new sell LO's price is less than or equal to the price of the best buy LO's price, we say that the *market has crossed* (to mean that the range of bid prices and the range of ask prices have intersected), which results in an immediate transaction that eats into the OB's Buy LOs.

Precisely, a new Sell LO (P, N) potentially transacts with (and hence, removes) the best Buy LOs on the OB.

$$\text{Removal: } [(P_i^{(b)}, \min(N_i^{(b)}, \max(0, N - \sum_{j=0}^{i-1} N_j^{(b)}))) \mid (i : P_i^{(b)} \geq P)] \qquad (10.1)$$

After this removal, it potentially adds the following LO to the asks side of the OB:

$$(P, \max(0, N - \sum_{i:P_i^{(b)} \geq P} N_i^{(b)})) \qquad (10.2)$$

Likewise, a new Buy MO (P, N) potentially transacts with (and hence, removes) the best Sell LOs on the OB

$$\text{Removal: } [(P_i^{(a)}, \min(N_i^{(a)}, \max(0, N - \sum_{j=0}^{i-1} N_j^{(a)}))) \mid (i : P_i^{(a)} \leq P)] \qquad (10.3)$$

After this removal, it potentially adds the following to the bids side of the OB:

$$(P, \max(0, N - \sum_{i:P_i^{(a)} \leq P} N_i^{(a)})) \qquad (10.4)$$

When a Market Order (MO) is submitted, things are simpler. A Sell Market Order of N shares will remove the best Buy LOs on the OB.

$$\text{Removal: } [(P_i^{(b)}, \min(N_i^{(b)}, \max(0, N - \sum_{j=0}^{i-1} N_j^{(b)}))) \mid 0 \leq i < m] \qquad (10.5)$$

The sales proceeds for this MO is:

$$\sum_{i=0}^{m-1} P_i^{(b)} \cdot (\min(N_i^{(b)}, \max(0, N - \sum_{j=0}^{i-1} N_j^{(b)}))) \qquad (10.6)$$

We note that if N is large, the sales proceeds for this MO can be significantly lower than the best possible sales proceeds $(= N \cdot P_0^{(b)})$, which happens only if $N \leq N_0^{(b)}$. Note also that if N is large, the new Best Bid Price (new value of $P_0^{(b)}$) can be significantly lower than the Best Bid Price before the MO was submitted (because the MO "eats into" a significant volume of Buy LOs on the OB). This "eating into" the Buy LOs on the OB and consequent lowering of the Best Bid Price (and hence, Mid Price) is known as *Price Impact* of an MO (more specifically, as the *Temporary Price Impact* of an MO). We use the word "temporary" because subsequent to this "eating into" the Buy LOs of the OB (and consequent, "hole", i.e., large Bid-Ask Spread), market participants will submit "replenishment LOs" (both

Buy LOs and Sell LOs) on the OB. These replenishments LOs would typically mitigate the Bid-Ask Spread and the eventual settlement of the Best Bid/Best Ask/Mid Prices constitutes what we call *Permanent Price Impact*—which refers to the changes in OB Best Bid/Best Ask/Mid prices relative to the corresponding prices before submission of the MO.

Likewise, a Buy Market Order of N shares will remove the best Sell LOs on the OB

$$\text{Removal: } [(P_i^{(a)}, \min(N_i^{(a)}, \max(0, N - \sum_{j=0}^{i-1} N_j^{(a)}))) \mid 0 \le i < n] \tag{10.7}$$

The purchase bill for this MO is:

$$\sum_{i=0}^{n-1} P_i^{(a)} \cdot (\min(N_i^{(a)}, \max(0, N - \sum_{j=0}^{i-1} N_j^{(a)}))) \tag{10.8}$$

If N is large, the purchase bill for this MO can be significantly higher than the best possible purchase bill $(= N \cdot P_0^{(a)})$, which happens only if $N \le N_0^{(a)}$. All that we wrote above in terms of Temporary and Permanent Price Impact naturally apply in the opposite direction for a Buy MO.

We refer to all of the above-described OB movements, including both temporary and permanent Price Impacts broadly as *Order Book Dynamics*. There is considerable literature on modeling Order Book Dynamics and some of these models can get fairly complex in order to capture various real-world nuances. Much of this literature is beyond the scope of this book. In this chapter, we will cover a few simple models for how a sell MO will move the OB's *Best Bid Price* (rather than a model for how it will move the entire OB). The model for how a buy MO will move the OB's *Best Ask Price* is naturally identical.

Now let's write some code that models how LOs and MOs interact with the OB. We write a class OrderBook that represents the Buy and Sell Limit Orders on the Order Book, which are each represented as a sorted sequence of the type DollarsAndShares, which is a dataclass we created to represent any pair of a dollar amount (dollar: float) and number of shares (shares: int). Sometimes, we use DollarsAndShares to represent an LO (pair of price and shares) as in the case of the sorted lists of Buy and Sell LOs. At other times, we use DollarsAndShares to represent the pair of total dollars transacted and total shares transacted when an MO is executed on the OB. The OrderBook maintains a price-descending sequence of PriceSizePairs for Buy LOs (descending_bids) and a price-ascending sequence of PriceSizePairs for Sell LOs (ascending_asks). We write the basic methods to get the OrderBook's highest bid price (method bid_price), lowest ask price (method ask_price), mid price (method mid_price), spread between the highest bid price and lowest ask price (method bid_ask_spread), and market depth (method market_depth).

```
@dataclass(frozen=True)
class DollarsAndShares:

    dollars: float
    shares: int

PriceSizePairs = Sequence[DollarsAndShares]

@dataclass(frozen=True)
class OrderBook:

    descending_bids: PriceSizePairs
    ascending_asks: PriceSizePairs

    def bid_price(self) -> float:
        return self.descending_bids[0].dollars

    def ask_price(self) -> float:
        return self.ascending_asks[0].dollars
```

```
def mid_price(self) -> float:
    return (self.bid_price() + self.ask_price()) / 2
def bid_ask_spread(self) -> float:
    return self.ask_price() - self.bid_price()
def market_depth(self) -> float:
    return self.ascending_asks[-1].dollars - \
        self.descending_bids[-1].dollars
```

Next, we want to write methods for LOs and MOs to interact with the OrderBook. Notice that each of Equation (10.1) (new Sell LO potentially removing some of the beginning of the Buy LOs on the OB), Equation (10.3) (new Buy LO potentially removing some of the beginning of the Sell LOs on the OB), Equation (10.5) (Sell MO removing some of the beginning of the Buy LOs on the OB) and Equation (10.7) (Buy MO removing some of the beginning of the Sell LOs on the OB) all perform a common core function—they "eat into" the most significant LOs (on the opposite side) on the OB. So we first write a @staticmethod eat_book for this common function.

eat_book takes as input a ps_pairs: PriceSizePairs (representing one side of the OB) and the number of shares: int to buy/sell. Notice eat_book's return type: Tuple[DollarsAndShares, PriceSizePairs]. The returned DollarsAndShares represents the pair of dollars transacted and the number of shares transacted (with number of shares transacted being less than or equal to the input shares). The returned PriceSizePairs represents the remainder of ps_pairs after the transacted number of shares have eaten into the input ps_pairs. eat_book first deletes (i.e. "eats up") as much of the *beginning* of the ps_pairs: PriceSizePairs data structure as it can (basically matching the input number of shares with an appropriate number of shares at the beginning of the ps_pairs: PriceSizePairs input). Note that the returned PriceSizePairs is a separate data structure, ensuring the immutability of the input ps_pairs: PriceSizePairs.

```
@staticmethod
def eat_book(
    ps_pairs: PriceSizePairs,
    shares: int
) -> Tuple[DollarsAndShares, PriceSizePairs]:
    rem_shares: int = shares
    dollars: float = 0.
    for i, d_s in enumerate(ps_pairs):
        this_price: float = d_s.dollars
        this_shares: int = d_s.shares
        dollars += this_price * min(rem_shares, this_shares)
        if rem_shares < this_shares:
            return (
                DollarsAndShares(dollars=dollars, shares=shares),
                [DollarsAndShares(
                    dollars=this_price,
                    shares=this_shares - rem_shares
                )] + list(ps_pairs[i+1:])
            )
        else:
            rem_shares -= this_shares
    return (
        DollarsAndShares(dollars=dollars, shares=shares - rem_shares),
        []
    )
```

Now we are ready to write the method sell_limit_order which takes Sell LO Price and Sell LO shares as input. As you can see in the code below, first it potentially removes (if it "crosses") an appropriate number of shares on the Buy LOs side of the OB (using the @staticmethod eat_book), and then potentially adds an appropriate number of shares at the Sell LO Price on the Sell LOs side of the OB. `sell_limit_order` returns a pair of `DollarsAndShares` type and `OrderBook` type. The returned `DollarsAndShares` represents the pair of dollars transacted and the number of shares transacted with the Buy LOs side of the OB (with number of shares transacted being less than or equal to the input `shares`). The returned `OrderBook` represents the new OB after potentially eating into the Buy LOs side of the OB and then potentially adding some shares at the Sell LO Price on the Sell LOs side of the OB. Note that the returned `OrderBook` is a newly-created data structure, ensuring the immutability of `self`. We urge you to read the code below carefully as there are many subtle details that are handled in the code.

```
from dataclasses import replace
    def sell_limit_order(self, price: float, shares: int) -> \
            Tuple[DollarsAndShares, OrderBook]:
        index: Optional[int] = next((i for i, d_s
                                    in enumerate(self.descending_bids)
                                    if d_s.dollars < price), None)
        eligible_bids: PriceSizePairs = self.descending_bids \
            if index is None else self.descending_bids[:index]
        ineligible_bids: PriceSizePairs = [] if index is None else \
            self.descending_bids[index:]

        d_s, rem_bids = OrderBook.eat_book(eligible_bids, shares)
        new_bids: PriceSizePairs = list(rem_bids) + list(ineligible_bids)
        rem_shares: int = shares - d_s.shares

        if rem_shares > 0:
            new_asks: List[DollarsAndShares] = list(self.ascending_asks)
            index1: Optional[int] = next((i for i, d_s
                                        in enumerate(new_asks)
                                        if d_s.dollars >= price), None)
            if index1 is None:
                new_asks.append(DollarsAndShares(
                    dollars=price,
                    shares=rem_shares
                ))
            elif new_asks[index1].dollars != price:
                new_asks.insert(index1, DollarsAndShares(
                    dollars=price,
                    shares=rem_shares
                ))
            else:
                new_asks[index1] = DollarsAndShares(
                    dollars=price,
                    shares=new_asks[index1].shares + rem_shares
                )
            return d_s, OrderBook(
                ascending_asks=new_asks,
                descending_bids=new_bids
            )
        else:
            return d_s, replace(
                self,
                descending_bids=new_bids
            )
```

Next, we write the easier method `sell_market_order` which takes as input the number of shares to be sold (as a market order). `sell_market_order` transacts with the appropriate number of shares on the Buy LOs side of the OB (removing those many shares from the Buy LOs side). It returns a pair of `DollarsAndShares` type and `OrderBook` type. The returned `DollarsAndShares` represents the pair of dollars transacted and the number of shares transacted (with number of shares transacted being less than or equal to the input `shares`). The returned `OrderBook` represents the remainder of the OB after the transacted number of shares have eaten into the Buy LOs side of the OB. Note that the returned OrderBook is a newly-created data structure, ensuring the immutability of self.

```
def sell_market_order(
    self,
    shares: int
) -> Tuple[DollarsAndShares, OrderBook]:
    d_s, rem_bids = OrderBook.eat_book(
        self.descending_bids,
        shares
    )
    return (d_s, replace(self, descending_bids=rem_bids))
```

We won't list the methods buy_limit_order and buy_market_order here as they are completely analogous (you can find the entire code for OrderBook in the file rl/chapter9/order_book.py). Now let us test out this code by creating a sample OrderBook and submitting some LOs and MOs to interact with the OrderBook.

```
bids: PriceSizePairs = [DollarsAndShares(
    dollars=x,
    shares=poisson(100. - (100 - x) * 10)
) for x in range(100, 90, -1)]
asks: PriceSizePairs = [DollarsAndShares(
    dollars=x,
    shares=poisson(100. - (x - 105) * 10)
) for x in range(105, 115, 1)]
ob0: OrderBook = OrderBook(descending_bids=bids, ascending_asks=asks)
```

The above code creates an OrderBook in the price range [91, 114] with a bid-ask spread of 5. Figure 10.2 depicts this OrderBook visually.

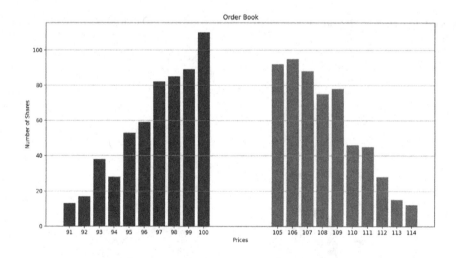

Figure 10.2 **Starting Order Book**

Let's submit a Sell LO that says we'd like to sell 40 shares as long as the transacted price is greater than or equal to 107. Our Sell LO should simply get added to the Sell LOs side of the OB.

```
d_s1, ob1 = ob0.sell_limit_order(107, 40)
```

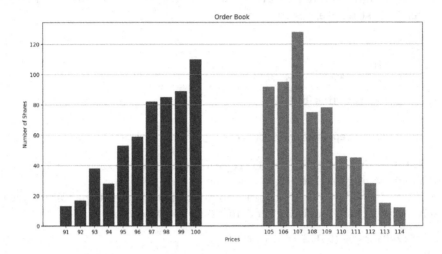

Figure 10.3 **Order Book after Sell LO**

The new `OrderBook` `ob1` has 40 more shares at the price level of 107, as depicted in Figure 10.3.

Now let's submit a Sell MO that says we'd like to sell 120 shares at the "best price". Our Sell MO should transact with 120 shares at "best prices" of 100 and 99 as well (since the OB does not have enough Buy LO shares at the price of 100).

```
d_s2, ob2 = ob1.sell_market_order(120)
```

The new `OrderBook` `ob2` has 120 less shares on the Buy LOs side of the OB, as depicted in Figure 10.4.

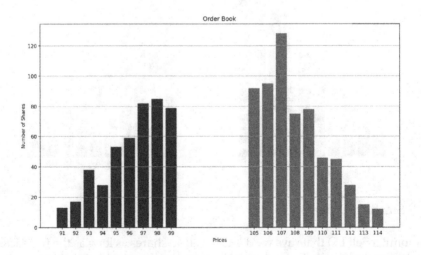

Figure 10.4 **Order Book after Sell MO**

Now let's submit a Buy LO that says we'd like to buy 80 shares as long as the transacted price is less than or equal to 100. Our Buy LO should get added to the Buy LOs side of the OB.

```
d_s3, ob3 = ob2.buy_limit_order(100, 80)
```

The new OrderBook ob3 has re-introduced a Buy LO at the price level of 100 (now with 80 shares), as depicted in Figure 10.5.

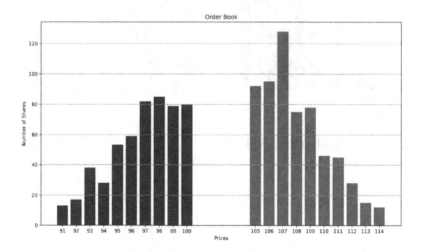

Figure 10.5 Order Book after Buy LO

Now let's submit a Sell LO that says we'd like to sell 60 shares as long as the transacted price is greater than or equal to 104. Our Sell LO should get added to the Sell LOs side of the OB.

```
d_s4, ob4 = ob3.sell_limit_order(104, 60)
```

The new OrderBook ob4 has introduced a Sell LO at a price of 104 with 60 shares, as depicted in Figure 10.6.

Now let's submit a Buy MO that says we'd like to buy 150 shares at the "best price". Our Buy MO should transact with 150 shares at "best prices" on the Sell LOs side of the OB.

```
d_s5, ob5 = ob4.buy_market_order(150)
```

The new OrderBook ob5 has 150 less shares on the Sell LOs side of the OB, wiping out all the shares at the price level of 104 and almost wiping out all the shares at the price level of 105, as depicted in Figure 10.7.

This has served as a good test of our code (transactions working as we'd like) and we encourage you to write more code of this sort to interact with the OrderBook, and to produce graphs of evolution of the OrderBook as this will help develop stronger intuition and internalize the concepts we've learnt above. All of the above code is in the file rl/chapter9/order_book.py.

Now we are ready to get started with the problem of Optimal Execution of a large-sized Market Order.

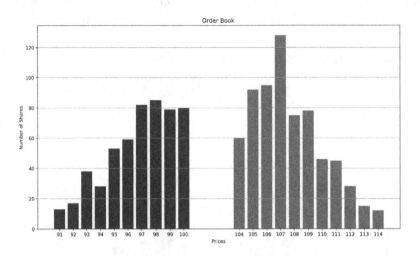

Figure 10.6 **Order Book after 2nd Sell LO**

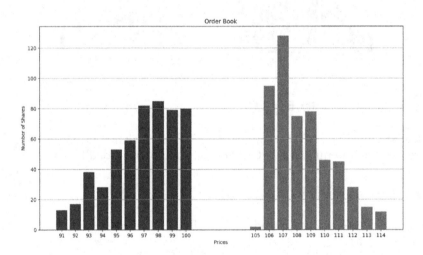

Figure 10.7 **Order Book after Buy MO**

10.2 OPTIMAL EXECUTION OF A MARKET ORDER

Imagine the following problem: You are a trader in a stock and your boss has instructed that you exit from trading in this stock because this stock doesn't meet your company's new investment objectives. You have to sell all of the N shares you own in this stock in the next T hours, but you have been instructed to accomplish the sale by submitting only Market Orders (not allowed to submit any Limit Orders because of the uncertainty in the time of execution of the sale with a Limit Order). You can submit sell market orders (of any size) at the start of each hour—so you have T opportunities to submit market orders of any size. Your goal is to maximize the Expected Total Utility of sales proceeds for all N shares over the T hours. Your task is to break up N into T appropriate chunks to maximize the Expected Total Utility objective. If you attempt to sell the N shares too fast (i.e., too many in the first few hours), as we've learnt above, each (MO) sale will eat a lot into the Buy LOs on the OB (Temporary Price Impact) which would result in transacting at prices below the best price (Best Bid Price). Moreover, you risk moving the *Best Bid Price* on the OB significantly lower (Permanent Price Impact) that would affect the sales proceeds for the next few sales you'd make. On the other hand, if you sell the N shares too slow (i.e., too few in the first few hours), you might transact at good prices but then you risk running out of time, which means you will have to dump a lot of shares with time running out which in turn would mean transacting at prices below the best price. Moreover, selling too slow exposes you to more uncertainty in market price movements over a longer time period, and more uncertainty in sales proceeds means the Expected Utility objective gets hurt. Thus, the precise timing and sizes in the breakup of shares is vital. You will need to have an estimate of the Temporary and Permanent Price Impact of your Market Orders, which can help you identify the appropriate number of shares to sell at the start of each hour.

Unsurprisingly, we can model this problem as a Market Decision Process control problem where the actions at each time step (each hour, in this case) are the number of shares sold at the time step and the rewards are the Utility of sales proceeds at each time step. To keep things simple and intuitive, we shall model *Price Impact* of Market Orders in terms of their effect on the *Best Bid Price* (rather than in terms of their effect on the entire OB). In other words, we won't be modeling the entire OB Price Dynamics, just the Best Bid Price Dynamics. We shall refer to the OB activity of an MO immediately "eating into the Buy LOs" (and hence, potentially transacting at prices lower than the best price) as the *Temporary* Price Impact. As mentioned earlier, this is followed by subsequent replenishment of both Buy and Sell LOs on the OB (stabilizing the OB)—we refer to any eventual (end of the hour) lowering of the Best Bid Price (relative to the Best Bid Price before the MO was submitted) as the *Permanent* Price Impact. Modeling the temporary and permanent Price Impacts separately helps us in deciding on the optimal actions (optimal shares to be sold at the start of each hour).

Now we develop some formalism to describe this problem precisely. As mentioned earlier, we make a number of simplifying assumptions in modeling the OB Dynamics for ease of articulation (without diluting the most important concepts). We index discrete time by $t = 0, 1, \ldots, T$. We denote P_t as the Best Bid Price on the OB at the start of time step t (for all $t = 0, 1, \ldots, T$) and N_t as the number of shares sold at time step t for all $t = 0, 1, \ldots, T-1$. We denote the number of shares remaining to be sold at the start of time step t as R_t for all $t = 0, 1, \ldots, T$. Therefore,

$$R_t = N - \sum_{i=0}^{t-1} N_i \text{ for all } t = 0, 1, \ldots, T$$

Note that:

$$R_0 = N$$
$$R_{t+1} = R_t - N_t \text{ for all } t = 0, 1, \ldots, T-1$$

Also note that we need to sell everything by time $t = T$ and so:

$$N_{T-1} = R_{T-1} \Rightarrow R_T = 0$$

The model of Best Bid Price Dynamics from one time step to the next is given by:

$$P_{t+1} = f_t(P_t, N_t, \epsilon_t) \text{ for all } t = 0, 1, \ldots, T-1$$

where f_t is an arbitrary function incorporating:

- The Permanent Price Impact of selling N_t shares.
- The Price-Impact-independent market-movement of the Best Bid Price from time t to time $t+1$.
- Noise ϵ_t, a source of randomness in Best Bid Price movements.

The sales proceeds from the sale at time step t, for all $t = 0, 1, \ldots, T-1$, is defined as:

$$N_t \cdot Q_t = N_t \cdot (P_t - g_t(P_t, N_t))$$

where g_t is a function modeling the Temporary Price Impact (i.e., the N_t MO "eating into" the Buy LOs on the OB). Q_t should be interpreted as the average Buy LO price transacted against by the N_t MO at time t.

Lastly, we denote the Utility (of Sales Proceeds) function as $U(\cdot)$.

As mentioned previously, solving for the optimal number of shares to be sold at each time step can be modeled as a discrete-time finite-horizon Markov Decision Process, which we describe below in terms of the order of MDP activity at each time step $t = 0, 1, \ldots, T-1$ (the MDP horizon is time T meaning all states at time T are terminal states). We follow the notational style of finite-horizon MDPs that should now be familiar from previous chapters.

Order of Events at time step t for all $t = 0, 1, \ldots, T-1$:

- Observe *State* $s_t := (P_t, R_t) \in \mathcal{S}_t$
- Perform *Action* $a_t := N_t \in \mathcal{A}_t$
- Receive *Reward* $r_{t+1} := U(N_t \cdot Q_t) = U(N_t \cdot (P_t - g_t(P_t, N_t)))$
- Experience Price Dynamics $P_{t+1} = f_t(P_t, N_t, \epsilon_t)$ and set $R_{t+1} = R_t - N_t$ so as to obtain the next state $s_{t+1} = (P_{t+1}, R_{t+1}) \in \mathcal{S}_{t+1}$.

Note that we have intentionally not specified if P_t, R_t, N_t are integers or real numbers, or if constrained to be non-negative etc. Those precise specifications will be customized to the nuances/constraints of the specific Optimal Order Execution problem we'd be solving. Be default, we shall assume that $P_t \in \mathbb{R}^+$ and $N_t, R_t \in \mathbb{Z}_{\geq 0}$ (as these represent realistic trading situations), although we do consider special cases later in the chapter where $P_t, R_t \in \mathbb{R}$ (unconstrained real numbers for analytical tractability).

The goal is to find the Optimal Policy $\pi^* = (\pi_0^*, \pi_1^*, \ldots, \pi_{T-1}^*)$ (defined as $\pi_t^*((P_t, R_t)) = N_t^*$ that maximizes:

$$\mathbb{E}[\sum_{t=0}^{T-1} \gamma^t \cdot U(N_t \cdot Q_t)]$$

where γ is the discount factor to account for the fact that future utility of sales proceeds can be modeled to be less valuable than today's.

Now let us write some code to solve this MDP. We write a class `OptimalOrderExecution` which models a fairly generic MDP for Optimal Order Execution as described above, and solves the Control problem with Approximate Value Iteration using the backward induction algorithm that we implemented in Chapter 6. Let us start by taking a look at the attributes (inputs) in `OptimalOrderExecution`:

- `shares` refers to the total number of shares N to be sold over T time steps.
- `time_steps` refers to the number of time steps T.
- `avg_exec_price_diff` refers to the time-sequenced functions g_t that return the reduction in the average price obtained by the Market Order at time t due to eating into the Buy LOs. g_t takes as input the type `PriceAndShares` that represents a pair of price: float and shares: int (in this case, the price is P_t and the shares is the MO size N_t at time t). As explained earlier, the sales proceeds at time t is: $N_t \cdot (P_t - g_t(P_t, N_t))$.

- `price_dynamics` refers to the time-sequenced functions f_t that represent the price dynamics: $P_{t+1} \sim f_t(P_t, N_t)$. f_t outputs a probability distribution of prices for P_{t+1}.
- `utility_func` refers to the Utility of Sales Proceeds function, incorporating any risk-aversion.
- `discount_factor` refers to the discount factor γ.
- `func_approx` refers to the `ValueFunctionApprox` type to be used to approximate the Value Function for each time step (since we are doing backward induction).
- `initial_price_distribution` refers to the probability distribution of prices P_0 at time 0, which is used to generate the samples of states at each of the time steps (needed in the approximate backward induction algorithm).

```
from rl.approximate_dynamic_programming import ValueFunctionApprox
@dataclass(frozen=True)
class PriceAndShares:
    price: float
    shares: int

@dataclass(frozen=True)
class OptimalOrderExecution:
    shares: int
    time_steps: int
    avg_exec_price_diff: Sequence[Callable[[PriceAndShares], float]]
    price_dynamics: Sequence[Callable[[PriceAndShares], Distribution[float]]]
    utility_func: Callable[[float], float]
    discount_factor: float
    func_approx: ValueFunctionApprox[PriceAndShares]
    initial_price_distribution: Distribution[float]
```

The two key things we need to perform the backward induction are:

- A method `get_mdp` that given a time step t, produces the `MarkovDecisionProcess` object representing the transitions from time t to time $t + 1$. The class `OptimalExecutionMDP` within `get_mdp` implements the abstract methods `step` and `actions` of the abstract class `MarkovDecisionProcess`. The code should be fairly self-explanatory—just a couple of things to point out here. Firstly, the input `p_r: NonTerminal[PriceAndShares]` to the `step` method represents the state (P_t, R_t) at time t, and the variable `p_s: PriceAndShares` represents the pair of (P_t, N_t), which serves as input to `avg_exec_price_diff` and `price_dynamics` (attributes of `OptimalOrderExecution`). Secondly, note that the `actions` method returns an `Iterator` on a single int at time $t = T - 1$ because of the constraint $N_{T-1} = R_{T-1}$.
- A method `get_states_distribution` that returns the probability distribution of states (P_t, R_t) at time t (of type `SampledDistribution[NonTerminal[PriceAndShares]]`). The code here is similar to the `get_states_distribiution` method of `AssetAllocDiscrete` in Chapter 8 (essentially, walking forward from time 0 to time t by sampling from the state-transition probability distribution and also sampling from uniform choices over all actions at each time step).

```
def get_mdp(self, t: int) -> MarkovDecisionProcess[PriceAndShares, int]:
    utility_f: Callable[[float], float] = self.utility_func
    price_diff: Sequence[Callable[[PriceAndShares], float]] = \
        self.avg_exec_price_diff
    dynamics: Sequence[Callable[[PriceAndShares], Distribution[float]]] = \
        self.price_dynamics
    steps: int = self.time_steps

    class OptimalExecutionMDP(MarkovDecisionProcess[PriceAndShares, int]):

        def step(
```

```
            self,
            p_r: NonTerminal[PriceAndShares],
            sell: int
        ) -> SampledDistribution[Tuple[State[PriceAndShares],
                                       float]]:
            def sr_sampler_func(
                p_r=p_r,
                sell=sell
            ) -> Tuple[State[PriceAndShares], float]:
                p_s: PriceAndShares = PriceAndShares(
                    price=p_r.state.price,
                    shares=sell
                )
                next_price: float = dynamics[t](p_s).sample()
                next_rem: int = p_r.state.shares - sell
                next_state: PriceAndShares = PriceAndShares(
                    price=next_price,
                    shares=next_rem
                )
                reward: float = utility_f(
                    sell * (p_r.state.price - price_diff[t](p_s))
                )
                return (NonTerminal(next_state), reward)
            return SampledDistribution(
                sampler=sr_sampler_func,
                expectation_samples=100
            )

        def actions(self, p_s: NonTerminal[PriceAndShares]) -> \
                Iterator[int]:
            if t == steps - 1:
                return iter([p_s.state.shares])
            else:
                return iter(range(p_s.state.shares + 1))

    return OptimalExecutionMDP()

def get_states_distribution(self, t: int) -> \
        SampledDistribution[NonTerminal[PriceAndShares]]:

    def states_sampler_func() -> NonTerminal[PriceAndShares]:
        price: float = self.initial_price_distribution.sample()
        rem: int = self.shares
        for i in range(t):
            sell: int = Choose(range(rem + 1)).sample()
            price = self.price_dynamics[i](PriceAndShares(
                price=price,
                shares=rem
            )).sample()
            rem -= sell
        return NonTerminal(PriceAndShares(
            price=price,
            shares=rem
        ))

    return SampledDistribution(states_sampler_func)
```

Finally, we produce the Optimal Value Function and Optimal Policy for each time step with the following method `backward_induction_vf_and_pi`:

```
from rl.approximate_dynamic_programming import back_opt_vf_and_policy

def backward_induction_vf_and_pi(
    self
) -> Iterator[Tuple[ValueFunctionApprox[PriceAndShares],
                    DeterministicPolicy[PriceAndShares, int]]]:

    mdp_f0_mu_triples: Sequence[Tuple[
        MarkovDecisionProcess[PriceAndShares, int],
        ValueFunctionApprox[PriceAndShares],
```

```
        SampledDistribution[NonTerminal[PriceAndShares]]
    ]] = [(
        self.get_mdp(i),
        self.func_approx,
        self.get_states_distribution(i)
    ) for i in range(self.time_steps)]

    num_state_samples: int = 10000
    error_tolerance: float = 1e-6

    return back_opt_vf_and_policy(
        mdp_f0_mu_triples=mdp_f0_mu_triples,
        gamma=self.discount_factor,
        num_state_samples=num_state_samples,
        error_tolerance=error_tolerance
    )
```

The above code is in the file rl/chapter9/optimal_order_execution.py. We encourage you to create a few different instances of `OptimalOrderExecution` by varying its inputs (try different temporary and permanent price impact functions, different utility functions, impose a few constraints etc.). Note that the above code has been written with an educational motivation rather than an efficient-computation motivation, so the convergence of the backward induction ADP algorithm is going to be slow. How do we know that the above code is correct? Well, we need to create a simple special case that yields a closed-form solution that we can compare the Optimal Value Function and Optimal Policy produced by `OptimalOrderExecution` against. This will be the subject of the following subsection.

10.2.1 Simple Linear Price Impact Model with No Risk-Aversion

Now we consider a special case of the above-described MDP—a simple linear Price Impact model with no risk-aversion. Furthermore, for analytical tractability, we assume N, N_t, P_t are all unconstrained continuous-valued (i.e., taking values $\in \mathbb{R}$).

We assume simple linear price dynamics as follows:

$$P_{t+1} = f_t(P_t, N_t, \epsilon) = P_t - \alpha \cdot N_t + \epsilon_t$$

where $\alpha \in \mathbb{R}$ and ϵ_t for all $t = 0, 1, \ldots, T - 1$ are independent and identically distributed (i.i.d.) with $\mathbb{E}[\epsilon_t | N_t, P_t] = 0$. Therefore, the Permanent Price Impact (as an Expectation) is $\alpha \cdot N_t$.

As for the Temporary Price Impact, we know that g_t needs to be a non-decreasing function of N_t. We assume a simple linear form for g_t as follows:

$$g_t(P_t, N_t) = \beta \cdot N_t \text{ for all } t = 0, 1, \ldots, T - 1$$

for some constant $\beta \in \mathbb{R}_{\geq 0}$. So, $Q_t = P_t - \beta N_t$. As mentioned above, we assume no risk-aversion, i.e., the Utility function $U(\cdot)$ is assumed to be the identity function. Also, we assume that the MDP discount factor $\gamma = 1$.

Note that all of these assumptions are far too simplistic and hence, an unrealistic model of the real-world, but starting with this simple model helps build good intuition and enables us to develop more realistic models by incrementally adding complexity/nuances from this simple base model.

As ever, in order to solve the Control problem, we define the Optimal Value Function and invoke the Bellman Optimality Equation. We shall use the standard notation for discrete-time finite-horizon MDPs that we are now very familiar with.

Denote the Value Function for policy π at time t (for all $t = 0, 1, \ldots T - 1$) as:

$$V_t^\pi((P_t, R_t)) = \mathbb{E}_\pi[\sum_{i=t}^{T-1} N_i \cdot (P_i - \beta \cdot N_i) | (P_t, R_t)]$$

Denote the Optimal Value Function at time t (for all $t = 0, 1, \ldots, T - 1$) as:

$$V_t^*((P_t, R_t)) = \max_\pi V_t^\pi((P_t, R_t))$$

The Optimal Value Function satisfies the finite-horizon Bellman Optimality Equation for all $t = 0, 1, \ldots, T - 2$, as follows:

$$V_t^*((P_t, R_t)) = \max_{N_t}\{N_t \cdot (P_t - \beta \cdot N_t) + \mathbb{E}[V_{t+1}^*((P_{t+1}, R_{t+1}))]\}$$

and

$$V_{T-1}^*((P_{T-1}, R_{T-1})) = N_{T-1} \cdot (P_{T-1} - \beta \cdot N_{T-1}) = R_{T-1} \cdot (P_{T-1} - \beta \cdot R_{T-1})$$

From the above, we can infer:

$$V_{T-2}^*((P_{T-2}, R_{T-2})) = \max_{N_{T-2}}\{N_{T-2} \cdot (P_{T-2} - \beta \cdot N_{T-2}) + \mathbb{E}[R_{T-1} \cdot (P_{T-1} - \beta \cdot R_{T-1})]\}$$

$$= \max_{N_{T-2}}\{N_{T-2} \cdot (P_{T-2} - \beta \cdot N_{T-2}) + \mathbb{E}[(R_{T-2} - N_{T-2})(P_{T-1} - \beta \cdot (R_{T-2} - N_{T-2}))]\}$$

$$= \max_{N_{T-2}}\{N_{T-2} \cdot (P_{T-2} - \beta \cdot N_{T-2}) + (R_{T-2} - N_{T-2}) \cdot (P_{T-2} - \alpha \cdot N_{T-2} - \beta \cdot (R_{T-2} - N_{T-2}))\}$$

This simplifies to:

$$V_{T-2}^*((P_{T-2}, R_{T-2})) = \max_{N_{T-2}}\{R_{T-2} \cdot P_{T-2} - \beta \cdot R_{T-2}^2 + (\alpha - 2\beta)(N_{T-2}^2 - N_{T-2} \cdot R_{T-2})\} \quad (10.9)$$

For the case $\alpha \geq 2\beta$, noting that $N_{T-2} \leq R_{T-2}$, we have the trivial solution:

$$N_{T-2}^* = 0 \text{ or } N_{T-2}^* = R_{T-2}$$

Substituting either of these two values for N_{T-2}^* in the right-hand-side of Equation (10.9) gives:

$$V_{T-2}^*((P_{T-2}, R_{T-2})) = R_{T-2} \cdot (P_{T-2} - \beta \cdot R_{T-2})$$

Continuing backwards in time in this manner (for the case $\alpha \geq 2\beta$) gives:

$$N_t^* = 0 \text{ or } N_t^* = R_t \text{ for all } t = 0, 1, \ldots, T - 1$$

$$V_t^*((P_t, R_t)) = R_t \cdot (P_t - \beta \cdot R_t) \text{ for all } t = 0, 1, \ldots, T - 1$$

So the solution for the case $\alpha \geq 2\beta$ is to sell all N shares at any one of the time steps $t = 0, 1, \ldots, T - 1$ (and none in the other time steps), and the Optimal Expected Total Sale Proceeds is $N \cdot (P_0 - \beta \cdot N)$

For the case $\alpha < 2\beta$, differentiating the term inside the max in Equation (10.9) with respect to N_{T-2}, and setting it to 0 gives:

$$(\alpha - 2\beta) \cdot (2N_{T-2}^* - R_{T-2}) = 0 \Rightarrow N_{T-2}^* = \frac{R_{T-2}}{2}$$

Substituting this solution for N^*_{T-2} in Equation (10.9) gives:

$$V^*_{T-2}((P_{T-2}, R_{T-2})) = R_{T-2} \cdot P_{T-2} - R^2_{T-2} \cdot (\frac{\alpha + 2\beta}{4})$$

Continuing backwards in time in this manner gives:

$$N^*_t = \frac{R_t}{T-t} \text{ for all } t = 0, 1, \ldots, T-1$$

$$V^*_t((P_t, R_t)) = R_t \cdot P_t - \frac{R^2_t}{2} \cdot (\frac{2\beta + \alpha \cdot (T-t-1)}{T-t}) \text{ for all } t = 0, 1, \ldots, T-1$$

Rolling forward in time, we see that $N^*_t = \frac{N}{T}$, i.e., splitting the N shares uniformly across the T time steps. Hence, the Optimal Policy is a constant deterministic function (i.e., independent of the *State*). Note that a uniform split makes intuitive sense because Price Impact and Market Movement are both linear and additive, and don't interact. This optimization is essentially equivalent to minimizing $\sum_{t=1}^{T} N^2_t$ with the constraint: $\sum_{t=1}^{T} N_t = N$. The Optimal Expected Total Sales Proceeds is equal to:

$$N \cdot P_0 - \frac{N^2}{2} \cdot (\alpha + \frac{2\beta - \alpha}{T})$$

Implementation Shortfall is the technical term used to refer to the reduction in Total Sales Proceeds relative to the maximum possible sales proceeds ($= N \cdot P_0$). So, in this simple linear model, the Implementation Shortfall from Price Impact is $\frac{N^2}{2} \cdot (\alpha + \frac{2\beta - \alpha}{T})$. Note that the Implementation Shortfall is non-zero even if one had infinite time available ($T \to \infty$) for the case of $\alpha > 0$. If Price Impact were purely temporary ($\alpha = 0$, i.e., Price fully snapped back), then the Implementation Shortfall is zero if one had infinite time available.

So now let's customize the class `OptimalOrderExecution` to this simple linear price impact model, and compare the Optimal Value Function and Optimal Policy produced by `OptimalOrderExecution` against the above-derived closed-form solutions. We write code below to create an instance of `OptimalOrderExecution` with time steps $T = 5$, total number of shares to be sold $N = 100$, linear temporary price impact with $\alpha = 0.03$, linear permanent price impact with $\beta = 0.03$, utility function as the identity function (no risk-aversion), and discount factor $\gamma = 1$. We set the standard deviation for the price dynamics probability distribution to 0 to speed up the calculation. Since we know the closed-form solution for the Optimal Value Function, we provide some assistance to `OptimalOrderExecution` by setting up a linear function approximation with two features: $P_t \cdot R_t$ and R^2_t. The task of `OptimalOrderExecution` is to infer the correct coefficients of these features for each time step. If the coefficients match that of the closed-form solution, it provides a great degree of confidence that our code is working correctly.

```
num_shares: int = 100
num_time_steps: int = 5
alpha: float = 0.03
beta: float = 0.05
init_price_mean: float = 100.0
init_price_stdev: float = 10.0

price_diff = [lambda p_s: beta * p_s.shares for _ in range(num_time_steps)]
dynamics = [lambda p_s: Gaussian(
    mu=p_s.price - alpha * p_s.shares,
```

```
    sigma=0.
) for _ in range(num_time_steps)]
ffs = [
    lambda p_s: p_s.state.price * p_s.state.shares,
    lambda p_s: float(p_s.state.shares * p_s.state.shares)
]
fa: FunctionApprox = LinearFunctionApprox.create(feature_functions=ffs)
init_price_distrib: Gaussian = Gaussian(
    mu=init_price_mean,
    sigma=init_price_stdev
)

ooe: OptimalOrderExecution = OptimalOrderExecution(
    shares=num_shares,
    time_steps=num_time_steps,
    avg_exec_price_diff=price_diff,
    price_dynamics=dynamics,
    utility_func=lambda x: x,
    discount_factor=1,
    func_approx=fa,
    initial_price_distribution=init_price_distrib
)
it_vf: Iterator[Tuple[ValueFunctionApprox[PriceAndShares],
                    DeterministicPolicy[PriceAndShares, int]]] = \
    ooe.backward_induction_vf_and_pi()
```

Next, we evaluate this Optimal Value Function and Optimal Policy on a particular state for all time steps, and compare that against the closed-form solution. The state we use for evaluation is as follows:

```
state: PriceAndShares = PriceAndShares(
    price=init_price_mean,
    shares=num_shares
)
```

The code to evaluate the obtained Optimal Value Function and Optimal Policy on the above state is as follows:

```
for t, (vf, pol) in enumerate(it_vf):
    print(f"Time {t:d}")
    print()
    opt_sale: int = pol.action_for(state)
    val: float = vf(NonTerminal(state))
    print(f"Optimal Sales = {opt_sale:d}, Opt Val = {val:.3f}")
    print()
    print("Optimal Weights below:")
    print(vf.weights.weights)
    print()
```

With 100,000 state samples for each time step and only 10 state transition samples (since the standard deviation of ϵ is set to be very small), this prints the following:

```
Time 0

Optimal Sales = 20, Opt Val = 9779.976

Optimal Weights below:
[ 0.9999948  -0.02199718]

Time 1

Optimal Sales = 20, Opt Val = 9762.479

Optimal Weights below:
[ 0.9999935  -0.02374564]

Time 2

Optimal Sales = 20, Opt Val = 9733.324

Optimal Weights below:
[ 0.99999335 -0.02666098]

Time 3

Optimal Sales = 20, Opt Val = 9675.013

Optimal Weights below:
[ 0.99999316 -0.03249182]

Time 4

Optimal Sales = 20, Opt Val = 9500.000

Optimal Weights below:
[ 1.    -0.05]
```

Now let's compare these results against the closed-form solution.

```python
for t in range(num_time_steps):
    print(f"Time {t:d}")
    print()
    left: int = num_time_steps - t
    opt_sale_anal: float = num_shares / num_time_steps
    wt1: float = 1
    wt2: float = -(2 * beta + alpha * (left - 1)) / (2 * left)
    val_anal: float = wt1 * state.price * state.shares + \
        wt2 * state.shares * state.shares

    print(f"Optimal Sales = {opt_sale_anal:.3f}, Opt Val = {val_anal:.3f}")
    print(f"Weight1 = {wt1:.3f}")
    print(f"Weight2 = {wt2:.3f}")
    print()
```

This prints the following:

```
Time 0

Optimal Sales = 20.000, Opt Val = 9780.000
Weight1 = 1.000
Weight2 = -0.022


Time 1

Optimal Sales = 20.000, Opt Val = 9762.500
Weight1 = 1.000
Weight2 = -0.024


Time 2

Optimal Sales = 20.000, Opt Val = 9733.333
Weight1 = 1.000
Weight2 = -0.027


Time 3

Optimal Sales = 20.000, Opt Val = 9675.000
Weight1 = 1.000
Weight2 = -0.033


Time 4

Optimal Sales = 20.000, Opt Val = 9500.000
Weight1 = 1.000
Weight2 = -0.050
```

We need to point out here that the general case of optimal order execution involving modeling of the entire Order Book's dynamics will have to deal with a large state space. This means the ADP algorithm will suffer from the curse of dimensionality, which means we will need to employ RL algorithms.

10.2.2 Paper by Bertsimas and Lo on Optimal Order Execution

A paper by Bertsimas and Lo on Optimal Order Execution (Bertsimas and Lo 1998) considered a special case of the simple Linear Impact model we sketched above. Specifically, they assumed no risk-aversion (Utility function is identity function) and assumed that the Permanent Price Impact parameter α is equal to the Temporary Price Impact Parameter β. In the same paper, Bertsimas and Lo then extended this Linear Impact Model to include dependence on a serially-correlated variable X_t as follows:

$$P_{t+1} = P_t - (\beta \cdot N_t + \theta \cdot X_t) + \epsilon_t$$

$$X_{t+1} = \rho \cdot X_t + \eta_t$$

$$Q_t = P_t - (\beta \cdot N_t + \theta \cdot X_t)$$

where ϵ_t and η_t are each independent and identically distributed random variables with mean zero for all $t = 0, 1, \ldots, T - 1$, ϵ_t and η_t are also independent of each other for all

$t = 0, 1, \ldots, T - 1$. X_t can be thought of as a market factor affecting P_t linearly. Applying the finite-horizon Bellman Optimality Equation on the Optimal Value Function (and the same backward-recursive approach as before) yields:

$$N_t^* = \frac{R_t}{T - t} + h(t, \beta, \theta, \rho) \cdot X_t$$

$$V_t^*((P_t, R_t, X_t)) = R_t \cdot P_t - (\text{quadratic in } (R_t, X_t) + \text{constant})$$

Essentially, the serial-correlation predictability ($\rho \neq 0$) alters the uniform-split strategy.

In the same paper, Bertsimas and Lo presented a more realistic model called *Linear-Percentage Temporary* (abbreviated as LPT) Price Impact model, whose salient features include:

- Geometric random walk: consistent with real data, and avoids non-positive prices.
- Fractional Price Impact $\frac{g_t(P_t, N_t)}{P_t}$ doesn't depend on P_t (this is validated by real data).
- Purely Temporary Price Impact, i.e., the price P_t snaps back after the Temporary Price Impact (no Permanent effect of Market Orders on future prices).

The specific model is:

$$P_{t+1} = P_t \cdot e^{Z_t}$$

$$X_{t+1} = \rho \cdot X_t + \eta_t$$

$$Q_t = P_t \cdot (1 - \beta \cdot N_t - \theta \cdot X_t)$$

where Z_t are independent and identically distributed random variables with mean μ_Z and variance σ_Z^2, η_t are independent and identically distributed random variables with mean zero for all $t = 0, 1, \ldots, T - 1$, Z_t and η_t are independent of each other for all $t = 0, 1, \ldots, T - 1$. X_t can be thought of as a market factor affecting P_t multiplicatively. With the same derivation methodology as before, we get the solution:

$$N_t^* = c_t^{(1)} + c_t^{(2)} R_t + c_t^{(3)} X_t$$

$$V_t^*((P_t, R_t, X_t)) = e^{\mu_Z + \frac{\sigma_Z^2}{2}} \cdot P_t \cdot (c_t^{(4)} + c_t^{(5)} R_t + c_t^{(6)} X_t + c_t^{(7)} R_t^2 + c_t^{(8)} X_t^2 + c_t^{(9)} R_t X_t)$$

where $c_t^{(k)}, 1 \leq k \leq 9$, are constants (independent of P_t, R_t, X_t).

As an exercise, we recommend implementing the above (LPT) model by customizing `OptimalOrderExecution` to compare the obtained Optimal Value Function and Optimal Policy against the closed-form solution (you can find the exact expressions for the $c_t^{(k)}$ coefficients in the Bertsimas and Lo paper).

10.2.3 Incorporating Risk-Aversion and Real-World Considerations

Bertsimas and Lo ignored risk-aversion for the purpose of analytical tractability. Although there was value in obtaining closed-form solutions, ignoring risk-aversion makes their model unrealistic. We have discussed in detail in Chapter 7 about the fact that traders are wary of the risk of uncertain revenues and would be willing to trade away some expected revenues for lower variance of revenues. This calls for incorporating risk-aversion in the maximization objective. Almgren and Chriss wrote an important paper (Almgren and Chriss 2000) where they work in this Risk-Aversion framework. They consider our simple linear price impact model and incorporate risk-aversion by maximizing $E[Y] - \lambda \cdot Var[Y]$ where Y is the total (uncertain) sales proceeds $\sum_{t=0}^{T-1} N_t \cdot Q_t$ and λ controls the degree of risk-aversion. The incorporation of risk-aversion affects the time-trajectory of N_t^*. Clearly, if $\lambda = 0$, we get the usual uniform-split strategy: $N_t^* = \frac{N}{T}$. The other extreme assumption is to minimize $Var[Y]$ which yields: $N_0^* = N$ (sell everything immediately because the only thing we want to avoid is uncertainty of sales proceeds). In their paper, Almgren and Chriss go on to derive the *Efficient Frontier* for this problem (analogous to the Efficient Frontier Portfolio Theory we outline in Appendix B). They also derive solutions for specific utility functions.

To model a real-world trading situation, the first step is to start with the MDP we described earlier with an appropriate model for the price dynamics $f_t(\cdot)$ and the temporary price impact $g_t(\cdot)$ (incorporating potential time-heterogeneity, non-linear price dynamics and non-linear impact). The OptimalOrderExecution class we wrote above allows us to incorporate all of the above. We can also model various real-world "frictions" such as discrete prices, discrete number of shares, constraints on prices and number of shares, as well as trading fees. To make the model truer to reality and more sophisticated, we can introduce various market factors in the *State* which would invariably lead to bloating of the State Space. We would also need to capture *Cross-Asset Market Impact*. As a further step, we could represent the entire Order Book (or a compact summary of the size/shape of the Order book) as part of the state, which leads to further bloating of the state space. All of this makes ADP infeasible and one would need to employ Reinforcement Learning algorithms. More importantly, we'd need to write a realistic Order Book Dynamics simulator capturing all of the above real-world considerations that an RL algorithm would learn from. There are a lot of practical and technical details involved in writing a real-world simulator and we won't be covering those details in this book. It suffices for here to say that the simulator would essentially be a sampling model that has learnt the Order Book Dynamics from market data (supervised learning of the Order Book Dynamics). Using such a simulator and with a deep learning-based function approximation of the Value Function, we can solve a practical Optimal Order Execution problem with Reinforcement Learning. We refer you to a couple of papers for further reading on this:

- Paper by Nevmyvaka, Feng, Kearns in 2006 (Nevmyvaka, Feng, and Kearns 2006)
- Paper by Vyetrenko and Xu in 2019 (Vyetrenko and Xu 2019)

Designing real-world simulators for Order Book Dynamics and using Reinforcement Learning for Optimal Order Execution is an exciting area for future research as well as engineering design. We hope this section has provided sufficient foundations for you to dig into this topic further.

10.3 OPTIMAL MARKET-MAKING

Now we move on to the second problem of this chapter involving trading on an Order Book—the problem of Optimal Market-Making. A market-maker is a company/individual which/who regularly quotes bid and ask prices in a financial asset (which, without loss of generality, we will refer to as a "stock"). The market-maker typically holds some inventory in the stock, always looking to buy at one's quoted bid price and sell at one's quoted ask price, thus looking to make money off the spread between one's quoted ask price and one's quoted bid price. The business of a market-maker is similar to that of a car dealer who maintains an inventory of cars and who will offer purchase and sales prices, looking to make a profit off the price spread and ensuring that the inventory of cars doesn't get too big. In this section, we consider the business of a market-maker who quotes one's bid prices by submitting Buy LOs on an OB and quotes one's ask prices by submitting Sell LOs on the OB. Market-makers are known as *liquidity providers* in the market because they make shares of the stock available for trading on the OB (both on the buy side and sell side). In general, anyone who submits LOs can be thought of as a market liquidity provider. Likewise, anyone who submits MOs can be thought of as a market *liquidity taker* (because an MO takes shares out of the volume that was made available for trading on the OB).

There is typically a fairly complex interplay between liquidity providers (including market-makers) and liquidity takers. Modeling OB dynamics is about modeling this complex interplay, predicting arrivals of MOs and LOs, in response to market events and in response to observed activity on the OB. In this section, we view the OB from the perspective of a single market-maker who aims to make money with Buy/Sell LOs of appropriate bid-ask spread and with appropriate volume of shares (specified in their submitted LOs). The market-maker is likely to be successful if she can do a good job of forecasting OB Dynamics and dynamically adjusting her Buy/Sell LOs on the OB. The goal of the market-maker is to maximize one's *Utility of Gains* at the end of a suitable horizon of time.

The core intuition in the decision of how to set the price and shares in the market-maker's Buy and Sell LOs is as follows: If the market-maker's bid-ask spread is too narrow, they will have more frequent transactions but smaller gains per transaction (more likelihood of their LOs being transacted against by an MO or an opposite-side LO). On the other hand, if the market-maker's bid-ask spread is too wide, they will have less frequent transactions but larger gains per transaction (less likelihood of their LOs being transacted against by an MO or an opposite-side LO). Also of great importance is the fact that a market-maker needs to carefully manage potentially large inventory buildup (either on the long side or the short side) so as to avoid scenarios of consequent unfavorable forced liquidation upon reaching the horizon time. Inventory buildup can occur if the market participants consistently transact against mostly one side of the market-maker's submitted LOs. With this high-level intuition, let us make these concepts of market-making precise. We start by developing some notation to help articulate the problem of Optimal Market-Making clearly. We will re-use some of the notation and terminology we had developed for the problem of Optimal Order Execution. As ever, for ease of exposition, we will simplify the setting for the Optimal Market-Making problem.

Assume there are a finite number of time steps indexed by $t = 0, 1, \ldots, T$. Assume the market-maker always shows a bid price and ask price (at each time t) along with the associated bid shares and ask shares on the OB. Also assume, for ease of exposition, that the market-maker can add or remove bid/ask shares from the OB *costlessly*. We use the following notation:

- Denote $W_t \in \mathbb{R}$ as the market-maker's trading account value at time t.
- Denote $I_t \in \mathbb{Z}$ as the market-maker's inventory of shares at time t (assume $I_0 = 0$). Note that the inventory can be positive or negative (negative means the market-maker is short a certain number of shares).
- Denote $S_t \in \mathbb{R}^+$ as the OB Mid Price at time t (assume a stochastic process for S_t).
- Denote $P_t^{(b)} \in \mathbb{R}^+$ as the market-maker's Bid Price at time t.
- Denote $N_t^{(b)} \in \mathbb{Z}^+$ as the market-maker's Bid Shares at time t.
- Denote $P_t^{(a)} \in \mathbb{R}^+$ as the market-maker's Ask Price at time t.
- Denote $N_t^{(a)} \in \mathbb{Z}^+$ as the market-maker's Ask Shares at time t.
- We refer to $\delta_t^{(b)} = S_t - P_t^{(b)}$ as the market-maker's Bid Spread (relative to OB Mid).
- We refer to $\delta_t^{(a)} = P_t^{(a)} - S_t$ as the market-maker's Ask Spread (relative to OB Mid).
- We refer to $\delta_t^{(b)} + \delta_t^{(a)} = P_t^{(a)} - P_t^{(b)}$ as the market-maker's Bid-Ask Spread.
- Random variable $X_t^{(b)} \in \mathbb{Z}_{\geq 0}$ refers to the total number of market-maker's Bid Shares that have been transacted against (by MOs or by Sell LOs) up to time t ($X_t^{(b)}$ is often

referred to as the cumulative "hits" up to time t, as in "the market-maker's buy offer has been *hit*").

- Random variable $X_t^{(a)} \in \mathbb{Z}_{\geq 0}$ refers to the total number of market-maker's Ask Shares that have been transacted against (by MOs or by Buy LOs) up to time t ($X_t^{(a)}$ is often referred to as the cumulative "lifts" up to time t, as in "the market-maker's sell offer has been *lifted*").

With this notation in place, we can write the trading account balance equation for all $t = 0, 1, \ldots, T - 1$ as follows:

$$W_{t+1} = W_t + P_t^{(a)} \cdot (X_{t+1}^{(a)} - X_t^{(a)}) - P_t^{(b)} \cdot (X_{t+1}^{(b)} - X_t^{(b)}) \qquad (10.10)$$

Note that since the inventory I_0 at time 0 is equal to 0, the inventory I_t at time t is given by the equation:

$$I_t = X_t^{(b)} - X_t^{(a)}$$

The market-maker's goal is to maximize (for an appropriately shaped concave utility function $U(\cdot)$) the sum of the trading account value at time T and the value of the inventory of shares held at time T, i.e., we maximize:

$$\mathbb{E}[U(W_T + I_T \cdot S_T)]$$

As we alluded to earlier, this problem can be cast as a discrete-time finite-horizon Markov Decision Process (with discount factor $\gamma = 1$). Following the usual notation for discrete-time finite-horizon MDPs, the order of activity for the MDP at each time step $t = 0, 1, \ldots, T - 1$ is as follows:

- Observe *State* $(S_t, W_t, I_t) \in \mathcal{S}_t$.
- Perform *Action* $(P_t^{(b)}, N_t^{(b)}, P_t^{(a)}, N_t^{(a)}) \in \mathcal{A}_t$.
- Random number of bid shares hit at time step t (this is equal to $X_{t+1}^{(b)} - X_t^{(b)}$).
- Random number of ask shares lifted at time step t (this is equal to $X_{t+1}^{(a)} - X_t^{(a)}$),
- Update of W_t to W_{t+1}.
- Update of I_t to I_{t+1}.
- Stochastic evolution of S_t to S_{t+1}.
- Receive *Reward* R_{t+1}, where

$$R_{t+1} := \begin{cases} 0 & \text{for } 1 \leq t + 1 \leq T - 1 \\ U(W_T + I_T \cdot S_T) & \text{for } t + 1 = T \end{cases}$$

The goal is to find an *Optimal Policy* $\pi^* = (\pi_0^*, \pi_1^*, \ldots, \pi_{T-1}^*)$, where

$$\pi_t^*((S_t, W_t, I_t)) = (P_t^{(b)}, N_t^{(b)}, P_t^{(a)}, N_t^{(a)})$$

that maximizes:

$$\mathbb{E}[\sum_{t=1}^{T} R_t] = \mathbb{E}[R_T] = \mathbb{E}[U(W_T + I_T \cdot S_T)]$$

10.3.1 Avellaneda-Stoikov Continuous-Time Formulation

A landmark paper by Avellaneda and Stoikov (Avellaneda and Stoikov 2008) formulated this optimal market-making problem in its continuous-time version. Their formulation is conducive to analytical tractability and they came up with a simple, clean and intuitive solution. In this subsection, we go over their formulation and in the next subsection, we show the derivation of their solution. We adapt our discrete-time notation above to their continuous-time setting.

$[(X_t^{(b)}|0 \leq t < T]$ and $[X_t^{(a)}|0 \leq t < T]$ are assumed to be continuous-time *Poisson processes* with the *hit rate per unit of time* and the *lift rate per unit of time* denoted as $\lambda_t^{(b)}$ and $\lambda_t^{(a)}$, respectively. Hence, we can write the following:

$$dX_t^{(b)} \sim Poisson(\lambda_t^{(b)} \cdot dt)$$

$$dX_t^{(a)} \sim Poisson(\lambda_t^{(a)} \cdot dt)$$

$$\lambda_t^{(b)} = f^{(b)}(\delta_t^{(b)})$$

$$\lambda_t^{(a)} = f^{(a)}(\delta_t^{(a)})$$

for decreasing functions $f^{(b)}(\cdot)$ and $f^{(a)}(\cdot)$.

$$dW_t = P_t^{(a)} \cdot dX_t^{(a)} - P_t^{(b)} \cdot dX_t^{(b)}$$

$$I_t = X_t^{(b)} - X_t^{(a)} \text{ (note: } I_0 = 0)$$

Since infinitesimal Poisson random variables $dX_t^{(b)}$ (shares hit in time interval from t to $t + dt$) and $dX_t^{(a)}$ (shares lifted in time interval from t to $t + dt$) are Bernoulli random variables (shares hit/lifted within time interval of duration dt will be 0 or 1), $N_t^{(b)}$ and $N_t^{(a)}$ (number of shares in the submitted LOs for the infinitesimal time interval from t to $t + dt$) can be assumed to be 1.

This simplifies the *Action* at time t to be just the pair:

$$(\delta_t^{(b)}, \delta_t^{(a)})$$

OB Mid Price Dynamics is assumed to be scaled Brownian motion:

$$dS_t = \sigma \cdot dz_t$$

for some $\sigma \in \mathbb{R}^+$.

The Utility function is assumed to be: $U(x) = -e^{-\gamma x}$ where $\gamma > 0$ is the risk-aversion parameter (this Utility function is essentially the CARA Utility function devoid of associated constants).

10.3.2 Solving the Avellaneda-Stoikov Formulation

The following solution is as presented in the Avellaneda-Stoikov paper. We can express the Avellaneda-Stoikov continuous-time formulation as a Hamilton-Jacobi-Bellman (HJB) formulation (note: for reference, the general HJB formulation is covered in Appendix D).

We denote the Optimal Value function as $V^*(t, S_t, W_t, I_t)$. Note that unlike Section 5.13 in Chapter 5 where we denoted the Optimal Value Function as a time-indexed sequence $V_t^*(\cdot)$, here we make t an explicit functional argument of V^* and each of S_t, W_t, I_t also as separate functional arguments of V^* (instead of the typical approach of making the state, as a tuple, a single functional argument). This is because in the continuous-time setting, we are interested in the time-differential of the Optimal Value Function and we also want to represent the dependency of the Optimal Value Function on each of S_t, W_t, I_t as explicit separate dependencies. Appendix D provides the derivation of the general HJB formulation (Equation (D.1) in Appendix D)—this general HJB Equation specializes here to the following:

$$\max_{\delta_t^{(b)}, \delta_t^{(a)}} \mathbb{E}[dV^*(t, S_t, W_t, I_t)] = 0 \text{ for } t < T$$

$$V^*(T, S_T, W_T, I_T) = -e^{-\gamma \cdot (W_T + I_T \cdot S_T)}$$

An infinitesimal change dV^* to $V^*(t, S_t, W_t, I_t)$ is comprised of 3 components:

- Due to pure movement in time (dependence of V^* on t).
- Due to randomness in OB Mid-Price (dependence of V^* on S_t).
- Due to randomness in hitting/lifting the market-maker's Bid/Ask (dependence of V^* on $\lambda_t^{(b)}$ and $\lambda_t^{(a)}$). Note that the probability of being hit in interval from t to $t + dt$ is $\lambda_t^{(b)} \cdot dt$ and probability of being lifted in interval from t to $t + dt$ is $\lambda_t^{(a)} \cdot dt$, upon which the trading account value W_t changes appropriately and the inventory I_t increments/decrements by 1.

With this, we can expand $dV^*(t, S_t, W_t, I_t)$ and rewrite HJB as:

$$\max_{\delta_t^{(b)}, \delta_t^{(a)}} \{ \frac{\partial V^*}{\partial t} \cdot dt + \mathbb{E}[\sigma \cdot \frac{\partial V^*}{\partial S_t} \cdot dz_t + \frac{\sigma^2}{2} \cdot \frac{\partial^2 V^*}{\partial S_t^2} \cdot (dz_t)^2]$$

$$+ \lambda_t^{(b)} \cdot dt \cdot V^*(t, S_t, W_t - S_t + \delta_t^{(b)}, I_t + 1)$$

$$+ \lambda_t^{(a)} \cdot dt \cdot V^*(t, S_t, W_t + S_t + \delta_t^{(a)}, I_t - 1)$$

$$+ (1 - \lambda_t^{(b)} \cdot dt - \lambda_t^{(a)} \cdot dt) \cdot V^*(t, S_t, W_t, I_t)$$

$$- V^*(t, S_t, W_t, I_t) \} = 0$$

Next, we want to convert the HJB Equation to a Partial Differential Equation (PDE). We can simplify the above HJB equation with a few observations:

- $\mathbb{E}[dz_t] = 0$.
- $\mathbb{E}[(dz_t)^2] = dt$.
- Organize the terms involving $\lambda_t^{(b)}$ and $\lambda_t^{(a)}$ better with some algebra.
- Divide throughout by dt.

$$\max_{\delta_t^{(b)}, \delta_t^{(a)}} \{ \frac{\partial V^*}{\partial t} + \frac{\sigma^2}{2} \cdot \frac{\partial^2 V^*}{\partial S_t^2}$$

$$+ \lambda_t^{(b)} \cdot (V^*(t, S_t, W_t - S_t + \delta_t^{(b)}, I_t + 1) - V^*(t, S_t, W_t, I_t))$$

$$+ \lambda_t^{(a)} \cdot (V^*(t, S_t, W_t + S_t + \delta_t^{(a)}, I_t - 1) - V^*(t, S_t, W_t, I_t)) \} = 0$$

Next, note that $\lambda_t^{(b)} = f^{(b)}(\delta_t^{(b)})$ and $\lambda_t^{(a)} = f^{(a)}(\delta_t^{(a)})$, and apply the max only on the relevant terms:

$$\frac{\partial V^*}{\partial t} + \frac{\sigma^2}{2} \cdot \frac{\partial^2 V^*}{\partial S_t^2}$$

$$+ \max_{\delta_t^{(b)}} \{ f^{(b)}(\delta_t^{(b)}) \cdot (V^*(t, S_t, W_t - S_t + \delta_t^{(b)}, I_t + 1) - V^*(t, S_t, W_t, I_t)) \}$$

$$+ \max_{\delta_t^{(a)}} \{ f^{(a)}(\delta_t^{(a)}) \cdot (V^*(t, S_t, W_t + S_t + \delta_t^{(a)}, I_t - 1) - V^*(t, S_t, W_t, I_t)) \} = 0$$

This combines with the boundary condition:

$$V^*(T, S_T, W_T, I_T) = -e^{-\gamma \cdot (W_T + I_T \cdot S_T)}$$

Next, we make an *educated guess* for the functional form of $V^*(t, S_t, W_t, I_t)$:

$$V^*(t, S_t, W_t, I_t) = -e^{-\gamma \cdot (W_t + \theta(t, S_t, I_t))} \tag{10.11}$$

to reduce the problem to a Partial Differential Equation (PDE) in terms of $\theta(t, S_t, I_t)$. Substituting this guessed functional form into the above PDE for $V^*(t, S_t, W_t, I_t)$ gives:

$$\frac{\partial \theta}{\partial t} + \frac{\sigma^2}{2} \cdot \left(\frac{\partial^2 \theta}{\partial S_t^2} - \gamma \cdot \left(\frac{\partial \theta}{\partial S_t} \right)^2 \right)$$

$$+ \max_{\delta_t^{(b)}} \left\{ \frac{f^{(b)}(\delta_t^{(b)})}{\gamma} \cdot \left(1 - e^{-\gamma \cdot (\delta_t^{(b)} - S_t + \theta(t, S_t, I_t + 1) - \theta(t, S_t, I_t))} \right) \right\}$$

$$+ \max_{\delta_t^{(a)}} \left\{ \frac{f^{(a)}(\delta_t^{(a)})}{\gamma} \cdot \left(1 - e^{-\gamma \cdot (\delta_t^{(a)} + S_t + \theta(t, S_t, I_t - 1) - \theta(t, S_t, I_t))} \right) \right\} = 0$$

The boundary condition is:

$$\theta(T, S_T, I_T) = I_T \cdot S_T$$

It turns out that $\theta(t, S_t, I_t + 1) - \theta(t, S_t, I_t)$ and $\theta(t, S_t, I_t) - \theta(t, S_t, I_t - 1)$ are equal to financially meaningful quantities known as *Indifference Bid and Ask Prices*.

Indifference Bid Price $Q^{(b)}(t, S_t, I_t)$ is defined as follows:

$$V^*(t, S_t, W_t - Q^{(b)}(t, S_t, I_t), I_t + 1) = V^*(t, S_t, W_t, I_t) \tag{10.12}$$

$Q^{(b)}(t, S_t, I_t)$ is the price to buy a single share with a *guarantee of immediate purchase* that results in the Optimum Expected Utility staying unchanged.

Likewise, Indifference Ask Price $Q^{(a)}(t, S_t, I_t)$ is defined as follows:

$$V^*(t, S_t, W_t + Q^{(a)}(t, S_t, I_t), I_t - 1) = V^*(t, S_t, W_t, I_t) \tag{10.13}$$

$Q^{(a)}(t, S_t, I_t)$ is the price to sell a single share with a *guarantee of immediate sale* that results in the Optimum Expected Utility staying unchanged.

For convenience, we abbreviate $Q^{(b)}(t, S_t, I_t)$ as $Q_t^{(b)}$ and $Q^{(a)}(t, S_t, I_t)$ as $Q_t^{(a)}$. Next, we express $V^*(t, S_t, W_t - Q_t^{(b)}, I_t + 1) = V^*(t, S_t, W_t, I_t)$ in terms of θ:

$$-e^{-\gamma \cdot (W_t - Q_t^{(b)} + \theta(t, S_t, I_t + 1))} = -e^{-\gamma \cdot (W_t + \theta(t, S_t, I_t))}$$

$$\Rightarrow Q_t^{(b)} = \theta(t, S_t, I_t + 1) - \theta(t, S_t, I_t) \tag{10.14}$$

Likewise for $Q_t^{(a)}$, we get:

$$Q_t^{(a)} = \theta(t, S_t, I_t) - \theta(t, S_t, I_t - 1) \tag{10.15}$$

Using Equations (10.14) and (10.15), bring $Q_t^{(b)}$ and $Q_t^{(a)}$ in the PDE for θ:

$$\frac{\partial \theta}{\partial t} + \frac{\sigma^2}{2} \cdot \left(\frac{\partial^2 \theta}{\partial S_t^2} - \gamma \cdot \left(\frac{\partial \theta}{\partial S_t} \right)^2 \right) + \max_{\delta_t^{(b)}} g(\delta_t^{(b)}) + \max_{\delta_t^{(a)}} h(\delta_t^{(b)}) = 0$$

where $g(\delta_t^{(b)}) = \dfrac{f^{(b)}(\delta_t^{(b)})}{\gamma} \cdot (1 - e^{-\gamma \cdot (\delta_t^{(b)} - S_t + Q_t^{(b)})})$

and $h(\delta_t^{(a)}) = \dfrac{f^{(a)}(\delta_t^{(a)})}{\gamma} \cdot (1 - e^{-\gamma \cdot (\delta_t^{(a)} + S_t - Q_t^{(a)})})$

To maximize $g(\delta_t^{(b)})$, differentiate g with respect to $\delta_t^{(b)}$ and set to 0:

$$e^{-\gamma \cdot (\delta_t^{(b)*} - S_t + Q_t^{(b)})} \cdot (\gamma \cdot f^{(b)}(\delta_t^{(b)*}) - \frac{\partial f^{(b)}}{\partial \delta_t^{(b)}}(\delta_t^{(b)*})) + \frac{\partial f^{(b)}}{\partial \delta_t^{(b)}}(\delta_t^{(b)*}) = 0$$

$$\Rightarrow \delta_t^{(b)*} = S_t - P_t^{(b)*} = S_t - Q_t^{(b)} + \frac{1}{\gamma} \cdot \log{(1 - \gamma \cdot \frac{f^{(b)}(\delta_t^{(b)*})}{\frac{\partial f^{(b)}}{\partial \delta_t^{(b)}}(\delta_t^{(b)*})})} \tag{10.16}$$

To maximize $h(\delta_t^{(a)})$, differentiate h with respect to $\delta_t^{(a)}$ and set to 0:

$$e^{-\gamma \cdot (\delta_t^{(a)*} + S_t - Q_t^{(a)})} \cdot (\gamma \cdot f^{(a)}(\delta_t^{(a)*}) - \frac{\partial f^{(a)}}{\partial \delta_t^{(a)}}(\delta_t^{(a)*})) + \frac{\partial f^{(a)}}{\partial \delta_t^{(a)}}(\delta_t^{(a)*}) = 0$$

$$\Rightarrow \delta_t^{(a)*} = P_t^{(a)*} - S_t = Q_t^{(a)} - S_t + \frac{1}{\gamma} \cdot \log{(1 - \gamma \cdot \frac{f^{(a)}(\delta_t^{(a)*})}{\frac{\partial f^{(a)}}{\partial \delta_t^{(a)}}(\delta_t^{(a)*})})} \tag{10.17}$$

Equations (10.16) and (10.17) are implicit equations for $\delta_t^{(b)*}$ and $\delta_t^{(a)*}$, respectively. Now let us write the PDE in terms of the Optimal Bid and Ask Spreads:

$$\frac{\partial \theta}{\partial t} + \frac{\sigma^2}{2} \cdot (\frac{\partial^2 \theta}{\partial S_t^2} - \gamma \cdot (\frac{\partial \theta}{\partial S_t})^2)$$
$$+ \frac{f^{(b)}(\delta_t^{(b)*})}{\gamma} \cdot (1 - e^{-\gamma \cdot (\delta_t^{(b)*} - S_t + \theta(t, S_t, I_t + 1) - \theta(t, S_t, I_t))}) \tag{10.18}$$
$$+ \frac{f^{(a)}(\delta_t^{(a)*})}{\gamma} \cdot (1 - e^{-\gamma \cdot (\delta_t^{(a)*} + S_t + \theta(t, S_t, I_t - 1) - \theta(t, S_t, I_t))}) = 0$$

with boundary condition: $\theta(T, S_T, I_T) = I_T \cdot S_T$

How do we go about solving this? Here are the steps:

- Firstly, we solve PDE (10.18) for θ in terms of $\delta_t^{(b)*}$ and $\delta_t^{(a)*}$. In general, this would be a numerical PDE solution.
- Using Equations (10.14) and (10.15), and using the above-obtained θ in terms of $\delta_t^{(b)*}$ and $\delta_t^{(a)*}$, we get $Q_t^{(b)}$ and $Q_t^{(a)}$ in terms of $\delta_t^{(b)*}$ and $\delta_t^{(a)*}$.
- Then we substitute the above-obtained $Q_t^{(b)}$ and $Q_t^{(a)}$ (in terms of $\delta_t^{(b)*}$ and $\delta_t^{(a)*}$) in Equations (10.16) and (10.17).
- Finally, we solve the implicit equations for $\delta_t^{(b)*}$ and $\delta_t^{(a)*}$ (in general, numerically).

This completes the (numerical) solution to the Avellaneda-Stoikov continuous-time formulation for the Optimal Market-Making problem. Having been through all the heavy equations above, let's now spend some time on building intuition.

Define the *Indifference Mid Price* $Q_t^{(m)} = \frac{Q_t^{(b)} + Q_t^{(a)}}{2}$. To develop intuition for Indifference Prices, consider a simple case where the market-maker doesn't supply any bids or asks after time t. This means the trading account value W_T at time T must be the same as the trading account value at time t and the inventory I_T at time T must be the same as the inventory I_t at time t. This implies:

$$V^*(t, S_t, W_t, I_t) = \mathbb{E}[-e^{-\gamma \cdot (W_t + I_t \cdot S_T)}]$$

The process $dS_t = \sigma \cdot dz_t$ implies that $S_T \sim \mathcal{N}(S_t, \sigma^2 \cdot (T - t))$, and hence:

$$V^*(t, S_t, W_t, I_t) = -e^{-\gamma \cdot (W_t + I_t \cdot S_t - \frac{\gamma \cdot I_t^2 \cdot \sigma^2 \cdot (T-t)}{2})}$$

Hence,

$$V^*(t, S_t, W_t - Q_t^{(b)}, I_t + 1) = -e^{-\gamma \cdot (W_t - Q_t^{(b)} + (I_t + 1) \cdot S_t - \frac{\gamma \cdot (I_t+1)^2 \cdot \sigma^2 \cdot (T-t)}{2})}$$

But from Equation (10.12), we know that:

$$V^*(t, S_t, W_t, I_t) = V^*(t, S_t, W_t - Q_t^{(b)}, I_t + 1)$$

Therefore,

$$-e^{-\gamma \cdot (W_t + I_t \cdot S_t - \frac{\gamma \cdot I_t^2 \cdot \sigma^2 \cdot (T-t)}{2})} = -e^{-\gamma \cdot (W_t - Q_t^{(b)} + (I_t + 1) \cdot S_t - \frac{\gamma \cdot (I_t+1)^2 \cdot \sigma^2 \cdot (T-t)}{2})}$$

This implies:

$$Q_t^{(b)} = S_t - (2I_t + 1) \cdot \frac{\gamma \cdot \sigma^2 \cdot (T - t)}{2}$$

Likewise, we can derive:

$$Q_t^{(a)} = S_t - (2I_t - 1) \cdot \frac{\gamma \cdot \sigma^2 \cdot (T - t)}{2}$$

The formulas for the Indifference Mid Price and the Indifference Bid-Ask Price Spread are as follows:

$$Q_t^{(m)} = S_t - I_t \cdot \gamma \cdot \sigma^2 \cdot (T - t)$$
$$Q_t^{(a)} - Q_t^{(b)} = \gamma \cdot \sigma^2 \cdot (T - t)$$

These results for the simple case of no-market-making-after-time-t serve as approximations for our problem of optimal market-making. Think of $Q_t^{(m)}$ as a *pseudo mid price* for the market-maker, an adjustment to the OB mid price S_t that takes into account the magnitude and sign of I_t. If the market-maker is long inventory ($I_t > 0$), then $Q_t^{(m)} < S_t$, which makes intuitive sense since the market-maker is interested in reducing her risk of inventory buildup and so, would be be more inclined to sell than buy, leading her to show bid and ask prices whose average is lower than the OB mid price S_t. Likewise, if the market-maker is short inventory ($I_t < 0$), then $Q_t^{(m)} > S_t$ indicating inclination to buy rather than sell.

Armed with this intuition, we come back to optimal market-making, observing from Equations (10.16) and (10.17):

$$P_t^{(b)*} < Q_t^{(b)} < Q_t^{(m)} < Q_t^{(a)} < P_t^{(a)*}$$

Visualize this ascending sequence of prices $[P_t^{(b)*}, Q_t^{(b)}, Q_t^{(m)}, Q_t^{(a)}, P_t^{(a)*}]$ as jointly sliding up/down (relative to OB mid price S_t) as a function of the inventory I_t's magnitude and sign, and perceive $P_t^{(b)^*}, P_t^{(a)^*}$ in terms of their spreads to the *pseudo mid price* $Q_t^{(m)}$:

$$Q_t^{(b)} - P_t^{(m)^*} = \frac{Q_t^{(b)} + Q_t^{(a)}}{2} + \frac{1}{\gamma} \cdot \log\left(1 - \gamma \cdot \frac{f^{(b)}(\delta_t^{(b)^*})}{\frac{\partial f^{(b)}}{\partial \delta_t^{(b)}}(\delta_t^{(b)^*})}\right)$$

$$P_t^{(a)^*} - Q_t^{(m)} = \frac{Q_t^{(b)} + Q_t^{(a)}}{2} + \frac{1}{\gamma} \cdot \log\left(1 - \gamma \cdot \frac{f^{(a)}(\delta_t^{(a)^*})}{\frac{\partial f^{(a)}}{\partial \delta_t^{(a)}}(\delta_t^{(a)^*})}\right)$$

10.3.3 Analytical Approximation to the Solution to Avellaneda-Stoikov Formulation

The PDE (10.18) we derived above for θ and the associated implicit Equations (10.16) and (10.17) for $\delta_t^{(b)^*}, \delta_t^{(a)^*}$ are messy. So we make some assumptions, simplify, and derive analytical approximations(as presented in the Avellaneda-Stoikov paper). We start by assuming a fairly standard functional form for $f^{(b)}$ and $f^{(a)}$:

$$f^{(b)}(\delta) = f^{(a)}(\delta) = c \cdot e^{-k \cdot \delta}$$

This reduces Equations (10.16) and (10.17) to:

$$\delta_t^{(b)^*} = S_t - Q_t^{(b)} + \frac{1}{\gamma} \cdot \log\left(1 + \frac{\gamma}{k}\right) \tag{10.19}$$

$$\delta_t^{(a)^*} = Q_t^{(a)} - S_t + \frac{1}{\gamma} \cdot \log\left(1 + \frac{\gamma}{k}\right) \tag{10.20}$$

which means $P_t^{(b)^*}$ and $P_t^{(a)^*}$ are equidistant from $Q_t^{(m)}$. Substituting these simplified $\delta_t^{(b)^*}, \delta_t^{(a)^*}$ in Equation (10.18) reduces the PDE to:

$$\frac{\partial \theta}{\partial t} + \frac{\sigma^2}{2} \cdot \left(\frac{\partial^2 \theta}{\partial S_t^2} - \gamma \cdot \left(\frac{\partial \theta}{\partial S_t}\right)^2\right) + \frac{c}{k+\gamma} \cdot \left(e^{-k \cdot \delta_t^{(b)^*}} + e^{-k \cdot \delta_t^{(a)^*}}\right) = 0 \tag{10.21}$$

with boundary condition $\theta(T, S_T, I_T) = I_T \cdot S_T$

Note that this PDE (10.21) involves $\delta_t^{(b)^*}$ and $\delta_t^{(a)^*}$. However, Equations (10.19), (10.20), (10.14) and (10.15) enable expressing $\delta_t^{(b)^*}$ and $\delta_t^{(a)^*}$ in terms of $\theta(t, S_t, I_t - 1), \theta(t, S_t, I_t)$ and $\theta(t, S_t, I_t + 1)$. This gives us a PDE just in terms of θ. Solving that PDE for θ would give us not only $V^*(t, S_t, W_t, I_t)$ but also $\delta_t^{(b)^*}$ and $\delta_t^{(a)^*}$ (using Equations (10.19), (10.20), (10.14) and (10.15)). To solve the PDE, we need to make a couple of approximations.

First, we make a linear approximation for $e^{-k \cdot \delta_t^{(b)^*}}$ and $e^{-k \cdot \delta_t^{(a)^*}}$ in PDE (10.21) as follows:

$$\frac{\partial \theta}{\partial t} + \frac{\sigma^2}{2} \cdot \left(\frac{\partial^2 \theta}{\partial S_t^2} - \gamma \cdot \left(\frac{\partial \theta}{\partial S_t}\right)^2\right) + \frac{c}{k+\gamma} \cdot \left(1 - k \cdot \delta_t^{(b)^*} + 1 - k \cdot \delta_t^{(a)^*}\right) = 0 \tag{10.22}$$

Combining the Equations (10.19), (10.20), (10.14) and (10.15) gives us:

$$\delta_t^{(b)^*} + \delta_t^{(a)^*} = \frac{2}{\gamma} \cdot \log\left(1 + \frac{\gamma}{k}\right) + 2\theta(t, S_t, I_t) - \theta(t, S_t, I_t + 1) - \theta(t, S_t, I_t - 1)$$

With this expression for $\delta_t^{(b)*} + \delta_t^{(a)*}$, PDE (10.22) takes the form:

$$\frac{\partial \theta}{\partial t} + \frac{\sigma^2}{2} \cdot \left(\frac{\partial^2 \theta}{\partial S_t^2} - \gamma \cdot \left(\frac{\partial \theta}{\partial S_t}\right)^2\right) + \frac{c}{k+\gamma} \cdot \left(2 - \frac{2k}{\gamma} \cdot \log\left(1 + \frac{\gamma}{k}\right)\right)$$
$$- k \cdot (2\theta(t, S_t, I_t) - \theta(t, S_t, I_t + 1) - \theta(t, S_t, I_t - 1))) = 0 \qquad (10.23)$$

To solve PDE (10.23), we consider the following asymptotic expansion of θ in I_t:

$$\theta(t, S_t, I_t) = \sum_{n=0}^{\infty} \frac{I_t^n}{n!} \cdot \theta^{(n)}(t, S_t)$$

So we need to determine the functions $\theta^{(n)}(t, S_t)$ for all $n = 0, 1, 2, \ldots$
For tractability, we approximate this expansion to the first 3 terms:

$$\theta(t, S_t, I_t) \approx \theta^{(0)}(t, S_t) + I_t \cdot \theta^{(1)}(t, S_t) + \frac{I_t^2}{2} \cdot \theta^{(2)}(t, S_t)$$

We note that the Optimal Value Function V^* can depend on S_t only through the current *Value of the Inventory* (i.e., through $I_t \cdot S_t$), i.e., it cannot depend on S_t in any other way. This means $V^*(t, S_t, W_t, 0) = -e^{-\gamma(W_t + \theta^{(0)}(t, S_t))}$ is independent of S_t. This means $\theta^{(0)}(t, S_t)$ is independent of S_t. So, we can write it as simply $\theta^{(0)}(t)$, meaning $\frac{\partial \theta^{(0)}}{\partial S_t}$ and $\frac{\partial^2 \theta^{(0)}}{\partial S_t^2}$ are equal to 0. Therefore, we can write the approximate expansion for $\theta(t, S_t, I_t)$ as:

$$\theta(t, S_t, I_t) = \theta^{(0)}(t) + I_t \cdot \theta^{(1)}(t, S_t) + \frac{I_t^2}{2} \theta^{(2)}(t, S_t) \qquad (10.24)$$

Substituting this approximation Equation (10.24) for $\theta(t, S_t, I_t)$ in PDE (10.23), we get:

$$\frac{\partial \theta^{(0)}}{\partial t} + I_t \cdot \frac{\partial \theta^{(1)}}{\partial t} + \frac{I_t^2}{2} \cdot \frac{\partial \theta^{(2)}}{\partial t} + \frac{\sigma^2}{2} \cdot \left(I_t \cdot \frac{\partial^2 \theta^{(1)}}{\partial S_t^2} + \frac{I_t^2}{2} \cdot \frac{\partial^2 \theta^{(2)}}{\partial S_t^2}\right)$$
$$- \frac{\gamma \sigma^2}{2} \cdot \left(I_t \cdot \frac{\partial \theta^{(1)}}{\partial S_t} + \frac{I_t^2}{2} \cdot \frac{\partial \theta^{(2)}}{\partial S_t}\right)^2 + \frac{c}{k+\gamma} \cdot \left(2 - \frac{2k}{\gamma} \cdot \log\left(1 + \frac{\gamma}{k}\right) + k \cdot \theta^{(2)}\right) = 0 \qquad (10.25)$$

with boundary condition:

$$\theta^{(0)}(T) + I_T \cdot \theta^{(1)}(T, S_T) + \frac{I_T^2}{2} \cdot \theta^{(2)}(T, S_T) = I_T \cdot S_T$$

Now we separately collect terms involving specific powers of I_t, each yielding a separate PDE:

- Terms devoid of I_t (i.e., I_t^0)
- Terms involving I_t (i.e., I_t^1)
- Terms involving I_t^2

We start by collecting terms involving I_t:

$$\frac{\partial \theta^{(1)}}{\partial t} + \frac{\sigma^2}{2} \cdot \frac{\partial^2 \theta^{(1)}}{\partial S_t^2} = 0 \text{ with boundary condition } \theta^{(1)}(T, S_T) = S_T$$

The solution to this PDE is:

$$\theta^{(1)}(t, S_t) = S_t \qquad (10.26)$$

Next, we collect terms involving I_t^2:

$$\frac{\partial \theta^{(2)}}{\partial t} + \frac{\sigma^2}{2} \cdot \frac{\partial^2 \theta^{(2)}}{\partial S_t^2} - \gamma \cdot \sigma^2 \cdot \left(\frac{\partial \theta^{(1)}}{\partial S_t}\right)^2 = 0 \text{ with boundary condition } \theta^{(2)}(T, S_T) = 0$$

Noting that $\theta^{(1)}(t, S_t) = S_t$, we solve this PDE as:

$$\theta^{(2)}(t, S_t) = -\gamma \cdot \sigma^2 \cdot (T - t) \tag{10.27}$$

Finally, we collect the terms devoid of I_t

$$\frac{\partial \theta^{(0)}}{\partial t} + \frac{c}{k + \gamma} \cdot \left(2 - \frac{2k}{\gamma} \cdot \log\left(1 + \frac{\gamma}{k}\right) + k \cdot \theta^{(2)}\right) = 0 \text{ with boundary } \theta^{(0)}(T) = 0$$

Noting that $\theta^{(2)}(t, S_t) = -\gamma\sigma^2 \cdot (T - t)$, we solve as:

$$\theta^{(0)}(t) = \frac{c}{k + \gamma} \cdot \left(\left(2 - \frac{2k}{\gamma} \cdot \log\left(1 + \frac{\gamma}{k}\right)\right) \cdot (T - t) - \frac{k\gamma\sigma^2}{2} \cdot (T - t)^2\right) \tag{10.28}$$

This completes the PDE solution for $\theta(t, S_t, I_t)$ and hence, for $V^*(t, S_t, W_t, I_t)$. Lastly, we derive formulas for $Q_t^{(b)}, Q_t^{(a)}, Q_t^{(m)}, \delta_t^{(b)*}, \delta_t^{(a)*}$.

Using Equations (10.14) and (10.15), we get:

$$Q_t^{(b)} = \theta^{(1)}(t, S_t) + (2I_t + 1) \cdot \theta^{(2)}(t, S_t) = S_t - (2I_t + 1) \cdot \frac{\gamma \cdot \sigma^2 \cdot (T - t)}{2} \tag{10.29}$$

$$Q_t^{(a)} = \theta^{(1)}(t, S_t) + (2I_t - 1) \cdot \theta^{(2)}(t, S_t) = S_t - (2I_t - 1) \cdot \frac{\gamma \cdot \sigma^2 \cdot (T - t)}{2} \tag{10.30}$$

Using equations (10.19) and (10.20), we get:

$$\delta_t^{(b)*} = \frac{(2I_t + 1) \cdot \gamma \cdot \sigma^2 \cdot (T - t)}{2} + \frac{1}{\gamma} \cdot \log\left(1 + \frac{\gamma}{k}\right) \tag{10.31}$$

$$\delta_t^{(a)*} = \frac{(1 - 2I_t) \cdot \gamma \cdot \sigma^2 \cdot (T - t)}{2} + \frac{1}{\gamma} \cdot \log\left(1 + \frac{\gamma}{k}\right) \tag{10.32}$$

$$\text{Optimal Bid-Ask Spread } \delta_t^{(b)*} + \delta_t^{(a)*} = \gamma \cdot \sigma^2 \cdot (T - t) + \frac{2}{\gamma} \cdot \log\left(1 + \frac{\gamma}{k}\right) \tag{10.33}$$

$$\text{Optimal Pseudo-Mid } Q_t^{(m)} = \frac{Q_t^{(b)} + Q_t^{(a)}}{2} = \frac{P_t^{(b)*} + P_t^{(a)*}}{2} = S_t - I_t \cdot \gamma \cdot \sigma^2 \cdot (T - t) \tag{10.34}$$

Now let's get back to developing intuition. Think of $Q_t^{(m)}$ as *inventory-risk-adjusted* mid-price (adjustment to S_t). If the market-maker is long inventory ($I_t > 0$), $Q_t^{(m)} < S_t$ indicating inclination to sell rather than buy, and if market-maker is short inventory, $Q_t^{(m)} > S_t$ indicating inclination to buy rather than sell. Think of the interval $[P_t^{(b)*}, P_t^{(a)*}]$ as being around the pseudo mid-price $Q_t^{(m)}$ (rather than thinking of it as being around the OB mid-price S_t). The interval $[P_t^{(b)*}, P_t^{(a)*}]$ moves up/down in tandem with $Q_t^{(m)}$ moving up/down (as a function of inventory I_t). Note from Equation (10.33) that the Optimal Bid-Ask Spread $P_t^{(a)*} - P_t^{(b)*}$ is independent of inventory I_t.

A useful view is:

$$P_t^{(b)*} < Q_t^{(b)} < Q_t^{(m)} < Q_t^{(a)} < P_t^{(a)*}$$

with the spreads as follows:

$$\text{Outer Spreads } P_t^{(a)*} - Q_t^{(a)} = Q_t^{(b)} - P_t^{(b)*} = \frac{1}{\gamma} \cdot \log\left(1 + \frac{\gamma}{k}\right)$$

$$\text{Inner Spreads } Q_t^{(a)} - Q_t^{(m)} = Q_t^{(m)} - Q_t^{(b)} = \frac{\gamma \cdot \sigma^2 \cdot (T - t)}{2}$$

This completes the analytical approximation to the solution of the Avellaneda-Stoikov continuous-time formulation of the Optimal Market-Making problem.

10.3.4 Real-World Market-Making

Note that while the Avellaneda-Stoikov continuous-time formulation and solution is elegant and intuitive, it is far from a real-world model. Real-world OB dynamics are time-heterogeneous, non-linear and far more complex. Furthermore, there are all kinds of real-world frictions we need to capture, such as discrete time, discrete prices/number of shares in a bid/ask submitted by the market-maker, various constraints on prices and number of shares in the bid/ask, and fees to be paid by the market-maker. Moreover, we need to capture various market factors in the *State* and in the OB Dynamics. This invariably leads to the *Curse of Dimensionality* and *Curse of Modeling*. This takes us down the same path that we've now got all too familiar with—Reinforcement Learning algorithms. This means we need a simulator that captures all of the above factors, features and frictions. Such a simulator is basically a *Market-Data-learnt Sampling Model* of OB Dynamics. We won't be covering the details of how to build such a simulator as that is outside the scope of this book (a topic under the umbrella of supervised learning of market patterns and behaviors). Using this simulator and neural-networks-based function approximation of the Value Function (and/or of the Policy function), we can leverage the power of RL algorithms (to be covered in the following chapters) to solve the problem of optimal market-making in practice. There are a number of papers written on how to build practical and useful market simulators and using Reinforcement Learning for Optimal Market-Making. We refer you to two such papers here:

- A paper from University of Liverpool (Spooner et al. 2018)
- A paper from J.P.Morgan Research (Ganesh et al. 2019)

This topic of development of models for OB Dynamics and RL algorithms for practical market-making is an exciting area for future research as well as engineering design. We hope this section has provided sufficient foundations for you to dig into this topic further.

10.4   KEY TAKEAWAYS FROM THIS CHAPTER

- Foundations of Order Book, Limit Orders, Market Orders, Price Impact of large Market Orders, and complexity of Order Book Dynamics.
- Casting Order Book trading problems such as Optimal Order Execution and Optimal Market-Making as Markov Decision Processes, developing intuition by deriving closed-form solutions for highly simplified assumptions (e.g., Bertsimas-Lo,

Avellaneda-Stoikov formulations), developing a deeper understanding by implementing a backward-induction ADP algorithm, and then moving on to develop RL algorithms (and associated market simulator) to solve this problem in a real-world setting to overcome the Curse of Dimensionality and Curse of Modeling.

III

Reinforcement Learning Algorithms

CHAPTER 11

Monte-Carlo and Temporal-Difference for Prediction

11.1 OVERVIEW OF THE REINFORCEMENT LEARNING APPROACH

In Module I, we covered Dynamic Programming (DP) and Approximate Dynamic Programming (ADP) algorithms to solve the problems of Prediction and Control. DP and ADP algorithms assume that we have access to a *model* of the MDP environment (by *model*, we mean the transitions defined by $\mathcal{P}_R$—notation from Chapter 4—referring to probabilities of next state and reward, given current state and action). However, in real-world situations, we often do not have access to a model of the MDP environment and so, we'd need to access the actual (real) MDP environment directly. As an example, a robotics application might not have access to a model of a certain type of terrain to learn to walk on, and so we'd need to access the actual (physical) terrain. This means we'd need to *interact* with the real MDP environment. Note that the real MDP environment doesn't give us transition probabilities—it simple serves up a new state and reward when we take an action in a certain state. In other words, it gives us individual experiences of next state and reward, rather than the explicit probabilities of occurrence of next states and rewards. So, the natural question to ask is whether we can infer the Optimal Value Function/Optimal Policy without access to a model (in the case of Prediction—the question is whether we can infer the Value Function for a given policy). The answer to this question is *Yes* and the algorithms that achieve this are known as Reinforcement Learning algorithms.

But Reinforcement Learning is often a great option even in situations where we do have a model. In typical real-world problems, the state space is large and the transitions structure is complex, so transition probabilities are either hard to compute or impossible to store/compute (within practical storage/compute constraints). This means even if we could *theoretically* estimate a model from interactions with the real environment and then run a DP/ADP algorithm, it's typically intractable/infeasible in a typical real-world problem. Moreover, a typical real-world environment is not stationary (meaning the probabilities $\mathcal{P}_R$ change over time) and so, the $\mathcal{P}_R$ probabilities model would need to be re-estimated periodically (making it cumbersome to do DP/ADP). All of this points to the practical alternative of constructing a *sampling model* (a model that serves up samples of next state and reward)—this is typically much more feasible than estimating a *probabilities model* (i.e. a

DOI: 10.1201/9781003229193-11

model of explicit transition probabilities). A sampling model can then be used by a Reinforcement Learning algorithm (as we shall explain shortly).

So what we are saying is that practically we are left with one of the following two options:

1. The AI Agent interacts with the real environment and doesn't bother with either a model of explicit transition probabilities (*probabilities model*) or a model of transition samples (*sampling model*).
2. We create a sampling model (by learning from interaction with the real environment) and treat this sampling model as a simulated environment (meaning, the AI agent interacts with this simulated environment).

From the perspective of the AI agent, either way there is an environment interface that will serve up (at each time step) a single experience of (next state, reward) pair when the agent performs a certain action in a given state. So essentially, either way, our access is simply to a stream of individual experiences of next state and reward rather than their explicit probabilities. So, then the question is—at a conceptual level, how does RL go about solving Prediction and Control problems with just this limited access (access to only experiences and not explicit probabilities)? This will become clearer and clearer as we make our way through Module III, but it would be a good idea now for us to briefly sketch an intuitive overview of the RL approach (before we dive into the actual RL algorithms).

To understand the core idea of how RL works, we take you back to the start of the book where we went over how a baby learns to walk. Specifically, we'd like you to develop intuition for how humans and other animals learn to perform requisite tasks or behave in appropriate ways, so as to get trained to make suitable decisions. Humans/animals don't build a model of explicit probabilities in their minds in a way that a DP/ADP algorithm would require. Rather, their learning is essentially a sort of "trial and error" method—they try an action, receive an experience (i.e., next state and reward) from their environment, then take a new action, receive another experience, and so on ... and then over a period of time, they figure out which actions might be leading to good outcomes (producing good rewards) and which actions might be leading to poor outcomes (poor rewards). This learning process involves raising the priority of actions perceived as good, and lowering the priority of actions perceived as bad. Humans/animals don't quite link their actions to the immediate reward—they link their actions to the cumulative rewards (*Returns*) obtained after performing an action. Linking actions to cumulative rewards is challenging because multiple actions have significantly overlapping rewards sequences, and often rewards show up in a delayed manner. Indeed, learning by attributing good versus bad outcomes to specific past actions is the powerful part of human/animal learning. Humans/animals are essentially estimating a Q-Value Function and are updating their Q-Value function each time they receive a new experience (of essentially a pair of next state and reward). Exactly how humans/animals manage to estimate Q-Value functions efficiently is unclear (a big area of ongoing research), but RL algorithms have specific techniques to estimate the Q-Value function in an incremental manner by updating the Q-Value function in subtle ways after each experience of next state and reward received from either the real environment or simulated environment.

We should also point out another important feature of human/animal learning—it is the fact that humans/animals are good at generalizing their inferences from experiences, i.e., they can interpolate and extrapolate the linkages between their actions and the outcomes received from their environment. Technically, this translates to a suitable function approximation of the Q-Value function. So before we embark on studying the details of various RL algorithms, it's important to recognize that RL overcomes complexity (specifically, the

Curse of Dimensionality and Curse of Modeling, as we have alluded to in previous chapters) with a combination of:

1. Learning from individual experiences of next state and reward received after performing actions in specific states.
2. Good generalization ability of the Q-Value function with a suitable function approximation (indeed, recent progress in capabilities of deep neural networks have helped considerably).

This idea of solving the MDP Prediction and Control problems in this manner (learning from a stream of experiences data with appropriate generalization ability in the Q-Value function approximation) came from the Ph.D. thesis of Chris Watkins (Watkins 1989). As mentioned before, we consider the RL book by Sutton and Barto (Richard S. Sutton and Barto 2018) as the best source for a comprehensive study of RL algorithms as well as the best source for all references associated with RL (hence, we don't provide too many references in this book).

As mentioned in previous chapters, most RL algorithms are founded on the Bellman Equations and all RL Control algorithms are based on the fundamental idea of *Generalized Policy Iteration* that we have explained in Chapter 4. But the exact ways in which the Bellman Equations and Generalized Policy Iteration idea are utilized in RL algorithms differ from one algorithm to another, and they differ significantly from how the Bellman Equations/Generalized Policy Iteration idea is utilized in DP algorithms.

As has been our practice, we start with the Prediction problem (this chapter) and then cover the Control problem (next chapter).

11.2   RL FOR PREDICTION

We re-use a lot of the notation we had developed in Module I. As a reminder, Prediction is the problem of estimating the Value Function of an MDP for a given policy π. We know from Chapter 4 that this is equivalent to estimating the Value Function of the π-implied MRP. So in this chapter, we assume that we are working with an MRP (rather than an MDP) and we assume that the MRP is available in the form of an interface that serves up an individual experience of (next state, reward) pair, given current state. The interface might be a real environment or a simulated environment. We refer to the agent's receipt of an individual experience of (next state, reward), given current state, as an *atomic experience*. Interacting with this interface in succession (starting from a state S_0) gives us a *trace experience* consisting of alternating states and rewards as follows:

$$S_0, R_1, S_1, R_2, S_2, \ldots$$

Given a stream of atomic experiences or a stream of trace experiences, the RL Prediction problem is to estimate the *Value Function* $V : \mathcal{N} \to \mathbb{R}$ of the MRP defined as:

$$V(s) = \mathbb{E}[G_t | S_t = s] \text{ for all } s \in \mathcal{N}, \text{ for all } t = 0, 1, 2, \ldots$$

where the *Return* G_t for each $t = 0, 1, 2, \ldots$ is defined as:

$$G_t = \sum_{i=t+1}^{\infty} \gamma^{i-t-1} \cdot R_i = R_{t+1} + \gamma \cdot R_{t+2} + \gamma^2 \cdot R_{t+3} + \ldots = R_{t+1} + \gamma \cdot G_{t+1}$$

We use the above definition of *Return* even for a terminating trace experience (say terminating at $t = T$, i.e., $S_T \in \mathcal{T}$), by treating $R_i = 0$ for all $i > T$.

The RL prediction algorithms we will soon develop consume a stream of atomic experiences or a stream of trace experiences to learn the requisite Value Function. So we want the input to an RL Prediction algorithm to be either an `Iterable` of atomic experiences or an `Iterable` of trace experiences. Now let's talk about the representation (in code) of a single atomic experience and the representation of a single trace experience. We take you back to the code in Chapter 3 where we had set up a `@dataclass` `TransitionStep` that served as a building block in the method `simulate_reward` in the abstract class `MarkovRewardProcess`.

```
@dataclass(frozen=True)
class TransitionStep(Generic[S]):
    state: NonTerminal[S]
    next_state: State[S]
    reward: float
```

`TransitionStep[S]` represents a single atomic experience. `simulate_reward` produces an `Iterator[TransitionStep[S]]` (i.e., a stream of atomic experiences in the form of a sampling trace) but in general, we can represent a single trace experience as an `Iterable[TransitionStep[S]]` (i.e., a sequence *or* stream of atomic experiences). Therefore, we want the input to an RL prediction algorithm to be either:

- an `Iterable[TransitionStep[S]]` representing a stream of atomic experiences
- an `Iterable[Iterable[TransitionStep[S]]]` representing a stream of trace experiences

Let's add a method `reward_traces` to `MarkovRewardProcess` that produces an `Iterator` (stream) of the sampling traces produced by `simulate_reward`. [1] So then we'd be able to use the output of `reward_traces` as the `Iterable[Iterable[TransitionStep[S]]]` input to an RL Prediction algorithm. Note that the input `start_state_distribution` is the specification of the probability distribution of start states (state from which we start a sampling trace that can be used as a trace experience).

```
def reward_traces(
        self,
        start_state_distribution: Distribution[NonTerminal[S]]
) -> Iterable[Iterable[TransitionStep[S]]]:
    while True:
        yield self.simulate_reward(start_state_distribution)
```

11.3 MONTE-CARLO (MC) PREDICTION

Monte-Carlo (MC) Prediction is a very simple RL algorithm that performs supervised learning to predict the expected return from any state of an MRP (i.e., it estimates the Value Function of an MRP), given a stream of trace experiences. Note that we wrote the abstract class `FunctionApprox` in Chapter 6 for supervised learning that takes data in the form of (x, y) pairs where x is the predictor variable and $y \in \mathbb{R}$ is the response variable. For the Monte-Carlo prediction problem, the x-values are the encountered states across the stream of input trace experiences and the y-values are the associated returns on the trace experiences (starting from the corresponding encountered state). The following function (in the file rl/monte_carlo.py) `mc_prediction` takes as input an `Iterable` of trace experiences, with each trace experience represented as an `Iterable` of `TransitionSteps`. `mc_prediction` performs the requisite supervised learning in an incremental manner, by calling the method `iterate_updates` of `approx_0: ValueFunctionApprox[S]` on an `Iterator` of (state, return) pairs that are extracted from each trace experience. As a reminder, the

---

[1] `reward_traces` is defined in the file rl/markov_process.py.

method `iterate_updates` calls the method `update` of `FunctionApprox` iteratively (in this case, each call to `update` updates the `ValueFunctionApprox` for a single (state, return) data point). `mc_prediction` produces as output an `Iterator` of `ValueFunctionApprox[S]`, i.e., an updated function approximation of the Value Function at the end of each trace experience (note that function approximation updates can be done only at the end of trace experiences because the trace experience returns are available only at the end of trace experiences).

```
import MarkovRewardProcess as mp

def mc_prediction(
    traces: Iterable[Iterable[mp.TransitionStep[S]]],
    approx_0: ValueFunctionApprox[S],
    gamma: float,
    episode_length_tolerance: float = 1e-6
) -> Iterator[ValueFunctionApprox[S]]:
    episodes: Iterator[Iterator[mp.ReturnStep[S]]] = \
        (returns(trace, gamma, episode_length_tolerance) for trace in traces)
    f = approx_0
    yield f

    for episode in episodes:
        f = last(f.iterate_updates(
            [(step.state, step.return_)] for step in episode
        ))
        yield f
```

The core of the `mc_prediction` function above is the call to the `returns` function (detailed below and available in the file rl/returns.py). `returns` takes as input: `trace` representing a trace experience (`Iterable` of `TransitionStep`), the discount factor `gamma`, and an `episodes_length_tolerance` that determines how many time steps to cover in each trace experience when $\gamma < 1$ (as many steps as until γ^{steps} falls below `episodes_length_tolerance` or until the trace experience ends in a terminal state, whichever happens first). If $\gamma = 1$, each trace experience needs to end in a terminal state (else the `returns` function will loop forever).

The `returns` function calculates the returns G_t (accumulated discounted rewards) starting from each state S_t in the trace experience. [2] The key is to walk backwards from the end of the trace experience to the start (so as to reuse the calculated returns while walking backwards: $G_t = R_{t+1} + \gamma \cdot G_{t+1}$). Note the use of `iterate.accumulate` to perform this backwards-walk calculation, which in turn uses the `add_return` method in `TransitionStep` to create an instance of `ReturnStep`. The `ReturnStep` (as seen in the code below) class is derived from the `TransitionStep` class and includes the additional attribute named `return_`.

We add a method called `add_return` in `TransitionStep` so we can augment the attributes `state, reward, next_state` with the additional attribute `return_` that is computed as reward plus gamma times the `return_` from the next state. [3]

```
@dataclass(frozen=True)
class TransitionStep(Generic[S]):
    state: NonTerminal[S]
    next_state: State[S]
    reward: float

    def add_return(self, gamma: float, return_: float) -> ReturnStep[S]:
        return ReturnStep(
            self.state,
            self.next_state,
            self.reward,
            return_=self.reward + gamma * return_
        )
```

---

[2] returns is defined in the file rl/returns.py.

[3] `TransitionStep` and the add_return method are defined in the file rl/markov_process.py.

```
@dataclass(frozen=True)
class ReturnStep(TransitionStep[S]):
    return_: float

import itertools

import rl.iterate as iterate
import rl.markov_process as mp

def returns(
        trace: Iterable[mp.TransitionStep[S]],
        gamma: float,
        tolerance: float
) -> Iterator[mp.ReturnStep[S]]:
    trace = iter(trace)

    max_steps = round(math.log(tolerance) / math.log(gamma)) if gamma < 1 \
        else None
    if max_steps is not None:
        trace = itertools.islice(trace, max_steps * 2)

    *transitions, last_transition = list(trace)

    return_steps = iterate.accumulate(
        reversed(transitions),
        func=lambda next, curr: curr.add_return(gamma, next.return_),
        initial=last_transition.add_return(gamma, 0)
    )
    return_steps = reversed(list(return_steps))

    if max_steps is not None:
        return_steps = itertools.islice(return_steps, max_steps)

    return return_steps
```

We say that the trace experiences are *episodic traces* if each trace experience ends in a terminal state to signify that each trace experience is an episode, after whose termination we move on to the next episode. Trace experiences that do not terminate are known as *continuing traces*. We say that an RL problem is *episodic* if the input trace experiences are all *episodic* (likewise, we say that an RL problem is *continuing* if some of the input trace experiences are *continuing*).

Assume that the probability distribution of returns conditional on a state is modeled by a function approximation as a (state-conditional) normal distribution, whose mean (Value Function) we denote as $V(s; \boldsymbol{w})$ where s denotes a state for which the function approximation is being evaluated and $\boldsymbol{w}$ denotes the set of parameters in the function approximation (e.g., the weights in a neural network). Then, the loss function for supervised learning of the Value Function is the sum of squares of differences between observed returns and the Value Function estimate from the function approximation. For a state S_t visited at time t in a trace experience and its associated return G_t on the trace experience, the contribution to the loss function is:

$$\mathcal{L}_{(S_t, G_t)}(\boldsymbol{w}) = \frac{1}{2} \cdot (V(S_t; \boldsymbol{w}) - G_t)^2 \qquad (11.1)$$

Its gradient with respect to $\boldsymbol{w}$ is:

$$\nabla_{\boldsymbol{w}} \mathcal{L}_{(S_t, G_t)}(\boldsymbol{w}) = (V(S_t; \boldsymbol{w}) - G_t) \cdot \nabla_{\boldsymbol{w}} V(S_t; \boldsymbol{w})$$

We know that the change in the parameters (adjustment to the parameters) is equal to the negative of the gradient of the loss function, scaled by the learning rate (let's denote the learning rate as α). Then the change in parameters is:

$$\Delta \boldsymbol{w} = \alpha \cdot (G_t - V(S_t; \boldsymbol{w})) \cdot \nabla_{\boldsymbol{w}} V(S_t; \boldsymbol{w}) \qquad (11.2)$$

This is a standard formula for change in parameters in response to incremental atomic data for supervised learning when the response variable has a conditional normal distribution. But it's useful to see this formula in an intuitive manner for this specialization of incremental supervised learning to Reinforcement Learning parameter updates. We should interpret the change in parameters Δw as the product of three conceptual entities:

- *Learning Rate* α
- *Return Residual* of the observed return G_t relative to the estimated conditional expected return $V(S_t; w)$
- *Estimate Gradient* of the conditional expected return $V(S_t; w)$ with respect to the parameters w

This interpretation of the change in parameters as the product of these three conceptual entities: (Learning rate, Return Residual, Estimate Gradient) is important as this will be a repeated pattern in many of the RL algorithms we will cover.

Now we consider a simple case of Monte-Carlo Prediction where the MRP consists of a finite state space with the non-terminal states $\mathcal{N} = \{s_1, s_2, \ldots, s_m\}$. In this case, we represent the Value Function of the MRP in a data structure (dictionary) of (state, expected return) pairs. This is known as "Tabular" Monte-Carlo (more generally as Tabular RL to reflect the fact that we represent the calculated Value Function in a "table", i.e., dictionary). Note that in this case, Monte-Carlo Prediction reduces to a very simple calculation wherein for each state, we simply maintain the average of the trace experience returns from that state onwards (averaged over state visitations across trace experiences), and the average is updated in an incremental manner. Recall from Section 6.4 of Chapter 6 that this is exactly what's done in the Tabular class (in file rl/func_approx.py). We also recall from Section 6.4 of Chapter 6 that Tabular implements the interface of the abstract class FunctionApprox and so, we can perform Tabular Monte-Carlo Prediction by passing a Tabular instance as the approx0: FunctionApprox argument to the mc_prediction function above. The implementation of the update method in Tabular is exactly as we desire: it performs an incremental averaging of the trace experience returns obtained from each state onwards (over a stream of trace experiences).

Let us denote $V_n(s_i)$ as the estimate of the Value Function for a state s_i after the n-th occurrence of the state s_i (when doing Tabular Monte-Carlo Prediction) and let $Y_i^{(1)}, Y_i^{(2)}, \ldots, Y_i^{(n)}$ be the trace experience returns associated with the n occurrences of state s_i. Let us denote the count_to_weight_func attribute of Tabular as f. Then, the Tabular update at the n-th occurrence of state s_i (with its associated return $Y_i^{(n)}$) is as follows:

$$V_n(s_i) = (1 - f(n)) \cdot V_{n-1}(s_i) + f(n) \cdot Y_i^{(n)} = V_{n-1}(s_i) + f(n) \cdot (Y_i^{(n)} - V_{n-1}(s_i)) \quad (11.3)$$

Thus, we see that the update (change) to the Value Function for a state s_i is equal to $f(n)$ (weight for the latest trace experience return $Y_i^{(n)}$ from state s_i) times the difference between the latest trace experience return $Y_i^{(n)}$ and the current Value Function estimate $V_{n-1}(s_i)$. This is a good perspective as it tells us how to adjust the Value Function estimate in an intuitive manner. In the case of the default setting of count_to_weight_func as $f(n) = \frac{1}{n}$, we get:

$$V_n(s_i) = \frac{n-1}{n} \cdot V_{n-1}(s_i) + \frac{1}{n} \cdot Y_i^{(n)} = V_{n-1}(s_i) + \frac{1}{n} \cdot (Y_i^{(n)} - V_{n-1}(s_i)) \quad (11.4)$$

So if we have 9 occurrences of a state with an average trace experience return of 50 and if the 10th occurrence of the state gives a trace experience return of 60, then we consider $\frac{1}{10}$ of $60 - 50$ (equal to 1) and increase the Value Function estimate for the state from 50 to $50+1 = 51$. This illustrates how we move the Value Function estimate in the direction from the current estimate to the latest trace experience return, by a magnitude of $\frac{1}{n}$ of their gap.

Expanding the incremental updates across values of n in Equation (11.3), we get:

$$V_n(s_i) = f(n) \cdot Y_i^{(n)} + (1 - f(n)) \cdot f(n-1) \cdot Y_i^{(n-1)} + \dots$$

$$+ (1 - f(n)) \cdot (1 - f(n-1)) \cdots (1 - f(2)) \cdot f(1) \cdot Y_i^{(1)} \qquad (11.5)$$

In the case of the default setting of `count_to_weight_func` as $f(n) = \frac{1}{n}$, we get:

$$V_n(s_i) = \frac{1}{n} \cdot Y_i^{(n)} + \frac{n-1}{n} \cdot \frac{1}{n-1} \cdot Y_i^{(n-1)} + \dots + \frac{n-1}{n} \cdot \frac{n-2}{n-1} \cdots \frac{1}{2} \cdot \frac{1}{1} \cdot Y_i^{(1)} = \frac{\sum_{k=1}^{n} Y_i^{(k)}}{n}$$
$$(11.6)$$

which is an equally-weighted average of the trace experience returns from the state. From the Law of Large Numbers, we know that the sample average converges to the expected value, which is the core idea behind the Monte-Carlo method.

Note that the `Tabular` class as an implementation of the abstract class `FunctionApprox` is not just a software design happenstance—there is a formal mathematical specialization here that is vital to recognize. This tabular representation is actually a special case of linear function approximation by setting a feature function $\phi_i(\cdot)$ for each x_i as: $\phi_i(x_i) = 1$ and $\phi(x) = 0$ for each $x \neq x_i$ (i.e., $\phi_i(\cdot)$ is the indicator function for x_i, and the Φ matrix of Chapter 6 reduces to the identity matrix). So we can conceptualize Tabular Monte-Carlo Prediction as a linear function approximation with the feature functions equal to the indicator functions for each of the non-terminal states and the linear-approximation parameters w_i equal to the Value Function estimates for the corresponding non-terminal states.

With this perspective, more broadly, we can view Tabular RL as a special case of RL with Linear Function Approximation of the Value Function. Moreover, the `count_to_weight_func` attribute of `Tabular` plays the role of the learning rate (as a function of the number of iterations in stochastic gradient descent). This becomes clear if we write Equation (11.3) in terms of parameter updates: write $V_n(s_i)$ as parameter value $w_i^{(n)}$ to denote the n-th update to parameter w_i corresponding to state s_i, and write $f(n)$ as learning rate α_n for the n-th update to w_i.

$$w_i^{(n)} = w_i^{(n-1)} + \alpha_n \cdot (Y_i^{(n)} - w_i^{(n-1)})$$

So, the change in parameter w_i for state s_i is α_n times $Y_i^{(n)} - w_i^{(n-1)}$. We observe that $Y_i^{(n)} - w_i^{(n-1)}$ represents the gradient of the loss function for the data point $(s_i, Y_i^{(n)})$ in the case of linear function approximation with features as indicator variables (for each state). This is because the loss function for the data point $(s_i, Y_i^{(n)})$ is $\frac{1}{2} \cdot (Y_i^{(n)} - \sum_{j=1}^{m} \phi_j(s_i) \cdot w_j^{(n-1)})^2$ which reduces to $\frac{1}{2} \cdot (Y_i^{(n)} - w_i^{(n-1)})^2$, whose gradient in the direction of w_i is $Y_i^{(n)} - w_i^{(n-1)}$ and 0 in the other directions (for $j \neq i$). So we see that `Tabular` updates are basically a special case of `LinearFunctionApprox` updates if we set the features to be indicator functions for each of the states (with `count_to_weight_func` playing the role of the learning rate).

Now that you recognize that `count_to_weight_func` essentially plays the role of the learning rate and governs the importance given to the latest trace experience return relative to past trace experience returns, we want to point out that real-world situations are not stationary in the sense that the environment typically evolves over a period of time and so, RL algorithms have to appropriately adapt to the changing environment. The way to adapt effectively is to have an element of "forgetfulness" of the past because if one learns about the distant past far too strongly in a changing environment, our predictions (and eventually control) would not be effective. So, how does an RL algorithm "forget"? Well, one can "forget" through an appropriate time-decay of the weights when averaging trace experience returns. If we set a constant learning rate α (in Tabular, this would correspond to count_to_weight_func=lambda _: alpha), we'd obtain "forgetfulness" with lower weights for old data points and higher weights for recent data points. This is because with a constant learning rate α, Equation (11.5) reduces to:

$$V_n(s_i) = \alpha \cdot Y_i^{(n)} + (1 - \alpha) \cdot \alpha \cdot Y_i^{(n-1)} + \ldots + (1 - \alpha)^{n-1} \cdot \alpha \cdot Y_i^{(1)}$$
$$= \sum_{j=1}^{n} \alpha \cdot (1 - \alpha)^{n-j} \cdot Y_i^{(j)}$$

which means we have exponentially-decaying weights in the weighted average of the trace experience returns for any given state.

Note that for $0 < \alpha \leq 1$, the weights sum up to 1 as n tends to infinity, i.e.,

$$\lim_{n \to \infty} \sum_{j=1}^{n} \alpha \cdot (1 - \alpha)^{n-j} = \lim_{n \to \infty} 1 - (1 - \alpha)^n = 1$$

It's worthwhile pointing out that the Monte-Carlo algorithm we've implemented above is known as Each-Visit Monte-Carlo to refer to the fact that we include each occurrence of a state in a trace experience. So if a particular state appears 10 times in a given trace experience, we have 10 (state, return) pairs that are used to make the update (for just that state) at the end of that trace experience. This is in contrast to First-Visit Monte-Carlo in which only the first occurrence of a state in a trace experience is included in the set of (state, return) pairs used to make an update at the end of the trace experience. So First-Visit Monte-Carlo needs to keep track of whether a state has already been visited in a trace experience (repeat occurrences of states in a trace experience are ignored). We won't implement First-Visit Monte-Carlo in this book, and leave it to you as an exercise.

Now let's write some code to test our implementation of Monte-Carlo Prediction. To do so, we go back to a simple finite MRP example from Chapter 3—SimpleInventoryMRPFinite. The following code creates an instance of the MRP and computes its exact Value Function based on Equation (3.2).

```
from rl.chapter2.simple_inventory_mrp import SimpleInventoryMRPFinite

user_capacity = 2
user_poisson_lambda = 1.0
user_holding_cost = 1.0
user_stockout_cost = 10.0
user_gamma = 0.9

si_mrp = SimpleInventoryMRPFinite(
    capacity=user_capacity,
    poisson_lambda=user_poisson_lambda,
    holding_cost=user_holding_cost,
    stockout_cost=user_stockout_cost
)
si_mrp.display_value_function(gamma=user_gamma)
```

This prints the following:

```
{NonTerminal(state=InventoryState(on_hand=0, on_order=0)): -35.511,
 NonTerminal(state=InventoryState(on_hand=0, on_order=1)): -27.932,
 NonTerminal(state=InventoryState(on_hand=0, on_order=2)): -28.345,
 NonTerminal(state=InventoryState(on_hand=1, on_order=0)): -28.932,
 NonTerminal(state=InventoryState(on_hand=1, on_order=1)): -29.345,
 NonTerminal(state=InventoryState(on_hand=2, on_order=0)): -30.345}
```

Next, we run Monte-Carlo Prediction by first generating a stream of trace experiences (in the form of sampling traces) from the MRP, and then calling mc_prediction using Tabular with equal-weights-learning-rate (i.e., default count_to_weight_func of lambda n: 1.0 / n).

```
from rl.chapter2.simple_inventory_mrp import InventoryState
from rl.function_approx import Tabular
from rl.approximate_dynamic_programming import ValueFunctionApprox
from rl.distribution import Choose
from rl.iterate import last
from rl.monte_carlo import mc_prediction
from itertools import islice
from pprint import pprint

traces: Iterable[Iterable[TransitionStep[S]]] = \
        mrp.reward_traces(Choose(si_mrp.non_terminal_states))
it: Iterator[ValueFunctionApprox[InventoryState]] = mc_prediction(
    traces=traces,
    approx_0=Tabular(),
    gamma=user_gamma,
    episode_length_tolerance=1e-6
)
num_traces = 60000

last_func: ValueFunctionApprox[InventoryState] = last(islice(it, num_traces))
pprint({s: round(last_func.evaluate([s])[0], 3)
        for s in si_mrp.non_terminal_states})
```

This prints the following:

```
{NonTerminal(state=InventoryState(on_hand=1, on_order=1)): -29.341,
 NonTerminal(state=InventoryState(on_hand=2, on_order=0)): -30.349,
 NonTerminal(state=InventoryState(on_hand=0, on_order=0)): -35.52,
 NonTerminal(state=InventoryState(on_hand=0, on_order=1)): -27.931,
 NonTerminal(state=InventoryState(on_hand=0, on_order=2)): -28.355,
 NonTerminal(state=InventoryState(on_hand=1, on_order=0)): -28.93}
```

We see that the Value Function computed by Tabular Monte-Carlo Prediction with 60000 trace experiences is within 0.01 of the exact Value Function, for each of the states.

This completes the coverage of our first RL Prediction algorithm: Monte-Carlo Prediction. This has the advantage of being a very simple, easy-to-understand algorithm with an unbiased estimate of the Value Function. But Monte-Carlo can be slow to converge to the correct Value Function and another disadvantage of Monte-Carlo is that it requires entire trace experiences (or long-enough trace experiences when $\gamma < 1$). The next RL Prediction algorithm we cover (Temporal-Difference) overcomes these weaknesses.

11.4 TEMPORAL-DIFFERENCE (TD) PREDICTION

To understand Temporal-Difference (TD) Prediction, we start with its Tabular version as it is simple to understand (and then we can generalize to TD Prediction with Function Approximation). To understand Tabular TD prediction, we begin by taking another look at the Value Function update in Tabular Monte-Carlo (MC) Prediction with constant learning rate.

$$V(S_t) \leftarrow V(S_t) + \alpha \cdot (G_t - V(S_t)) \tag{11.7}$$

where S_t is the state visited at time step t in the current trace experience, G_t is the trace experience return obtained from time step t onwards, and α denotes the learning rate (based on count_to_weight_func attribute in the Tabular class). The key in moving from MC to TD is to take advantage of the recursive structure of the Value Function as given by the MRP Bellman Equation (Equation (3.1)). Although we only have access to individual experiences of next state S_{t+1} and reward R_{t+1}, and not the transition probabilities of next state and reward, we can approximate G_t as experience reward R_{t+1} plus γ times $V(S_{t+1})$ (where S_{t+1} is the experience's next state). The idea is to build upon (the term we use is *bootstrap*) the Value Function that is currently estimated. Clearly, this is a biased estimate of the Value Function meaning the update to the Value Function for S_t will be biased. But the bias disadvantage is outweighed by the reduction in variance (which we will discuss more about later), by speedup in convergence (bootstrapping is our friend here), and by the fact that we don't actually need entire/long-enough trace experiences (again, bootstrapping is our friend here). So, the update for Tabular TD Prediction is:

$$V(S_t) \leftarrow V(S_t) + \alpha \cdot (R_{t+1} + \gamma \cdot V(S_{t+1}) - V(S_t)) \tag{11.8}$$

Note how we've simply replaced G_t in Equation (11.7) (Tabular MC Prediction update) with $R_{t+1} + \gamma \cdot V(S_{t+1})$.

To facilitate understanding, for the remainder of the book, we shall interpret $V(S_{t+1})$ as being equal to 0 if $S_{t+1} \in \mathcal{T}$ (note: technically, this notation is incorrect because $V(\cdot)$ is a function with domain $\mathcal{N}$). Likewise, we shall interpret the function approximation notation $V(S_{t+1}; w)$ as being equal to 0 if $S_{t+1} \in \mathcal{T}$.

We refer to $R_{t+1} + \gamma \cdot V(S_{t+1})$ as the TD target and we refer to $\delta_t = R_{t+1} + \gamma \cdot V(S_{t+1}) - V(S_t)$ as the TD Error. The TD Error is the crucial quantity since it represents the "sample Bellman Error" and hence, the TD Error can be used to move $V(S_t)$ appropriately (as shown in the above adjustment to $V(S_t)$), which in turn has the effect of bridging the TD error (on an expected basis).

An important practical advantage of TD is that (unlike MC) we can use it in situations where we have incomplete trace experiences (happens often in real-world situations where experiments gets curtailed/disrupted) and also, we can use it in situations where we never reach a terminal state (*continuing* trace). The other appealing thing about TD is that it is learning (updating Value Function) after each atomic experience (we call it *continuous learning*) versus MC's learning at the end of trace experiences. This also means that TD can be run on *any* stream of atomic experiences, not just atomic experiences that are part of a trace experience. This is a major advantage as we can chop the available data and serve it any order, freeing us from the order in which the data arrives.

Now that we understand how TD Prediction works for the Tabular case, let's consider TD Prediction with Function Approximation. Here, each time we transition from a state S_t to state S_{t+1} with reward R_{t+1}, we make an update to the parameters of the function approximation. To understand how the parameters of the function approximation update, let's consider the loss function for TD. We start with the single-state loss function for MC (Equation (11.1)) and simply replace G_t with $R_{t+1} + \gamma \cdot V(S_{t+1}, w)$ as follows:

$$\mathcal{L}_{(S_t, S_{t+1}, R_{t+1})}(w) = \frac{1}{2} \cdot (V(S_t; w) - (R_{t+1} + \gamma \cdot V(S_{t+1}; w)))^2 \tag{11.9}$$

Unlike MC, in the case of TD, we don't take the gradient of this loss function. Instead we "cheat" in the gradient calculation by ignoring the dependency of $V(S_{t+1}; w)$ on w.

This "gradient with cheating" calculation is known as *semi-gradient*. Specifically, we pretend that the only dependency of the loss function on w is through $V(S_t; w)$. Hence, the semi-gradient calculation results in the following formula for change in parameters w:

$$\Delta w = \alpha \cdot (R_{t+1} + \gamma \cdot V(S_{t+1}; w) - V(S_t; w)) \cdot \nabla_w V(S_t; w) \qquad (11.10)$$

This looks similar to the formula for parameters update in the case of MC (with G_t replaced by $R_{t+1} + \gamma \cdot V(S_{t+1}; w)$). Hence, this has the same structure as MC in terms of conceptualizing the change in parameters as the product of the following 3 entities:

- *Learning Rate* α
- *TD Error* $\delta_t = R_{t+1} + \gamma \cdot V(S_{t+1}; w) - V(S_t; w)$
- *Estimate Gradient* of the conditional expected return $V(S_t; w)$ with respect to the parameters w

Now let's write some code to implement TD Prediction (with Function Approximation). Unlike MC which takes as input a stream of trace experiences, TD works with a more granular stream: a stream of *atomic experiences*. Note that a stream of trace experiences can be broken up into a stream of atomic experiences, but we could also obtain a stream of atomic experiences in other ways (not necessarily from a stream of trace experiences). Thus, the TD prediction algorithm we write below (td_prediction) takes as input an Iterable[TransitionStep[S]]. td_prediction produces an Iterator of ValueFunctionApprox[S], i.e., an updated function approximation of the Value Function after each atomic experience in the input atomic experiences stream. Similar to our implementation of MC, our implementation of TD is based on supervised learning on a stream of (x, y) pairs, but there are two key differences:

1. The update of the ValueFunctionApprox is done after each atomic experience, versus MC where the updates are done at the end of each trace experience.
2. The y-value depends on the Value Function estimate, as seen from the update Equation (11.10) above. This means we cannot use the iterate_updates method of FunctionApprox that MC Prediction uses. Rather, we need to directly use the rl.iterate.accumulate function (a wrapped version of itertools.accumulate). As seen in the code below, the accumulation is performed on the input transitions: Iterable[TransitionStep[S]] and the function governing the accumulation is the step function in the code below that calls the update method of ValueFunctionApprox. Note that the y-values passed to update involve a call to the estimated Value Function v for the next_state of each transition. However, since the next_state could be Terminal or NonTerminal, and since ValueFunctionApprox is valid only for non-terminal states, we use the extended_vf function we had implemented in Chapter 6 to handle the cases of the next state being Terminal or NonTerminal (with terminal states evaluating to the default value of 0).

```
import rl.iterate as iterate
import rl.markov_process as mp
from rl.approximate_dynamic_programming import ValueFunctionApprox
from rl.approximate_dynamic_programming import extended_vf

def td_prediction(
        transitions: Iterable[mp.TransitionStep[S]],
        approx_0: ValueFunctionApprox[S],
        gamma: float
) -> Iterator[ValueFunctionApprox[S]]:
    def step(
            v: ValueFunctionApprox[S],
            transition: mp.TransitionStep[S]
```

```
    ) -> ValueFunctionApprox[S]:
        return v.update([(
            transition.state,
            transition.reward + gamma * extended_vf(v, transition.next_state)
        )])

    return iterate.accumulate(transitions, step, initial=approx_0)
```

The above code is in the file rl/td.py.

Now let's write some code to test our implementation of TD Prediction. We test on the same SimpleInventoryMRPFinite that we had tested MC Prediction on. Let us see how close we can get to the true Value Function (that we had calculated above while testing MC Prediction). But first we need to write a function to construct a stream of atomic experiences (Iterator[TransitionStep[S]]) from a given FiniteMarkovRewardProcess (below code is in the file rl/chapter10/prediction_utils.py). Note the use of itertools.chain.from.iterable to chain together a stream of trace experiences (obtained by calling method reward_traces) into a stream of atomic experiences in the below function unit_experiences_from_episodes.

```
import itertools
from rl.distribution import Distribution, Choose
from rl.approximate_dynamic_programming import NTStateDistribution

def mrp_episodes_stream(
    mrp: MarkovRewardProcess[S],
    start_state_distribution: NTStateDistribution[S]
) -> Iterable[Iterable[TransitionStep[S]]]:
    return mrp.reward_traces(start_state_distribution)

def fmrp_episodes_stream(
    fmrp: FiniteMarkovRewardProcess[S]
) -> Iterable[Iterable[TransitionStep[S]]]:
    return mrp_episodes_stream(fmrp, Choose(fmrp.non_terminal_states))

def unit_experiences_from_episodes(
    episodes: Iterable[Iterable[TransitionStep[S]]],
    episode_length: int
) -> Iterable[TransitionStep[S]]:
    return itertools.chain.from_iterable(
        itertools.islice(episode, episode_length) for episode in episodes
    )
```

Effective use of Tabular TD Prediction requires us to create an appropriate learning rate schedule by suitably lowering the learning rate as a function of the number of occurrences of a state in the atomic experiences stream (learning rate schedule specified by count_to_weight_func attribute of Tabular class). We write below (code in the file rl/function_approx.py) the following learning rate schedule:

$$\alpha_n = \frac{\alpha}{1 + (\frac{n-1}{H})^\beta} \tag{11.11}$$

where α_n is the learning rate to be used at the n-th Value Function update for a given state, α is the initial learning rate (i.e. $\alpha = \alpha_1$), H (we call it "half life") is the number of updates for the learning rate to decrease to half the initial learning rate (if β is 1), and β is the exponent controlling the curvature of the decrease in the learning rate. We shall often set $\beta = 0.5$.

```
def learning_rate_schedule(
    initial_learning_rate: float,
    half_life: float,
    exponent: float
) -> Callable[[int], float]:
    def lr_func(n: int) -> float:
        return initial_learning_rate * (1 + (n - 1) / half_life) ** -exponent
    return lr_func
```

With these functions available, we can now write code to test our implementation of TD Prediction. We use the same instance si_mrp: SimpleInventoryMRPFinite that we had created above when testing MC Prediction. We use the same number of episodes (60000) we had used when testing MC Prediction. We set initial learning rate $\alpha = 0.03$, half life $H = 1000$ and exponent $\beta = 0.5$. We set the episode length (number of atomic experiences in a single trace experience) to be 100 (about the same as with the settings we had for testing MC Prediction). We use the same discount factor $\gamma = 0.9$.

```
import rl.iterate as iterate
import rl.td as td
import itertools
from pprint import pprint
from rl.chapter10.prediction_utils import fmrp_episodes_stream
from rl.chapter10.prediction_utils import unit_experiences_from_episodes
from rl.function_approx import learning_rate_schedule

episode_length: int = 100
initial_learning_rate: float = 0.03
half_life: float = 1000.0
exponent: float = 0.5
gamma: float = 0.9

episodes: Iterable[Iterable[TransitionStep[S]]] = \
    fmrp_episodes_stream(si_mrp)
td_experiences: Iterable[TransitionStep[S]] = \
    unit_experiences_from_episodes(
        episodes,
        episode_length
    )
learning_rate_func: Callable[[int], float] = learning_rate_schedule(
    initial_learning_rate=initial_learning_rate,
    half_life=half_life,
    exponent=exponent
)
td_vfs: Iterator[ValueFunctionApprox[S]] = td.td_prediction(
    transitions=td_experiences,
    approx_0=Tabular(count_to_weight_func=learning_rate_func),
    gamma=gamma
)

num_episodes = 60000

final_td_vf: ValueFunctionApprox[S] = \
    iterate.last(itertools.islice(td_vfs, episode_length * num_episodes))
pprint({s: round(final_td_vf(s), 3) for s in si_mrp.non_terminal_states})
```

This prints the following:

```
{NonTerminal(state=InventoryState(on_hand=0, on_order=0)): -35.529,
 NonTerminal(state=InventoryState(on_hand=0, on_order=1)): -27.868,
 NonTerminal(state=InventoryState(on_hand=0, on_order=2)): -28.344,
 NonTerminal(state=InventoryState(on_hand=1, on_order=0)): -28.935,
 NonTerminal(state=InventoryState(on_hand=1, on_order=1)): -29.386,
 NonTerminal(state=InventoryState(on_hand=2, on_order=0)): -30.305}
```

Thus, we see that our implementation of TD prediction with the above settings fetches us an estimated Value Function within 0.065 of the true Value Function after 60,000 episodes.

As ever, we encourage you to play with various settings for MC Prediction and TD prediction to develop some intuition for how the results change as you change the settings. You can play with the code in the file rl/chapter10/simple_inventory_mrp.py.

11.5   TD VERSUS MC

It is often claimed that TD is the most significant and innovative idea in the development of the field of Reinforcement Learning. The key to TD is that it blends the advantages of Dynamic Programming (DP) and Monte-Carlo (MC). Like DP, TD updates the Value Function estimate by bootstrapping from the Value Function estimate of the next state experienced (essentially, drawing from Bellman Equation). Like MC, TD learns from experiences without requiring access to transition probabilities (MC and TD updates are *experience updates* while DP updates are *transition-probabilities-averaged-updates*). So TD overcomes curse of dimensionality and curse of modeling (computational limitation of DP), and also has the advantage of not requiring entire trace experiences (practical limitation of MC).

The TD idea has its origins in a seminal book by Harry Klopf (Klopf and Data Sciences Laboratory 1972) that greatly influenced Richard Sutton and Andrew Barto to pursue the TD idea further, after which they published several papers on TD, much of whose content is covered in their RL book (Richard S. Sutton and Barto 2018).

11.5.1   TD Learning Akin to Human Learning

Perhaps the most attractive thing about TD (versus MC) is that it is akin to how humans learn. Let us illustrate this point with how a soccer player learns to improve her game in the process of playing many soccer games. Let's simplify the soccer game to a "golden-goal" soccer game, i.e., the game ends when a team scores a goal. The reward in such a soccer game is +1 for scoring (and winning), 0 if the opponent scores, and also 0 for the entire duration of the game before the goal is scored. The soccer player (who is learning) has her *State* comprising of her position/velocity/posture etc., the other players' positions/velocity etc., the soccer ball's position/velocity etc. The *Action*s of the soccer player are her physical movements, including the ways to dribble/kick the ball. If the soccer player learns in an MC style (a single episode is a single soccer game), then the soccer player analyzes (at the end of the game) all possible states and actions that occurred during the game and assesses how the actions in each state might have affected the final outcome of the game. You can see how laborious and difficult this actions-reward linkage would be, and you might even argue that it's impossible to disentangle the effects of various actions during the game on the goal that was eventually scored. In any case, you should recognize that this is absolutely not how a soccer player would analyze and learn. Rather, a soccer player learns *during the game*—she is continuously evaluating how her actions change the probability of scoring the goal (which is essentially the Value Function). If a pass to her teammate did not result in a goal but greatly increased the chances of scoring a goal, then the action of passing the ball to one's teammate in that state is a good action, boosting the action's Q-value immediately, and she will likely try that action (or a similar action) again, meaning actions with better Q-values are prioritized, which drives towards better and quicker goal-scoring opportunities, and likely eventually results in a goal. Such goal-scoring (based on active learning during the game, cutting out poor actions and promoting good actions) would be hailed by commentators as "success from continuous and eager learning" on the part of the soccer player. This is essentially TD learning.

If you think about career decisions and relationship decisions in our lives, MC-style learning is quite infeasible because we simply don't have sufficient "episodes" (for certain decisions, our entire life might be a single episode), and waiting to analyze and adjust until the end of an episode might be far too late in our lives. Rather, we learn and adjust our evaluations of situations constantly in a TD-like manner. Think about various important decisions we make in our lives and you will see that we learn by perpetual adjustment of estimates and we are efficient in the use of limited experiences we obtain in our lives.

11.5.2   Bias, Variance and Convergence

Now let's talk about bias and variance of the MC and TD prediction estimates, and their convergence properties.

Say we are at state S_t at time step t on a trace experience, and G_t is the return from that state S_t onwards on this trace experience. G_t is an unbiased estimate of the true value function for state S_t, which is a big advantage for MC when it comes to convergence, even with function approximation of the Value Function. On the other hand, the TD Target $R_{t+1} + \gamma \cdot V(S_{t+1}; w)$ is a biased estimate of the true value function for state S_t. There is considerable literature on formal proofs of TD Prediction convergence and we won't cover it in detail here, but here's a qualitative summary: Tabular TD Prediction converges to the true value function in the mean for constant learning rate, and converges to the true value function if the following stochastic approximation conditions are satisfied for the learning rate schedule $\alpha_n, n = 1, 2, \ldots$, where the index n refers to the n-th occurrence of a particular state whose Value Function is being updated:

$$\sum_{n=1}^{\infty} \alpha_n = \infty \text{ and } \sum_{n=1}^{\infty} \alpha_n^2 < \infty$$

The stochastic approximation conditions above are known as the Robbins-Monro schedule and apply to a general class of iterative methods used for root-finding or optimization when data is noisy. The intuition here is that the steps should be large enough (first condition) to eventually overcome any unfavorable initial values or noisy data and yet the steps should eventually become small enough (second condition) to ensure convergence. Note that in Equation (11.11), exponent $\beta = 1$ satisfies the Robbins-Monro conditions. In particular, our default choice of count_to_weight_func=lambda n: 1.0 / n in Tabular satisfies the Robbins-Monro conditions, but our other common choice of constant learning rate does not satisfy the Robbins-Monro conditions. However, we want to emphasize that the Robbins-Monro conditions are typically not that useful in practice because it is not a statement of speed of convergence and it is not a statement on closeness to the true optima (in practice, the goal is typically simply to get fairly close to the true answer reasonably quickly).

The bad news with TD (due to the bias in its update) is that TD Prediction with function approximation does not always converge to the true value function. Most TD Prediction convergence proofs are for the Tabular case, however some proofs are for the case of linear function approximation of the Value Function.

The flip side of MC's bias advantage over TD is that the TD Target $R_{t+1} + \gamma \cdot V(S_{t+1}; w)$ has much lower variance than G_t because G_t depends on many random state transitions and random rewards (on the remainder of the trace experience) whose variances accumulate, whereas the TD Target depends on only the next random state transition S_{t+1} and the next random reward R_{t+1}.

As for speed of convergence and efficiency in use of limited set of experiences data, we still don't have formal proofs on whether MC is better or TD is better. More importantly, because MC and TD have significant differences in their usage of data, nature of updates, and frequency of updates, it is not even clear how to create a level-playing field when comparing MC and TD for speed of convergence or for efficiency in usage of limited experiences data. The typical comparisons between MC and TD are done with constant learning rates, and it's been determined that practically TD learns faster than MC with constant learning rates.

A popular simple problem in the literature (when comparing RL prediction algorithms) is a random walk MRP with states $\{0, 1, 2, \ldots, B\}$ with 0 and B as the terminal states (think of these as terminating barriers of a random walk) and the remaining states as the non-terminal states. From any non-terminal state i, we transition to state $i + 1$ with probability

p and to state $i - 1$ with probability $1 - p$. The reward is 0 upon each transition, except if we transition from state $B - 1$ to terminal state B which results in a reward of 1. It's quite obvious that for $p = 0.5$ (symmetric random walk), the Value Function is given by: $V(i) = \frac{i}{B}$ for all $0 < i < B$. We'd like to analyze how MC and TD converge, if at all, to this Value Function, starting from a neutral initial Value Function of $V(i) = 0.5$ for all $0 < i < B$. The following code sets up this random walk MRP.

```python
from rl.distribution import Categorical
class RandomWalkMRP(FiniteMarkovRewardProcess[int]):
    barrier: int
    p: float

    def __init__(
        self,
        barrier: int,
        p: float
    ):
        self.barrier = barrier
        self.p = p
        super().__init__(self.get_transition_map())

    def get_transition_map(self) -> \
            Mapping[int, Categorical[Tuple[int, float]]]:
        d: Dict[int, Categorical[Tuple[int, float]]] = {
            i: Categorical({
                (i + 1, 0. if i < self.barrier - 1 else 1.): self.p,
                (i - 1, 0.): 1 - self.p
            }) for i in range(1, self.barrier)
        }
        return d
```

The above code is in the file rl/chapter10/random_walk_mrp.py. Next, we generate a stream of trace experiences from the MRP, use the trace experiences stream to perform MC Prediction, split the trace experiences stream into a stream of atomic experiences so as to perform TD Prediction, run MC and TD Prediction with a variety of learning rate choices, and plot the root-mean-squared-errors (RMSE) of the Value Function averaged across the non-terminal states as a function of episode batches (i.e., visualize how the RMSE of the Value Function evolves as the MC/TD algorithm progresses). This is done by calling the function compare_mc_and_td which is in the file rl/chapter10/prediction_utils.py.

Figure 11.1 depicts the convergence for our implementations of MC and TD Prediction for constant learning rates of $\alpha = 0.01$ (darker curves) and $\alpha = 0.05$ (lighter curves). We produced this figure by using data from 700 episodes generated from the random walk MRP with barrier $B = 10$, $p = 0.5$ and discount factor $\gamma = 1$ (a single episode refers to a single trace experience that terminates either at state 0 or at state B). We plotted the RMSE after each batch of 7 episodes, hence each of the 4 curves shown in the Figure have 100 RMSE data points plotted. Firstly, we clearly see that MC has significantly more variance as evidenced by the choppy MC RMSE progression curves. Secondly, we note that $\alpha = 0.01$ is a fairly small learning rate and so, the progression of RMSE is quite slow on the darker curves. On the other hand, notice the quick learning for $\alpha = 0.05$ (lighter curves). MC RMSE curve is not just choppy, it's evident that it progresses quite quickly in the first few episode batches (relative to the corresponding TD) but is slow after the first few episode batches (relative to the corresponding TD). This results in TD reaching fairly small RMSE quicker than the corresponding MC (this is especially stark for TD with $\alpha = 0.005$, i.e. the dashed lighter curve in the figure). This behavior of TD outperforming the comparable MC (with constant learning rate) is typical for MRP problems.

Lastly, it's important to recognize that MC is not very sensitive to the initial Value Function while TD is more sensitive to the initial Value Function. We encourage you to play with

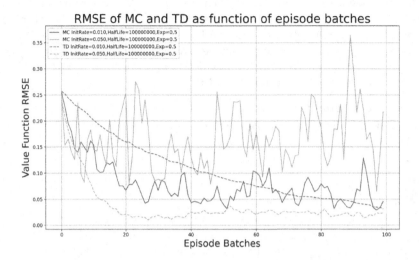

Figure 11.1 MC and TD Convergence for Random Walk MRP

the initial Value Function for this random walk example and evaluate how it affects MC and TD convergence speed.

More generally, we encourage you to play with the compare_mc_and_td function on other choices of MRP (ones we have created earlier in this book such as the inventory examples, or make up your own MRPs) so you can develop good intuition for how MC and TD Prediction algorithms converge for a variety of choices of learning rate schedules, initial Value Function choices, choices of discount factor etc.

11.5.3   Fixed-Data Experience Replay on TD versus MC

We have talked a lot about *how* TD learns versus *how* MC learns. In this subsection, we turn our focus to *what* TD learns and *what* MC learns, to shed light on a profound conceptual difference between TD and MC. We illuminate this difference with a special setting—we are given a fixed finite set of trace experiences (versus usual settings considered in this chapter so far where we had an "endless" stream of trace experiences). The agent is allowed to tap into this fixed finite set of traces experiences endlessly, i.e., the MC or TD Prediction RL agent can indeed consume an endless stream of experiences, but all of that stream of experiences must ultimately be sourced from the given fixed finite set of trace experiences. This means we'd end up tapping into trace experiences (or its component atomic experiences) repeatedly. We call this technique of re-using experiences data encountered previously as *Experience Replay*. We will cover this Experience Replay technique in more detail in Chapter 13, but for now, we shall uncover the key conceptual difference between *what* MC and TD learn by running the algorithms on an *Experience Replay* of a fixed finite set of trace experiences.

So let us start by setting up this experience replay with some code. Firstly, we represent the given input data of the fixed finite set of trace experiences as the data type:

Sequence[Sequence[Tuple[S, float]]]

The outer Sequence refers to the sequence of trace experiences, and the inner Sequence refers to the sequence of (state, reward) pairs in a trace experience (to represent the alternating sequence of states and rewards in a trace experience). The first function we write is to convert this data set into a:

```
Sequence[Sequence[TransitionStep[S]]]
```

which is consumable by MC and TD Prediction algorithms (since their interfaces work with the `TransitionStep[S]` data type). The following function does this job:

```python
def get_fixed_episodes_from_sr_pairs_seq(
    sr_pairs_seq: Sequence[Sequence[Tuple[S, float]]],
    terminal_state: S
) -> Sequence[Sequence[TransitionStep[S]]]:
    return [[TransitionStep(
        state=NonTerminal(s),
        reward=r,
        next_state=NonTerminal(trace[i+1][0])
        if i < len(trace) - 1 else Terminal(terminal_state)
    ) for i, (s, r) in enumerate(trace)] for trace in sr_pairs_seq]
```

We'd like MC Prediction to run on an endless stream of `Sequence[TransitionStep[S]]` sourced from the fixed finite data set produced by `get_fixed_episodes_from_sr_pairs_seq`. So we write the following function to generate an endless stream by repeatedly randomly (uniformly) sampling from the fixed finite set of trace experiences:

```python
import numpy as np

def get_episodes_stream(
    fixed_episodes: Sequence[Sequence[TransitionStep[S]]]
) -> Iterator[Sequence[TransitionStep[S]]]:
    num_episodes: int = len(fixed_episodes)
    while True:
        yield fixed_episodes[np.random.randint(num_episodes)]
```

As we know, TD works with atomic experiences rather than trace experiences. So we need the following function to split the fixed finite set of trace experiences into a fixed finite set of atomic experiences:

```python
import itertools

def fixed_experiences_from_fixed_episodes(
    fixed_episodes: Sequence[Sequence[TransitionStep[S]]]
) -> Sequence[TransitionStep[S]]:
    return list(itertools.chain.from_iterable(fixed_episodes))
```

We'd like TD Prediction to run on an endless stream of `TransitionStep[S]` from the fixed finite set of atomic experiences produced by `fixed_experiences_from_fixed_episodes`. So we write the following function to generate an endless stream by repeatedly randomly (uniformly) sampling from the fixed finite set of atomic experiences:

```python
def get_experiences_stream(
    fixed_experiences: Sequence[TransitionStep[S]]
) -> Iterator[TransitionStep[S]]:
    num_experiences: int = len(fixed_experiences)
    while True:
        yield fixed_experiences[np.random.randint(num_experiences)]
```

Ok—now we are ready to run MC and TD Prediction algorithms on an experience replay of the given input of a fixed finite set of trace experiences. It is quite obvious what the MC Prediction algorithm would learn. MC Prediction is simply supervised learning of a data set of states and their associated returns, and here we have a fixed finite set of states (across the trace experiences) and the corresponding trace experience returns associated with each of those states. Hence, MC Prediction should return a Value Function comprising of the average returns seen in the fixed finite data set for each of the states in the data set. So let us first write a function to explicitly calculate the average returns, and then we can confirm that MC Prediction will give the same answer.

```python
from rl.returns import returns
from rl.markov_process import ReturnStep
def get_return_steps_from_fixed_episodes(
    fixed_episodes: Sequence[Sequence[TransitionStep[S]]],
    gamma: float
) -> Sequence[ReturnStep[S]]:
    return list(itertools.chain.from_iterable(returns(episode, gamma, 1e-8)
                                        for episode in fixed_episodes))

def get_mean_returns_from_return_steps(
    returns_seq: Sequence[ReturnStep[S]]
) -> Mapping[NonTerminal[S], float]:
    def by_state(ret: ReturnStep[S]) -> S:
        return ret.state.state

    sorted_returns_seq: Sequence[ReturnStep[S]] = sorted(
        returns_seq,
        key=by_state
    )
    return {NonTerminal(s): np.mean([r.return_ for r in l])
            for s, l in itertools.groupby(
                sorted_returns_seq,
                key=by_state
            )}
```

To facilitate comparisons, we will do all calculations on the following simple hand-entered input data set:

```python
given_data: Sequence[Sequence[Tuple[str, float]]] = [
    [('A', 2.), ('A', 6.), ('B', 1.), ('B', 2.)],
    [('A', 3.), ('B', 2.), ('A', 4.), ('B', 2.), ('B', 0.)],
    [('B', 3.), ('B', 6.), ('A', 1.), ('B', 1.)],
    [('A', 0.), ('B', 2.), ('A', 4.), ('B', 4.), ('B', 2.), ('B', 3.)],
    [('B', 8.), ('B', 2.)]
]
```

The following code runs get_mean_returns_from_return_steps on this simple input data set.

```python
from pprint import pprint
gamma: float = 0.9

fixed_episodes: Sequence[Sequence[TransitionStep[str]]] = \
    get_fixed_episodes_from_sr_pairs_seq(
        sr_pairs_seq=given_data,
        terminal_state='T'
    )

returns_seq: Sequence[ReturnStep[str]] = \
    get_return_steps_from_fixed_episodes(
        fixed_episodes=fixed_episodes,
        gamma=gamma
    )

mean_returns: Mapping[NonTerminal[str], float] = \
    get_mean_returns_from_return_steps(returns_seq)

pprint(mean_returns)
```

This prints:

```
{NonTerminal(state='B'): 5.190378571428572,
 NonTerminal(state='A'): 8.261809999999999}
```

Now let's run MC Prediction with experience-replayed 100,000 trace experiences with equal weighting for each of the (state, return) pairs, i.e., with count_to_weights_func attribute of Tabular set to the function lambda n: 1.0 / n:

```python
import rl.monte_carlo as mc
import rl.iterate as iterate

def mc_prediction(
    episodes_stream: Iterator[Sequence[TransitionStep[S]]],
    gamma: float,
    num_episodes: int
) -> Mapping[NonTerminal[S], float]:
    return iterate.last(itertools.islice(
        mc.mc_prediction(
            traces=episodes_stream,
            approx_0=Tabular(),
            gamma=gamma,
            episode_length_tolerance=1e-10
        ),
        num_episodes
    )).values_map

num_mc_episodes: int = 100000

episodes: Iterator[Sequence[TransitionStep[str]]] = \
    get_episodes_stream(fixed_episodes)

mc_pred: Mapping[NonTerminal[str], float] = mc_prediction(
    episodes_stream=episodes,
    gamma=gamma,
    num_episodes=num_mc_episodes
)

pprint(mc_pred)
```

This prints:

```
{NonTerminal(state='A'): 8.262643843836214,
 NonTerminal(state='B'): 5.191276907315868}
```

So, as expected, it ties out within the standard error for 100,000 trace experiences. Now let's move on to TD Prediction. Let's run TD Prediction on experience-replayed 1,000,000 atomic experiences with a learning rate schedule having an initial learning rate of 0.01, decaying with a half life of 10000, and with an exponent of 0.5.

```python
import rl.td as td
from rl.function_approx import learning_rate_schedule, Tabular

def td_prediction(
    experiences_stream: Iterator[TransitionStep[S]],
    gamma: float,
    num_experiences: int
) -> Mapping[NonTerminal[S], float]:
    return iterate.last(itertools.islice(
        td.td_prediction(
            transitions=experiences_stream,
            approx_0=Tabular(count_to_weight_func=learning_rate_schedule(
                initial_learning_rate=0.01,
                half_life=10000,
                exponent=0.5
            )),
            gamma=gamma
        ),
        num_experiences
    )).values_map
```

```
num_td_experiences: int = 1000000

fixed_experiences: Sequence[TransitionStep[str]] = \
    fixed_experiences_from_fixed_episodes(fixed_episodes)

experiences: Iterator[TransitionStep[str]] = \
    get_experiences_stream(fixed_experiences)

td_pred: Mapping[NonTerminal[str], float] = td_prediction(
    experiences_stream=experiences,
    gamma=gamma,
    num_experiences=num_td_experiences
)

pprint(td_pred)
```

This prints:

```
{NonTerminal(state='A'): 9.899838136517303,
 NonTerminal(state='B'): 7.444114569419306}
```

We note that this Value Function is vastly different from the Value Function produced by MC Prediction. Is there a bug in our code, or perhaps a more serious conceptual problem? Nope—there is neither a bug here nor a more serious problem. This is exactly what TD Prediction on Experience Replay on a fixed finite data set is meant to produce. So, what Value Function does this correspond to? It turns out that TD Prediction drives towards a Value Function of an MRP that is *implied* by the fixed finite set of given experiences. By the term *implied*, we mean the maximum likelihood estimate for the transition probabilities $\mathcal{P}_R$, estimated from the given fixed finite data, i.e.,

$$\mathcal{P}_R(s, r, s') = \frac{\sum_{i=1}^{N} \mathbb{I}_{S_i=s, R_{i+1}=r, S_{i+1}=s'}}{\sum_{i=1}^{N} \mathbb{I}_{S_i=s}} \tag{11.12}$$

where the fixed finite set of atomic experiences are $[(S_i, R_{i+1}, S_{i+1})|1 \leq i \leq N]$, and $\mathbb{I}$ denotes the indicator function.

So let's write some code to construct this MRP based on the above formula.

```
from rl.distribution import Categorical
from rl.markov_process import FiniteMarkovRewardProcess
def finite_mrp(
    fixed_experiences: Sequence[TransitionStep[S]]
) -> FiniteMarkovRewardProcess[S]:
    def by_state(tr: TransitionStep[S]) -> S:
        return tr.state.state

    d: Mapping[S, Sequence[Tuple[S, float]]] = \
        {s: [(t.next_state.state, t.reward) for t in l] for s, l in
         itertools.groupby(
             sorted(fixed_experiences, key=by_state),
             key=by_state
         )}
    mrp: Dict[S, Categorical[Tuple[S, float]]] = \
        {s: Categorical({x: y / len(l) for x, y in
                         collections.Counter(l).items()})
         for s, l in d.items()}
    return FiniteMarkovRewardProcess(mrp)
```

Now let's print its Value Function.

```
fmrp: FiniteMarkovRewardProcess[str] = finite_mrp(fixed_experiences)
fmrp.display_value_function(gamma)
```

This prints:

```
{NonTerminal(state='A'): 9.958, NonTerminal(state='B'): 7.545}
```

So our TD Prediction algorithm doesn't exactly match the Value Function of the data-implied MRP, but it gets close. It turns out that a variation of our TD Prediction algorithm exactly matches the Value Function of the data-implied MRP. We won't implement this variation in this chapter, but will describe it briefly here. The variation is as follows:

- The Value Function is not updated after each atomic experience, rather the Value Function is updated at the end of each *batch of atomic experiences*.
- Each batch of atomic experiences consists of a single occurrence of each atomic experience in the given fixed finite data set.
- The updates to the Value Function to be performed at the end of each batch are accumulated in a buffer after each atomic experience and the buffer's contents are used to update the Value Function only at the end of the batch. Specifically, this means that the right-hand-side of Equation (11.10) is calculated at the end of each atomic experience and these calculated values are accumulated in the buffer until the end of the batch, at which point the buffer's contents are used to update the Value Function.

This variant of the TD Prediction algorithm is known as *Batch Updating* and more broadly, RL algorithms that update the Value Function at the end of a batch of experiences are referred to as *Batch Methods*. This contrasts with *Incremental Methods*, which are RL algorithms that update the Value Function after each atomic experience (in the case of TD) or at the end of each trace experience (in the case of MC). The MC and TD Prediction algorithms we implemented earlier in this chapter are Incremental Methods. We will cover Batch Methods in detail in Chapter 13.

Although our TD Prediction algorithm is an Incremental Method, it did get fairly close to the Value Function of the data-implied MRP. So let us ignore the nuance that our TD Prediction algorithm didn't exactly match the Value Function of the data-implied MRP and instead focus on the fact that our MC Prediction algorithm and our TD Prediction algorithm drove towards two very different Value Functions. The MC Prediction algorithm learns a "fairly naive" Value Function—one that is based on the mean of the observed returns (for each state) in the given fixed finite data. The TD Prediction algorithm is learning something "deeper"—it is (implicitly) constructing an MRP based on the given fixed finite data (Equation (11.12)), and then (implicitly) calculating the Value Function of the constructed MRP. The mechanics of the TD Prediction algorithms don't actually construct the MRP and calculate the Value Function of the MRP—rather, the TD Prediction algorithm directly drives towards the Value Function of the data-implied MRP. However, the fact that it gets to this "more nuanced" Value Function means that it is (implicitly) trying to infer a transitions structure from the given data, and hence, we say that it is learning something "deeper" than what MC is learning. This has practical implications. Firstly, this learning facet of TD means that it exploits any Markov property in the environment and so, TD algorithms are more efficient (learn faster than MC) in Markov environments. On the other hand, the naive nature of MC (not exploiting any Markov property in the environment) is advantageous (more effective than TD) in non-Markov environments.

We encourage you to try Experience Replay on larger input data sets, and to code up Batch Method variants of MC and TD prediction algorithms. As a starting point, the experience replay code for this chapter is in the file rl/chapter10/mc_td_experience_replay.py.

11.5.4 Bootstrapping and Experiencing

We summarize MC, TD and DP in terms of whether they bootstrap (or not) and in terms of whether they experience interactions with an real/simulated environment (or not).

- Bootstrapping: By "bootstrapping", we mean that an update to the Value Function utilizes a current or prior estimate of the Value Function. MC *does not bootstrap* since its Value Function updates use actual trace experience returns and not any current or prior estimates of the Value Function. On the other hand, TD and DP *do bootstrap*.
- Experiencing: By "experiencing", we mean that the algorithm uses experiences obtained by interacting with a real or simulated environment, rather than performing expectation calculations with a model of transition probabilities (the latter doesn't require interactions with an environment and hence, doesn't "experience"). MC and TD *do experience*, while DP *does not experience*.

We illustrate this perspective of bootstrapping (or not) and experiencing (or not) with some very popular diagrams that we are borrowing from lecture slides from David Silver's RL course and from teaching content prepared by Richard Sutton.

The first diagram is Figure 11.2, known as the MC *backup* diagram for an MDP (although we are covering Prediction in this chapter, these concepts also apply to MDP Control). The root of the tree is the state whose Value Function we want to update. The remaining nodes of the tree are the future states that might be visited and future actions that might be taken. The branching on the tree is due to the probabilistic transitions of the MDP and the multiple choices of actions that might be taken at each time step. The nodes marked as "T" are the terminal states. The highlighted path on the tree from the root node (current state) to a terminal state indicates a particular trace experience used by the MC algorithm. The highlighted path is the set of future states/actions used in updating the Value Function of the current state (root node). We say that the Value Function is "backed up" along this highlighted path (to mean that the Value Function update calculation propagates from the bottom of the highlighted path to the top, since the trace experience return is calculated as accumulated rewards from the bottom to the top, i.e., from the end of the trace experience to the beginning of the trace experience). This is why we refer to such diagrams as *backup* diagrams. Since MC "experiences", it only considers a single child node from any node (rather than all the child nodes, which would be the case if we considered all probabilistic transitions or considered all action choices). So the backup is narrow (doesn't go wide across the tree). Since MC does not "bootstrap", it doesn't use the Value Function estimate from its child/grandchild node (next time step's state/action)—instead, it utilizes the rewards at all future states/actions along the entire trace experience. So the backup works deep into the tree (is not shallow as would be the case in "bootstrapping"). In summary, the MC backup is narrow and deep.

The next diagram is Figure 11.3, known as the TD *backup* diagram for an MDP. Again, the highlighting applies to the future states/actions used in updating the Value Function of the current state (root node). The Value Function is "backed up" along this highlighted portion of the tree. Since TD "experiences", it only considers a single child node from any node (rather than all the child nodes, which would be the case if we considered all probabilistic transitions or considered all actions choices). So the backup is narrow (doesn't go wide across the tree). Since TD "bootstraps", it uses the Value Function estimate from its child/grandchild node (next time step's state/action) and doesn't utilize rewards at states/actions beyond the next time step's state/action. So the backup is shallow (doesn't work deep into the tree). In summary, the TD backup is narrow and shallow.

The next diagram is Figure 11.4, known as the DP *backup* diagram for an MDP. Again, the highlighting applies to the future states/actions used in updating the Value Function of the

Monte Carlo (Supervised Learning) (MC)

$$V(S_t) \leftarrow V(S_t) + \alpha\big[G_t - V(S_t)\big]$$

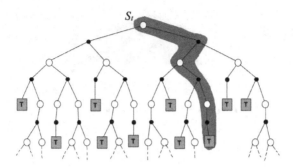

Figure 11.2 MC Backup Diagram (Image Credit: David Silver's RL Course)

Simplest TD Method

$$V(S_t) \leftarrow V(S_t) + \alpha\big[R_{t+1} + \gamma V(S_{t+1}) - V(S_t)\big]$$

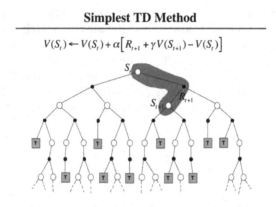

Figure 11.3 TD Backup Diagram (Image Credit: David Silver's RL Course)

current state (root node). The Value Function is "backed up" along this highlighted portion of the tree. Since DP does not "experience" and utilizes the knowledge of probabilities of all next states and considers all choices of actions (in the case of Control), it considers all child nodes (all choices of actions) and all grandchild nodes (all probabilistic transitions to next states) from the root node (current state). So the backup goes wide across the tree. Since DP "bootstraps", it uses the Value Function estimate from its children/grandchildren nodes (next time step's states/actions) and doesn't utilize rewards at states/actions beyond the next time step's states/actions. So the backup is shallow (doesn't work deep into the tree). In summary, the DP backup is wide and shallow.

This perspective of shallow versus deep (for "bootstrapping" or not) and of narrow versus wide (for "experiencing" or not) is a great way to visualize and internalize the core ideas within MC, TD and DP, and it helps us compare and contrast these methods in a simple and intuitive manner. We must thank Rich Sutton for this excellent pedagogical contribution. This brings us to the next diagram (Figure 11.5) which provides a unified view of RL in a single picture. The top of this figure shows methods that "bootstrap"

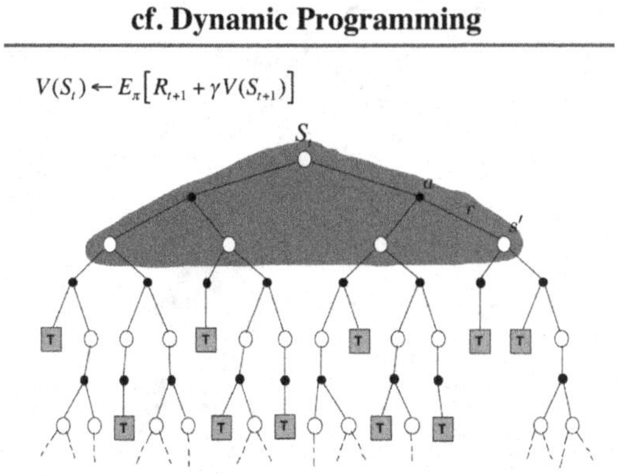

cf. Dynamic Programming

$$V(S_t) \leftarrow E_\pi\big[R_{t+1} + \gamma V(S_{t+1})\big]$$

Figure 11.4 DP Backup Diagram (Image Credit: David Silver's RL Course)

(including TD and DP) and the bottom of this figure shows methods that do not "boot-strap" (including MC and methods known as "Exhaustive Search" that go both deep into the tree and wide across the tree—we shall cover some of these methods in a later chapter). Therefore the vertical dimension of this figure refers to the depth of the backup. The left of this figure shows methods that "experience" (including TD and MC) and the right of this figure shows methods than do not "experience" (including DP and "Exhaustive Search"). Therefore, the horizontal dimension of this figure refers to the width of the backup.

Unified View

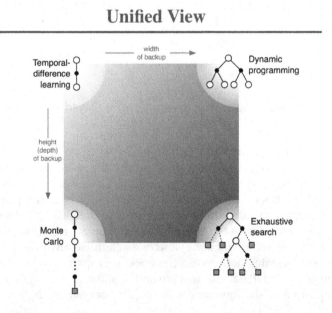

Figure 11.5 Unified View of RL (Image Credit: Sutton-Barto's RL Book)

11.6   TD(λ) PREDICTION

Now that we've seen the contrasting natures of TD and MC (and their respective pros and cons), it's natural to wonder if we could design an RL Prediction algorithm that combines the features of TD and MC and perhaps fetch us a blend of their respective benefits. It turns out this is indeed possible, and is the subject of this section—an innovative approach to RL Prediction known as TD(λ). λ is a continuous-valued parameter in the range $[0, 1]$ such that $\lambda = 0$ corresponds to the TD approach and $\lambda = 1$ corresponds to the MC approach. Tuning λ between 0 and 1 allows us to span the spectrum from the TD approach to the MC approach, essentially a blended approach known as the TD(λ) approach. The TD(λ) approach for RL Prediction gives us the TD(λ) Prediction algorithm. To get to the TD(λ) Prediction algorithm (in this section), we start with the TD Prediction algorithm we wrote earlier, generalize it to a multi-time-step bootstrapping prediction algorithm, extend that further to an algorithm known as the λ-Return Prediction algorithm, after which we shall be ready to present the TD(λ) Prediction algorithm.

11.6.1 n-Step Bootstrapping Prediction Algorithm

In this subsection, we generalize the bootstrapping approach of TD to multi-time-step bootstrapping (which we refer to as n-step bootstrapping). We start with Tabular Prediction as it is very easy to explain and understand. To understand n-step bootstrapping, let us take another look at the TD update equation for the Tabular case (Equation (11.8)):

$$V(S_t) \leftarrow V(S_t) + \alpha \cdot (R_{t+1} + \gamma \cdot V(S_{t+1}) - V(S_t))$$

The basic idea was that we replaced G_t (in the case of the MC update) with the estimate $R_{t+1} + \gamma \cdot V(S_{t+1})$, by using the current estimate of the Value Function of the state that is 1 time step ahead on the trace experience. It's then natural to extend this idea to instead use the current estimate of the Value Function for the state that is 2 time steps ahead on the trace experience, which would yield the following update:

$$V(S_t) \leftarrow V(S_t) + \alpha \cdot (R_{t+1} + \gamma \cdot R_{t+2} + \gamma^2 \cdot V(S_{t+2}) - V(S_t))$$

We can generalize this to an update that uses the current estimate of the Value Function for the state that is $n \geq 1$ time steps ahead on the trace experience, as follows:

$$V(S_t) \leftarrow V(S_t) + \alpha \cdot (G_{t,n} - V(S_t)) \tag{11.13}$$

where $G_{t,n}$ (known as n-step bootstrapped return) is defined as:

$$G_{t,n} = \sum_{i=t+1}^{t+n} \gamma^{i-t-1} \cdot R_i + \gamma^n \cdot V(S_{t+n})$$

$$= R_{t+1} + \gamma \cdot R_{t+2} + \gamma^2 \cdot R_{t+3} + \ldots + \gamma^{n-1} \cdot R_{t+n} + \gamma^n \cdot V(S_{t+n})$$

If the trace experience terminates at $t = T$, i.e., $S_T \in \mathcal{T}$, the above equation applies only for t, n such that $t + n < T$. Essentially, each n-step bootstrapped return $G_{t,n}$ is an approximation of the full return G_t, by truncating G_t at n steps and adjusting for the remainder with the Value Function estimate $V(S_{t+n})$ for the state S_{t+n}. If $t + n \geq T$, then there is no need for a truncation and the n-step bootstrapped return $G_{t,n}$ is equal to the full return G_t.

It is easy to generalize this n-step bootstrapping Prediction algorithm to the case of Function Approximation for the Value Function. The update Equation (11.13) generalizes to:

$$\Delta \boldsymbol{w} = \alpha \cdot (G_{t,n} - V(S_t; \boldsymbol{w})) \cdot \nabla_{\boldsymbol{w}} V(S_t; \boldsymbol{w}) \tag{11.14}$$

where the n-step bootstrapped return $G_{t,n}$ is now defined in terms of the function approximation for the Value Function (rather than the tabular Value Function), as follows:

$$G_{t,n} = \sum_{i=t+1}^{t+n} \gamma^{i-t-1} \cdot R_i + \gamma^n \cdot V(S_{t+n}; \boldsymbol{w})$$

$$= R_{t+1} + \gamma \cdot R_{t+2} + \gamma^2 \cdot R_{t+3} + \ldots + \gamma^{n-1} \cdot R_{t+n} + \gamma^n \cdot V(S_{t+n}; \boldsymbol{w})$$

The nuances we outlined above for when the trace experience terminates naturally apply here as well.

Equation (11.14) looks similar to the parameters update equations for the MC and TD Prediction algorithms we covered earlier, in terms of conceptualizing the change in parameters as the product of the following 3 entities:

- *Learning Rate* α.
- *n-step Bootstrapped Error* $G_{t,n} - V(S_t; \boldsymbol{w})$.
- *Estimate Gradient* of the conditional expected return $V(S_t; \boldsymbol{w})$ with respect to the parameters $\boldsymbol{w}$.

n serves as a parameter taking us across the spectrum from TD to MC. $n = 1$ is the case of TD while sufficiently large n is the case of MC. If a trace experience is of length T (i.e., $S_T \in \mathcal{T}$), then $n \geq T$ will not have any bootstrapping (since the bootstrapping target goes beyond the length of the trace experience) and hence, this makes it identical to MC.

We note that for large n, the update to the Value Function for state S_t visited at time t happens in a delayed manner (after n steps, at time $t + n$), which is unlike the TD algorithm we had developed earlier where the update happens at the very next time step. We won't be implementing this n-step bootstrapping Prediction algorithm and leave it as an exercise for you to implement (re-using some of the functions/classes we have developed so far in this book). A key point to note for your implementation: The input won't be an `Iterable` of atomic experiences (like in the case of the TD Prediction algorithm we implemented), rather it will be an `Iterable` of trace experiences (i.e., the input will be the same as for our MC Prediction algorithm: `Iterable[Iterable[TransitionStep[S]]]`) since we need multiple future rewards in the trace to perform an update to the current state.

11.6.2 λ-Return Prediction Algorithm

Now we extend the n-step bootstrapping Prediction Algorithm to the λ-Return Prediction Algorithm. The idea behind this extension is really simple: Since the target for each n (in n-step bootstrapping) is $G_{t,n}$, a valid target can also be a weighted-average target:

$$\sum_{n=1}^{N} u_n \cdot G_{t,n} + u \cdot G_t \text{ where } u + \sum_{n=1}^{N} u_n = 1$$

Note that any of the u_n or u can be 0, as long as they all sum up to 1. The λ-Return target is a special case of the weights u_n and u, and applies to episodic problems (i.e., where every trace experience terminates). For a given state S_t with the episode terminating at time T (i.e., $S_T \in \mathcal{T}$), the weights for the λ-Return target are as follows:

$$u_n = (1 - \lambda) \cdot \lambda^{n-1} \text{ for all } n = 1, \ldots, T - t - 1, u_n = 0 \text{ for all } n \geq T - t \text{ and } u = \lambda^{T-t-1}$$

We denote the λ-Return target as $G_t^{(\lambda)}$, defined as:

$$G_t^{(\lambda)} = (1 - \lambda) \cdot \sum_{n=1}^{T-t-1} \lambda^{n-1} \cdot G_{t,n} + \lambda^{T-t-1} \cdot G_t \qquad (11.15)$$

Thus, the update Equation is:

$$\Delta w = \alpha \cdot (G_t^{(\lambda)} - V(S_t; w)) \cdot \nabla_w V(S_t; w) \tag{11.16}$$

We note that for $\lambda = 0$, the λ-Return target reduces to the TD (1-step bootstrapping) target and for $\lambda = 1$, the λ-Return target reduces to the MC target G_t. The λ parameter gives us a smooth way of tuning from TD ($\lambda = 0$) to MC ($\lambda = 1$).

Note that for $\lambda > 0$, Equation (11.16) tells us that the parameters w of the function approximation can be updated only at the end of an episode (the term *episode* refers to a terminating trace experience). Updating w according to Equation (11.16) for all states $S_t, t = 0, \ldots, T - 1$, *at the end of each episode* gives us the *Offline λ-Return Prediction* algorithm. The term *Offline* refers to the fact that we have to wait till the end of an episode to make an update to the parameters w of the function approximation (rather than making parameter updates after each time step in the episode, which we refer to as an *Online* algorithm). Online algorithms are appealing because the Value Function update for an atomic experience could be utilized immediately by the updates for the next few atomic experiences, and so it facilitates continuous/fast learning. So the natural question to ask here is if we can turn the Offline λ-return Prediction algorithm outlined above to an Online version. An online version is indeed possible (it's known as the TD(λ) Prediction algorithm) and is the topic of the remaining subsections of this section. But before we begin the coverage of the (Online) TD(λ) Prediction algorithm, let's wrap up this subsection with an implementation of this Offline version (i.e., the λ-Return Prediction algorithm).

```python
import rl.markov_process as mp
import numpy as np
from rl.approximate_dynamic_programming import ValueFunctionApprox

def lambda_return_prediction(
        traces: Iterable[Iterable[mp.TransitionStep[S]]],
        approx_0: ValueFunctionApprox[S],
        gamma: float,
        lambd: float
) -> Iterator[ValueFunctionApprox[S]]:
    func_approx: ValueFunctionApprox[S] = approx_0
    yield func_approx

    for trace in traces:
        gp: List[float] = [1.]
        lp: List[float] = [1.]
        predictors: List[NonTerminal[S]] = []
        partials: List[List[float]] = []
        weights: List[List[float]] = []
        trace_seq: Sequence[mp.TransitionStep[S]] = list(trace)
        for t, tr in enumerate(trace_seq):
            for i, partial in enumerate(partials):
                partial.append(
                    partial[-1] +
                    gp[t - i] * (tr.reward - func_approx(tr.state)) +
                    (gp[t - i] * gamma * extended_vf(func_approx, tr.next_state)
                     if t < len(trace_seq) - 1 else 0.)
                )
                weights[i].append(
                    weights[i][-1] * lambd if t < len(trace_seq)
                    else lp[t - i]
                )
            predictors.append(tr.state)
            partials.append([tr.reward +
                            (gamma * extended_vf(func_approx, tr.next_state)
                             if t < len(trace_seq) - 1 else 0.)])
            weights.append([1. - (lambd if t < len(trace_seq) else 0.)])
            gp.append(gp[-1] * gamma)
            lp.append(lp[-1] * lambd)
```

```
    responses: Sequence[float] = [np.dot(p, w) for p, w in
                                  zip(partials, weights)]
    for p, r in zip(predictors, responses):
        func_approx = func_approx.update([(p, r)])
    yield func_approx
```

The above code is in the file rl/td_lambda.py.

Note that this λ-Return Prediction algorithm is not just Offline, it is also a highly ineffi-cient algorithm because of the two loops within each trace experience. However, it serves as a pedagogical benefit before moving on to the (efficient) Online TD(λ) Prediction algo-rithm.

11.6.3 Eligibility Traces

Now we are ready to start developing the TD(λ) Prediction algorithm. The TD(λ) Predic-tion algorithm is founded on the concept of *Eligibility Traces*. So we start by introducing the concept of Eligibility traces (first for the Tabular case, then generalize to Function Approx-imations), then go over the TD(λ) Prediction algorithm (based on Eligibility traces), and finally explain why the TD(λ) Prediction algorithm is essentially the *Online* version of the *Offline* λ-Return Prediction algorithm we've implemented above.

We begin the story of Eligibility Traces with the concept of a (for lack of a better term) *Memory* function. Assume that we have an event happening at specific points in time, say at times $t_1, t_2, \ldots, t_n \in \mathbb{R}_{\geq 0}$ with $t_1 < t_2 < \ldots < t_n$, and we'd like to construct a *Memory* function $M : \mathbb{R}_{\geq 0} \to \mathbb{R}_{\geq 0}$ such that the *Memory* function (at any point in time t) remembers the number of times the event has occurred up to time t, but also has an element of "for-getfulness" in the sense that recent occurrences of the event are remembered better than older occurrences of the event. So the function M needs to have an element of memory-decay in remembering the count of the occurrences of the events. In other words, we want the function M to produce a time-decayed count of the event occurrences. We do this by constructing the function M as follows (for some decay-parameter $\theta \in [0,1]$):

$$M(t) = \begin{cases} \mathbb{I}_{t=t_1} & \text{if } t \leq t_1, \text{ else} \\ M(t_i) \cdot \theta^{t-t_i} + \mathbb{I}_{t=t_{i+1}} & \text{if } t_i < t \leq t_{i+1} \text{ for any } 1 \leq i < n, \text{ else} \\ M(t_n) \cdot \theta^{t-t_n} & \text{otherwise (i.e., if } t > t_n) \end{cases} \quad (11.17)$$

where $\mathbb{I}$ denotes the indicator function.

This means the memory function has an uptick of 1 each time the event occurs (at time t_i, for each $i = 1, 2, \ldots n$), but then decays by a factor of $\theta^{\Delta t}$ over any interval Δt where the event doesn't occur. Thus, the memory function captures the notion of frequency of the events as well as the recency of the events.

Let's write some code to plot this function in order to visualize it and gain some intu-ition.

```
def plot_memory_function(theta: float, event_times: List[float]) -> None:
    step: float = 0.01
    x_vals: List[float] = [0.0]
    y_vals: List[float] = [0.0]
    for t in event_times:
        rng: Sequence[int] = range(1, int(math.floor((t - x_vals[-1]) / step)))
        x_vals += [x_vals[-1] + i * step for i in rng]
        y_vals += [y_vals[-1] * theta ** (i * step) for i in rng]
        x_vals.append(t)
        y_vals.append(y_vals[-1] * theta ** (t - x_vals[-1]) + 1.0)
    plt.plot(x_vals, y_vals)
```

```
plt.grid()
plt.xticks([0.0] + event_times)
plt.xlabel("Event Timings", fontsize=15)
plt.ylabel("Memory Funtion Values", fontsize=15)
plt.title("Memory Function (Frequency and Recency)", fontsize=25)
plt.show()
```

Let's run this for $\theta = 0.8$ and an arbitrary sequence of event times:

```
theta = 0.8
event_times = [2.0, 3.0, 4.0, 7.0, 9.0, 14.0, 15.0, 21.0]
plot_memory_function(theta, event_times)
```

This produces the graph in Figure 11.6.

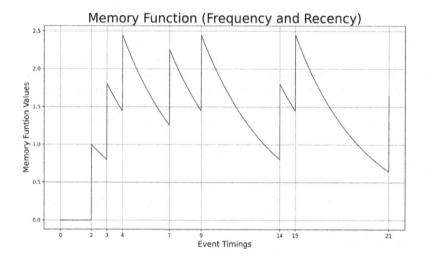

Figure 11.6 **Memory Function (Frequency and Recency)**

The above code is in the file rl/chapter10/memory_function.py.

This memory function is actually quite useful as a model for a variety of modeling situations in the broader world of Applied Mathematics where we want to combine the notions of frequency and recency. However, here we want to use this memory function as a way to model *Eligibility Traces* for the tabular case, which in turn will give us the tabular TD(λ) Prediction algorithm (online version of the offline tabular λ-Return Prediction algorithm we covered earlier).

Now we are ready to define Eligibility Traces for the Tabular case. We assume a finite state space with the set of non-terminal states $\mathcal{N} = \{s_1, s_2, \ldots, s_m\}$. Eligibility Trace for each state $s \in \mathcal{N}$ is defined as the Memory function $M(\cdot)$ with $\theta = \gamma \cdot \lambda$ (i.e., the product of the discount factor and the TD-λ parameter) and the event timings are the time steps at which the state s occurs in a trace experience. Thus, we define Eligibility Traces for a given trace experience at any time step t (of the trace experience) as a function $E_t : \mathcal{N} \to \mathbb{R}_{\geq 0}$ as follows:

$$E_0(s) = \mathbb{I}_{S_0 = s}, \text{ for all } s \in \mathcal{N}$$

$$E_t(s) = \gamma \cdot \lambda \cdot E_{t-1}(s) + \mathbb{I}_{S_t = s}, \text{ for all } s \in \mathcal{N}, \text{ for all } t = 1, 2, \ldots$$

where $\mathbb{I}$ denotes the indicator function.

Then, the Tabular TD(λ) Prediction algorithm performs the following updates to the Value Function at each time step t in each trace experience:

$$V(s) \leftarrow V(s) + \alpha \cdot (R_{t+1} + \gamma \cdot V(S_{t+1}) - V(S_t)) \cdot E_t(s), \text{ for all } s \in \mathcal{N}$$

Note the similarities and differences relative to the TD update we have seen earlier. Firstly, this is an online algorithm since we make an update at each time step in a trace experience. Secondly, we update the Value Function *for all* states at each time step (unlike TD Prediction which updates the Value Function only for the particular state that is visited at that time step). Thirdly, the change in the Value Function for each state $s \in \mathcal{N}$ is proportional to the TD-Error $\delta_t = R_{t+1} + \gamma \cdot V(S_{t+1}) - V(S_t)$, much like in the case of the TD update. However, here the TD-Error is multiplied by the eligibility trace $E_t(s)$ for each state s at each time step t. So, we can compactly write the update as:

$$V(s) \leftarrow V(s) + \alpha \cdot \delta_t \cdot E_t(s), \text{ for all } s \in \mathcal{N} \tag{11.18}$$

where α is the learning rate.

This is it—this is the Tabular TD(λ) Prediction algorithm! Now the question is—how is this linked to the Tabular λ-Return Prediction algorithm? It turns out that if we made all the updates of Equation (11.18) in an offline manner (at the end of each trace experience), then the sum of the changes in the Value Function for any specific state $s \in \mathcal{N}$ over the course of the entire trace experience is equal to the change in the Value Function for s in the Tabular λ-Return Prediction algorithm as a result of its offline update for state s. Concretely,

Theorem 11.6.1.

$$\sum_{t=0}^{T-1} \alpha \cdot \delta_t \cdot E_t(s) = \sum_{t=0}^{T-1} \alpha \cdot (G_t^{(\lambda)} - V(S_t)) \cdot \mathbb{I}_{S_t=s}, \text{ for all } s \in \mathcal{N}$$

where $\mathbb{I}$ denotes the indicator function.

Proof. We begin the proof with the following important identity:

$$
\begin{aligned}
G_t^{(\lambda)} - V(S_t) = -V(S_t) \quad &+(1-\lambda) \cdot \lambda^0 \cdot (R_{t+1} + \gamma \cdot V(S_{t+1})) \\
&+(1-\lambda) \cdot \lambda^1 \cdot (R_{t+1} + \gamma \cdot R_{t+2} + \gamma^2 \cdot V(S_{t+2})) \\
&+(1-\lambda) \cdot \lambda^2 \cdot (R_{t+1} + \gamma \cdot R_{t+2} + \gamma^2 \cdot R_{t+3} + \gamma^3 \cdot V(S_{t+3})) \\
&+ \ldots \\
= -V(S_t) \quad &+(\gamma\lambda)^0 \cdot (R_{t+1} + \gamma \cdot V(S_{t+1}) - \gamma\lambda \cdot V(S_{t+1})) \\
&+(\gamma\lambda)^1 \cdot (R_{t+2} + \gamma \cdot V(S_{t+2}) - \gamma\lambda \cdot V(S_{t+2})) \\
&+(\gamma\lambda)^2 \cdot (R_{t+3} + \gamma \cdot V(S_{t+3}) - \gamma\lambda \cdot V(S_{t+3})) \\
&+ \ldots \\
= \quad &\quad (\gamma\lambda)^0 \cdot (R_{t+1} + \gamma \cdot V(S_{t+1}) - V(S_t)) \\
&+(\gamma\lambda)^1 \cdot (R_{t+2} + \gamma \cdot V(S_{t+2}) - V(S_{t+1})) \\
&+(\gamma\lambda)^2 \cdot (R_{t+3} + \gamma \cdot V(S_{t+3}) - V(S_{t+2})) \\
&+ \ldots \\
= \delta_t + \gamma\lambda \cdot \delta_{t+1} &+ (\gamma\lambda)^2 \cdot \delta_{t+2} + \ldots
\end{aligned}
$$

$$\tag{11.19}$$

Now assume that a specific non-terminal state s appears at time steps $t_1, t_2, \ldots, t_n$. Then,

$$\sum_{t=0}^{T-1} \alpha \cdot (G_t^{(\lambda)} - V(S_t)) \cdot \mathbb{I}_{S_t=s} = \sum_{i=1}^{n} \alpha \cdot (G_{t_i}^{(\lambda)} - V(S_{t_i}))$$

$$= \sum_{i=1}^{n} \alpha \cdot (\delta_{t_i} + \gamma\lambda \cdot \delta_{t_i+1} + (\gamma\lambda)^2 \cdot \delta_{t_i+2} + \ldots)$$

$$= \sum_{t=0}^{T-1} \alpha \cdot \delta_t \cdot E_t(s)$$

$\square$

If we set $\lambda = 0$ in this Tabular TD(λ) Prediction algorithm, we note that $E_t(s)$ reduces to $\mathbb{I}_{S_t=s}$ and so, the Tabular TD(λ) prediction algorithm's update for $\lambda = 0$ at each time step t reduces to:

$$V(S_t) \leftarrow V(S_t) + \alpha \cdot \delta_t$$

which is exactly the update of the Tabular TD Prediction algorithm. Therefore, TD algorithms are often referred to as TD(0).

If we set $\lambda = 1$ in this Tabular TD(λ) Prediction algorithm with episodic traces (i.e., all trace experiences terminating), Theorem 11.6.1 tells us that the sum of all changes in the Value Function for any specific state $s \in \mathcal{N}$ over the course of the entire trace experience ($= \sum_{t=0}^{T-1} \alpha \cdot \delta_t \cdot E_t(s)$) is equal to the change in the Value Function for s in the Every-Visit MC Prediction algorithm as a result of its offline update for state s ($= \sum_{t=0}^{T-1} \alpha \cdot (G_t - V(S_t) \cdot \mathbb{I}_{S_t=s})$). Hence, TD(1) is considered to be "equivalent" to Every-Visit MC.

To clarify, TD(λ) Prediction is an online algorithm and hence, not exactly equivalent to the offline λ-Return Prediction algorithm. However, if we modified the TD(λ) Prediction algorithm to be offline, then they are equivalent. The offline version of TD(λ) Prediction would not make the updates to the Value Function at each time step—rather, it would accumulate the changes to the Value Function (as prescribed by the TD(λ) update formula) in a buffer, and then at the end of the trace experience, it would update the Value Function with the contents of the buffer.

However, as explained earlier, online update are desirable because the changes to the Value Function at each time step can be immediately usable for the next time steps' updates and so, it promotes rapid learning without having to wait for a trace experience to end. Moreover, online algorithms can be used in situations where we don't have a complete episode.

With an understanding of Tabular TD(λ) Prediction in place, we can generalize TD(λ) Prediction to the case of function approximation in a straightforward manner. In the case of function approximation, the data type of eligibility traces will be the same data type as that of the parameters w in the function approximation (so here we denote eligibility traces at time t of a trace experience as simply E_t rather than as a function of states as we had done for the Tabular case above). We initialize E_0 at the start of each trace experience to $\nabla_w V(S_0; w)$. Then, for each time step $t > 0$, E_t is calculated recursively in terms of the previous time step's value E_{t-1}, which is then used to update the parameters of the Value Function approximation, as follows:

$$E_t = \gamma\lambda \cdot E_{t-1} + \nabla_w V(S_t; w)$$
$$\Delta w = \alpha \cdot (R_{t+1} + \gamma \cdot V(S_{t+1}; w) - V(S_t; w)) \cdot E_t$$

The update to the parameters w can be expressed more succinctly as:

$$\Delta w = \alpha \cdot \delta_t \cdot E_t$$

where δ_t now denotes the TD Error based on the function approximation for the Value Function.

The idea of Eligibility Traces has its origins in a seminal book by Harry Klopf (Klopf and Data Sciences Laboratory 1972) that greatly influenced Richard Sutton and Andrew Barto to pursue the idea of Eligibility Traces further, after which they published several papers on Eligibility Traces, much of whose content is covered in their RL book (Richard S. Sutton and Barto 2018).

11.6.4 Implementation of the TD(λ) Prediction Algorithm

You'd have observed that the TD(λ) update is not as simple as the MC and TD updates, where we were able to use the FunctionApprox interface in a straightforward manner. For TD(λ), it might appear that we can't quite use the FunctionApprox interface and would need to write custom-code for its implementation. However, by noting that the FunctionApprox method objective_gradient is quite generic and that FunctionApprox and Gradient support methods __add__ and __mul__ (vector space operations), we can actually implement the TD(λ) in terms of the FunctionApprox interface.

The function td_lambda_prediction below takes as input an Iterable of trace experiences (traces), an initial FunctionApprox (approx_0), and the γ and λ parameters. At the start of each trace experience, we need to initialize the eligibility traces to 0. The data type of the eligibility traces is the Gradient type and so we invoke the zero method for Gradient(func_approx) in order to initialize the eligibility traces to 0. Then, at every time step in every trace experience, we first set the predictor variable x_t to be the state and the response variable y_t to be the TD target. Then we need to update the eligibility traces el_tr and update the function approximation func_approx using the updated el_tr.

Thankfully, the __mul__ method of Gradient class enables us to conveniently multiply el_tr with $\gamma \cdot \lambda$ and then, it also enables us to multiply the updated el_tr with the prediction error $\mathbb{E}_M[y|x_t] - y_t = V(S_t; w) - (R_{t+1} + \gamma \cdot V(S_{t+1}; w))$ (in the code as func_approx(x) - y), which is then used (as a Gradient type) to update the internal parameters of the func_approx. The __add__ method of Gradient enables us to add $\nabla_w V(S_t; w)$ (as a Gradient type) to el_tr * gamma * lambd. The only seemingly difficult part is calculating $\nabla_w V(S_t; w)$. The FunctionApprox interface provides us with a method objective_gradient to calculate the gradient of any specified objective (call it $Obj(x, y)$). But here we have to calculate the gradient of the prediction of the function approximation. Thankfully, the interface of objective_gradient is fairly generic and we actually have a choice of constructing $Obj(x, y)$ to be whatever function we want (not necessarily a minimizing Objective Function). We specify $Obj(x, y)$ in terms of the obj_deriv_out_func argument, which as a reminder, represents $\frac{\partial Obj(x,y)}{\partial Out(x)}$. Note that we have assumed a Gaussian distribution for the returns conditioned on the state. So we can set $Out(x)$ to be the function approximation's prediction $V(S_t; w)$ and we can set $Obj(x, y) = Out(x)$, meaning obj_deriv_out_func ($\frac{\partial Obj(x,y)}{\partial Out(x)}$) is a function returning the constant value of 1 (as seen in the code below).

```
import rl.markov_process as mp
import numpy as np
from rl.function_approx import Gradient
from rl.approximate_dynamic_programming import ValueFunctionApprox

def td_lambda_prediction(
        traces: Iterable[Iterable[mp.TransitionStep[S]]],
```

```
        approx_0: ValueFunctionApprox[S],
        gamma: float,
        lambd: float
) -> Iterator[ValueFunctionApprox[S]]:
    func_approx: ValueFunctionApprox[S] = approx_0
    yield func_approx

    for trace in traces:
        el_tr: Gradient[ValueFunctionApprox[S]] = Gradient(func_approx).zero()
        for step in trace:
            x: NonTerminal[S] = step.state
            y: float = step.reward + gamma * \
                extended_vf(func_approx, step.next_state)
            el_tr = el_tr * (gamma * lambd) + func_approx.objective_gradient(
                xy_vals_seq=[(x, y)],
                obj_deriv_out_fun=lambda x1, y1: np.ones(len(x1))
            )
            func_approx = func_approx.update_with_gradient(
                el_tr * (func_approx(x) - y)
            )
            yield func_approx
```

The above code is in the file rl/td_lambda.py.

Let's use the same instance si_mrp: SimpleInventoryMRPFinite that we had created above when testing MC and TD Prediction. We use the same number of episodes (60000) we had used when testing MC Prediction. Just like in the case of testing TD prediction, we set initial learning rate $\alpha = 0.03$, half life $H = 1000$ and exponent $\beta = 0.5$. We set the episode length (number of atomic experiences in a single trace experience) to be 100 (same as with the settings we had for testing TD Prediction and consistent with MC Prediction as well). We use the same discount factor $\gamma = 0.9$. Let's set $\lambda = 0.3$.

```
import rl.iterate as iterate
import rl.td_lambda as td_lambda
import itertools
from pprint import pprint
from rl.chapter10.prediction_utils import fmrp_episodes_stream
from rl.function_approx import learning_rate_schedule

gamma: float = 0.9
episode_length: int = 100
initial_learning_rate: float = 0.03
half_life: float = 1000.0
exponent: float = 0.5
lambda_param = 0.3

episodes: Iterable[Iterable[TransitionStep[S]]] = \
    fmrp_episodes_stream(si_mrp)
curtailed_episodes: Iterable[Iterable[TransitionStep[S]]] = \
    (itertools.islice(episode, episode_length) for episode in episodes)
learning_rate_func: Callable[[int], float] = learning_rate_schedule(
    initial_learning_rate=initial_learning_rate,
    half_life=half_life,
    exponent=exponent
)
td_lambda_vfs: Iterator[ValueFunctionApprox[S]] = td_lambda.td_lambda_prediction(
    traces=curtailed_episodes,
    approx_0=Tabular(count_to_weight_func=learning_rate_func),
    gamma=gamma,
    lambd=lambda_param
)

num_episodes = 60000

final_td_lambda_vf: ValueFunctionApprox[S] = \
    iterate.last(itertools.islice(td_lambda_vfs, episode_length * num_episodes))
pprint({s: round(final_td_lambda_vf(s), 3) for s in si_mrp.non_terminal_states})
```

This prints the following:

```
{NonTerminal(state=InventoryState(on_hand=0, on_order=0)): -35.545,
 NonTerminal(state=InventoryState(on_hand=0, on_order=1)): -27.97,
 NonTerminal(state=InventoryState(on_hand=0, on_order=2)): -28.396,
 NonTerminal(state=InventoryState(on_hand=1, on_order=0)): -28.943,
 NonTerminal(state=InventoryState(on_hand=1, on_order=1)): -29.506,
 NonTerminal(state=InventoryState(on_hand=2, on_order=0)): -30.339}
```

Thus, we see that our implementation of TD(λ) Prediction with the above settings fetches us an estimated Value Function fairly close to the true Value Function. As ever, we encourage you to play with various settings for TD(λ) Prediction to develop an intuition for how the results change as you change the settings, and particularly as you change the λ parameter. You can play with the code in the file rl/chapter10/simple_inventory_mrp.py.

11.7   KEY TAKEAWAYS FROM THIS CHAPTER

- Bias-Variance tradeoff of TD versus MC.
- MC learns the statistical mean of the observed returns while TD learns something "deeper"—it implicitly estimates an MRP from the observed data and produces the Value Function of the implicitly-estimated MRP.
- Understanding TD versus MC versus DP from the perspectives of "bootstrapping" and "experiencing" (Figure 11.5 provides a great view).
- "Equivalence" of λ-Return Prediction and TD(λ) Prediction, hence TD is equivalent to TD(0) and MC is "equivalent" to TD(1).

CHAPTER 12

Monte-Carlo and Temporal-Difference for Control

In chapter 11, we covered MC and TD algorithms to solve the *Prediction* problem. In this chapter, we cover MC and TD algorithms to solve the *Control* problem. As a reminder, MC and TD algorithms are Reinforcement Learning algorithms that only have access to an individual experience (at a time) of next state and reward when the AI agent performs an action in a given state. The individual experience could be the result of an interaction with a real environment or could be served by a simulated environment (as explained at the state of Chapter 11). It also pays to remind that RL algorithms overcome the Curse of Dimensionality and the Curse of Modeling by learning an appropriate function approximation of the Value Function from a stream of individual experiences. Hence, large-scale Control problems that are typically seen in the real-world are often tackled by RL.

12.1   REFRESHER ON *GENERALIZED POLICY ITERATION* (GPI)

We shall soon see that all RL Control algorithms are based on the fundamental idea of *Generalized Policy Iteration* (introduced initially in Chapter 5), henceforth abbreviated as GPI. The exact ways in which the GPI idea is utilized in RL algorithms differs from one algorithm to another, and they differ significantly from how the GPI idea is utilized in DP algorithms. So before we get into RL Control algorithms, it's important to ground on the abstract concept of GPI. We now ask you to re-read Section 5.11 in Chapter 5.

To summarize, the key concept in GPI is that we can evaluate the Value Function for a policy with *any* Policy Evaluation method, and we can improve a policy with *any* Policy Improvement method (not necessarily the methods used in the classical Policy Iteration DP algorithm). The word *any* does not simply mean alternative algorithms for Policy Evaluation and/or Policy Improvements—the word *any* also refers to the fact that we can do a "partial" Policy Evaluation or a "partial" Policy Improvement. The word "partial" is used quite generically here—any set of calculations that simply take us *towards* a complete Policy Evaluation or *towards* a complete Policy Improvement qualify. This means GPI allows us to switch from Policy Evaluation to Policy Improvements without doing a complete Policy Evaluation or complete Policy Improvement (for instance, we don't have to take Policy Evaluation calculations all the way to convergence). Figure 12.1 illustrates Generalized Policy Iteration as the shorter-length arrows (versus the longer-length arrows seen in Figure

DOI: 10.1201/9781003229193-12

5.3 for the usual Policy Iteration algorithm). Note how these shorter-length arrows don't go all the way to either the "value function line" or the "policy line" but the shorter-length arrows do go some part of the way towards the line they are meant to go towards at that stage in the algorithm.

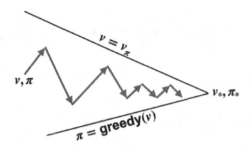

Figure 12.1 **Progression Lines of Value Function and Policy in Generalized Policy Iteration (Image Credit: Coursera Course on Fundamentals of RL)**

As has been our norm in the book so far, our approach to RL Control algorithms is to first cover the simple case of Tabular RL Control algorithms to illustrate the core concepts in a simple and intuitive manner. In many Tabular RL Control algorithms (especially Tabular TD Control), GPI consists of the Policy Evaluation step for just a single state (versus for all states in usual Policy Iteration) and the Policy Improvement step is also done for just a single state. So essentially these RL Control algorithms are an alternating sequence of single-state policy evaluation and single-state policy improvement (where the single-state is the state produced by sampling or the state that is encountered in a real-world environment interaction). Similar to the case of Prediction, we first cover Monte-Carlo (MC) Control and then move on to Temporal-Difference (TD) Control.

12.2   GPI WITH EVALUATION AS MONTE-CARLO

Let us think about how to do MC Control based on the GPI idea. The natural idea that emerges is to do Policy Evaluation with MC (this is basically MC Prediction), followed by greedy Policy Improvement, then MC Policy Evaluation with the improved policy, and so on … . This seems like a reasonable idea, but there is a problem with doing greedy Policy Improvement. The problem is that the Greedy Policy Improvement calculation (Equation 12.1) requires a model of the state transition probability function $\mathcal{P}$ and the reward function $\mathcal{R}$, which is not available in an RL interface.

$$\pi'_D(s) = \arg\max_{a \in \mathcal{A}}\{\mathcal{R}(s,a) + \gamma \sum_{s' \in \mathcal{N}} \mathcal{P}(s,a,s') \cdot V^\pi(s')\} \text{ for all } s \in \mathcal{N} \qquad (12.1)$$

However, we note that Equation 12.1 can be written more succinctly as:

$$\pi'_D(s) = \arg\max_{a \in \mathcal{A}} Q^\pi(s,a) \text{ for all } s \in \mathcal{N} \qquad (12.2)$$

This view of Greedy Policy Improvement is valuable because instead of doing Policy Evaluation for calculating V^π (MC Prediction), we can instead do Policy Evaluation to calculate Q^π (with MC Prediction for the Q-Value Function). With this modification to Policy Evaluation, we can keep alternating between Policy Evaluation and Policy Improvement until convergence to obtain the Optimal Value Function and Optimal Policy. Indeed, this is a valid MC Control algorithm. However, this algorithm is not practical as each Policy Evaluation (MC Prediction) typically takes very long to converge (as we have noted in Chapter 11) and the number of iterations of Evaluation and Improvement until GPI convergence will also be large. More importantly, this algorithm simply modifies the Policy Iteration DP/ADP algorithm by replacing DP/ADP Policy Evaluation with MC Q-Value Policy Evaluation—hence, we simply end up with a slower version of the Policy Iteration DP/ADP algorithm. Instead, we seek an MC Control Algorithm that switches from Policy Evaluation to Policy Improvement without requiring Policy Evaluation to converge (this is essentially the GPI idea).

So the natural GPI idea here would be to do the usual MC Prediction updates (of the Q-Value estimate) at the end of an episode, then improve the policy at the end of that episode, then perform MC Prediction updates (with the improved policy) at the end of the next episode, and so on ... . Let's see what this algorithm looks like. Equation 12.2 tells us that all we need to perform the requisite greedy action (from the improved policy) at any time step in any episode is an estimate of the Q-Value Function. For ease of understanding, for now, let us just restrict ourselves to the case of Tabular Every-Visit MC Control with equal weights for each of the *Return* data points obtained for any (state, action) pair. In this case, we can simply perform the following two updates at the end of each episode for each (S_t, A_t) pair encountered in the episode (note that at each time step t, A_t is based on the greedy policy derived from the current estimate of the Q-Value function):

$$Count(S_t, A_t) \leftarrow Count(S_t, A_t) + 1$$
$$Q(S_t, A_t) \leftarrow Q(S_t, A_t) + \frac{1}{Count(S_t, A_t)} \cdot (G_t - Q(S_t, A_t)) \tag{12.3}$$

It's important to note that $Count(S_t, A_t)$ is accumulated over the set of all episodes seen thus far. Note that the estimate $Q(S_t, A_t)$ is not an estimate of the Q-Value Function for a single policy—rather, it keeps updating as we encounter new greedy policies across the set of episodes.

So is this now our first Tabular RL Control algorithm? Not quite—there is yet another problem. This problem is more subtle and we illustrate the problem with a simple example. Let's consider a specific state (call it s) and assume that there are only two allowable actions a_1 and a_2 for state s. Let's say the true Q-Value Function for state s is: $Q_{true}(s, a_1) = 2, Q_{true}(s, a_2) = 5$. Let's say we initialize the Q-Value Function estimate as: $Q(s, a_1) = Q(s, a_2) = 0$. When we encounter state s for the first time, the action to be taken is arbitrary between a_1 and a_2 since they both have the same Q-Value estimate (meaning both a_1 and a_2 yield the same max value for $Q(s, a)$ among the two choices for a). Let's say we arbitrarily pick a_1 as the action choice and let's say for this first encounter of state s (with the arbitrarily picked action a_1), the return obtained is 3. So $Q(s, a_1)$ updates to the value 3. So when the state s is encountered for the second time, we see that $Q(s, a_1) = 3$ and $Q(s, a_2) = 0$ and so, action a_1 will be taken according to the greedy policy implied by the estimate of the Q-Value Function. Let's say we now obtain a return of -1, updating $Q(s, a_1)$ to $\frac{3-1}{2} = 1$. When s is encountered for the third time, yet again action a_1 will be taken according to the greedy policy implied by the estimate of the Q-Value Function. Let's say we now obtain a return of 2, updating $Q(s, a_1)$ to $\frac{3-1+2}{3} = \frac{4}{3}$. We see that as long as the returns associated with a_1 are not negative enough to make the estimate $Q(s, a_1)$ negative, a_2 is "locked out" by a_1 because the first few occurrences of a_1 happen to yield an average return greater than the initialization of $Q(s, a_2)$. Even if a_2 was chosen, it is possible that the first few occurrences of a_2 yield an average return smaller than the average return obtained on the first few occurrences of a_1, in which case a_2 could still get locked-out prematurely.

This problem goes beyond MC Control and applies to the broader problem of RL Control—updates can get biased by initial random occurrences of returns (or return estimates), which in turn could prevent certain actions from being sufficiently chosen (thus, disallowing accurate estimates of the Q-Values for those actions). While we do want to *exploit* actions that seem to be fetching higher returns, we also want to adequately *explore* all possible actions so we can obtain an accurate-enough estimate of their Q-Values. This is essentially the Explore-Exploit dilemma of the famous Multi-Armed Bandit Problem. In Chapter 15, we will cover the Multi-Armed Bandit problem in detail, along with a variety of techniques to solve the Multi-Armed Bandit problem (which are essentially creative ways of resolving the Explore-Exploit dilemma). We will see in Chapter 15 that a simple way of resolving the Explore-Exploit dilemma is with a method known as ϵ-greedy, which essentially means we must be greedy ("exploit") a certain $(1 - \epsilon)$ fraction of the time and for the remaining (ϵ) fraction of the time, we explore all possible actions. The term "certain fraction of the time" refers to probabilities of choosing actions, which means an ϵ-greedy policy (generated from a Q-Value Function estimate) will be a stochastic policy. For the sake of simplicity, in this book, we will employ the ϵ-greedy method to resolve the Explore-Exploit dilemma in all RL Control algorithms involving the Explore-Exploit dilemma (although you must understand that we can replace the ϵ-greedy method by the other methods we shall cover in Chapter 15 in any of the RL Control algorithms where we run into the Explore-Exploit dilemma). So we need to tweak the Tabular MC Control algorithm described above to perform Policy Improvement with the ϵ-greedy method. The formal definition of the ϵ-greedy stochastic policy π' (obtained from the current estimate of the Q-Value Function) for a Finite MDP (since we are focused on Tabular RL Control) is as follows:

$$\text{Improved Stochastic Policy } \pi'(s, a) = \begin{cases} \frac{\epsilon}{|\mathcal{A}|} + 1 - \epsilon & \text{if } a = \arg\max_{b \in \mathcal{A}} Q(s, b) \\ \frac{\epsilon}{|\mathcal{A}|} & \text{otherwise} \end{cases}$$

where $\mathcal{A}$ denotes the set of allowable actions and $\epsilon \in [0, 1]$ is the specification of the degree of exploration.

This says that with probability $1 - \epsilon$, we select the action that maximizes the Q-Value Function estimate for a given state, and with probability ϵ, we uniform-randomly select each of the allowable actions (including the maximizing action). Hence, the maximizing action is chosen with probability $\frac{\epsilon}{|\mathcal{A}|} + 1 - \epsilon$. Note that if ϵ is zero, π' reduces to the deterministic greedy policy π'_D that we had defined earlier. So the greedy policy can be considered to be a special case of ϵ-greedy policy with $\epsilon = 0$.

But we haven't yet actually proved that an ϵ-greedy policy is indeed an improved policy. We do this in the theorem below. Note that in the following theorem's proof, we re-use the notation and inductive-proof approach used in the Policy Improvement Theorem (Theorem 5.6.1) in Chapter 5. So it would be a good idea to re-read the proof of Theorem 5.6.1 in Chapter 5 before reading the following theorem's proof.

Theorem 12.2.1. *For a Finite MDP, if π is a policy such that for all $s \in \mathcal{N}, \pi(s, a) \geq \frac{\epsilon}{|\mathcal{A}|}$ for all $a \in \mathcal{A}$, then the ϵ-greedy policy π' obtained from Q^π is an improvement over π, i.e., $V^{\pi'}(s) \geq V^\pi(s)$ for all $s \in \mathcal{N}$.*

Proof. We've previously learnt that for any policy π', if we apply the Bellman Policy Operator $B^{\pi'}$ repeatedly (starting with V^π), we converge to $V^{\pi'}$. In other words,

$$\lim_{i \to \infty} (B^{\pi'})^i (V^\pi) = V^{\pi'}$$

So the proof is complete if we prove that:

$$(\boldsymbol{B}^{\pi'})^{i+1}(\boldsymbol{V}^{\pi}) \geq (\boldsymbol{B}^{\pi'})^{i}(\boldsymbol{V}^{\pi}) \text{ for all } i = 0, 1, 2, \ldots$$

In plain English, this says we need to prove that repeated application of $\boldsymbol{B}^{\pi'}$ produces a non-decreasing sequence of Value Functions $[(\boldsymbol{B}^{\pi'})^{i}(\boldsymbol{V}^{\pi})|i = 0, 1, 2, \ldots]$.

We prove this by induction. The base case of the proof by induction is to show that $\boldsymbol{B}^{\pi'}(\boldsymbol{V}^{\pi}) \geq \boldsymbol{V}^{\pi}$

$$\begin{aligned}
\boldsymbol{B}^{\pi'}(\boldsymbol{V}^{\pi})(s) &= (\boldsymbol{\mathcal{R}}^{\pi'} + \gamma \cdot \boldsymbol{\mathcal{P}}^{\pi'} \cdot \boldsymbol{V}^{\pi})(s) \\
&= \boldsymbol{\mathcal{R}}^{\pi'}(s) + \gamma \cdot \sum_{s' \in \mathcal{N}} \boldsymbol{\mathcal{P}}^{\pi'}(s, s') \cdot \boldsymbol{V}^{\pi}(s') \\
&= \sum_{a \in \mathcal{A}} \pi'(s, a) \cdot (\mathcal{R}(s, a) + \gamma \cdot \sum_{s' \in \mathcal{N}} \mathcal{P}(s, a, s') \cdot \boldsymbol{V}^{\pi}(s')) \\
&= \sum_{a \in \mathcal{A}} \pi'(s, a) \cdot Q^{\pi}(s, a) \\
&= \sum_{a \in \mathcal{A}} \frac{\epsilon}{|\mathcal{A}|} \cdot Q^{\pi}(s, a) + (1 - \epsilon) \cdot \max_{a \in \mathcal{A}} Q^{\pi}(s, a) \\
&\geq \sum_{a \in \mathcal{A}} \frac{\epsilon}{|\mathcal{A}|} \cdot Q^{\pi}(s, a) + (1 - \epsilon) \cdot \sum_{a \in \mathcal{A}} \frac{\pi(s, a) - \frac{\epsilon}{|\mathcal{A}|}}{1 - \epsilon} \cdot Q^{\pi}(s, a) \\
&= \sum_{a \in \mathcal{A}} \pi(s, a) \cdot Q^{\pi}(s, a) \\
&= \boldsymbol{V}^{\pi}(s) \text{ for all } s \in \mathcal{N}
\end{aligned}$$

The line with the inequality above is due to the fact that for any fixed $s \in \mathcal{N}$, $\max_{a \in \mathcal{A}} Q^{\pi}(s, a) \geq \sum_{a \in \mathcal{A}} w_a \cdot Q^{\pi}(s, a)$ (maximum Q-Value greater than or equal to a weighted average of all Q-Values, for a given state) with the weights $w_a = \frac{\pi(s,a) - \frac{\epsilon}{|\mathcal{A}|}}{1 - \epsilon}$ such that $\sum_{a \in \mathcal{A}} w_a = 1$ and $0 \leq w_a \leq 1$ for all $a \in \mathcal{A}$.

This completes the base case of the proof by induction.

The induction step is easy and is proved as a consequence of the monotonicity property of the $\boldsymbol{B}^{\pi}$ operator (for any π), which is defined as follows:

$$\text{Monotonicity Property of } \boldsymbol{B}^{\pi} : \boldsymbol{X} \geq \boldsymbol{Y} \Rightarrow \boldsymbol{B}^{\pi}(\boldsymbol{X}) \geq \boldsymbol{B}^{\pi}(\boldsymbol{Y})$$

Note that we proved the monotonicity property of the $\boldsymbol{B}^{\pi}$ operator in Chapter 5. A straightforward application of this monotonicity property provides the induction step of the proof:

$$(\boldsymbol{B}^{\pi'})^{i+1}(\boldsymbol{V}^{\pi}) \geq (\boldsymbol{B}^{\pi'})^{i}(\boldsymbol{V}^{\pi}) \Rightarrow (\boldsymbol{B}^{\pi'})^{i+2}(\boldsymbol{V}^{\pi}) \geq (\boldsymbol{B}^{\pi'})^{i+1}(\boldsymbol{V}^{\pi}) \text{ for all } i = 0, 1, 2, \ldots$$

This completes the proof. □

We note that for any ϵ-greedy policy π, we do ensure the condition that for all $s \in \mathcal{N}$, $\pi(s, a) \geq \frac{\epsilon}{|\mathcal{A}|}$ for all $a \in \mathcal{A}$. So we just need to ensure that this condition holds true for the initial choice of π (in the GPI with MC algorithm). An easy way to ensure this is to choose the initial π to be a uniform choice over actions (for each state), i.e., for all $s \in \mathcal{N}$, $\pi(s, a) = \frac{1}{|\mathcal{A}|}$ for all $a \in \mathcal{A}$.

12.3 GLIE MONTE-CONTROL CONTROL

So to summarize, we've resolved two problems—firstly, we replaced the state-value function estimate with the action-value function estimate and secondly, we replaced greedy policy improvement with ϵ-greedy policy improvement. So our MC Control algorithm does GPI as follows:

- Do Policy Evaluation with the Q-Value Function with Q-Value updates at the end of each episode.
- Do Policy Improvement with an ϵ-greedy Policy (readily obtained from the Q-Value Function estimate at any time step for any episode).

So now we are ready to develop the details of the Monte-Control algorithm that we've been seeking. For ease of understanding, we first cover the Tabular version and then we will implement the generalized version with function approximation. Note that an ϵ-greedy policy enables adequate exploration of actions, but we will also need to do adequate exploration of states in order to achieve a suitable estimate of the Q-Value Function. Moreover, as our Control algorithm proceeds and the Q-Value Function estimate gets better and better, we reduce the amount of exploration and eventually (as the number of episodes tend to infinity), we want to have ϵ (degree of exploration) tend to zero. In fact, this behavior has a catchy acronym associated with it, which we define below:

Definition 12.3.1. We refer to *Greedy In The Limit with Infinite Exploration* (abbreviated as GLIE) as the behavior that has the following two properties:

1. All state-action pairs are explored infinitely many times, i.e., for all $s \in \mathcal{N}$, for all $a \in \mathcal{A}$, and $Count_k(s, a)$ denoting the number of occurrences of (s, a) pairs after k episodes:

$$\lim_{k \to \infty} Count_k(s, a) = \infty$$

2. The policy converges to a greedy policy, i.e., for all $s \in \mathcal{N}$, for all $a \in \mathcal{A}$, and $\pi_k(s, a)$ denoting the ϵ-greedy policy obtained from the Q-Value Function estimate after k episodes:

$$\lim_{k \to \infty} \pi_k(s, a) = \mathbb{I}_{a = \arg\max_{b \in \mathcal{A}} Q(s, b)}$$

A simple way by which our method of using the ϵ-greedy policy (for policy improvement) can be made GLIE is by reducing ϵ as a function of number of episodes k as follows:

$$\epsilon_k = \frac{1}{k}$$

So now we are ready to describe the Tabular MC Control algorithm we've been seeking. We ensure that this algorithm has GLIE behavior and so, we refer to it as *GLIE Tabular Monte-Carlo Control*. The following is the outline of the procedure for each episode (terminating trace experience) in the algorithm:

- Generate the trace experience (episode) with actions sampled from the ϵ-greedy policy π obtained from the estimate of the Q-Value Function that is available at the start of the trace experience. Also, sample the first state of the trace experience from a uniform distribution of states in $\mathcal{N}$. This ensures infinite exploration of both states and actions. Let's denote the contents of this trace experience as:

$$S_0, A_0, R_1, S_1, A_1, \ldots, R_T, S_T$$

and define the trace experience return G_t associated with (S_t, A_t) as:

$$G_t = \sum_{i=t+1}^{T} \gamma^{i-t-1} \cdot R_i = R_{t+1} + \gamma \cdot R_{t+2} + \gamma^2 \cdot R_{t+3} + \ldots \gamma^{T-t-1} \cdot R_T$$

- For each state S_t and action A_t in the trace experience, perform the following updates at the end of the trace experience:

$$Count(S_t, A_t) \leftarrow Count(S_t, A_t) + 1$$

$$Q(S_t, A_t) \leftarrow Q(S_t, A_t) + \frac{1}{Count(S_t, A_t)} \cdot (G_t - Q(S_t, A_t))$$

- Let's say this trace experience is the k-th trace experience in the sequence of trace experiences. Then, at the end of the trace experience, set:

$$\epsilon \leftarrow \frac{1}{k}$$

We state the following important theorem without proof.

Theorem 12.3.1. *The above-described GLIE Tabular Monte-Carlo Control algorithm converges to the Optimal Action-Value function: $Q(s, a) \rightarrow Q^*(s, a)$ for all $s \in \mathcal{N}$, for all $a \in \mathcal{A}$. Hence, GLIE Tabular Monte-Carlo Control converges to an Optimal (Deterministic) Policy π^*.*

The extension from Tabular to Function Approximation of the Q-Value Function is straightforward. The update (change) in the parameters w of the Q-Value Function Approximation $Q(s, a; w)$ is as follows:

$$\Delta w = \alpha \cdot (G_t - Q(S_t, A_t; w)) \cdot \nabla_w Q(S_t, A_t; w) \qquad (12.4)$$

where α is the learning rate in the stochastic gradient descent and G_t is the trace experience return from state S_t upon taking action A_t at time t on a trace experience.

Now let us write some code to implement the above description of GLIE Monte-Carlo Control, generalized to handle Function Approximation of the Q-Value Function. As you shall see in the code below, there are a couple of other generalizations from the algorithm outline described above. Let us start by understanding the various arguments to the below function `glie_mc_control`.

- `mdp: MarkovDecisionProcess[S, A]`—This represents the interface to an abstract Markov Decision Process. Note that this interface doesn't provide any access to the transition probabilities or reward function. The core functionality available through this interface are the two @abstractmethods `step` and `actions`. The `step` method only allows us to access an individual experience of the next state and reward pair given the current state and action (since it returns an abstract `Distribution` object). The `actions` method gives us the allowable actions for a given state.
- `states: NTStateDistribution[S]`—This represents an arbitrary distribution of the non-terminal states, which in turn allows us to sample the starting state (from this distribution) for each trace experience.
- `approx_0: QValueFunctionApprox[S, A]`—This represents the initial function approximation of the Q-Value function (that is meant to be updated, in an immutable manner, through the course of the algorithm).
- `gamma: float`—This represents the discount factor to be used in estimating the Q-Value Function.

- epsilon_as_func_of_episodes: Callable[[int], float]—This represents the extent of exploration (ϵ) as a function of the number of trace experiences done so far (allowing us to generalize from our default choice of $\epsilon(k) = \frac{1}{k}$).
- episode_length_tolerance: float—This represents the *tolerance* that determines the trace experience length T (the minimum T such that $\gamma^T < tolerance$).

glie_mc_control produces a generator (Iterator) of Q-Value Function estimates at the end of each trace experience. The code is fairly self-explanatory. The method simulate_actions of mdp: MarkovDecisionProcess creates a single sampling trace (i.e., a trace experience). At the end of each trace experience, the update method of FunctionApprox updates the Q-Value Function (creates a new Q-Value Function without mutating the currrent Q-Value Function) using each of the returns (and associated state-actions pairs) from the trace experience. The ϵ-greedy policy is derived from the Q-Value Function estimate by using the function epsilon_greedy_policy that is shown below and is quite self-explanatory.

```
from rl.markov_decision_process import epsilon_greedy_policy, TransitionStep
from rl.approximate_dynamic_programming import QValueFunctionApprox
from rl.approximate_dynamic_programming import NTStateDistribution
def glie_mc_control(
    mdp: MarkovDecisionProcess[S, A],
    states: NTStateDistribution[S],
    approx_0: QValueFunctionApprox[S, A],
    gamma: float,
    epsilon_as_func_of_episodes: Callable[[int], float],
    episode_length_tolerance: float = 1e-6
) -> Iterator[QValueFunctionApprox[S, A]]:
    q: QValueFunctionApprox[S, A] = approx_0
    p: Policy[S, A] = epsilon_greedy_policy(q, mdp, 1.0)
    yield q

    num_episodes: int = 0
    while True:
        trace: Iterable[TransitionStep[S, A]] = \
            mdp.simulate_actions(states, p)
        num_episodes += 1
        for step in returns(trace, gamma, episode_length_tolerance):
            q = q.update([((step.state, step.action), step.return_)])
        p = epsilon_greedy_policy(
            q,
            mdp,
            epsilon_as_func_of_episodes(num_episodes)
        )
        yield q
```

The implementation of epsilon_greedy_policy is as follows:

```
from rl.policy import DeterministicPolicy, Policy, RandomPolicy
def greedy_policy_from_qvf(
    q: QValueFunctionApprox[S, A],
    actions: Callable[[NonTerminal[S]], Iterable[A]]
) -> DeterministicPolicy[S, A]:
    def optimal_action(s: S) -> A:
        _, a = q.argmax((NonTerminal(s), a) for a in actions(NonTerminal(s)))
        return a
    return DeterministicPolicy(optimal_action)

def epsilon_greedy_policy(
    q: QValueFunctionApprox[S, A],
    mdp: MarkovDecisionProcess[S, A],
    epsilon: float = 0.0
) -> Policy[S, A]:
    def explore(s: S, mdp=mdp) -> Iterable[A]:
```

```
        return mdp.actions(NonTerminal(s))
    return RandomPolicy(Categorical(
        {UniformPolicy(explore): epsilon,
         greedy_policy_from_qvf(q, mdp.actions): 1 - epsilon}
    ))
```

The above code is in the file rl/monte_carlo.py.

Note that `epsilon_greedy_policy` returns an instance of the class `RandomPolicy`. `RandomPolicy` creates a policy that randomly selects one of several specified policies (in this case, we need to select between the greedy policy of type `DeterministicPolicy` and the `UniformPolicy`). The implementation of `RandomPolicy` is shown below and you can find its code in the file rl/policy.py.

```
@dataclass(frozen=True)
class RandomPolicy(Policy[S, A]):
    policy_choices: Distribution[Policy[S, A]]

    def act(self, state: NonTerminal[S]) -> Distribution[A]:
        policy: Policy[S, A] = self.policy_choices.sample()
        return policy.act(state)
```

Now let us test `glie_mc_control` on the simple inventory MDP we wrote in Chapter 4.

```
from rl.chapter3.simple_inventory_mdp_cap import SimpleInventoryMDPCap
```

```
capacity: int = 2
poisson_lambda: float = 1.0
holding_cost: float = 1.0
stockout_cost: float = 10.0
gamma: float = 0.9

si_mdp: SimpleInventoryMDPCap = SimpleInventoryMDPCap(
    capacity=capacity,
    poisson_lambda=poisson_lambda,
    holding_cost=holding_cost,
    stockout_cost=stockout_cost
)
```

First, let's run Value Iteration so we can determine the true Optimal Value Function and Optimal Policy

```
from rl.dynamic_programming import value_iteration_result
true_opt_vf, true_opt_policy = value_iteration_result(fmdp, gamma=gamma)
print("True Optimal Value Function")
pprint(true_opt_vf)
print("True Optimal Policy")
print(true_opt_policy)
```

This prints:

```
True Optimal Value Function
{NonTerminal(state=InventoryState(on_hand=0, on_order=0)): -34.894855194671294,
 NonTerminal(state=InventoryState(on_hand=0, on_order=1)): -27.66095964467877,
 NonTerminal(state=InventoryState(on_hand=0, on_order=2)): -27.99189950444479,
 NonTerminal(state=InventoryState(on_hand=1, on_order=0)): -28.66095964467877,
 NonTerminal(state=InventoryState(on_hand=1, on_order=1)): -28.99189950444479,
 NonTerminal(state=InventoryState(on_hand=2, on_order=0)): -29.991899504444792}
True Optimal Policy
For State InventoryState(on_hand=0, on_order=0): Do Action 1
For State InventoryState(on_hand=0, on_order=1): Do Action 1
For State InventoryState(on_hand=0, on_order=2): Do Action 0
For State InventoryState(on_hand=1, on_order=0): Do Action 1
For State InventoryState(on_hand=1, on_order=1): Do Action 0
For State InventoryState(on_hand=2, on_order=0): Do Action 0
```

Now let's run GLIE MC Control with the following parameters:

```
from rl.function_approx import Tabular
from rl.distribution import Choose
from rl.chapter3.simple_inventory_mdp_cap import InventoryState

episode_length_tolerance: float = 1e-5
epsilon_as_func_of_episodes: Callable[[int], float] = lambda k: k ** -0.5
initial_learning_rate: float = 0.1
half_life: float = 10000.0
exponent: float = 1.0

initial_qvf_dict: Mapping[Tuple[NonTerminal[InventoryState], int], float] = {
    (s, a): 0. for s in si_mdp.non_terminal_states for a in si_mdp.actions(s)
}
learning_rate_func: Callable[[int], float] = learning_rate_schedule(
    initial_learning_rate=initial_learning_rate,
    half_life=half_life,
    exponent=exponent
)
qvfs: Iterator[QValueFunctionApprox[InventoryState, int] = glie_mc_control(
    mdp=si_mdp,
    states=Choose(si_mdp.non_terminal_states),
    approx_0=Tabular(
        values_map=initial_qvf_dict,
        count_to_weight_func=learning_rate_func
    ),
    gamma=gamma,
    epsilon_as_func_of_episodes=epsilon_as_func_of_episodes,
    episode_length_tolerance=episode_length_tolerance
)
```

Now let's fetch the final estimate of the Optimal Q-Value Function after `num_episodes` have run, and extract from it the estimate of the Optimal State-Value Function and the Optimal Policy.

```
from rl.distribution import Constant
from rl.dynamic_programming import V
import itertools
import rl.iterate as iterate

num_episodes = 10000
final_qvf: QValueFunctionApprox[InventoryState, int] = \
    iterate.last(itertools.islice(qvfs, num_episodes))

def get_vf_and_policy_from_qvf(
    mdp: FiniteMarkovDecisionProcess[S, A],
    qvf: QValueFunctionApprox[S, A]
) -> Tuple[V[S], FiniteDeterministicPolicy[S, A]]:
    opt_vf: V[S] = {
        s: max(qvf((s, a)) for a in mdp.actions(s))
        for s in mdp.non_terminal_states
    }
    opt_policy: FiniteDeterministicPolicy[S, A] = \
        FiniteDeterministicPolicy({
            s.state: qvf.argmax((s, a) for a in mdp.actions(s))[1]
            for s in mdp.non_terminal_states
        })
    return opt_vf, opt_policy

opt_vf, opt_policy = get_vf_and_policy_from_qvf(
    mdp=si_mdp,
    qvf=final_qvf
)
print(f"GLIE MC Optimal Value Function with {num_episodes:d} episodes")
pprint(opt_vf)
print(f"GLIE MC Optimal Policy with {num_episodes:d} episodes")
print(opt_policy)
```

This prints:

```
GLIE MC Optimal Value Function with 10000 episodes
{NonTerminal(state=InventoryState(on_hand=0, on_order=0)): -34.76212336633032,
 NonTerminal(state=InventoryState(on_hand=0, on_order=1)): -27.90668364332291,
 NonTerminal(state=InventoryState(on_hand=0, on_order=2)): -28.306190508518398,
 NonTerminal(state=InventoryState(on_hand=1, on_order=0)): -28.548284937363526,
 NonTerminal(state=InventoryState(on_hand=1, on_order=1)): -28.864409885059185,
 NonTerminal(state=InventoryState(on_hand=2, on_order=0)): -30.23156422557605}
GLIE MC Optimal Policy with 10000 episodes
For State InventoryState(on_hand=0, on_order=0): Do Action 1
For State InventoryState(on_hand=0, on_order=1): Do Action 1
For State InventoryState(on_hand=0, on_order=2): Do Action 0
For State InventoryState(on_hand=1, on_order=0): Do Action 1
For State InventoryState(on_hand=1, on_order=1): Do Action 0
For State InventoryState(on_hand=2, on_order=0): Do Action 0
```

We see that this reasonably converges to the true Value Function (and reaches the true Optimal Policy) as produced by Value Iteration.

The code above is in the file rl/chapter11/simple_inventory_mdp_cap.py. Also see the helper functions in rl/chapter11/control_utils.py which you can use to run your own experiments and tests for RL Control algorithms.

12.4   SARSA

Just like in the case of RL Prediction, the natural idea is to replace MC Control with TD Control using the TD Target $R_{t+1} + \gamma \cdot Q(S_{t+1}, A_{t+1}; w)$ as a biased estimate of G_t when updating $Q(S_t, A_t; w)$. This means the parameters update in Equation (12.4) gets modified to the following parameters update:

$$\Delta w = \alpha \cdot (R_{t+1} + \gamma \cdot Q(S_{t+1}, A_{t+1}; w) - Q(S_t, A_t; w)) \cdot \nabla_w Q(S_t, A_t; w) \qquad (12.5)$$

Unlike MC Control where updates are made at the end of each trace experience (i.e., episode), a TD control algorithm can update at the end of each atomic experience. This means the Q-Value Function Approximation is updated after each atomic experience (*continuous learning*), which in turn means that the ϵ-greedy policy will be (automatically) updated at the end of each atomic experience. At each time step t in a trace experience, the current ϵ-greedy policy is used to sample A_t from S_t and is also used to sample A_{t+1} from S_{t+1}. Note that in MC Control, the same ϵ-greedy policy is used to sample all the actions from their corresponding states in the trace experience, and so in MC Control, we were able to generate the entire trace experience with the currently available ϵ-greedy policy. However, here in TD Control, we need to generate a trace experience incrementally since the action to be taken from a state depends on the just-updated ϵ-greedy policy (that is derived from the just-updated Q-Value Function).

Just like in the case of RL Prediction, the disadvantage of the TD Target being a biased estimate of the return is compensated by a reduction in the variance of the return estimate. Also, TD Control offers a better speed of convergence (as we shall soon illustrate). Most importantly, TD Control offers the ability to use in situations where we have incomplete trace experiences (happens often in real-world situations where experiments gets curtailed/disrupted) and also, we can use it in situations where we never reach a terminal state (*continuing trace*).

Note that Equation (12.5) has the entities

- **State** S_t

- **Action** A_t

- **Reward** R_t

- **State** S_{t+1}

- **Action** A_{t+1}

which prompted this TD Control algorithm to be named SARSA (for **State-Action-Reward-State-Action**). Following our convention from Chapter 4, we depict the SARSA algorithm in Figure 12.2 with states as elliptical-shaped nodes, actions as rectangular-shaped nodes, and the edges as samples from transition probability distribution and ϵ-greedy policy distribution.

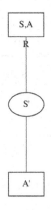

Figure 12.2 **Visualization of SARSA Algorithm**

Now let us write some code to implement the above-described SARSA algorithm. Let us start by understanding the various arguments to the below function `glie_sarsa`.

- `mdp: MarkovDecisionProcess[S, A]`—This represents the interface to an abstract Markov Decision Process. We want to remind that this interface doesn't provide any access to the transition probabilities or reward function. The core functionality available through this interface are the two `@abstractmethods` `step` and `actions`. The `step` method only allows us to access a sample of the next state and reward pair given the current state and action (since it returns an abstract `Distribution` object). The `actions` method gives us the allowable actions for a given state.
- `states: NTStateDistribution[S]`—This represents an arbitrary distribution of the non-terminal states, which in turn allows us to sample the starting state (from this distribution) for each trace experience.
- `approx_0: QValueFunctionApprox[S, A]`—This represents the initial function approximation of the Q-Value function (that is meant to be updated, after each atomic experience, in an immutable manner, through the course of the algorithm).
- `gamma: float`—This represents the discount factor to be used in estimating the Q-Value Function.
- `epsilon_as_func_of_episodes: Callable[[int], float]`—This represents the extent of exploration (ϵ) as a function of the number of episodes.

- `max_episode_length: int`—This represents the number of time steps at which we would curtail a trace experience and start a new one. As we've explained, TD Control doesn't require complete trace experiences, and so we can do as little or as large a number of time steps in a trace experience (`max_episode_length` gives us that control).

`glie_sarsa` produces a generator (`Iterator`) of Q-Value Function estimates at the end of each atomic experience. The `while True` loops over trace experiences. The inner `while` loops over time steps—each of these steps involves the following:

- Given the current `state` and `action`, we obtain a sample of the pair of `next_state` and `reward` (using the `sample` method of the `Distribution` obtained from `mdp.step(state, action)`.
- Obtain the `next_action` from `next_state` using the function `epsilon_greedy_action` which utilizes the ϵ-greedy policy derived from the current Q-Value Function estimate (referenced by q).
- Update the Q-Value Function based on Equation (12.5) (using the `update` method of q: `QValueFunctionApprox[S, A]`). Note that this is an immutable update since we produce an `Iterable` (generator) of the Q-Value Function estimate after each time step.

Before the code for `glie_sarsa`, let's understand the code for `epsilon_greedy_action` which returns an action sampled from the ϵ-greedy policy probability distribution that is derived from the Q-Value Function estimate, given as input a non-terminal state, a Q-Value Function estimate, the set of allowable actions, and ϵ.

```
from operator import itemgetter
from Distribution import Categorical
def epsilon_greedy_action(
    q: QValueFunctionApprox[S, A],
    nt_state: NonTerminal[S],
    actions: Set[A],
    epsilon: float
) -> A:
    greedy_action: A = max(
        ((a, q((nt_state, a))) for a in actions),
        key=itemgetter(1)
    )[0]
    return Categorical(
        {a: epsilon / len(actions) +
         (1 - epsilon if a == greedy_action else 0.) for a in actions}
    ).sample()
def glie_sarsa(
    mdp: MarkovDecisionProcess[S, A],
    states: NTStateDistribution[S],
    approx_0: QValueFunctionApprox[S, A],
    gamma: float,
    epsilon_as_func_of_episodes: Callable[[int], float],
    max_episode_length: int
) -> Iterator[QValueFunctionApprox[S, A]]:
    q: QValueFunctionApprox[S, A] = approx_0
    yield q
    num_episodes: int = 0
    while True:
        num_episodes += 1
        epsilon: float = epsilon_as_func_of_episodes(num_episodes)
        state: NonTerminal[S] = states.sample()
        action: A = epsilon_greedy_action(
            q=q,
            nt_state=state,
            actions=set(mdp.actions(state)),
```

```
        epsilon=epsilon
    )
    steps: int = 0
    while isinstance(state, NonTerminal) and steps < max_episode_length:
        next_state, reward = mdp.step(state, action).sample()
        if isinstance(next_state, NonTerminal):
            next_action: A = epsilon_greedy_action(
                q=q,
                nt_state=next_state,
                actions=set(mdp.actions(next_state)),
                epsilon=epsilon
            )
            q = q.update([(
                (state, action),
                reward + gamma * q((next_state, next_action))
            )])
            action = next_action
        else:
            q = q.update([((state, action), reward)])
        yield q
        steps += 1
        state = next_state
```

The above code is in the file rl/td.py.

Let us test this on the simple inventory MDP we tested GLIE MC Control on (we use the same si_mdp: SimpleInventoryMDPCap object and the same parameter values that were set up earlier when testing GLIE MC Control).

```
from rl.chapter3.simple_inventory_mdp_cap import InventoryState

max_episode_length: int = 100
epsilon_as_func_of_episodes: Callable[[int], float] = lambda k: k ** -0.5
initial_learning_rate: float = 0.1
half_life: float = 10000.0
exponent: float = 1.0
gamma: float = 0.9

initial_qvf_dict: Mapping[Tuple[NonTerminal[InventoryState], int], float] = {
    (s, a): 0. for s in si_mdp.non_terminal_states for a in si_mdp.actions(s)
}
learning_rate_func: Callable[[int], float] = learning_rate_schedule(
    initial_learning_rate=initial_learning_rate,
    half_life=half_life,
    exponent=exponent
)
qvfs: Iterator[QValueFunctionApprox[InventoryState, int]] = glie_sarsa(
    mdp=si_mdp,
    states=Choose(si_mdp.non_terminal_states),
    approx_0=Tabular(
        values_map=initial_qvf_dict,
        count_to_weight_func=learning_rate_func
    ),
    gamma=gamma,
    epsilon_as_func_of_episodes=epsilon_as_func_of_episodes,
    max_episode_length=max_episode_length
)
```

Now let's fetch the final estimate of the Optimal Q-Value Function after num_episodes * max_episode_length updates of the Q-Value Function, and extract from it the estimate of the Optimal State-Value Function and the Optimal Policy (using the function get_vf_and_policy_from_qvf that we had written earlier).

```
import itertools
import rl.iterate as iterate

num_updates = num_episodes * max_episode_length
```

```
final_qvf: QValueFunctionApprox[InventoryState, int] = \
    iterate.last(itertools.islice(qvfs, num_updates))
opt_vf, opt_policy = get_vf_and_policy_from_qvf(
    mdp=si_mdp,
    qvf=final_qvf
)
print(f"GLIE SARSA Optimal Value Function with {num_updates:d} updates")
pprint(opt_vf)
print(f"GLIE SARSA Optimal Policy with {num_updates:d} updates")
print(opt_policy)
```

This prints:

```
GLIE SARSA Optimal Value Function with 1000000 updates
{NonTerminal(state=InventoryState(on_hand=0, on_order=0)): -35.05830797041331,
 NonTerminal(state=InventoryState(on_hand=0, on_order=1)): -27.8507256742493,
 NonTerminal(state=InventoryState(on_hand=0, on_order=2)): -27.735579652721434,
 NonTerminal(state=InventoryState(on_hand=1, on_order=0)): -28.984534974043097,
 NonTerminal(state=InventoryState(on_hand=1, on_order=1)): -29.325829885558885,
 NonTerminal(state=InventoryState(on_hand=2, on_order=0)): -30.236704327526777}
GLIE SARSA Optimal Policy with 1000000 updates
For State InventoryState(on_hand=0, on_order=0): Do Action 1
For State InventoryState(on_hand=0, on_order=1): Do Action 1
For State InventoryState(on_hand=0, on_order=2): Do Action 0
For State InventoryState(on_hand=1, on_order=0): Do Action 1
For State InventoryState(on_hand=1, on_order=1): Do Action 0
For State InventoryState(on_hand=2, on_order=0): Do Action 0
```

We see that this reasonably converges to the true Value Function (and reaches the true Optimal Policy) as produced by Value Iteration (whose results were displayed when we tested GLIE MC Control).

The code above is in the file rl/chapter11/simple_inventory_mdp_cap.py. Also see the helper functions in rl/chapter11/control_utils.py which you can use to run your own experiments and tests for RL Control algorithms.

For Tabular GLIE MC Control, we stated a theorem for theoretical guarantee of convergence to the true Optimal Value Function (and hence, true Optimal Policy). Is there something analogous for Tabular GLIE SARSA? This answers in the affirmative with the added condition that we reduce the learning rate according to the Robbins-Monro schedule. We state the following theorem without proof.

Theorem 12.4.1. *Tabular SARSA converges to the Optimal Action-Value function, $Q(s, a) \rightarrow Q^*(s, a)$ (hence, converges to an Optimal Deterministic Policy π^*), under the following conditions:*

- *GLIE schedule of policies $\pi_t(s, a)$*

- *Robbins-Monro schedule of step-sizes α_t:*

$$\sum_{t=1}^{\infty} \alpha_t = \infty$$

$$\sum_{t=1}^{\infty} \alpha_t^2 < \infty$$

Now let's compare GLIE MC Control and GLIE SARSA. This comparison is analogous to the comparison in Section 11.5.2 in Chapter 11 regarding their bias, variance and convergence properties. GLIE SARSA carries a biased estimate of the Q-Value Function compared to the unbiased estimate of GLIE MC Control. On the flip side, the TD Target $R_{t+1} + \gamma \cdot Q(S_{t+1}, A_{t+1}; w)$ has much lower variance than G_t because G_t depends on many random state transitions and random rewards (on the remainder of the trace experience) whose variances accumulate, whereas the TD Target depends on only the next random state transition S_{t+1} and the next random reward R_{t+1}. The bad news with GLIE SARSA (due to the bias in its update) is that with function approximation, it does not always converge to the Optimal Value Function/Policy.

As mentioned in Chapter 11, because MC and TD have significant differences in their usage of data, nature of updates, and frequency of updates, it is not even clear how to create a level-playing field when comparing MC and TD for speed of convergence or for efficiency in usage of limited experiences data. The typical comparisons between MC and TD are done with constant learning rates, and it's been determined that practically GLIE SARSA learns faster than GLIE MC Control with constant learning rates. We illustrate this by running GLIE MC Control and GLIE SARSA on SimpleInventoryMDPCap, and plot the root-mean-squared-errors (RMSE) of the Q-Value Function estimates as a function of batches of episodes (i.e., visualize how the RMSE of the Q-Value Function evolves as the two algorithms progress). This is done by calling the function compare_mc_sarsa_ql which is in the file rl/chapter11/control_utils.py.

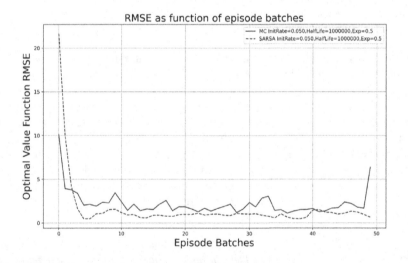

Figure 12.3 **GLIE MC Control and GLIE SARSA Convergence for SimpleInventoryMDPCap**

Figure 12.3 depicts the convergence for our implementations of GLIE MC Control and GLIE SARSA for a constant learning rate of $\alpha = 0.05$. We produced this figure by using data from 500 episodes generated from the same SimpleInventoryMDPCap object we had created earlier (with same discount factor $\gamma = 0.9$). We plotted the RMSE after each batch of 10 episodes, hence both curves shown in the figure have 50 RMSE data points plotted. Firstly, we clearly see that MC Control has significantly more variance as evidenced by the choppy MC Control RMSE progression curve. Secondly, we note that the MC Control RMSE curve progresses quite quickly in the first few episode batches but is slow to converge after the first few episode batches (relative to the progression of SARSA). This results in SARSA reaching fairly small RMSE quicker than MC Control. This behavior of GLIE SARSA outperforming the comparable GLIE MC Control (with constant learning rate) is typical in most MDP Control problems.

Lastly, it's important to recognize that MC Control is not very sensitive to the initial Value Function while SARSA is more sensitive to the initial Value Function. We encourage you to play with the initial Value Function for this `SimpleInventoryMDPCap` example and evaluate how it affects the convergence speeds.

More generally, we encourage you to play with the `compare_mc_sarsa_ql` function on other MDP choices (ones we have created earlier in this book, or make up your own MDPs) so you can develop good intuition for how GLIE MC Control and GLIE SARSA algorithms converge for a variety of choices of learning rate schedules, initial Value Function choices, choices of discount factor etc.

12.5   SARSA(λ)

Much like how we extended TD Prediction to TD(λ) Prediction, we can extend SARSA to SARSA(λ), which gives us a way to tune the spectrum from MC Control to SARSA using the λ parameter. Recall that in order to develop TD(λ) Prediction from TD Prediction, we first developed the n-step TD Prediction Algorithm, then the Offline λ-Return TD Algorithm, and finally the Online TD(λ) Algorithm. We develop an analogous progression from SARSA to SARSA(λ).

So the first thing to do is to extend SARSA to 2-step-bootstrapped SARSA, whose update is as follows:

$$\Delta w = \alpha \cdot (R_{t+1} + \gamma \cdot R_{t+2} + \gamma^2 \cdot Q(S_{t+2}, A_{t+2}; w) - Q(S_t, A_t; w)) \cdot \nabla_w Q(S_t, A_t; w)$$

Generalizing this to n-step-bootstrapped SARSA, the update would then be as follows:

$$\Delta w = \alpha \cdot (G_{t,n} - Q(S_t, A_t; w)) \cdot \nabla_w Q(S_t, A_t; w)$$

where the n-step-bootstrapped Return $G_{t,n}$ is defined as:

$$G_{t,n} = \sum_{i=t+1}^{t+n} \gamma^{i-t-1} \cdot R_i + \gamma^n \cdot Q(S_{t+n}, A_{t+n}; w)$$

$$= R_{t+1} + \gamma \cdot R_{t+2} + \ldots + \gamma^{n-1} \cdot R_{t+n} + \gamma^n \cdot Q(S_{t+n}, A_{t+n}; w)$$

Instead of $G_{t,n}$, a valid target is a weighted-average target:

$$\sum_{n=1}^{N} u_n \cdot G_{t,n} + u \cdot G_t \text{ where } u + \sum_{n=1}^{N} u_n = 1$$

Any of the u_n or u can be 0, as long as they all sum up to 1. The λ-Return target is a special case of weights u_n and u, defined as follows:

$$u_n = (1 - \lambda) \cdot \lambda^{n-1} \text{ for all } n = 1, \ldots, T - t - 1$$

$$u_n = 0 \text{ for all } n \geq T - t \text{ and } u = \lambda^{T-t-1}$$

We denote the λ-Return target as $G_t^{(\lambda)}$, defined as:

$$G_t^{(\lambda)} = (1 - \lambda) \cdot \sum_{n=1}^{T-t-1} \lambda^{n-1} \cdot G_{t,n} + \lambda^{T-t-1} \cdot G_t$$

Then, the Offline λ-Return SARSA Algorithm makes the following updates (performed at the end of each trace experience) for each (S_t, A_t) encountered in the trace experience:

$$\Delta w = \alpha \cdot (G_t^{(\lambda)} - Q(S_t, A_t; w)) \cdot \nabla_w Q(S_t, A_t; w)$$

Finally, we create the SARSA(λ) Algorithm, which is the online "version" of the above λ-Return SARSA Algorithm. The calculations/updates at each time step t for each trace experience are as follows:

$$\delta_t = R_{t+1} + \gamma \cdot Q(S_{t+1}, A_{t+1}; w) - Q(S_t, A_t; w)$$
$$E_t = \gamma\lambda \cdot E_{t-1} + \nabla_w Q(S_t, A_t; w)$$
$$\Delta w = \alpha \cdot \delta_t \cdot E_t$$

with the eligibility traces initialized at time 0 for each trace experience as $E_0 = \nabla_w V(S_0; w)$. Note that just like in SARSA, the ϵ-greedy policy improvement is automatic from the updated Q-Value Function estimate after each time step.

We leave the implementation of SARSA(λ) in Python code as an exercise for you to do.

12.6 OFF-POLICY CONTROL

All control algorithms face a tension between wanting to learn Q-Values contingent on *subsequent optimal behavior* versus wanting to explore all actions. This almost seems contradictory because the quest for exploration deters one from optimal behavior. Our approach so far of pursuing an ϵ-greedy policy (to be thought of as an *almost optimal* policy) is a hack to resolve this tension. A cleaner approach is to use two separate policies for the two separate goals of wanting to be optimal and wanting to explore. The first policy is the one that we learn about (which eventually becomes the optimal policy)—we call this policy the *Target Policy* (to signify the "target" of Control). The second policy is the one that behaves in an exploratory manner, so we can obtain sufficient data for all actions, enabling us to adequately estimate the Q-Value Function—we call this policy the *Behavior Policy*.

In SARSA, at a given time step, we are in a current state S, take action A, after which we obtain the reward R and next state S', upon which we take the next action A'. The action A taken from the current state S is meant to come from an exploratory policy (behavior policy) so that for each state S, we have adequate occurrences of all actions in order to accurately estimate the Q-Value Function. The action A' taken from the next state S' is meant to come from the target policy as we aim for *subsequent optimal behavior* ($Q^*(S, A)$ requires optimal behavior subsequent to taking action A). However, in the SARSA algorithm, the behavior policy producing A from S and the target policy producing A' from S' are, in fact, the same policy—the ϵ-greedy policy. Algorithms such as SARSA in which the behavior policy is the same as the target policy are referred to as On-Policy Algorithms to indicate the fact that the behavior used to generate data (experiences) does not deviate from the policy we are aiming for (target policy, which drives towards the optimal policy).

The separation of behavior policy and target policy as two separate policies gives us algorithms that are known as Off-Policy Algorithms to indicate the fact that the behavior policy is allowed to "deviate off" from the target policy. This separation enables us to construct more general and more powerful RL algorithms. We will use the notation π for the target policy and the notation μ for the behavior policy—therefore, we say that Off-Policy algorithms estimate the Value Function for target policy π while following behavior policy μ. Off-Policy algorithms can be very valuable in real-world situations where we can learn the target policy π by observing humans or other AI agents who follow a behavior policy μ. Another great practical benefit is to be able to re-use prior experiences that were generated from old policies, say $\pi_1, \pi_2, \ldots$ Yet another powerful benefit is that we can learn multiple policies $\mu_1, \mu_2, \ldots$ while following one behavior policy π. Let's now make the concept of Off-Policy Learning concrete by covering the most basic (and most famous) Off-Policy Control Algorithm, which goes by the name of Q-Learning.

12.6.1 Q-Learning

The best way to understand the (Off-Policy) Q-Learning algorithm is to tweak SARSA to make it Off-Policy. Instead of having both the action A and the next action A' being generated by the same ϵ-greedy policy, we generate (i.e., sample) action A (from state S) using an exploratory behavior policy μ and we generate the next action A' (from next state S') using the target policy π. The behavior policy can be any policy as long as it is exploratory enough to be able to obtain sufficient data for all actions (in order to obtain an adequate estimate of the Q-Value Function). Note that in SARSA, when we roll over to the next (new) time step, the new time step's state S is set to be equal to the previous time step's next state S' and the new time step's action A is set to be equal to the previous time step's next action A'. However, in Q-Learning, we only set the new time step's state S to be equal to the previous time step's next state S'. The action A for the new time step will be generated using the behavior policy μ, and won't be equal to the previous time step's next action A' (that would have been generated using the target policy π).

This Q-Learning idea of two separate policies—behavior policy and target policy—is fairly generic, and can be used in algorithms beyond solving the Control problem. However, here we are interested in Q-Learning for Control and so, we want to ensure that the target policy eventually becomes the optimal policy. One straightforward way to accomplish this is to make the target policy equal to the deterministic greedy policy derived from the Q-Value Function estimate at every step. Thus, the update for Q-Learning Control algorithm is as follows:

$$\Delta w = \alpha \cdot \delta_t \cdot \nabla_w Q(S_t, A_t; w)$$

where

$$\delta_t = R_{t+1} + \gamma \cdot Q(S_{t+1}, \arg\max_{a \in \mathcal{A}} Q(S_{t+1}, a; w); w) - Q(S_t, A_t; w)$$
$$= R_{t+1} + \gamma \cdot \max_{a \in \mathcal{A}} Q(S_{t+1}, a; w) - Q(S_t, A_t; w)$$

Following our convention from Chapter 4, we depict the Q-Learning algorithm in Figure 12.4 with states as elliptical-shaped nodes, actions as rectangular-shaped nodes, and the edges as samples from transition probability distribution and action choices.

Although we have highlighted some attractive features of Q-Learning (on account of being Off-Policy), it turns out that Q-Learning when combined with function approximation of the Q-Value Function leads to convergence issues (more on this later). However, Tabular Q-Learning converges under the usual appropriate conditions. There is considerable literature on convergence of Tabular Q-Learning and we won't go over those convergence theorems in this book—here it suffices to say that the convergence proofs for Tabular Q-Learning require infinite exploration of all (state, action) pairs and appropriate stochastic approximation conditions for step sizes.

Now let us write some code for Q-Learning. The function q_learning below is quite similar to the function glie_sarsa we wrote earlier. Here are the differences:

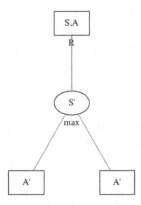

Figure 12.4 **Visualization of Q-Learning Algorithm**

- q_learning takes as input the argument policy_from_q: PolicyFromQType, which is a function with two arguments—a Q-Value Function and a MarkovDecisionProcess object—and returns the policy derived from the Q-Value Function. Thus, q_learning takes as input a general behavior policy whereas glie_sarsa uses the ϵ-greedy policy as its behavior (and target) policy. However, you should note that the typical descriptions of Q-Learning in the RL literature specialize the behavior policy to be the ϵ-greedy policy (we've simply chosen to describe and implement Q-Learning in its more general form of using an arbitrary user-specified behavior policy).
- glie_sarsa takes as input epsilon_as_func_of_episodes: Callable[[int], float] whereas q_learning doesn't require this argument (Q-Learning can converge even if its behavior policy has an unchanging ϵ, and any ϵ specification in q_learning would be built into the policy_from_q argument).
- As explained above, in q_learning, the action from the state is obtained using the specified behavior policy policy_from_q and the "next action" from the next_state is implicitly obtained using the deterministic greedy policy derived from the Q-Value Function estimate q. In glie_sarsa, both action and next_action were obtained from the ϵ-greedy policy.
- As explained above, in q_learning, as we move to the next time step, we set state to be equal to the previous time step's next_state whereas in glie_sarsa, we not only do this but we also set action to be equal to the previous time step's next_action.

```
PolicyFromQType = Callable[
    [QValueFunctionApprox[S, A], MarkovDecisionProcess[S, A]],
    Policy[S, A]
]
def q_learning(
    mdp: MarkovDecisionProcess[S, A],
    policy_from_q: PolicyFromQType,
    states: NTStateDistribution[S],
    approx_0: QValueFunctionApprox[S, A],
    gamma: float,
    max_episode_length: int
) -> Iterator[QValueFunctionApprox[S, A]]:
    q: QValueFunctionApprox[S, A] = approx_0
    yield q
    while True:
        state: NonTerminal[S] = states.sample()
        steps: int = 0
        while isinstance(state, NonTerminal) and steps < max_episode_length:
            policy: Policy[S, A] = policy_from_q(q, mdp)
```

```
action: A = policy.act(state).sample()
next_state, reward = mdp.step(state, action).sample()
next_return: float = max(
    q((next_state, a))
    for a in mdp.actions(next_state)
) if isinstance(next_state, NonTerminal) else 0.
q = q.update([((state, action), reward + gamma * next_return)])
yield q
steps += 1
state = next_state
```

The above code is in the file rl/td.py. Much like how we tested GLIE SARSA on `SimpleInventoryMDPCap`, the code in the file rl/chapter11/simple_inventory_mdp_cap.py also tests Q-Learning on `SimpleInventoryMDPCap`. We encourage you to leverage the helper functions in rl/chapter11/control_utils.py to run your own experiments and tests for Q-Learning. In particular, the functions for Q-Learning in rl/chapter11/control_utils.py employ the common practice of using the ϵ-greedy policy as the behavior policy.

12.6.2 Windy Grid

Now we cover an interesting Control problem that is quite popular in the RL literature—how to navigate a "Windy Grid". We have added some bells and whistles to this problem to make it more interesting. We want to evaluate SARSA and Q-Learning on this problem. Here's the detailed description of this problem:

We are given a grid comprising of cells arranged in the form of m rows and n columns, defined as $\mathcal{G} = \{(i, j) \mid 0 \le i < m, 0 \le j < n\}$. A subset of $\mathcal{G}$ (denoted $\mathcal{B}$) are uninhabitable cells known as *blocks*. A subset of $\mathcal{G}-\mathcal{B}$ (denoted $\mathcal{T}$) is known as the set of goal cells. We have to find a least-cost path from each of the cells in $\mathcal{G} - \mathcal{B} - \mathcal{T}$ to any of the cells in $\mathcal{T}$. At each step, we are required to make a move to a non-block cell (we cannot remain stationary). Right after we make our move, a random vertical wind could move us one cell up or down, unless limited by a block or limited by the boundary of the grid.

Each column has its own random wind specification given by two parameters $0 \le p_1 \le 1$ and $0 \le p_2 \le 1$ with $p_1 + p_2 \le 1$. The wind blows downwards with probability p_1, upwards with probability p_2, and there is no wind with probability $1 - p_1 - p_2$. If the wind makes us bump against a block or against the boundary of the grid, we incur a cost of $b \in \mathbb{R}^+$ in addition to the usual cost of 1 for each move we make. Thus, here the cost includes not just the time spent on making the moves, but also the cost of bumping against blocks or against the boundary of the grid (due to the wind). Minimizing the expected total cost amounts to finding our way to a goal state in a manner that combines minimization of the number of moves with the minimization of the hurt caused by bumping (assume discount factor of 1 when minimizing this expected total cost). If the wind causes us to bump against a wall or against a boundary, we bounce and settle in the cell we moved to just before being blown by the wind (note that the wind blows immediately after we make a move). The wind will never move us by more than one cell between two successive moves, and the wind is never horizontal. Note also that if we move to a goal cell, the process ends immediately without any wind-blow following the movement to the goal cell. The random wind for all the columns is specified as a sequence $[(p_{1,j}, p_{2,j}) \mid 0 \le j < n]$.

Let us model this problem of minimizing the expected total cost while reaching a goal cell as a Finite Markov Decision Process.

State Space $\mathcal{S} = \mathcal{G} - \mathcal{B}$, Non-Terminal States $\mathcal{N} = \mathcal{G} - \mathcal{B} - \mathcal{T}$, Terminal States are $\mathcal{T}$.

We denote the set of all possible moves {UP, DOWN, LEFT, RIGHT} as:
$\mathcal{A} = \{(1, 0), (-1, 0), (0, -1), (0, 1)\}$.

The actions $\mathcal{A}(s)$ for a given non-terminal state $s \in \mathcal{N}$ is defined as: $\{a \mid a \in \mathcal{A}, s+a \in \mathcal{S}\}$ where $+$ denotes element-wise addition of integer 2-tuples.

For all $(s_r, s_c) \in \mathcal{N}$, for all $(a_r, a_c) \in \mathcal{A}((s_r, s_c))$, if $(s_r + a_r, s_c + a_c) \in \mathcal{T}$, then:

$$\mathcal{P}_R((s_r, s_c), (a_r, a_c), -1, (s_r + a_r, s_c + a_c)) = 1$$

For all $(s_r, s_c) \in \mathcal{N}$, for all $(a_r, a_c) \in \mathcal{A}((s_r, s_c))$, if $(s_r + a_r, s_c + a_c) \in \mathcal{N}$, then:

$$\mathcal{P}_R((s_r, s_c), (a_r, a_c), -1 - b, (s_r + a_r, s_c + a_c))$$
$$= p_{1,s_c+a_c} \cdot \mathbb{I}_{(s_r+a_r-1,s_c+a_c) \notin \mathcal{S}} + p_{2,s_c+a_c} \cdot \mathbb{I}_{(s_r+a_r+1,s_c+a_c) \notin \mathcal{S}}$$

$$\mathcal{P}_R((s_r, s_c), (a_r, a_c), -1, (s_r + a_r - 1, s_c + a_c)) = p_{1,s_c+a_c} \cdot \mathbb{I}_{(s_r+a_r-1,s_c+a_c) \in \mathcal{S}}$$

$$\mathcal{P}_R((s_r, s_c), (a_r, a_c), -1, (s_r + a_r + 1, s_c + a_c)) = p_{2,s_c+a_c} \cdot \mathbb{I}_{(s_r+a_r+1,s_c+a_c) \in \mathcal{S}}$$

$$\mathcal{P}_R((s_r, s_c), (a_r, a_c), -1, (s_r + a_r, s_c + a_c)) = 1 - p_{1,s_c+a_c} - p_{2,s_c+a_c}$$

Discount Factor $\gamma = 1$

Now let's write some code to model this problem with the above MDP spec, and run Value Iteration, SARSA and Q-Learning as three different ways of solving this MDP Control problem.

We start with the problem specification in the form of a Python class WindyGrid and write some helper functions before getting into the MDP creation and DP/RL algorithms.

```python
'''
Cell specifies (row, column) coordinate
'''
Cell = Tuple[int, int]
CellSet = Set[Cell]
Move = Tuple[int, int]
'''
WindSpec specifies a random vertical wind for each column.
Each random vertical wind is specified by a (p1, p2) pair
where p1 specifies probability of Downward Wind (could take you
one step lower in row coordinate unless prevented by a block or
boundary) and p2 specifies probability of Upward Wind (could take
you onw step higher in column coordinate unless prevented by a
block or boundary). If one bumps against a block or boundary, one
incurs a bump cost and doesn't move. The remaining probability
1- p1 - p2 corresponds to No Wind.
'''
WindSpec = Sequence[Tuple[float, float]]

possible_moves: Mapping[Move, str] = {
    (-1, 0): 'D',
    (1, 0): 'U',
    (0, -1): 'L',
    (0, 1): 'R'
}

@dataclass(frozen=True)
class WindyGrid:

    rows: int   # number of grid rows
    columns: int   # number of grid columns
    blocks: CellSet   # coordinates of block cells
    terminals: CellSet   # coordinates of goal cells
    wind: WindSpec   # spec of vertical random wind for the columns
    bump_cost: float   # cost of bumping against block or boundary

    @staticmethod
    def add_move_to_cell(cell: Cell, move: Move) -> Cell:
        return cell[0] + move[0], cell[1] + move[1]

    def is_valid_state(self, cell: Cell) -> bool:
        '''
        checks if a cell is a valid state of the MDP
        '''
```

```
        return 0 <= cell[0] < self.rows and 0 <= cell[1] < self.columns \
            and cell not in self.blocks
    def get_all_nt_states(self) -> CellSet:
        '''
        returns all the non-terminal states
        '''
        return {(i, j) for i in range(self.rows) for j in range(self.columns)
                if (i, j) not in set.union(self.blocks, self.terminals)}
    def get_actions_and_next_states(self, nt_state: Cell) \
            -> Set[Tuple[Move, Cell]]:
        '''
        given a non-terminal state, returns the set of all possible
        (action, next_state) pairs
        '''
        temp: Set[Tuple[Move, Cell]] = {(a, WindyGrid.add_move_to_cell(
            nt_state,
            a
        )) for a in possible_moves}
        return {(a, s) for a, s in temp if self.is_valid_state(s)}
```

Next, we write a method to calculate the transition probabilities. The code below should be self-explanatory and mimics the description of the problem above and the mathematical specification of the transition probabilities given above.

```
from rl.distribution import Categorical
    def get_transition_probabilities(self, nt_state: Cell) \
            -> Mapping[Move, Categorical[Tuple[Cell, float]]]:
        '''
        given a non-terminal state, return a dictionary whose
        keys are the valid actions (moves) from the given state
        and the corresponding values are the associated probabilities
        (following that move) of the (next_state, reward) pairs.
        The probabilities are determined from the wind probabilities
        of the column one is in after the move. Note that if one moves
        to a goal cell (terminal state), then one ends up in that
        goal cell with 100% probability (i.e., no wind exposure in a
        goal cell).
        '''
        d: Dict[Move, Categorical[Tuple[Cell, float]]] = {}
        for a, (r, c) in self.get_actions_and_next_states(nt_state):
            if (r, c) in self.terminals:
                d[a] = Categorical({((r, c), -1.): 1.})
            else:
                down_prob, up_prob = self.wind[c]
                stay_prob: float = 1. - down_prob - up_prob
                d1: Dict[Tuple[Cell, float], float] = \
                    {((r, c), -1.): stay_prob}
                if self.is_valid_state((r - 1, c)):
                    d1[((r - 1, c), -1.)] = down_prob
                if self.is_valid_state((r + 1, c)):
                    d1[((r + 1, c), -1.)] = up_prob
                d1[((r, c), -1. - self.bump_cost)] = \
                    down_prob * (1 - self.is_valid_state((r - 1, c))) + \
                    up_prob * (1 - self.is_valid_state((r + 1, c)))
                d[a] = Categorical(d1)
        return d
```

Next, we write a method to create the `MarkovDecisionProcess` for the Windy Grid.

```
from rl.markov_decision_process import FiniteMarkovDecisionProcess
    def get_finite_mdp(self) -> FiniteMarkovDecisionProcess[Cell, Move]:
        '''
        returns the FiniteMarkovDecision object for this windy grid problem
        '''
```

```
    return FiniteMarkovDecisionProcess(
        {s: self.get_transition_probabilities(s) for s in
         self.get_all_nt_states()}
    )
```

Next, we write methods for Value Iteration, SARSA and Q-Learning

```python
from rl.markov_decision_process import FiniteDeterministicPolicy
from rl.dynamic_programming import value_iteration_result, V
from rl.chapter11.control_utils import glie_sarsa_finite_learning_rate
from rl.chapter11.control_utils import q_learning_finite_learning_rate
from rl.chapter11.control_utils import get_vf_and_policy_from_qvf
    def get_vi_vf_and_policy(self) -> \
            Tuple[V[Cell], FiniteDeterministicPolicy[Cell, Move]]:
        '''
        Performs the Value Iteration DP algorithm returning the
        Optimal Value Function (as a V[Cell]) and the Optimal Policy
        (as a FiniteDeterministicPolicy[Cell, Move])
        '''
        return value_iteration_result(self.get_finite_mdp(), gamma=1.)
    def get_glie_sarsa_vf_and_policy(
        self,
        epsilon_as_func_of_episodes: Callable[[int], float],
        learning_rate: float,
        num_updates: int
    ) -> Tuple[V[Cell], FiniteDeterministicPolicy[Cell, Move]]:
        qvfs: Iterator[QValueFunctionApprox[Cell, Move]] = \
            glie_sarsa_finite_learning_rate(
                fmdp=self.get_finite_mdp(),
                initial_learning_rate=learning_rate,
                half_life=1e8,
                exponent=1.0,
                gamma=1.0,
                epsilon_as_func_of_episodes=epsilon_as_func_of_episodes,
                max_episode_length=int(1e8)
            )
        final_qvf: QValueFunctionApprox[Cell, Move] = \
            iterate.last(itertools.islice(qvfs, num_updates))
        return get_vf_and_policy_from_qvf(
            mdp=self.get_finite_mdp(),
            qvf=final_qvf
        )

    def get_q_learning_vf_and_policy(
        self,
        epsilon: float,
        learning_rate: float,
        num_updates: int
    ) -> Tuple[V[Cell], FiniteDeterministicPolicy[Cell, Move]]:
        qvfs: Iterator[QValueFunctionApprox[Cell, Move]] = \
            q_learning_finite_learning_rate(
                fmdp=self.get_finite_mdp(),
                initial_learning_rate=learning_rate,
                half_life=1e8,
                exponent=1.0,
                gamma=1.0,
                epsilon=epsilon,
                max_episode_length=int(1e8)
            )
        final_qvf: QValueFunctionApprox[Cell, Move] = \
            iterate.last(itertools.islice(qvfs, num_updates))
        return get_vf_and_policy_from_qvf(
            mdp=self.get_finite_mdp(),
            qvf=final_qvf
        )
```

The above code is in the file rl/chapter11/windy_grid.py. Note that this file also contains some helpful printing functions that pretty-prints the grid, along with the calculated Optimal Value Functions and Optimal Policies. The method `print_wind_and_bumps` prints the column wind probabilities and the cost of bumping into a block/boundary. The method `print_vf_and_policy` prints a given Value Function and a given Deterministic Policy—this method can be used to print the Optimal Value Function and Optimal Policy produced by Value Iteration, by SARSA and by Q-Learning. In the printing of a deterministic policy, "X" represents a block, "T" represents a terminal cell, and the characters "L", "R", "D", "U" represent "Left", "Right", "Down", "Up" moves, respectively.

Now let's run our code on a small instance of a Windy Grid.

```python
wg = WindyGrid(
    rows=5,
    columns=5,
    blocks={(0, 1), (0, 2), (0, 4), (2, 3), (3, 0), (4, 0)},
    terminals={(3, 4)},
    wind=[(0., 0.9), (0.0, 0.8), (0.7, 0.0), (0.8, 0.0), (0.9, 0.0)],
    bump_cost=4.0
)
wg.print_wind_and_bumps()
vi_vf_dict, vi_policy = wg.get_vi_vf_and_policy()
print("Value Iteration\n")
wg.print_vf_and_policy(
    vf_dict=vi_vf_dict,
    policy=vi_policy
)
epsilon_as_func_of_episodes: Callable[[int], float] = lambda k: 1. / k
learning_rate: float = 0.03
num_updates: int = 100000
sarsa_vf_dict, sarsa_policy = wg.get_glie_sarsa_vf_and_policy(
    epsilon_as_func_of_episodes=epsilon_as_func_of_episodes,
    learning_rate=learning_rate,
    num_updates=num_updates
)
print("SARSA\n")
wg.print_vf_and_policy(
    vf_dict=sarsa_vf_dict,
    policy=sarsa_policy
)
epsilon: float = 0.2
ql_vf_dict, ql_policy = wg.get_q_learning_vf_and_policy(
    epsilon=epsilon,
    learning_rate=learning_rate,
    num_updates=num_updates
)
print("Q-Learning\n")
wg.print_vf_and_policy(
    vf_dict=ql_vf_dict,
    policy=ql_policy
)
```

This prints the following:

```
Column 0: Down Prob = 0.00, Up Prob = 0.90
Column 1: Down Prob = 0.00, Up Prob = 0.80
Column 2: Down Prob = 0.70, Up Prob = 0.00
Column 3: Down Prob = 0.80, Up Prob = 0.00
Column 4: Down Prob = 0.90, Up Prob = 0.00
Bump Cost = 4.00

Value Iteration
```

```
          0     1     2     3     4
4 XXXXX  5.25  2.02  1.10  1.00
3 XXXXX  8.53  5.20  1.00  0.00
2  9.21  6.90  8.53 XXXXX  1.00
1  8.36  9.21  8.36 12.16 11.00
0 10.12 XXXXX XXXXX 17.16 XXXXX
```

```
   0 1 2 3 4
4  X R R R D
3  X R R R T
2  R U U X U
1  R U L L U
0  U X X U X
```

SARSA

```
          0     1     2     3     4
4 XXXXX  5.47  2.02  1.08  1.00
3 XXXXX  8.78  5.37  1.00  0.00
2  9.14  7.03  8.29 XXXXX  1.00
1  8.51  9.16  8.27 11.92 12.58
0 10.05 XXXXX XXXXX 16.48 XXXXX
```

```
   0 1 2 3 4
4  X R R R D
3  X R R R T
2  R U U X U
1  R U L L U
0  U X X U X
```

Q-Learning

```
          0     1     2     3     4
4 XXXXX  5.45  2.02  1.09  1.00
3 XXXXX  8.09  5.12  1.00  0.00
2  8.78  6.76  7.92 XXXXX  1.00
1  8.31  8.85  8.09 11.52 10.93
0  9.85 XXXXX XXXXX 16.16 XXXXX
```

```
   0 1 2 3 4
4  X R R R D
3  X R R R T
2  R U U X U
1  R U L L U
0  U X X U X
```

Value Iteration should be considered as the benchmark since it calculates the Optimal Value Function within the default tolerance of 1e-5. We see that both SARSA and Q-Learning get fairly close to the Optimal Value Function after only 100,000 updates (i.e.,

100,000 moves across various episodes). We also see that both SARSA and Q-Learning obtain the true Optimal Policy, consistent with Value Iteration.

Now let's explore SARSA and Q-Learning's speed of convergence to the Optimal Value Function.

We first run GLIE SARSA and Q-Learning for the above settings of bump cost = 4.0, GLIE SARSA $\epsilon(k) = \frac{1}{k}$, Q-Learning $\epsilon = 0.2$. Figure 12.5 depicts the trajectory of Root-Mean-Squared-Error (RMSE) of the Q-Values relative to the Q-Values obtained by Value Iteration. The RMSE is plotted as a function of progressive batches of 10 episodes. We can see that GLIE SARSA and Q-Learning have roughly the same convergence trajectory.

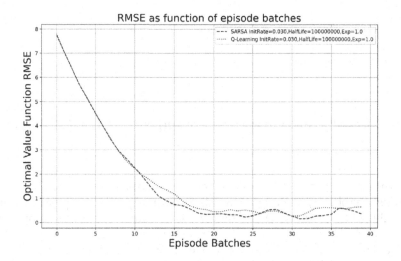

Figure 12.5   GLIE SARSA and Q-Learning Convergence for Windy Grid (Bump Cost = 4)

Now let us set the bump cost to a very high value of 100,000. Figure 12.6 depicts the convergence trajectory for bump cost of 100,000. We see that Q-Learning converges much faster than GLIE SARSA (we kept GLIE SARSA $\epsilon(k) = \frac{1}{k}$ and Q-Learning $\epsilon = 0.2$). So why does Q-Learning do better? Q-Learning has two advantages over GLIE SARSA here: Firstly, its behavior policy is exploring at the constant amount of 20% whereas GLIE SARSA's exploration declines to 10% after just the 10th episode. This means Q-Learning gets sufficient data quicker than GLIE SARSA for the entire set of (state, action) pairs. Secondly, Q-Learning's target policy is greedy, versus GLIE SARSA's declining-ϵ-greedy. This means GLIE SARSA's Optimal Q-Value estimation is compromised due to the exploration of actions in its target policy (rather than a pure exploitation with max over actions, as is the case with Q-Learning). Thus, the separation between behavior policy and target policy in Q-Learning fetches it the best of both worlds and enables it to perform better than GLIE SARSA in this example.

SARSA is a more "conservative" algorithm in the sense that if there is a risk of a large negative reward close to the optimal path, SARSA will tend to avoid that dangerous optimal path and only slowly learn to use that optimal path when ϵ (exploration) reduces sufficiently. Q-Learning, on the other hand, will tend to take that risk while exploring and learns fast through "big failures". This provides us with a guide on when to use SARSA and when to use Q-Learning. Roughly speaking, use SARSA if you are training your AI agent with interaction with the real environment where you care about time and money

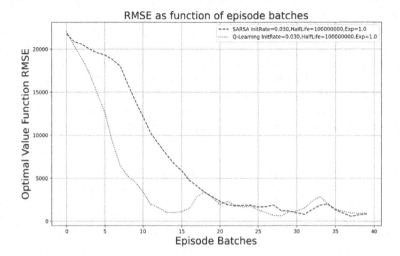

Figure 12.6 GLIE SARSA and Q-Learning Convergence for Windy Grid (Bump Cost = 100,000)

consumed while doing the training with real environment-interaction (e.g., you don't want to risk damaging a robot by walking it towards an optimal path in the proximity of physical danger). On the other hand, use Q-Learning if you are training your AI agent with a simulated environment where large negative rewards don't cause actual time/money losses, but these large negative rewards help the AI agent learn quickly. In a financial trading example, if you are training your RL agent in a real trading environment, you'd want to use SARSA as Q-Learning can potentially incur big losses while SARSA (although slower in learning) will avoid real trading losses during the process of learning. On the other hand, if you are training your RL agent in a simulated trading environment, Q-Learning is the way to go as it will learn fast by incuring "paper trading" losses as part of the process of executing risky trades.

Note that Q-Learning (and Off-policy Learning in general) has higher per-sample variance than SARSA, which could lead to problems in convergence, especially when we employ function approximation for the Q-Value Function. Q-Learning has been shown to be particularly problematic in converging when using neural networks for its Q-Value function approximation.

The SARSA algorithm was introduced in a paper by Rummery and Niranjan (Rummery and Niranjan 1994). The Q-Learning algorithm was introduced in the Ph.D. thesis of Chris Watkins (Watkins 1989).

12.6.3 Importance Sampling

Now that we've got a good grip of Off-Policy Learning through the Q-Learning algorithm, we show a very different (arguably simpler) method of doing Off-Policy Learning. This method is known as Importance Sampling, a fairly general technique (beyond RL) for estimating properties of a particular probability distribution, while only having access to samples of a different probability distribution. Specializing this technique to Off-Policy Control, we estimate the Value Function for the target policy (probability distribution of interest) while having access to samples generated from the probability distribution of the behavior policy. Specifically, Importance Sampling enables us to calculate $\mathbb{E}_{X \sim P}[f(X)]$ (where P is the probability distribution of interest), given samples from probability distribution Q, as follows:

$$
\begin{aligned}
\mathbb{E}_{X \sim P}[f(X)] &= \sum P(X) \cdot f(X) \\
&= \sum Q(X) \cdot \frac{P(X)}{Q(X)} \cdot f(X) \\
&= \mathbb{E}_{X \sim Q}[\frac{P(X)}{Q(X)} \cdot f(X)]
\end{aligned}
$$

So basically, the function $f(X)$ of samples X are scaled by the ratio of the probabilities $P(X)$ and $Q(X)$.

Let's employ this Importance Sampling method for Off-Policy Monte Carlo Prediction, where we need to estimate the Value Function for policy π while only having access to trace experience returns generated using policy μ. The idea is straightforward—we simply weight the returns G_t according to the similarity between policies π and μ, by multiplying importance sampling corrections along whole episodes. Let us define ρ_t as the product of the ratio of action probabilities (on the two policies π and μ) from time t to time $T - 1$ (assume episode ends at time T). Specifically,

$$
\rho_t = \frac{\pi(S_t, A_t)}{\mu(S_t, A_t)} \cdot \frac{\pi(S_{t+1}, A_{t+1})}{\mu(S_{t+1}, A_{t+1})} \cdots \frac{\pi(S_{T-1}, A_{T-1})}{\mu(S_{T-1}, A_{T-1})}
$$

We've learnt in Chapter 11 that the learning rate α (treated as an update step-size) serves as a weight to the update target (in the case of MC, the update target is the return G_t). So all we have to do is to scale the step-size α for the update for time t by ρ_t. Hence, the MC Prediction update is tweaked to be the following when doing Off-Policy with Importance Sampling:

$$
\Delta \boldsymbol{w} = \alpha \cdot \rho_t \cdot (G_t - V(S_t; \boldsymbol{w})) \cdot \nabla_{\boldsymbol{w}} V(S_t; \boldsymbol{w})
$$

For MC Control, we make the analogous tweak to the update for the Q-Value Function, as follows:

$$
\Delta \boldsymbol{w} = \alpha \cdot \rho_t \cdot (G_t - Q(S_t, A_t; \boldsymbol{w})) \cdot \nabla_{\boldsymbol{w}} Q(S_t, A_t; \boldsymbol{w})
$$

Note that we cannot use this method if μ is zero when π is non-zero (since μ is in the denominator).

A key disadvantage of Off-Policy MC with Importance Sampling is that it dramatically increases the variance of the Value Function estimate. To contain the variance, we can use TD targets (instead of trace experience returns) generated from μ to evaluate the Value Function for π. For Off-Policy TD Prediction, we essentially weight TD target $R + \gamma \cdot V(S'; \boldsymbol{w})$ with importance sampling. Here we only need a single importance sampling correction, as follows:

$$
\Delta \boldsymbol{w} = \alpha \cdot \frac{\pi(S_t, A_t)}{\mu(S_t, A_t)} \cdot (R_{t+1} + \gamma \cdot V(S_{t+1}; \boldsymbol{w}) - V(S_t; \boldsymbol{w})) \cdot \nabla_{\boldsymbol{w}} V(S_t; \boldsymbol{w})
$$

For TD Control, we do the analogous update for the Q-Value Function:

$$
\Delta \boldsymbol{w} = \alpha \cdot \frac{\pi(S_t, A_t)}{\mu(S_t, A_t)} \cdot (R_{t+1} + \gamma \cdot Q(S_{t+1}, A_{t+1}; \boldsymbol{w}) - Q(S_t, A_t; \boldsymbol{w})) \cdot \nabla_{\boldsymbol{w}} Q(S_t, A_t; \boldsymbol{w})
$$

This has much lower variance than MC importance sampling. A key advantage of TD importance sampling is that policies only need to be similar over a single time step.

Since the modifications from On-Policy algorithms to Off-Policy algorithms based on Importance Sampling are just a small tweak of scaling the update by importance sampling corrections, we won't implement the Off-Policy Importance Sampling algorithms in Python code. However, we encourage you to implement the Prediction and Control MC and TD Off-Policy algorithms (based on Importance Sampling) described above.

12.7 CONCEPTUAL LINKAGE BETWEEN DP AND TD ALGORITHMS

It's worthwhile placing RL algorithms in terms of their conceptual relationship to DP algorithms. Let's start with the Prediction problem, whose solution is based on the Bellman Expectation Equation. Figure 12.8 depicts TD Prediction, which is the sample backup version of Policy Evaluation, depicted in Figure 12.7 as a full backup DP algorithm.

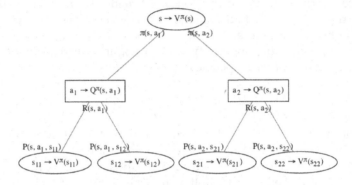

Figure 12.7 **Policy Evaluation (DP Algorithm with Full Backup)**

Figure 12.8 **TD Prediction (RL Algorithm with Sample Backup)**

Likewise, Figure 12.10 depicts SARSA, which is the sample backup version of Q-Policy Iteration (Policy Iteration on Q-Value), depicted in Figure 12.9 as a full backup DP algorithm.

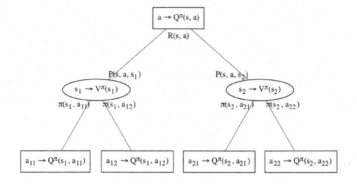

Figure 12.9 Q-Policy Iteration (DP Algorithm with Full Backup)

Figure 12.10 SARSA (RL Algorithm with Sample Backup)

Finally, Figure 12.12 depicts Q-Learning, which is the sample backup version of Q-Value Iteration (Value Iteration on Q-Value), depicted in Figure 12.11 as a full backup DP algorithm.

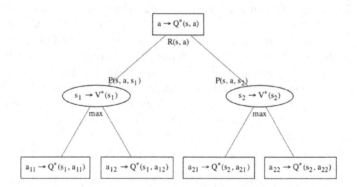

Figure 12.11 Q-Value Iteration (DP Algorithm with Full Backup)

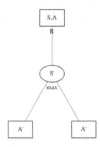

Figure 12.12 Q-Learning (RL Algorithm with Sample Backup)

The table in Figure 12.13 summarizes these RL algorithms, along with their corresponding DP algorithms, showing the expectation targets of the DP algorithms' updates along with the corresponding sample targets of the RL algorithms' updates.

Full Backup (DP)	Sample Backup (TD)
Policy Evaluation's $V(S)$ update: $\mathbb{E}[R + \gamma V(S')\|S]$	TD Learning's $V(S)$ update: sample $R + \gamma V(S')$
Q-Policy Iteration's $Q(S, A)$ update: $\mathbb{E}[R + \gamma Q(S', A')\|S, A]$	SARSA's $Q(S, A)$ update: sample $R + \gamma Q(S', A')$
Q-Value Iteration's $Q(S, A)$ update: $\mathbb{E}[R + \gamma \max_{a'} Q(S', a')\|S, A]$	Q-Learning's $Q(S, A)$ update: sample $R + \gamma \max_{a'} Q(S', a')$

Figure 12.13 Relationship between DP and RL Algorithms

12.8 CONVERGENCE OF RL ALGORITHMS

Now we provide an overview of convergence of RL Algorithms. Let us start with RL Prediction. Figure 12.14 provides the overview of RL Prediction. As you can see, Monte-Carlo Prediction has convergence guarantees, whether On-Policy or Off-Policy, whether Tabular or with Function Approximation (even with non-linear Function Approximation). However, Temporal-Difference Prediction can have convergence issues—the core reason for this is that the TD update is not a true gradient update (as we've explained in Chapter 11, it is a *semi-gradient* update). As you can see, although we have convergence guarantees for On-Policy TD Prediction with linear function approximation, there is no convergence guarantee for On-Policy TD Prediction with non-linear function approximation. The situation is even worse for Off-Policy TD Prediction—there is no convergence guarantee even for linear function approximation.

We want to highlight a confluence pattern in RL Algorithms where convergence problems arise. As a rule of thumb, if we do all of the following three, then we run into convergence problems.

- Bootstrapping, i.e., updating with a target that involves the current Value Function estimate (as is the case with Temporal-Difference)
- Off-Policy
- Function Approximation of the Value Function

Hence, [Bootstrapping, Off-Policy, Function Approximation] is known as the *Deadly Triad*, a term emphasized and popularized by Richard Sutton in a number of publications

On/Off Policy	Algorithm	Tabular	Linear	Non-Linear
On-Policy	MC	✓	✓	✓
	TD(0)	✓	✓	✗
	TD(λ)	✓	✓	✗
Off-Policy	MC	✓	✓	✓
	TD(0)	✓	✗	✗
	TD(λ)	✓	✗	✗

Figure 12.14 Convergence of RL Prediction Algorithms

and lectures. We should highlight that the *Deadly Triad* phenomenon is not a theorem—rather, it should be viewed as a rough pattern and as a rule of thumb. So to achieve convergence, we avoid at least one of the above three. We have seen that each of [Bootstrapping, Off-Policy, Function Approximation] provides benefits, but when all three come together, we run into convergence problems. The fundamental problem is that semi-gradient bootstrapping does not follow the gradient of *any* objective function and this causes TD to diverge when running off-policy and when using function approximations.

Function Approximation is typically unavoidable in real-world problems because of the size of real-world problems. So we are looking at avoiding semi-gradient bootstrapping or avoiding off-policy. Note that semi-gradient bootstrapping can be mitigated by tuning the TD λ parameter to a high-enough value. However, if we want to get around this problem in a fundamental manner, we can avoid the core issue of semi-gradient bootstrapping by instead doing a *true gradient* with a method known as *Gradient Temporal-Difference* (or *Gradient TD*, for short). We will cover Gradient TD in detail in Chapter 13, but for now, we want to simply share that Gradient TD updates the value function approximation's parameters with the actual gradient (not semi-gradient) of an appropriate loss function and the gradient formula involves bootstrapping. Thus, it avails of the advantages of bootstrapping without the disadvantages of semi-gradient (which we cheekily referred to as "cheating" in Chapter 11). Figure 12.15 expands upon Figure 12.14 by incorporating convergence properties of Gradient TD.

On/Off Policy	Algorithm	Tabular	Linear	Non-Linear
On-Policy	MC	✓	✓	✓
	TD	✓	✓	✗
	Gradient TD	✓	✓	✓
Off-Policy	MC	✓	✓	✓
	TD	✓	✗	✗
	Gradient TD	✓	✓	✓

Figure 12.15 Convergence of RL Prediction Algorithms, including Gradient TD

Now let's move on to convergence of Control Algorithms. Figure 12.16 provides the picture. (✓) means it doesn't quite hit the Optimal Value Function, but bounces around near the Optimal Value Function. Gradient Q-Learning is the adaptation of Q-Learning with Gradient TD. So this method is Off-Policy, is bootstrapped, but avoids semi-gradient. This enables it to converge for linear function approximations. However, it diverges when used with non-linear function approximations. So, for Control, even with Gradient TD, the deadly triad still exists for a combination of [Bootstrapping, Off-Policy, Non-Linear Function Approximation]. In Chapter 13, we shall cover the DQN algorithm which is an innovative and practically effective method for getting around the deadly triad for RL Control.

Algorithm	Tabular	Linear	Non-Linear
MC Control	✓	(✓)	✗
SARSA	✓	(✓)	✗
Q-Learning	✓	✗	✗
Gradient Q-Learning	✓	✓	✗

Figure 12.16 Convergence of RL Control Algorithms

12.9   KEY TAKEAWAYS FROM THIS CHAPTER

- RL Control is based on the idea of Generalized Policy Iteration (GPI):
    - Policy Evaluation with Q-Value Function (instead of State-Value Function V)
    - Improved Policy needs to be exploratory, e.g., ϵ-greedy
- On-Policy versus Off-Policy (e.g., SARSA versus Q-Learning)
- Deadly Triad := [Bootstrapping, Off-Policy, Function Approximation]

CHAPTER 13

Batch RL, Experience-Replay, DQN, LSPI, Gradient TD

In Chapters 11 and 12, we covered the basic RL algorithms for Prediction and Control, respectively. Specifically, we covered the basic Monte-Carlo (MC) and Temporal-Difference (TD) techniques. We want to highlight two key aspects of these basic RL algorithms:

1. The experiences data arrives in the form of a single unit of experience at a time (single unit is a *trace experience* for MC and an *atomic experience* for TD), the unit of experience is used by the algorithm for Value Function learning, and then that unit of experience is not used later in the algorithm (essentially, that unit of experience, once consumed, is *not re-consumed* for further learning later in the algorithm). It doesn't have to be this way—one can develop RL algorithms that re-use experience data—this approach is known as *Experience-Replay* (in fact, we saw a glimpse of Experience-Replay in Section 11.5.3 of Chapter 11).

2. Learning occurs in a *granularly incremental* manner, by updating the Value Function after each unit of experience. It doesn't have to be this way—one can develop RL algorithms that take an entire batch of experiences (or in fact, all of the experiences that one could possibly get), and learn the Value Function directly for that entire batch of experiences. A key idea here is that if we know in advance what experiences data we have (or will have), and if we collect and organize all of that data, then we could directly (i.e., not incrementally) estimate the Value Function for *that* experiences data set. This approach to RL is known as *Batch RL* (versus the basic RL algorithms we covered in the previous chapters that can be termed as *Incremental RL*).

Thus, we have a choice or doing Experience-Replay or not, and we have a choice of doing Batch RL or Incremental RL. In fact, some of the interesting and practically effective algorithms combine both the ideas of Experience-Replay and Batch RL. This chapter starts with the coverage of Batch RL and Experience-Replay. Then, we cover some key algorithms (including Deep Q-Networks and Least Squares Policy Iteration) that effectively leverage Batch RL and/or Experience-Replay. Next, we look deeper into the issue of the *Deadly Triad* (that we had alluded to in Chapter 12) by viewing Value Functions as Vectors (we had done this in Chapter 5), understand Value Function Vector transformations with a balance of geometric intuition and mathematical rigor, providing insights into convergence issues for a variety of traditional loss functions used to develop RL algorithms. Finally, this treatment of Value Functions as Vectors leads us in the direction of overcoming the Deadly Triad by defining an appropriate loss function, calculating whose gradient provides a more robust set of RL algorithms known as Gradient Temporal Difference (abbreviated, as Gradient TD).

13.1 BATCH RL AND EXPERIENCE-REPLAY

Let us understand Incremental RL versus Batch RL in the context of fixed finite experiences data. To make things simple and easy to understand, we first focus on understanding the difference for the case of MC Prediction (i.e., to calculate the Value Function of an MRP using Monte-Carlo). In fact, we had covered this setting in Section 11.5.3 of Chapter 11.

To refresh this setting, specifically we have access to a fixed finite sequence/stream of MRP trace experiences (i.e., `Iterable[Iterable[TransitionStep[S]]]`), which we know can be converted to returns-augmented data of the form `Iterable[Iterable[ReturnStep[S]]]` (using the returns function[1]). Flattening this data to `Iterable[ReturnStep[S]]` and extracting from it the (state, return) pairs gives us the fixed, finite training data for MC Prediction, that we denote as follows:

$$\mathcal{D} = [(S_i, G_i)|1 \leq i \leq n]$$

We've learnt in Chapter 11 that we can do an Incremental MC Prediction estimation $V(s; \boldsymbol{w})$ by updating $\boldsymbol{w}$ after each MRP trace experience with the gradient calculation $\nabla_{\boldsymbol{w}}\mathcal{L}(\boldsymbol{w})$ for each data pair (S_i, G_i), as follows:

$$\mathcal{L}_{(S_i, G_i)}(\boldsymbol{w}) = \frac{1}{2} \cdot (V(S_i; \boldsymbol{w}) - G_i)^2$$

$$\nabla_{\boldsymbol{w}}\mathcal{L}_{(S_i, G_i)}(\boldsymbol{w}) = (V(S_i; \boldsymbol{w}) - G_i) \cdot \nabla_{\boldsymbol{w}}V(S_i; \boldsymbol{w})$$

$$\Delta\boldsymbol{w} = \alpha \cdot (G_i - V(S_i; \boldsymbol{w})) \cdot \nabla_{\boldsymbol{w}}V(S_i; \boldsymbol{w})$$

The Incremental MC Prediction algorithm performs n updates in sequence for data pairs $(S_i, G_i), i = 1, 2, \ldots, n$ using the update method of `FunctionApprox`. We note that Incremental RL makes inefficient use of available training data $\mathcal{D}$ because we essentially "discard" each of these units of training data after it's used to perform an update. We want to make efficient use of the given data with Batch RL. Batch MC Prediction aims to estimate the MRP Value Function $V(s; \boldsymbol{w}^*)$ such that

$$\boldsymbol{w}^* = \arg\min_{\boldsymbol{w}} \frac{1}{2n} \cdot \sum_{i=1}^{n}(V(S_i; \boldsymbol{w}) - G_i)^2$$

$$= \arg\min_{\boldsymbol{w}} \mathbb{E}_{(S,G)\sim\mathcal{D}}[\frac{1}{2} \cdot (V(S; \boldsymbol{w}) - G)^2]$$

This, in fact, is the `solve` method of `FunctionApprox` on training data $\mathcal{D}$. This approach is called Batch RL because we first collect and store the entire set (batch) of data $\mathcal{D}$ available to us, and then we find the best possible parameters $\boldsymbol{w}^*$ fitting this data $\mathcal{D}$. Note that unlike Incremental RL, here we are not updating the MRP Value Function estimate while the data arrives—we simply store the data as it arrives and start the MRP Value Function estimation procedure once we are ready with the entire (batch) data $\mathcal{D}$ in storage. As we know from the implementation of the `solve` method of `FunctionApprox`, finding the best possible parameters $\boldsymbol{w}^*$ from the batch $\mathcal{D}$ involves calling the update method of `FunctionApprox` with

---

[1] returns is defined in the file rl/returns.py

repeated use of the available data pairs (S, G) in the stored data set $\mathcal{D}$. Each of these updates to the parameters w is as follows:

$$\Delta w = \alpha \cdot \frac{1}{n} \cdot \sum_{i=1}^{n} (G_i - V(S_i; w)) \cdot \nabla_w V(S_i; w)$$

Note that unlike Incremental MC where each update to w uses data from a single trace experience, each update to w in Batch MC uses all of the trace experiences data (all of the batch data). If we keep doing these updates repeatedly, we will ultimately converge to the desired MRP Value Function $V(s; w^*)$. The repeated use of the available data in $\mathcal{D}$ means that we are doing Batch MC Prediction using *Experience-Replay*. So we see that this makes more efficient use of the available training data $\mathcal{D}$ due to the re-use of the data pairs in $\mathcal{D}$.

The code for this Batch MC Prediction algorithm (batch_mc_prediction) is shown below.[2] From the input trace experiences (traces in the code below), we first create the set of ReturnStep transitions that span across the set of all input trace experiences (return_steps in the code below). This involves calculating the return associated with each state encountered in traces (across all trace experiences). From return_steps, we create the (state, return) pairs that constitute the fixed, finite training data $\mathcal{D}$, which is then passed to the solve method of approx: ValueFunctionApprox[S].

```
import rl.markov_process as mp
from rl.returns import returns
from rl.approximate_dynamic_programming import ValueFunctionApprox
import itertools

def batch_mc_prediction(
    traces: Iterable[Iterable[mp.TransitionStep[S]]],
    approx: ValueFunctionApprox[S],
    gamma: float,
    episode_length_tolerance: float = 1e-6,
    convergence_tolerance: float = 1e-5
) -> ValueFunctionApprox[S]:
    '''traces is a finite iterable'''
    return_steps: Iterable[mp.ReturnStep[S]] = \
        itertools.chain.from_iterable(
            returns(trace, gamma, episode_length_tolerance) for trace in traces
        )
    return approx.solve(
        [(step.state, step.return_) for step in return_steps],
        convergence_tolerance
    )
```

Now let's move on to Batch TD Prediction. Here we have fixed, finite experiences data $\mathcal{D}$ available as:

$$\mathcal{D} = [(S_i, R_i, S_i') | 1 \le i \le n]$$

where (R_i, S_i') is the pair of reward and next state from a state S_i. So, Experiences Data $\mathcal{D}$ is presented in the form of a fixed, finite number of atomic experiences. This is represented in code as an Iterable[TransitionStep[S]].

Just like Batch MC Prediction, here in Batch TD Prediction, we first collect and store the data as it arrives, and once we are ready with the batch of data $\mathcal{D}$ in storage, we start the MRP Value Function estimation procedure. The parameters w are updated with repeated use of the atomic experiences in the stored data $\mathcal{D}$. Each of these updates to the parameters w is as follows:

$$\Delta w = \alpha \cdot \frac{1}{n} \cdot \sum_{i=1}^{n} (R_i + \gamma \cdot V(S_i'; w) - V(S_i; w)) \cdot \nabla_w V(S_i; w)$$

---

[2]batch_mc_prediction is defined in the file rl/monte_carlo.py.

Note that unlike Incremental TD where each update to w uses data from a single atomic experience, each update to w in Batch TD uses all of the atomic experiences data (all of the batch data). The repeated use of the available data in $\mathcal{D}$ means that we are doing Batch TD Prediction using *Experience-Replay*. So we see that this makes more efficient use of the available training data $\mathcal{D}$ due to the re-use of the data pairs in $\mathcal{D}$.

The code for this Batch TD Prediction algorithm (batch_td_prediction) is shown below.[3] We create a Sequence[TransitionStep] from the fixed, finite-length input atomic experiences $\mathcal{D}$ (transitions in the code below), and call the update method of FunctionApprox repeatedly, passing the data $\mathcal{D}$ (now in the form of a Sequence[TransitionStep]) to each invocation of the update method (using the function itertools.repeat). This repeated invocation of the update method is done by using the function iterate.accumulate. This is done until convergence (convergence based on the done function in the code below), at which point we return the converged FunctionApprox.

```
import rl.markov_process as mp
from rl.approximate_dynamic_programming import ValueFunctionApprox, extended_vf
import rl.iterate as iterate
import itertools
import numpy as np
def batch_td_prediction(
    transitions: Iterable[mp.TransitionStep[S]],
    approx_0: ValueFunctionApprox[S],
    gamma: float,
    convergence_tolerance: float = 1e-5
) -> ValueFunctionApprox[S]:
    '''transitions is a finite iterable'''
    def step(
        v: ValueFunctionApprox[S],
        tr_seq: Sequence[mp.TransitionStep[S]]
    ) -> ValueFunctionApprox[S]:
        return v.update([(
            tr.state, tr.reward + gamma * extended_vf(v, tr.next_state)
        ) for tr in tr_seq])

    def done(
        a: ValueFunctionApprox[S],
        b: ValueFunctionApprox[S],
        convergence_tolerance=convergence_tolerance
    ) -> bool:
        return b.within(a, convergence_tolerance)

    return iterate.converged(
        iterate.accumulate(
            itertools.repeat(list(transitions)),
            step,
            initial=approx_0
        ),
        done=done
```

Likewise, we can do Batch TD(λ) Prediction. Here we are given a fixed, finite number of trace experiences

$$\mathcal{D} = [(S_{i,0}, R_{i,1}, S_{i,1}, R_{i,2}, S_{i,2}, \ldots, R_{i,T_i}, S_{i,T_i})|1 \leq i \leq n]$$

For trace experience i, for each time step t in the trace experience, we calculate the eligibility traces as follows:

$$E_{i,t} = \gamma\lambda \cdot E_{i,t-1} + \nabla_w V(S_{i,t}; w) \text{ for all } t = 1, 1, \ldots T_i - 1$$

with the eligiblity traces initialized at time 0 for trace experience i as $E_{i,0} = \nabla_w V(S_{i,0}; w)$.

---

[3]batch_td_prediction is defined in the file rl/td.py.

Then, each update to the parameters w is as follows:

$$\Delta w = \alpha \cdot \frac{1}{n} \cdot \sum_{i=1}^{n} \frac{1}{T_i} \cdot \sum_{t=0}^{T_i-1} (R_{i,t+1} + \gamma \cdot V(S_{i,t+1}; w) - V(S_{i,t}; w)) \cdot E_{i,t} \qquad (13.1)$$

13.2 A GENERIC IMPLEMENTATION OF EXPERIENCE-REPLAY

Before we proceed to more algorithms involving Experience-Replay and/or Batch RL, it is vital to recognize that the concept of Experience-Replay stands on its own, independent of its use in Batch RL. In fact, Experience-Replay is a much broader concept, beyond its use in RL. The idea of Experience-Replay is that we have a stream of data coming in and instead of consuming it in an algorithm as soon as it arrives, we store each unit of incoming data in memory (which we shall call *Experience-Replay-Memory*, abbreviated as ER-Memory), and use samples of data from ER-Memory (with replacement) for our algorithm's needs. Thus, we are routing the incoming stream of data to ER-Memory and sourcing data needed for our algorithm from ER-Memory (by sampling with replacement). This enables re-use of the incoming data stream. It also gives us flexibility to sample an arbitrary number of data units at a time, so our algorithm doesn't need to be limited to using a single unit of data at a time. Lastly, we organize the data in ER-Memory in such a manner that we can assign different sampling weights to different units of data, depending on the arrival time of the data. This is quite useful for many algorithms that wish to give more importance to recently arrived data and de-emphasize/forget older data.

Let us now write some code to implement all of these ideas described above. The code below uses an arbitrary data type `T`, which means that the unit of data being handled with Experience-Replay could be any data structure (specifically, not limited to the `TransitionStep` data type that we care about for RL with Experience-Replay).

The attribute `saved_transitions: List[T]` is the data structure storing the incoming units of data, with the most recently arrived unit of data at the end of the list (since we append to the list). The attribute `time_weights_func` lets the user specify a function from the reverse-time-stamp of a unit of data to the sampling weight to assign to that unit of data ("reverse-time-stamp" means the most recently-arrived unit of data has a time-index of 0, although physically it is stored at the end of the list, rather than at the start). The attribute `weights` simply stores the sampling weights of all units of data in `saved_transitions`, and the attribute `weights_sum` stores the sum of the `weights` (the attributes `weights` and `weights_sum` are there purely for computational efficiency to avoid too many calls to `time_weights_func` and avoidance of summing a long list of weights, which is required to normalize the weights to sum up to 1).

`add_data` appends an incoming unit of data (`transition: T`) to `self.saved_transitions` and updates `self.weights` and `self.weights_sum`. `sample_mini_batches` returns a sample of specified size `mini_batch_size`, using the sampling weights in `self.weights`. We also have a method `replay` that takes as input an `Iterable` of transitions and a `mini_batch_size`, and returns an `Iterator` of `mini_batch_sized` data units. As long as the input `transitions: Iterable[T]` is not exhausted, `replay` appends each unit of data in `transitions` to `self.saved_transitions` and then yields a `mini_batch_sized` sample of data. Once `transitions: Iterable[T]` is exhausted, it simply yields the samples of data. The `Iterator` generated by `replay` can be piped to any algorithm that expects an `Iterable` of the units of data as input, essentially enabling us to replace the pipe carrying an input data stream with a pipe carrying the data stream sourced from ER-Memory.

```
T = TypeVar('T')
```

```python
class ExperienceReplayMemory(Generic[T]):
    saved_transitions: List[T]
    time_weights_func: Callable[[int], float]
    weights: List[float]
    weights_sum: float

    def __init__(
        self,
        time_weights_func: Callable[[int], float] = lambda _: 1.0,
    ):
        self.saved_transitions = []
        self.time_weights_func = time_weights_func
        self.weights = []
        self.weights_sum = 0.0

    def add_data(self, transition: T) -> None:
        self.saved_transitions.append(transition)
        weight: float = self.time_weights_func(len(self.saved_transitions) - 1)
        self.weights.append(weight)
        self.weights_sum += weight

    def sample_mini_batch(self, mini_batch_size: int) -> Sequence[T]:
        num_transitions: int = len(self.saved_transitions)
        return Categorical(
            {tr: self.weights[num_transitions - 1 - i] / self.weights_sum
                for i, tr in enumerate(self.saved_transitions)}
        ).sample_n(min(mini_batch_size, num_transitions))

    def replay(
        self,
        transitions: Iterable[T],
        mini_batch_size: int
    ) -> Iterator[Sequence[T]]:

        for transition in transitions:
            self.add_data(transition)
            yield self.sample_mini_batch(mini_batch_size)

        while True:
            yield self.sample_mini_batch(mini_batch_size)
```

The code above is in the file rl/experience_replay.py. We encourage you to implement Batch MC Prediction and Batch TD Prediction using this ExperienceReplayMemory class.

13.3 LEAST-SQUARES RL PREDICTION

We've seen how Batch RL Prediction is an iterative process of weight updates until convergence—the MRP Value Function is updated with repeated use of the fixed, finite (batch) data that is made available. However, if we assume that the MRP Value Function approximation $V(s; w)$ is a linear function approximation (linear in a set of feature functions of the state space), then we can solve for the MRP Value Function with direct and simple linear algebra operations (i.e., without the need for iterative weight updates until convergence). Let us see how.

We define a sequence of feature functions $\phi_j : \mathcal{N} \to \mathbb{R}, j = 1, 2, \ldots, m$ and we assume the parameters w is a weights vector $w = (w_1, w_2, \ldots, w_m) \in \mathbb{R}^m$. Therefore, the MRP Value Function is approximated as:

$$V(s; w) = \sum_{j=1}^{m} \phi_j(s) \cdot w_j = \phi(s)^T \cdot w \text{ for all } s \in \mathcal{N}$$

where $\phi(s) \in \mathbb{R}^m$ is the feature vector for state s.

The direct solution of the MRP Value Function using simple linear algebra operations is known as Least-Squares (abbreviated as LS) solution. We start with Batch MC Prediction for the case of linear function approximation, which is known as Least-Squares Monte-Carlo (abbreviated as LSMC).

13.3.1 Least-Squares Monte-Carlo (LSMC)

For the case of linear function approximation, the loss function for Batch MC Prediction with data $[(S_i, G_i)|1 \leq i \leq n]$ is:

$$\mathcal{L}(\boldsymbol{w}) = \frac{1}{2n} \cdot \sum_{i=1}^{n} (\sum_{j=1}^{m} \phi_j(S_i) \cdot w_j - G_i)^2 = \frac{1}{2n} \cdot \sum_{i=1}^{n} (\phi(S_i)^T \cdot \boldsymbol{w} - G_i)^2$$

We set the gradient of this loss function to 0, and solve for $\boldsymbol{w}^*$. This yields:

$$\sum_{i=1}^{n} \phi(S_i) \cdot (\phi(S_i)^T \cdot \boldsymbol{w}^* - G_i) = 0$$

We can calculate the solution $\boldsymbol{w}^*$ as $\boldsymbol{A}^{-1} \cdot \boldsymbol{b}$, where the $m \times m$ Matrix $\boldsymbol{A}$ is accumulated at each data pair (S_i, G_i) as:

$$\boldsymbol{A} \leftarrow \boldsymbol{A} + \phi(S_i) \cdot \phi(S_i)^T \text{ (i.e., outer-product of } \phi(S_i) \text{ with itself)}$$

and the m-Vector $\boldsymbol{b}$ is accumulated at each data pair (S_i, G_i) as:

$$\boldsymbol{b} \leftarrow \boldsymbol{b} + \phi(S_i) \cdot G_i$$

To implement this algorithm, we can simply call `batch_mc_prediction` that we had written earlier by setting the argument `approx` as `LinearFunctionApprox` and by setting the attribute `direct_solve` in `approx: LinearFunctionApprox[S]` as `True`. If you read the code under `direct_solve=True` branch in the `solve` method, you will see that it indeed performs the above-described linear algebra calculations. The inversion of the matrix $\boldsymbol{A}$ is $O(m^3)$ complexity. However, we can speed up the algorithm to be $O(m^2)$ with a different implementation—we can maintain the inverse of $\boldsymbol{A}$ after each (S_i, G_i) update to $\boldsymbol{A}$ by applying the Sherman-Morrison formula for incremental inverse (Sherman and Morrison 1950). The Sherman-Morrison incremental inverse for $\boldsymbol{A}$ is as follows:

$$(\boldsymbol{A} + \phi(S_i) \cdot \phi(S_i)^T)^{-1} = \boldsymbol{A}^{-1} - \frac{\boldsymbol{A}^{-1} \cdot \phi(S_i) \cdot \phi(S_i)^T \cdot \boldsymbol{A}^{-1}}{1 + \phi(S_i)^T \cdot \boldsymbol{A}^{-1} \cdot \phi(S_i)}$$

with $\boldsymbol{A}^{-1}$ initialized to $\frac{1}{\epsilon} \cdot \boldsymbol{I}_m$, where $\boldsymbol{I}_m$ is the $m \times m$ identity matrix, and $\epsilon \in \mathbb{R}^+$ is a small number provided as a parameter to the algorithm. $\frac{1}{\epsilon}$ should be considered to be a proxy for the step-size α which is not required for least-squares algorithms. If ϵ is too small, the sequence of inverses of $\boldsymbol{A}$ can be quite unstable and if ϵ is too large, the learning is slowed.

This brings down the computational complexity of this algorithm to $O(m^2)$. We won't implement the Sherman-Morrison incremental inverse for LSMC, but in the next subsection we shall implement it for Least-Squares Temporal Difference (LSTD).

13.3.2 Least-Squares Temporal-Difference (LSTD)

For the case of linear function approximation, the loss function for Batch TD Prediction with data $[(S_i, R_i, S'_i)|1 \leq i \leq n]$ is:

$$\mathcal{L}(\boldsymbol{w}) = \frac{1}{2n} \cdot \sum_{i=1}^{n} (\phi(S_i)^T \cdot \boldsymbol{w} - (R_i + \gamma \cdot \phi(S'_i)^T \cdot \boldsymbol{w}))^2$$

We set the semi-gradient of this loss function to 0, and solve for $\boldsymbol{w}^*$. This yields:

$$\sum_{i=1}^{n} \phi(S_i) \cdot (\phi(S_i)^T \cdot \boldsymbol{w}^* - (S_i + \gamma \cdot \phi(S_i')^T \cdot \boldsymbol{w}^*)) = 0$$

We can calculate the solution $\boldsymbol{w}^*$ as $\boldsymbol{A}^{-1} \cdot \boldsymbol{b}$, where the $m \times m$ Matrix $\boldsymbol{A}$ is accumulated at each atomic experience (S_i, R_i, S_i') as:

$$\boldsymbol{A} \leftarrow \boldsymbol{A} + \phi(S_i) \cdot (\phi(S_i) - \gamma \cdot \phi(S_i'))^T \text{ (note the Outer-Product)}$$

and the m-Vector $\boldsymbol{b}$ is accumulated at each atomic experience (S_i, R_i, S_i') as:

$$\boldsymbol{b} \leftarrow \boldsymbol{b} + \phi(S_i) \cdot R_i$$

With Sherman-Morrison incremental inverse, we can reduce the computational complexity from $O(m^3)$ to $O(m^2)$, as follows:

$$(\boldsymbol{A} + \phi(S_i) \cdot (\phi(S_i) - \gamma \cdot \phi(S_i))^T)^{-1} = \boldsymbol{A}^{-1} - \frac{\boldsymbol{A}^{-1} \cdot \phi(S_i) \cdot (\phi(S_i) - \gamma \cdot \phi(S_i'))^T \cdot \boldsymbol{A}^{-1}}{1 + (\phi(S_i) - \gamma \cdot \phi(S_i'))^T \cdot \boldsymbol{A}^{-1} \cdot \phi(S_i)}$$

with $\boldsymbol{A}^{-1}$ initialized to $\frac{1}{\epsilon} \cdot \boldsymbol{I}_m$, where $\boldsymbol{I}_m$ is the $m \times m$ identity matrix, and $\epsilon \in \mathbb{R}^+$ is a small number provided as a parameter to the algorithm.

This algorithm is known as the Least-Squares Temporal-Difference (LSTD) algorithm and is due to Bradtke and Barto (Bradtke and Barto 1996).

Now let's write some code to implement this LSTD algorithm. The arguments transitions, feature_functions, gamma and epsilon of the function least_squares_td below are quite self-explanatory. This is a batch method with direct calculation of the estimated Value Function from batch data (rather than iterative weight updates), so least_squares_td returns the estimated Value Function of type LinearFunctionApprox[NonTerminal[S]], rather than an Iterator over the updated function approximations (as was the case in Incremental RL algorithms).

The code below should be fairly self-explanatory. a_inv refers to $\boldsymbol{A}^{-1}$ which is updated with the Sherman-Morrison incremental inverse method. b_vec refers to the $\boldsymbol{b}$ vector. phi1 refers to $\phi(S_i)$, phi2 refers to $\phi(S_i) - \gamma \cdot \phi(S_i')$ (except when S_i' is a terminal state, in which case phi2 is simply $\phi(S_i)$). The temporary variable temp refers to $(\boldsymbol{A}^{-1})^T \cdot (\phi(S_i) - \gamma \cdot \phi(S_i'))$ and is used both in the numerator and denominator in the Sherman-Morrison formula to update $\boldsymbol{A}^{-1}$.

```
from rl.function_approx import LinearFunctionApprox
import rl.markov_process as mp
import numpy as np
def least_squares_td(
        transitions: Iterable[mp.TransitionStep[S]],
        feature_functions: Sequence[Callable[[NonTerminal[S]], float]],
        gamma: float,
        epsilon: float
) -> LinearFunctionApprox[NonTerminal[S]]:
    ''' transitions is a finite iterable '''
    num_features: int = len(feature_functions)
    a_inv: np.ndarray = np.eye(num_features) / epsilon
    b_vec: np.ndarray = np.zeros(num_features)
    for tr in transitions:
        phi1: np.ndarray = np.array([f(tr.state) for f in feature_functions])
        if isinstance(tr.next_state, NonTerminal):
            phi2 = phi1 - gamma * np.array([f(tr.next_state)
                                            for f in feature_functions])
```

```
      else:
          phi2 = phi1
      temp: np.ndarray = a_inv.T.dot(phi2)
      a_inv = a_inv - np.outer(a_inv.dot(phi1), temp) / (1 + phi1.dot(temp))
      b_vec += phi1 * tr.reward
  opt_wts: np.ndarray = a_inv.dot(b_vec)
  return LinearFunctionApprox.create(
      feature_functions=feature_functions,
      weights=Weights.create(opt_wts)
  )
```

The code above is in the file rl/td.py.

Now let's test this on transitions data sampled from the RandomWalkMRP example we had constructed in Chapter 11. As a reminder, this MRP consists of a random walk across states $\{0, 1, 2, \dots, B\}$ with 0 and B as the terminal states (think of these as terminating barriers of a random walk) and the remaining states as the non-terminal states. From any non-terminal state i, we transition to state $i + 1$ with probability p and to state $i - 1$ with probability $1 - p$. The reward is 0 upon each transition, except if we transition from state $B - 1$ to terminal state B which results in a reward of 1. The code for RandomWalkMRP is in the file rl/chapter10/random_walk_mrp.py.

First, we set up a RandomWalkMRP object with $B = 20, p = 0.55$ and calculate its true Value Function (so we can later compare against Incremental TD and LSTD methods).

```
from rl.chapter10.random_walk_mrp import RandomWalkMRP
import nump as np

this_barrier: int = 20
this_p: float = 0.55
random_walk: RandomWalkMRP = RandomWalkMRP(
    barrier=this_barrier,
    p=this_p
)
gamma = 1.0
true_vf: np.ndarray = random_walk.get_value_function_vec(gamma=gamma)
```

Let's say we have access to only 10,000 transitions (each transition is an object of the type TransitionStep). First, we generate these 10,000 sampled transitions from the RandomWalkMRP object we created above.

```
from rl.approximate_dynamic_programming import NTStateDistribution
from rl.markov_process import TransitionStep
import itertools

num_transitions: int = 10000
nt_states: Sequence[NonTerminal[int]] = random_walk.non_terminal_states
start_distribution: NTStateDistribution[int] = Choose(set(nt_states))
traces: Iterable[Iterable[TransitionStep[int]]] = \
    random_walk.reward_traces(start_distribution)
transitions: Iterable[TransitionStep[int]] = \
    itertools.chain.from_iterable(traces)
td_transitions: Iterable[TransitionStep[int]] = \
    itertools.islice(transitions, num_transitions)
```

Before running LSTD, let's run Incremental Tabular TD on the 10,000 transitions in td_transitions and obtain the resultant Value Function (td_vf in the code below). Since there are only 10,000 transitions, we use an aggressive initial learning rate of 0.5 to promote fast learning, but we let this high learning rate decay quickly so the learning stabilizes.

```
from rl.function_approx import Tabular
import rl.iterate as iterate

initial_learning_rate: float = 0.5
```

```
half_life: float = 1000
exponent: float = 0.5
approx0: Tabular[NonTerminal[int]] = Tabular(
    count_to_weight_func=learning_rate_schedule(
        initial_learning_rate=initial_learning_rate,
        half_life=half_life,
        exponent=exponent
    )
)
td_func: Tabular[NonTerminal[int]] = \
    iterate.last(itertools.islice(
        td_prediction(
            transitions=td_transitions,
            approx_0=approx0,
            gamma=gamma
        ),
        num_transitions
    ))
td_vf: np.ndarray = td_func.evaluate(nt_states)
```

Finally, we run the LSTD algorithm on 10,000 transitions. Note that the Value Function of RandomWalkMRP, for $p \neq 0.5$, is non-linear as a function of the integer states. So we use non-linear features that can approximate arbitrary non-linear shapes—a good choice is the set of (orthogonal) Laguerre Polynomials. In the code below, we use the first 5 Laguerre Polynomials (i.e., upto degree 4 polynomial) as the feature functions for the linear function approximation of the Value Function. Then we invoke the LSTD algorithm we wrote above to calculate the LinearFunctionApprox based on this batch of 10,000 transitions.

```
from rl.chapter12.laguerre import laguerre_state_features
from rl.function_approx import LinearFunctionApprox

num_polynomials: int = 5
features: Sequence[Callable[[NonTerminal[int]], float]] = \
    laguerre_state_features(num_polynomials)
lstd_transitions: Iterable[TransitionStep[int]] = \
    itertools.islice(transitions, num_transitions)
epsilon: float = 1e-4

lstd_func: LinearFunctionApprox[NonTerminal[int]] = \
    least_squares_td(
        transitions=lstd_transitions,
        feature_functions=features,
        gamma=gamma,
        epsilon=epsilon
    )
lstd_vf: np.ndarray = lstd_func.evaluate(nt_states)
```

Figure 13.1 depicts how the LSTD Value Function estimate (for 10,000 transitions) lstd_vf compares against Incremental Tabular TD Value Function estimate (for 10,000 transitions) td_vf and against the true value function true_vf (obtained using the linear-algebra-solver-based calculation of the MRP Value Function). We encourage you to modify the parameters used in the code above to see how it alters the results—specifically play around with this_barrier, this_p, gamma, num_transitions, the learning rate trajectory for Incremental Tabular TD, the number of Laguerre polynomials, and epsilon. The above code is in the file rl/chapter12/random_walk_lstd.py.

13.3.3 LSTD(λ)

Likewise, we can do LSTD(λ) using Eligibility Traces. Here we are given a fixed, finite number of trace experiences

$$\mathcal{D} = [(S_{i,0}, R_{i,1}, S_{i,1}, R_{i,2}, S_{i,2}, \ldots, R_{i,T_i}, S_{i,T_i}) | 1 \leq i \leq n]$$

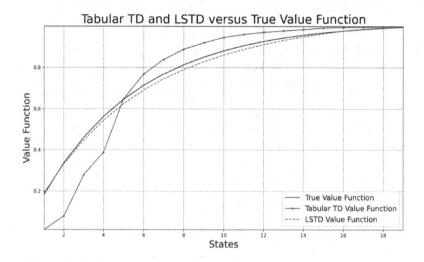

Figure 13.1 **LSTD and Tabular TD Value Functions**

Denote the Eligibility Traces of trace experience i at time t as $E_{i,t}$. Note that the eligibility traces accumulate $\nabla_{w} V(s; w) = \phi(s)$ in each trace experience. When accumulating, the previous time step's eligibility traces is discounted by $\lambda\gamma$. By setting the right-hand-side of Equation (13.1) to 0 (i.e., setting the update to w over all atomic experiences data to 0), we get:

$$\sum_{i=1}^{n} \frac{1}{T_i} \cdot \sum_{t=0}^{T_i-1} E_{i,t} \cdot (\phi(S_{i,t})^{T} \cdot w^{*} - (R_{i,t+1} + \gamma \cdot \phi(S_{i,t+1})^{T} \cdot w^{*})) = 0$$

We can calculate the solution w^* as $A^{-1} \cdot b$, where the $m \times m$ Matrix A is accumulated at each atomic experience $(S_{i,t}, R_{i,t+1}, S_{i,t+1})$ as:

$$A \leftarrow A + \frac{1}{T_i} \cdot E_{i,t} \cdot (\phi(S_{i,t}) - \gamma \cdot \phi(S_{i,t+1}))^{T} \text{ (note the Outer-Product)}$$

and the m-Vector b is accumulated at each atomic experience $(S_{i,t}, R_{i,t+1}, S_{i,t+1})$ as:

$$b \leftarrow b + \frac{1}{T_i} \cdot E_{i,t} \cdot R_{i,t+1}$$

With Sherman-Morrison incremental inverse, we can reduce the computational complexity from $O(m^3)$ to $O(m^2)$.

13.3.4 Convergence of Least-Squares Prediction

Before we move on to Least-Squares for the Control problem, we want to point out that the convergence behavior of Least-Squares Prediction algorithms are identical to their counter-part Incremental RL Prediction algorithms, with the exception that Off-Policy LSMC does not have convergence guarantees. Figure 13.2 shows the updated summary table for convergence of RL Prediction algorithms (that we had displayed at the end of Chapter 12) to now also include Least-Squares Prediction algorithms.

This ends our coverage of Least-Squares Prediction. Before we move on to Least-Squares Control, we need to cover Incremental RL Control with Experience-Replay as it serves as a stepping stone towards Least-Squares Control.

On/Off Policy	Algorithm	Tabular	Linear	Non-Linear
On-Policy	MC	✓	✓	✓
	LSMC	✓	✓	-
	TD	✓	✓	✗
	LSTD	✓	✓	-
	Gradient TD	✓	✓	✓
Off-Policy	MC	✓	✓	✓
	LSMC	✓	✗	-
	TD	✓	✗	✗
	LSTD	✓	✗	-
	Gradient TD	✓	✓	✓

Figure 13.2 Convergence of RL Prediction Algorithms

13.4 Q-LEARNING WITH EXPERIENCE-REPLAY

In this section, we cover Off-Policy Incremental TD Control with Experience-Replay. Specifically, we revisit the Q-Learning algorithm we covered in Chapter 12, but we tweak that algorithm such that the transitions used to make the Q-Learning updates are sourced from an experience-replay memory, rather than from a behavior policy derived from the current Q-Value estimate. While investigating the challenges with Off-Policy TD methods with deep learning function approximation, researchers identified two challenges:

1) The sequences of states made available to deep learning through trace experiences are highly correlated, whereas deep learning algorithms are premised on data samples being independent.
2) The data distribution changes as the RL algorithm learns new behaviors, whereas deep learning algorithms are premised on a fixed underlying distribution (i.e., stationary).

Experience-Replay serves to smooth the training data distribution over many past behaviors, effectively resolving the correlation issue as well as the non-stationary issue. Hence, Experience-Replay is a powerful idea for Off-Policy TD Control. The idea of using Experience-Replay for Off-Policy TD Control is due to the Ph.D. thesis of Long Lin (Lin 1993).

To make this idea of Q-Learning with Experience-Replay clear, we make a few changes to the q_learning function we had written in Chapter 12 with the following function q_learning_experience_replay.

```
from rl.markov_decision_process import TransitionStep
from rl.approximate_dynamic_programming import QValueFunctionApprox
from rl.approximate_dynamic_programming import NTStateDistribution
from rl.experience_replay import ExperienceReplayMemory

PolicyFromQType = Callable[
    [QValueFunctionApprox[S, A], MarkovDecisionProcess[S, A]],
    Policy[S, A]
]

def q_learning_experience_replay(
    mdp: MarkovDecisionProcess[S, A],
    policy_from_q: PolicyFromQType,
    states: NTStateDistribution[S],
    approx_0: QValueFunctionApprox[S, A],
    gamma: float,
    max_episode_length: int,
```

```
        mini_batch_size: int,
        weights_decay_half_life: float
) -> Iterator[QValueFunctionApprox[S, A]]:
        exp_replay: ExperienceReplayMemory[TransitionStep[S, A]] = \
            ExperienceReplayMemory(
                time_weights_func=lambda t: 0.5 ** (t / weights_decay_half_life),
            )
        q: QValueFunctionApprox[S, A] = approx_0
        yield q
        while True:
            state: NonTerminal[S] = states.sample()
            steps: int = 0
            while isinstance(state, NonTerminal) and steps < max_episode_length:
                policy: Policy[S, A] = policy_from_q(q, mdp)
                action: A = policy.act(state).sample()
                next_state, reward = mdp.step(state, action).sample()
                exp_replay.add_data(TransitionStep(
                    state=state,
                    action=action,
                    next_state=next_state,
                    reward=reward
                ))
                trs: Sequence[TransitionStep[S, A]] = \
                    exp_replay.sample_mini_batch(mini_batch_size)
                q = q.update(
                    [(
                        (tr.state, tr.action),
                        tr.reward + gamma * (
                            max(q((tr.next_state, a))
                                for a in mdp.actions(tr.next_state))
                            if isinstance(tr.next_state, NonTerminal) else 0.)
                    ) for tr in trs],
                )
                yield q
                steps += 1
                state = next_state
```

The key difference between the q_learning algorithm we wrote in Chapter 12 and this q_learning_experience_replay algorithm is that here we have an experience-replay memory (using the ExperienceReplayMemory class we had implemented earlier). In the q_learning algorithm, the (state, action, next_state, reward) 4-tuple comprising TransitionStep (that is used to perform the Q-Learning update) was the result of action being sampled from the behavior policy (derived from the current estimate of the Q-Value Function, e.g., ϵ-greedy), and then the next_state and reward being generated from the (state, action) pair using the step method of mdp. Here in q_learning_experience_replay, we don't use this 4-tuple TransitionStep to perform the update—rather, we append this 4-tuple to the ExperienceReplayMemory (using the add_data method), then we sample mini_batch_sized TransitionSteps from the ExperienceReplayMemory (giving more sampling weightage to the more recently added TransitionSteps), and use those 4-tuple TransitionSteps to perform the Q-Learning update. Note that these sampled TransitionSteps might be from old behavior policies (derived from old estimates of the Q-Value estimate). The key is that this algorithm re-uses atomic experiences that were previously prepared by the algorithm, which also means that it re-uses behavior policies that were previously constructed by the algorithm.

The argument mini_batch_size refers to the number of TransitionSteps to be drawn from the ExperienceReplayMemory at each step. The argument weights_decay_half_life refers to the half life of an exponential decay function for the weights used in the sampling of the TransitionSteps (the most recently added TransitionStep has the highest weight). With this understanding, the code should be self-explanatory.

The above code is in the file rl/td.py.

13.4.1 Deep Q-Networks (DQN) Algorithm

DeepMind developed an innovative and practically effective RL Control algorithm based on Q-Learning with Experience-Replay—an algorithm they named as Deep Q-Networks (abberviated as DQN). Apart from reaping the above-mentioned benefits of Experience-Replay for Q-Learning with a Deep Neural Network approximating the Q-Value function, they also benefited from employing a second Deep Neural Network (let us call the main DNN as the Q-Network, referring to its parameters at w, and the second DNN as the target network, referring to its parameters as w^-). The parameters w^- of the target network are infrequently updated to be made equal to the parameters w of the Q-network. The purpose of the Q-Network is to evaluate the Q-Value of the current state s and the purpose of the target network is to evaluate the Q-Value of the next state s', which in turn is used to obtain the Q-Learning target (note that the Q-Value of the current state is $Q(s, a; w)$ and the Q-Learning target is $r + \gamma \cdot \max_{a'} Q(s', a'; w^-)$ for a given atomic experience (s, a, r, s')).

Deep Learning is premised on the fact that the supervised learning targets (response values y corresponding to predictor values x) are pre-generated fixed values. This is not the case in TD learning where the targets are dependent on the Q-Values. As Q-Values are updated at each step, the targets also get updated, and this correlation between the current state's Q-Value estimate and the target value typically leads to oscillations or divergence of the Q-Value estimate. By infrequently updating the parameters w^- of the target network (providing the target values) to be made equal to the parameters w of the Q-network (which are updated at each iteration), the targets in the Q-Learning update are essentially kept fixed. This goes a long way in resolving the core issue of correlation between the current state's Q-Value estimate and the target values, helping considerably with convergence of the Q-Learning algorithm. Thus, DQN reaps the benefits of not just Experience-Replay in Q-Learning (which we articulated earlier), but also the benefits of having "fixed" targets. DNN utilizes a parameter C such that the updating of w^- to be made equal to w is done once every C updates to w (updates to w are based on the usual Q-Learning update equation).

We won't implement the DQN algorithm in Python code—however, we sketch the outline of the algorithm, as follows:

At each time t for each episode:

- Given state S_t, take action A_t according to ϵ-greedy policy extracted from Q-network values $Q(S_t, a; w)$.
- Given state S_t and action A_t, obtain reward R_{t+1} and next state S_{t+1} from the environment.
- Append atomic experience $(S_t, A_t, R_{t+1}, S_{t+1})$ in experience-replay memory $\mathcal{D}$.
- Sample a random mini-batch of atomic experiences $(s_i, a_i, r_i, s'_i) \sim \mathcal{D}$.
- Using this mini-batch of atomic experiences, update the Q-network parameters w with the Q-learning targets based on "frozen" parameters w^- of the target network.

$$\Delta w = \alpha \cdot \sum_i (r_i + \gamma \cdot \max_{a'_i} Q(s'_i, a'_i; w^-) - Q(s_i, a_i; w)) \cdot \nabla_w Q(s_i, a_i; w)$$

- $S_t \leftarrow S_{t+1}$
- Once every C time steps, set $w^- \leftarrow w$.

To learn more about the effectiveness of DQN for Atari games, see the Original DQN Paper (Mnih et al. 2013) and the DQN Nature Paper (Mnih et al. 2015) that DeepMind has published.

Now we are ready to cover Batch RL Control (specifically Least-Squares TD Control), which combines the ideas of Least-Squares TD Prediction and Q-Learning with Experience-Replay.

13.5 LEAST-SQUARES POLICY ITERATION (LSPI)

Having seen Least-Squares Prediction, the natural question is whether we can extend the Least-Squares (batch with linear function approximation) methodology to solve the Control problem. For On-Policy MC Control and On-Policy TD Control, we take the usual route of Generalized Policy Iteration (GPI) with:

1. Policy Evaluation as Least-Squares Q-Value Prediction. Specifically, the Q-Value for a policy π is approximated as:

$$Q^\pi(s,a) \approx Q(s,a;\boldsymbol{w}) = \boldsymbol{\phi}(s,a)^T \cdot \boldsymbol{w} \text{ for all } s \in \mathcal{N}, \text{ for all } a \in \mathcal{A}$$

with a direct linear-algebraic solve for the linear function approximation weights $\boldsymbol{w}$ using batch experiences data generated using policy π.

2. ϵ-Greedy Policy Improvement.

In this section, we focus on Off-Policy Control with Least-Squares TD. This algorithm is known as Least-Squares Policy Iteration, abbreviated as LSPI, developed by Lagoudakis and Parr (Lagoudakis and Parr 2003). LSPI has been an important go-to algorithm in the history of RL Control because of its simplicity and effectiveness. The basic idea of LSPI is that it does Generalized Policy Iteration (GPI) in the form of *Q-Learning with Experience-Replay*, with the key being that instead of doing the usual Q-Learning update after each atomic experience, we do *batch Q-Learning* for the Policy Evaluation phase of GPI. We spend the rest of this section describing LSPI in detail and then implementing it in Python code.

The input to LSPI is a fixed finite data set $\mathcal{D}$, consisting of a set of (s, a, r, s') atomic experiences, i.e., a set of `rl.markov_decision_process.TransitionStep` objects, and the task of LSPI is to determine the Optimal Q-Value Function (and hence, Optimal Policy) based on this experiences data set $\mathcal{D}$ using an experience-replayed, batch Q-Learning technique described below. Assume $\mathcal{D}$ consists of n atomic experiences, indexed as $i = 1, 2, \ldots n$, with atomic experience i denoted as (s_i, a_i, r_i, s'_i).

In LSPI, each iteration of GPI involves access to:

- The experiences data set $\mathcal{D}$.
- A *Deterministic Target Policy* (call it π_D), that is made available from the previous iteration of GPI.

Given $\mathcal{D}$ and π_D, the goal of each iteration of GPI is to solve for weights $\boldsymbol{w}^*$ that minimizes:

$$\mathcal{L}(\boldsymbol{w}) = \sum_{i=1}^{n} (Q(s_i, a_i; \boldsymbol{w}) - (r_i + \gamma \cdot Q(s'_i, \pi_D(s'_i); \boldsymbol{w})))^2$$

$$= \sum_{i=1}^{n} (\boldsymbol{\phi}(s_i, a_i)^T \cdot \boldsymbol{w} - (r_i + \gamma \cdot \boldsymbol{\phi}(s'_i, \pi_D(s'_i))^T \cdot \boldsymbol{w}))^2$$

The solution for the weights $\boldsymbol{w}^*$ is attained by setting the semi-gradient of $\mathcal{L}(\boldsymbol{w})$ to 0, i.e.,

$$\sum_{i=1}^{n} \boldsymbol{\phi}(s_i, a_i) \cdot (\boldsymbol{\phi}(s_i, a_i)^T \cdot \boldsymbol{w}^* - (r_i + \gamma \cdot \boldsymbol{\phi}(s'_i, \pi_D(s'_i))^T \cdot \boldsymbol{w}^*)) = 0 \qquad (13.2)$$

We can calculate the solution $\boldsymbol{w}^*$ as $\boldsymbol{A}^{-1} \cdot \boldsymbol{b}$, where the $m \times m$ Matrix $\boldsymbol{A}$ is accumulated for each `TransitionStep` (s_i, a_i, r_i, s'_i) as:

$$\boldsymbol{A} \leftarrow \boldsymbol{A} + \boldsymbol{\phi}(s_i, a_i) \cdot (\boldsymbol{\phi}(s_i, a_i) - \gamma \cdot \boldsymbol{\phi}(s'_i, \pi_D(s'_i)))^T$$

and the m-Vector b is accumulated at each atomic experience (s_i, a_i, r_i, s_i') as:

$$b \leftarrow b + \phi(s_i, a_i) \cdot r_i$$

With Sherman-Morrison incremental inverse, we can reduce the computational complexity from $O(m^3)$ to $O(m^2)$.

This solved w^* defines an updated Q-Value Function as follows:

$$Q(s, a; w^*) = \phi(s, a)^T \cdot w^* = \sum_{j=1}^{m} \phi_j(s, a) \cdot w_j^*$$

This defines an updated, improved deterministic policy π_D' (serving as the *Deterministic Target Policy* for the next iteration of GPI):

$$\pi_D'(s) = \arg\max_a Q(s, a; w^*) \text{ for all } s \in \mathcal{N}$$

This least-squares solution of w^* (Prediction) is known as Least-Squares Temporal Difference for Q-Value, abbreviated as *LSTDQ*. Thus, LSPI is GPI with LSTDQ and greedy policy improvements. Note how LSTDQ in each iteration re-uses the same data $\mathcal{D}$, i.e., LSPI does experience-replay.

We should point out here that the LSPI algorithm we described above should be considered as the *standard variant* of LSPI. However, we can design several other variants of LSPI, in terms of how the experiences data is sourced and used. Firstly, we should note that the experiences data $\mathcal{D}$ essentially provides the behavior policy for Q-Learning (along with the consequent reward and next state transition). In the *standard variant* we described above, since $\mathcal{D}$ is provided from an external source, the behavior policy that generates this data $\mathcal{D}$ must come from an external source. It doesn't have to be this way—we could generate the experiences data from a behavior policy derived from the Q-Value estimates produced by LSTDQ (e.g., ϵ-greedy policy). This would mean the experiences data used in the algorithm is not a fixed, finite data set, rather a variable, incrementally-produced data set. Even if the behavior policy was external, the data set $\mathcal{D}$ might not be a fixed finite data set—rather, it could be made available as an on-demand, variable data stream. Furthermore, in each iteration of GPI, we could use a subset of the experiences data made available until that point of time (rather than the approach of the standard variant of LSPI that uses all of the available experiences data). If we choose to sample a subset of the available experiences data, we might give more sampling-weightage to the more recently generated data. This would especially be the case if the experiences data was being generated from a policy derived from the Q-Value estimates produced by LSTDQ. In this case, we would leverage the ExperienceReplayMemory class we'd written earlier.

Next, we write code to implement the *standard variant* of LSPI we described above. First, we write a function to implement LSTDQ. As described above, the inputs to LSTDQ are the experiences data $\mathcal{D}$ (transitions in the code below) and a deterministic target policy π_D (target_policy in the code below). Since we are doing a linear function approximation, the input also includes a set of features, described as functions of state and action (feature_functions in the code below). Lastly, the inputs also include the discount factor γ and the numerical control parameter ϵ. The code below should be fairly self-explanatory, as it is a straightforward extension of LSTD (implemented in function least_squares_td earlier). The key differences are that this is an estimate of the Action-Value (Q-Value) function, rather than the State-Value Function, and the target used in the least-squares calculation is the Q-Learning target (produced by the target_policy).

```python
def least_squares_tdq(
    transitions: Iterable[TransitionStep[S, A]],
    feature_functions: Sequence[Callable[[Tuple[NonTerminal[S], A]], float]],
    target_policy: DeterministicPolicy[S, A],
    gamma: float,
    epsilon: float
) -> LinearFunctionApprox[Tuple[NonTerminal[S], A]]:
    '''transitions is a finite iterable'''
    num_features: int = len(feature_functions)
    a_inv: np.ndarray = np.eye(num_features) / epsilon
    b_vec: np.ndarray = np.zeros(num_features)
    for tr in transitions:
        phi1: np.ndarray = np.array([f((tr.state, tr.action))
                                     for f in feature_functions])
        if isinstance(tr.next_state, NonTerminal):
            phi2 = phi1 - gamma * np.array([
                f((tr.next_state, target_policy.action_for(tr.next_state.state)))
                for f in feature_functions])
        else:
            phi2 = phi1
        temp: np.ndarray = a_inv.T.dot(phi2)
        a_inv = a_inv - np.outer(a_inv.dot(phi1), temp) / (1 + phi1.dot(temp))
        b_vec += phi1 * tr.reward

    opt_wts: np.ndarray = a_inv.dot(b_vec)
    return LinearFunctionApprox.create(
        feature_functions=feature_functions,
        weights=Weights.create(opt_wts)
    )
```

Now we are ready to write the standard variant of LSPI. The code below is a straight-forward implementation of our description above, looping through the iterations of GPI, yielding the Q-Value `LinearFunctionApprox` after each iteration of GPI.

```python
def least_squares_policy_iteration(
    transitions: Iterable[TransitionStep[S, A]],
    actions: Callable[[NonTerminal[S]], Iterable[A]],
    feature_functions: Sequence[Callable[[Tuple[NonTerminal[S], A]], float]],
    initial_target_policy: DeterministicPolicy[S, A],
    gamma: float,
    epsilon: float
) -> Iterator[LinearFunctionApprox[Tuple[NonTerminal[S], A]]]:
    '''transitions is a finite iterable'''
    target_policy: DeterministicPolicy[S, A] = initial_target_policy
    transitions_seq: Sequence[TransitionStep[S, A]] = list(transitions)
    while True:
        q: LinearFunctionApprox[Tuple[NonTerminal[S], A]] = \
            least_squares_tdq(
                transitions=transitions_seq,
                feature_functions=feature_functions,
                target_policy=target_policy,
                gamma=gamma,
                epsilon=epsilon,
            )
        target_policy = greedy_policy_from_qvf(q, actions)
        yield q
```

The above code is in the file rl/td.py.

13.5.1 Saving Your Village from a Vampire

Now we consider a Control problem we'd like to test the above LSPI algorithm on. We call it the Vampire problem that can be described as a good old-fashioned bedtime story, as follows:

A village is visited by a vampire every morning who uniform-randomly eats 1 villager upon entering the village, then retreats to the hills, planning to come back the next morning. The villagers come up with a plan. They will poison a certain number of villagers each night until the vampire eats a poisoned villager the next morning, after which the vampire dies immediately (due to the poison in the villager the vampire ate). Unfortunately, all villagers who get poisoned also die the day after they are given the poison. If the goal of the villagers is to maximize the expected number of villagers at termination (termination is when either the vampire dies or all villagers die), what should be the optimal poisoning strategy? In other words, if there are n villagers on any day, how many villagers should be poisoned (as a function of n)?

It is straightforward to model this problem as an MDP. The *State* is the number of villagers at risk on any given night (if the vampire is still alive, the *State* is the number of villagers and if the vampire is dead, the *State* is 0, which is the only *Terminal State*). The *Action* is the number of villagers poisoned on any given night. The *Reward* is zero as long as the vampire is alive, and is equal to the number of villagers remaining if the vampire dies. Let us refer to the initial number of villagers as I. Thus,

$$\mathcal{S} = \{0, 1, \ldots, I\}, \mathcal{T} = \{0\}$$

$$\mathcal{A}(s) = \{0, 1, \ldots, s - 1\} \text{ where } s \in \mathcal{N}$$

For all $s \in \mathcal{N}$, for all $a \in \mathcal{A}(s)$,

$$\mathcal{P}_R(s, a, r, s') = \begin{cases} \frac{s-a}{s} & \text{if } r = 0 \text{ and } s' = s - a - 1 \\ \frac{a}{s} & \text{if } r = s - a \text{ and } s' = 0 \\ 0 & \text{otherwise} \end{cases}$$

It is rather straightforward to solve this with Dynamic Programming (say, Value Iteration) since we know the transition probabilities and rewards function and since the state and action spaces are finite. However, in a situation where we don't know the exact probabilities with which the vampire operates, and we only had access to observations on specific days, we can attempt to solve this problem with Reinforcement Learning (assuming we had access to observations of many vampires operating on many villages). In any case, our goal here is to test LSPI using this vampire problem as an example. So we write some code to first model this MDP as described above, solve it with value iteration (to obtain the benchmark, i.e., true Optimal Value Function and true Optimal Policy to compare against), then generate atomic experiences data from the MDP, and then solve this problem with LSPI using this stream of generated atomic experiences.

```python
from rl.markov_decision_process import TransitionStep
from rl.distribution import Categorical, Choose
from rl.function_approx import LinearFunctionApprox
from rl.policy import DeterministicPolicy, FiniteDeterministicPolicy
from rl.dynamic_programming import value_iteration_result, V
from rl.chapter11.control_utils import get_vf_and_policy_from_qvf
from rl.td import least_squares_policy_iteration
from numpy.polynomial.laguerre import lagval
import itertools
import rl.iterate as iterate
import numpy as np

class VampireMDP(FiniteMarkovDecisionProcess[int, int]):

    initial_villagers: int

    def __init__(self, initial_villagers: int):
        self.initial_villagers = initial_villagers
```

```python
        super().__init__(self.mdp_map())
    def mdp_map(self) -> \
            Mapping[int, Mapping[int, Categorical[Tuple[int, float]]]]:
        return {s: {a: Categorical(
            {(s - a - 1, 0.): 1 - a / s, (0, float(s - a)): a / s}
        ) for a in range(s)} for s in range(1, self.initial_villagers + 1)}
    def vi_vf_and_policy(self) -> \
            Tuple[V[int], FiniteDeterministicPolicy[int, int]]:
        return value_iteration_result(self, 1.0)
    def lspi_features(
        self,
        factor1_features: int,
        factor2_features: int
    ) -> Sequence[Callable[[Tuple[NonTerminal[int], int]], float]]:
        ret: List[Callable[[Tuple[NonTerminal[int], int]], float]] = []
        ident1: np.ndarray = np.eye(factor1_features)
        ident2: np.ndarray = np.eye(factor2_features)
        for i in range(factor1_features):
            def factor1_ff(x: Tuple[NonTerminal[int], int], i=i) -> float:
                return lagval(
                    float((x[0].state - x[1]) ** 2 / x[0].state),
                    ident1[i]
                )
            ret.append(factor1_ff)
        for j in range(factor2_features):
            def factor2_ff(x: Tuple[NonTerminal[int], int], j=j) -> float:
                return lagval(
                    float((x[0].state - x[1]) * x[1] / x[0].state),
                    ident2[j]
                )
            ret.append(factor2_ff)
        return ret
    def lspi_transitions(self) -> Iterator[TransitionStep[int, int]]:
        states_distribution: Choose[NonTerminal[int]] = \
            Choose(self.non_terminal_states)
        while True:
            state: NonTerminal[int] = states_distribution.sample()
            action: int = Choose(range(state.state)). sample()
            next_state, reward = self.step(state, action).sample()
            transition: TransitionStep[int, int] = TransitionStep(
                state=state,
                action=action,
                next_state=next_state,
                reward=reward
            )
            yield transition
    def lspi_vf_and_policy(self) -> \
            Tuple[V[int], FiniteDeterministicPolicy[int, int]]:
        transitions: Iterable[TransitionStep[int, int]] = itertools.islice(
            self.lspi_transitions(),
            20000
        )
        qvf_iter: Iterator[LinearFunctionApprox[Tuple[
            NonTerminal[int], int]]] = least_squares_policy_iteration(
                transitions=transitions,
                actions=self.actions,
                feature_functions=self.lspi_features(4, 4),
                initial_target_policy=DeterministicPolicy(
                    lambda s: int(s / 2)
                ),
                gamma=1.0,
                epsilon=1e-5
        )
        qvf: LinearFunctionApprox[Tuple[NonTerminal[int], int]] = \
            iterate.last(
```

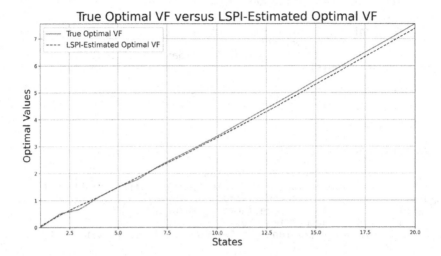

Figure 13.3 **True versus LSPI Optimal Value Function**

```
itertools.islice(
    qvf_iter,
    20
)
)
return get_vf_and_policy_from_qvf(self, qvf)
```

The above code should be self-explanatory. The main challenge with LSPI is that we need to construct features function of the state and action such that the Q-Value Function is linear in those features. In this case, since we simply want to test the correctness of our LSPI implementation, we define feature functions (in method lspi_feature above) based on our knowledge of the true optimal Q-Value Function from the Dynamic Programming solution. The atomic experiences comprising the experiences data $\mathcal{D}$ for LSPI to use is generated with a uniform distribution of non-terminal states and a uniform distribution of actions for a given state (in method lspi_transitions above).

Figure 13.3 shows the plot of the True Optimal Value Function (from Value Iteration) versus the LSPI-estimated Optimal Value Function.

Figure 13.4 shows the plot of the True Optimal Policy (from Value Iteration) versus the LSPI-estimated Optimal Policy.

The above code is in the file rl/chapter12/vampire.py. As ever, we encourage you to modify some of the parameters in this code (including choices of feature functions, nature and number of atomic transitions used, number of GPI iterations, choice of ϵ, and perhaps even a different dynamic for the vampire behavior), and see how the results change.

13.5.2 Least-Squares Control Convergence

We wrap up this section by including the convergence behavior of LSPI in the summary table for convergence of RL Control algorithms (that we had displayed at the end of Chapter 12). Figure 13.5 shows the updated summary table for convergence of RL Control algorithms to now also include LSPI. Note that (✓) means it doesn't quite hit the Optimal Value Function, but bounces around near the Optimal Value Function. But this is better than Q-Learning in the case of linear function approximation.

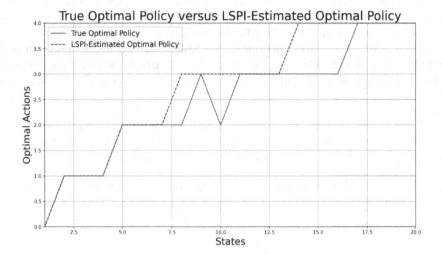

Figure 13.4 **True versus LSPI Optimal Policy**

Algorithm	Tabular	Linear	Non-Linear
MC Control	✓	(✓)	✗
SARSA	✓	(✓)	✗
Q-Learning	✓	✗	✗
LSPI	✓	(✓)	-
Gradient Q-Learning	✓	✓	✗

Figure 13.5 **Convergence of RL Control Algorithms**

13.6 RL FOR OPTIMAL EXERCISE OF AMERICAN OPTIONS

We learnt in Chapter 9 that the American Options Pricing problem is an Optimal Stopping problem and can be modeled as an MDP so that solving the Control problem of the MDP gives us the fair price of an American Option. We can solve it with Dynamic Programming or Reinforcement Learning, as appropriate.

In the financial trading industry, it has traditionally not been a common practice to explicitly view the American Options Pricing problem as an MDP. Specialized algorithms have been developed to price American Options. We now provide a quick overview of the common practice in pricing American Options in the financial trading industry. Firstly, we should note that the price of some American Options is equal to the price of the corresponding European Option, for which we have a closed-form solution under the assumption of a lognormal process for the underlying—this is the case for a plain-vanilla American call option whose price (as we proved in Chapter 9) is equal to the price of a plain-vanilla European call option. However, this is not the case for a plain-vanilla American put option. Secondly, we should note that if the payoff of an American option is dependent on only the current price of the underlying (and not on the past prices of the underlying)—in which case, we say that the option payoff is not "history-dependent"—and if the dimension of the state space is not large, then we can do a simple backward induction on a binomial tree (as we showed in Chapter 9). In practice, a more detailed data structure such as a trinomial tree or a lattice is often used for more accurate backward-induction calculations. However, if the payoff is history-dependent (i.e., payoff depends on past prices of the underlying) or if the payoff depends on the prices of several underlying assets, then the state space is too large for backward induction to handle. In such cases, the standard approach in the financial trading industry is to use the Longstaff-Schwartz pricing algorithm (Longstaff and Schwartz 2001). We won't cover the Longstaff-Schwartz pricing algorithm in detail in this book—it suffices to share here that the Longstaff-Schwartz pricing algorithm combines 3 ideas:

- The Pricing is based on a set of sampling traces of the underlying prices.
- Function approximation of the continuation value for in-the-money states.
- Backward-recursive determination of early exercise states.

The goal of this section is to explain how to price American Options with Reinforcement Learning, as an alternative to the Longstaff-Schwartz algorithm.

13.6.1 LSPI for American Options Pricing

A paper by Li, Szepesvari, Schuurmans (Li, Szepesvári, and Schuurmans 2009) showed that LSPI can be an attractive alternative to the Longstaff-Schwartz algorithm in pricing American Options. Before we dive into the details of pricing American Options with LSPI, let's review the MDP model for American Options Pricing.

- *State* is [Current Time, Relevant History of Underlying Security Prices].
- *Action* is Boolean: Exercise (i.e., Stop) or Continue.
- *Reward* is always 0, except upon Exercise (when the *Reward* is equal to the Payoff).
- *State*-transitions are based on the Underlying Securities' Risk-Neutral Process.

The key is to create a linear function approximation of the state-conditioned *continuation value* of the American Option (*continuation value* is the price of the American Option at the current state, conditional on not exercising the option at the current state, i.e., continuing to hold the option). Knowing the continuation value in any state enables us to compare the continuation value against the exercise value (i.e., payoff), thus providing us with the Optimal Stopping criteria (as a function of the state), which in turn enables us to determine the Price of the American Option. Furthermore, we can customize the LSPI algorithm to the nuances of the American Option Pricing problem, yielding a specialized version of LSPI. The key customization comes from the fact that there are only two actions. The action to exercise produces a (state-conditioned) reward (i.e., option payoff) and transition to a terminal state. The action to continue produces no reward and transitions to a new state at the next time step. Let us refer to these 2 actions as: $a = c$ (continue the option) and $a = e$ (exercise the option).

Since we know the exercise value in any state, we only need to create a linear function approximation for the continuation value, i.e., for the Q-Value $Q(s, c)$ for all non-terminal states s. If we denote the payoff in non-terminal state s as $g(s)$, then $Q(s, e) = g(s)$. So we write

$$\hat{Q}(s, a; w) = \begin{cases} \phi(s)^T \cdot w & \text{if } a = c \\ g(s) & \text{if } a = e \end{cases} \quad \text{for all } s \in \mathcal{N}$$

for feature functions $\phi(\cdot) = [\phi_i(\cdot) | i = 1, \ldots, m]$, which are feature functions of only state (and not action).

Each iteration of GPI in the LSPI algorithm starts with a deterministic target policy $\pi_D(\cdot)$ that is made available as a greedy policy derived from the previous iteration's LSTDQ-solved $\hat{Q}(s; a; w^*)$. The LSTDQ solution for w^* is based on pre-generated training data and with the Q-Learning target policy set to be π_D. Since we learn the Q-Value function for only $a = c$, the behavior policy μ generating experiences data for training is a constant function $\mu(s) = c$. Note also that for American Options, the reward for $a = c$ is 0. So each atomic experience for training is of the form $(s, c, 0, s')$. This means we can represent each atomic experience for training as a 2-tuple (s, s'). This reduces the LSPI Semi-Gradient Equation (13.2) to:

$$\sum_i \phi(s_i) \cdot (\phi(s_i)^T \cdot \boldsymbol{w}^* - \gamma \cdot \hat{Q}(s'_i, \pi_D(s'_i); \boldsymbol{w}^*)) = 0 \qquad (13.3)$$

We need to consider two cases for the term $\hat{Q}(s'_i, \pi_D(s'_i); \boldsymbol{w}^*)$:

- $C1$: If s'_i is non-terminal and $\pi_D(s'_i) = c$ (i.e., $\phi(s'_i)^T \cdot \boldsymbol{w} \geq g(s'_i)$): Substitute $\phi(s'_i)^T \cdot \boldsymbol{w}^*$ for $\hat{Q}(s'_i, \pi_D(s'_i); \boldsymbol{w}^*)$ in Equation (13.3)
- $C2$: If s'_i is a terminal state or $\pi_D(s'_i) = e$ (i.e., $g(s'_i) > \phi(s'_i)^T \cdot \boldsymbol{w}$): Substitute $g(s'_i)$ for $\hat{Q}(s'_i, \pi_D(s'_i); \boldsymbol{w}^*)$ in Equation (13.3)

So we can rewrite Equation (13.3) using indicator notation $\mathbb{I}$ for cases $C1, C2$ as:

$$\sum_i \phi(s_i) \cdot (\phi(s_i)^T \cdot \boldsymbol{w}^* - \mathbb{I}_{C1} \cdot \gamma \cdot \phi(s'_i)^T \cdot \boldsymbol{w}^* - \mathbb{I}_{C2} \cdot \gamma \cdot g(s'_i)) = 0$$

Factoring out $\boldsymbol{w}^*$, we get:

$$(\sum_i \phi(s_i) \cdot (\phi(s_i) - \mathbb{I}_{C1} \cdot \gamma \cdot \phi(s'_i))^T) \cdot \boldsymbol{w}^* = \gamma \cdot \sum_i \mathbb{I}_{C2} \cdot \phi(s_i) \cdot g(s'_i)$$

This can be written in the familiar vector-matrix notation as: $\boldsymbol{A} \cdot \boldsymbol{w}^* = \boldsymbol{b}$

$$\boldsymbol{A} = \sum_i \phi(s_i) \cdot (\phi(s_i) - \mathbb{I}_{C1} \cdot \gamma \cdot \phi(s'_i))^T$$

$$\boldsymbol{b} = \gamma \cdot \sum_i \mathbb{I}_{C2} \cdot \phi(s_i) \cdot g(s'_i)$$

The $m \times m$ Matrix $\boldsymbol{A}$ is accumulated at each atomic experience (s_i, s'_i) as:

$$\boldsymbol{A} \leftarrow \boldsymbol{A} + \phi(s_i) \cdot (\phi(s_i) - \mathbb{I}_{C1} \cdot \gamma \cdot \phi(s'_i))^T$$

The m-Vector $\boldsymbol{b}$ is accumulated at each atomic experience (s_i, s'_i) as:

$$\boldsymbol{b} \leftarrow \boldsymbol{b} + \gamma \cdot \mathbb{I}_{C2} \cdot \phi(s_i) \cdot g(s'_i)$$

With Sherman-Morrison incremental inverse of $\boldsymbol{A}$, we can reduce the time-complexity from $O(m^3)$ to $O(m^2)$.

This solved $\boldsymbol{w}^*$ updates the Q-Value Function Approximation to $\hat{Q}(s, a; \boldsymbol{w}^*)$. This defines an updated, improved deterministic policy π'_D (serving as the *Deterministic Target Policy* for the next iteration of GPI):

$$\pi'_D(s) = \arg\max_a \hat{Q}(s, a; \boldsymbol{w}^*) \text{ for all } s \in \mathcal{N}$$

Li, Szepesvari, Schuurmans (Li, Szepesvári, and Schuurmans 2009) recommend in their paper to use 7 feature functions, the first 4 Laguerre polynomials that are functions of the underlying price and 3 functions of time. Precisely, the feature functions they recommend are:

- $\phi_0(S_t) = 1$
- $\phi_1(S_t) = e^{-\frac{M_t}{2}}$
- $\phi_2(S_t) = e^{-\frac{M_t}{2}} \cdot (1 - M_t)$
- $\phi_3(S_t) = e^{-\frac{M_t}{2}} \cdot (1 - 2M_t + M_t^2/2)$
- $\phi_0^{(t)}(t) = sin(\frac{\pi(T-t)}{2T})$
- $\phi_1^{(t)}(t) = \log(T - t)$
- $\phi_2^{(t)}(t) = (\frac{t}{T})^2$

where $M_t = \frac{S_t}{K}$ (S_t is the current underlying price and K is the American Option strike), t is the current time, and T is the expiration time (i.e., $0 \leq t < T$).

13.6.2   Deep Q-Learning for American Options Pricing

LSPI is data-efficient and compute-efficient, but linearity is a limitation in the function approximation. The alternative is (incremental) Q-Learning with neural network function approximation, which we cover in this subsection. We employ the same set up as LSPI (including Experience Replay)—specifically, the function approximation is required only for continuation value. Precisely,

$$\hat{Q}(s, a; \boldsymbol{w}) = \begin{cases} f(s; \boldsymbol{w}) & \text{if } a = c \\ g(s) & \text{if } a = e \end{cases} \text{ for all } s \in \mathcal{N}$$

where $f(s; \boldsymbol{w})$ is the deep neural network function approximation.

The Q-Learning update for each atomic experience (s_i, s_i') is:

$$\Delta\boldsymbol{w} = \alpha \cdot (\gamma \cdot \hat{Q}(s_i', \pi(s_i'); \boldsymbol{w}) - f(s_i; \boldsymbol{w})) \cdot \nabla_{\boldsymbol{w}} f(s_i; \boldsymbol{w})$$

When s_i' is a non-terminal state, the update is:

$$\Delta\boldsymbol{w} = \alpha \cdot (\gamma \cdot \max(g(s_i'), f(s_i'; \boldsymbol{w})) - f(s_i; \boldsymbol{w})) \cdot \nabla_{\boldsymbol{w}} f(s_i; \boldsymbol{w})$$

When s_i' is a terminal state, the update is:

$$\Delta\boldsymbol{w} = \alpha \cdot (\gamma \cdot g(s_i') - f(s_i; \boldsymbol{w})) \cdot \nabla_{\boldsymbol{w}} f(s_i; \boldsymbol{w})$$

13.7   VALUE FUNCTION GEOMETRY

Now we look deeper into the issue of the *Deadly Triad* (that we had alluded to in Chapter 12) by viewing Value Functions as Vectors (we had done this in Chapter 5), understand Value Function Vector transformations with a balance of geometric intuition and mathematical rigor, providing insights into convergence issues for a variety of traditional loss functions used to develop RL algorithms. As ever, the best way to understand Vector transformations is to visualize it and so, we loosely refer to this topic as Value Function Geometry. The geometric intuition is particularly useful for linear function approximations. To promote intuition, we shall present this content for linear function approximations of the Value Function and stick to Prediction (rather than Control) although many of the concepts covered in this section are well-extensible to non-linear function approximations and to the Control problem.

This treatment was originally presented in the LSPI paper by Lagoudakis and Parr (Lagoudakis and Parr 2003) and has been covered in detail in the RL book by Sutton and

Barto (Richard S. Sutton and Barto 2018). This treatment of Value Functions as Vectors leads us in the direction of overcoming the Deadly Triad by defining an appropriate loss function, calculating whose gradient provides a more robust set of RL algorithms known as Gradient Temporal-Difference (abbreviated, as Gradient TD), which we shall cover in the next section.

Along with visual intuition, it is important to write precise notation for Value Function transformations and approximations. So we start with a set of formal definitions, keeping the setting fairly simple and basic for ease of understanding.

13.7.1   Notation and Definitions

Assume our state space is finite without any terminal states, i.e. $\mathcal{S} = \mathcal{N} = \{s_1, s_2, \ldots, s_n\}$. Assume our action space $\mathcal{A}$ consists of a finite number of actions. This coverage can be extended to infinite/continuous spaces, but we shall stick to this simple setting in this section. Also, as mentioned above, we restrict this coverage to the case of a fixed (potentially stochastic) policy denoted as $\pi : \mathcal{S} \times \mathcal{A} \to [0, 1]$. This means we are restricting to the case of the Prediction problem (although it's possible to extend some of this coverage to the case of Control).

We denote the Value Function for a policy π as $V^\pi : \mathcal{S} \to \mathbb{R}$. Consider the n-dimensional vector space $\mathbb{R}^n$, with each dimension corresponding to a state in $\mathcal{S}$. Think of a Value Function (typically denoted V): $\mathcal{S} \to \mathbb{R}$ as a vector in the $\mathbb{R}^n$ vector space. Each dimension's coordinate is the evaluation of the Value Function for that dimension's state. The coordinates of vector V^π for policy π are: $[V^\pi(s_1), V^\pi(s_2), \ldots, V^\pi(s_n)]$. Note that this treatment is the same as the treatment in our coverage of Dynamic Programming in Chapter 5.

Our interest is in identifying an appropriate function approximation of the Value Function V^π. For the function approximation, assume there are m feature functions $\phi_1, \phi_2, \ldots, \phi_m : \mathcal{S} \to \mathbb{R}$, with $\phi(s) \in \mathbb{R}^m$ denoting the feature vector for any state $s \in \mathcal{S}$. To keep things simple and to promote understanding of the concepts, we limit ourselves to linear function approximations. For linear function approximation of the Value Function with weights $w = (w_1, w_2, \ldots, w_m)$, we use the notation $V_w : \mathcal{S} \to \mathbb{R}$, defined as:

$$V_w(s) = \phi(s)^T \cdot w = \sum_{j=1}^{m} \phi_j(s) \cdot w_j \text{ for all } s \in \mathcal{S}.$$

Assuming independence of the feature functions, the m feature functions give us m independent vectors in the vector space $\mathbb{R}^n$. Feature function ϕ_j gives us the vector $[\phi_j(s_1), \phi_j(s_2), \ldots, \phi_j(s_n)] \in \mathbb{R}^n$. These m vectors are the m columns of the $n \times m$ matrix $\Phi = [\phi_j(s_i)], 1 \leq i \leq n, 1 \leq j \leq m$. The span of these m independent vectors is an m-dimensional vector subspace within this n-dimensional vector space, spanned by the set of all $w = (w_1, w_2, \ldots, w_m) \in \mathbb{R}^m$. The vector $V_w = \Phi \cdot w$ in this vector subspace has co-ordinates $[V_w(s_1), V_w(s_2), \ldots, V_w(s_n)]$. The vector V_w is fully specified by w (so we often say w to mean V_w). Our interest is in identifying an appropriate $w \in \mathbb{R}^m$ that represents an adequate linear function approximation $V_w = \Phi \cdot w$ of the Value Function V^π.

We denote the probability distribution of occurrence of states under policy π as $\mu_\pi : \mathcal{S} \to [0, 1]$. In accordance with the notation we used in Chapter 4, $\mathcal{R}(s, a)$ refers to the Expected Reward upon taking action a in state s, and $\mathcal{P}(s, a, s')$ refers to the probability of transition from state s to state s' upon taking action a. Define

$$\mathcal{R}^\pi(s) = \sum_{a \in \mathcal{A}} \pi(s, a) \cdot \mathcal{R}(s, a) \text{ for all } s \in \mathcal{S}$$

$$\mathcal{P}^\pi(s, s') = \sum_{a \in \mathcal{A}} \pi(s, a) \cdot \mathcal{P}(s, a, s') \text{ for all } s, s' in \mathcal{S}$$

to denote the Expected Reward and state transition probabilities respectively of the π-implied MRP.

$\mathcal{R}^\pi$ refers to vector $[\mathcal{R}^\pi(s_1), \mathcal{R}^\pi(s_2), \ldots, \mathcal{R}^\pi(s_n)]$ and $\mathcal{P}^\pi$ refers to matrix $[\mathcal{P}^\pi(s_i, s_{i'})], 1 \le i, i' \le n$. Denote $\gamma < 1$ (since there are no terminal states) as the MDP discount factor.

13.7.2 Bellman Policy Operator and Projection Operator

In Chapter 4, we introduced the Bellman Policy Operator B^π for policy π operating on any Value Function vector V. As a reminder,

$$B^\pi(V) = \mathcal{R}^\pi + \gamma \mathcal{P}^\pi \cdot V \text{ for any VF vector } V \in \mathbb{R}^n$$

Note that B^π is a linear operator in vector space $\mathbb{R}^n$. So we henceforth denote and treat B^π as an $n \times n$ matrix, representing the linear operator. We've learnt in Chapter 4 that V^π is the fixed point of B^π. Therefore, we can write:

$$B^\pi \cdot V^\pi = V^\pi$$

This means, if we start with an arbitrary Value Function vector V and repeatedly apply B^π, by Banach Fixed-Point Theorem 5.3.1, we will reach the fixed point V^π. We've learnt in Chapter 4 that this is, in fact, the Dynamic Programming Policy Evaluation algorithm. Note that Tabular Monte Carlo also converges to V^π (albeit slowly).

Next, we introduce the Projection Operator Π_Φ for the subspace spanned by the column vectors (feature functions) of Φ. We define $\Pi_\Phi(V)$ as the vector in the subspace spanned by the column vectors of Φ that represents the orthogonal projection of Value Function vector V on the Φ subspace. To make this precise, we first define "distance" $d(V_1, V_2)$ between Value Function vectors V_1, V_2, weighted by μ_π across the n dimensions of V_1, V_2. Specifically,

$$d(V_1, V_2) = \sum_{i=1}^{n} \mu_\pi(s_i) \cdot (V_1(s_i) - V_2(s_i))^2 = (V_1 - V_2)^T \cdot D \cdot (V_1 - V_2)$$

where D is the square diagonal matrix consisting of the diagonal elements $\mu_\pi(s_i), 1 \le i \le n$.

With this "distance" metric, we define $\Pi_\Phi(V)$ as the Value Function vector in the subspace spanned by the column vectors of Φ that is given by $\arg\min_w d(V, V_w)$. This is a weighted least squares regression with solution:

$$w^* = (\Phi^T \cdot D \cdot \Phi)^{-1} \cdot \Phi^T \cdot D \cdot V$$

Since $\Pi_\Phi(V) = \Phi \cdot w^*$, we henceforth denote and treat Projection Operator Π_Φ as the following $n \times n$ matrix:

$$\Pi_\Phi = \Phi \cdot (\Phi^T \cdot D \cdot \Phi)^{-1} \cdot \Phi^T \cdot D$$

13.7.3 Vectors of Interest in the Φ Subspace

In this section, we cover 4 Value Function vectors of interest in the Φ subspace, as candidate linear function approximations of the Value Function V^π. To lighten notation, we will refer to the Φ-subspace Value Function vectors by their corresponding weights w. All 4 of these Value Function vectors are depicted in Figure 13.6, an image we are borrowing from Sutton and Barto's RL book (Richard S. Sutton and Barto 2018). We spend the rest of this section going over these 4 Value Function vectors in detail.

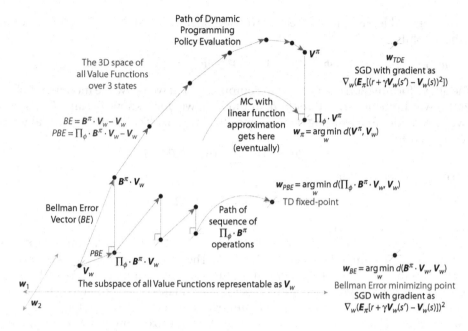

Figure 13.6 **Value Function Geometry (Image Credit: Sutton-Barto's RL Book)**

The first Value Function vector of interest in the Φ subspace is the Projection $\Pi_\Phi \cdot V^\pi$, denoted as $w_\pi = \arg\min_w d(V^\pi, V_w)$. This is the linear function approximation of the Value Function V^π we seek because it is the Value Function vector in the Φ subspace that is "closest" to V^π. Monte-Carlo with linear function approximation will (slowly) converge to w_π. Figure 13.6 provides the visualization. We've learnt that Monte-Carlo can be slow to converge, so we seek function approximations in the Φ subspace that are based on Temporal-Difference (TD), i.e., bootstrapped methods. The remaining three Value Function vectors in the Φ subspace are based on TD methods.

We denote the second Value Function vector of interest in the Φ subspace as w_{BE}. The acronym BE stands for *Bellman Error*. To understand this, consider the application of the Bellman Policy Operator B^π on a Value Function vector V_w in the Φ subspace. Applying B^π on V_w typically throws V_w out of the Φ subspace. The idea is to find a Value Function vector V_w in the Φ subspace such that the "distance" between V_w and $B^\pi \cdot V_w$ is minimized, i.e. we minimize the "error vector" $BE = B^\pi \cdot V_w - V_w$ (Figure 13.6 provides the visualization). Hence, we say we are minimizing the *Bellman Error* (or simply that we are minimizing BE), and we refer to w_{BE} as the Value Function vector in the Φ subspace for which BE is minimized. Formally, we define it as:

$$w_{BE} = \arg\min_w d(B^\pi \cdot V_w, V_w)$$
$$= \arg\min_w d(V_w, \mathcal{R}^\pi + \gamma \mathcal{P}^\pi \cdot V_w)$$
$$= \arg\min_w d(\Phi \cdot w, \mathcal{R}^\pi + \gamma \mathcal{P}^\pi \cdot \Phi \cdot w)$$
$$= \arg\min_w d(\Phi \cdot w - \gamma \mathcal{P}^\pi \cdot \Phi \cdot w, \mathcal{R}^\pi)$$
$$= \arg\min_w d((\Phi - \gamma \mathcal{P}^\pi \cdot \Phi) \cdot w, \mathcal{R}^\pi)$$

This is a weighted least-squares linear regression of $\mathcal{R}^\pi$ against $\Phi - \gamma \mathcal{P}^\pi \cdot \Phi$ with weights μ_π, whose solution is:

$$w_{BE} = ((\Phi - \gamma \mathcal{P}^\pi \cdot \Phi)^T \cdot D \cdot (\Phi - \gamma \mathcal{P}^\pi \cdot \Phi))^{-1} \cdot (\Phi - \gamma \mathcal{P}^\pi \cdot \Phi)^T \cdot D \cdot \mathcal{R}^\pi$$

The above formulation can be used to compute w_{BE} if we know the model probabilities $\mathcal{P}^\pi$ and reward function $\mathcal{R}^\pi$. But often, in practice, we don't know $\mathcal{P}^\pi$ and $\mathcal{R}^\pi$, in which case we seek model-free learning of w_{BE}, specifically with a TD (bootstrapped) algorithm.

Let us refer to

$$(\Phi - \gamma \mathcal{P}^\pi \cdot \Phi)^T \cdot D \cdot (\Phi - \gamma \mathcal{P}^\pi \cdot \Phi)$$

as matrix A and let us refer to

$$(\Phi - \gamma \mathcal{P}^\pi \cdot \Phi)^T \cdot D \cdot \mathcal{R}^\pi$$

as vector b so that $w_{BE} = A^{-1} \cdot b$.

Following policy π, each time we perform an individual transition from s to s' getting reward r, we get a sample estimate of A and b. The sample estimate of A is the outer-product of vector $\phi(s) - \gamma \cdot \phi(s')$ with itself. The sample estimate of b is scalar r times vector $\phi(s) - \gamma \cdot \phi(s')$. We average these sample estimates across many such individual transitions. However, this requires m (the number of features) to be not too large.

If m is large or if we are doing non-linear function approximation or off-policy, then we seek a gradient-based TD algorithm. We defined w_{BE} as the vector in the Φ subspace for which the Bellman Error is minimized. But Bellman Error for a state is the expectation of the TD error δ for that state when following policy π. So we want to do Stochastic Gradient Descent with the gradient of the square of expected TD error, as follows:

$$\begin{aligned} \Delta w &= -\alpha \cdot \frac{1}{2} \cdot \nabla_w (\mathbb{E}_\pi[\delta])^2 \\ &= -\alpha \cdot \mathbb{E}_\pi[r + \gamma \cdot \phi(s')^T \cdot w - \phi(s)^T \cdot w] \cdot \nabla_w \mathbb{E}_\pi[\delta] \\ &= \alpha \cdot (\mathbb{E}_\pi[r + \gamma \cdot \phi(s')^T \cdot w] - \phi(s)^T \cdot w) \cdot (\phi(s) - \gamma \cdot \mathbb{E}_\pi[\phi(s')]) \end{aligned}$$

This is called the *Residual Gradient* algorithm, due to Leemon Baird (Baird 1995). It requires two independent samples of s' transitioning from s. If we do have that, it converges to w_{BE} robustly (even for non-linear function approximations). But this algorithm is slow, and doesn't converge to a desirable place. Another issue is that w_{BE} is not learnable if we can only access the features, and not underlying states. These issues led researchers to consider alternative TD algorithms.

We denote the third Value Function vector of interest in the Φ subspace as w_{TDE} and define it as the vector in the Φ subspace for which the expected square of the TD error δ (when following policy π) is minimized. Formally,

$$w_{TDE} = \underset{w}{\arg\min} \sum_{s \in \mathcal{S}} \mu_\pi(s) \sum_{r,s'} \mathbb{P}_\pi(r, s'|s) \cdot (r + \gamma \cdot \phi(s')^T \cdot w - \phi(s)^T \cdot w)^2$$

To perform Stochastic Gradient Descent, we have to estimate the gradient of the expected square of TD error by sampling. The weight update for each gradient sample in the Stochastic Gradient Descent is:

$$\begin{aligned} \Delta w &= -\alpha \cdot \frac{1}{2} \cdot \nabla_w (r + \gamma \cdot \phi(s')^T \cdot w - \phi(s)^T \cdot w)^2 \\ &= \alpha \cdot (r + \gamma \cdot \phi(s')^T \cdot w - \phi(s)^T \cdot w) \cdot (\phi(s) - \gamma \cdot \phi(s')) \end{aligned}$$

This algorithm is called *Naive Residual Gradient*, due to Leemon Baird (Baird 1995). Naive Residual Gradient converges robustly, but again, not to a desirable place. So researchers had to look even further.

This brings us to the fourth (and final) Value Function vector of interest in the Φ subspace. We denote this Value Function vector as w_{PBE}. The acronym PBE stands for *Projected Bellman Error*. To understand this, first consider the composition of the Projection Operator Π_Φ and the Bellman Policy Operator B^π, i.e., $\Pi_\Phi \cdot B^\pi$ (we call this composed operator as the *Projected Bellman* operator). Visualize the application of this *Projected Bellman* operator on a Value Function vector V_w in the Φ subspace. Applying B^π on V_w typically throws V_w out of the Φ subspace and then further applying Π_Φ brings it back to the Φ subspace (call this resultant Value Function vector $V_{w'}$). The idea is to find a Value Function vector V_w in the Φ subspace for which the "distance" between V_w and $V_{w'}$ is minimized, i.e. we minimize the "error vector" $PBE = \Pi_\Phi \cdot B^\pi \cdot V_w - V_w$ (Figure 13.6 provides the visualization). Hence, we say we are minimizing the *Projected Bellman Error* (or simply that we are minimizing PBE), and we refer to w_{PBE} as the Value Function vector in the Φ subspace for which PBE is minimized. It turns out that the minimum of PBE is actually zero, i.e., $\Phi \cdot w_{PBE}$ is a fixed point of operator $\Pi_\Phi \cdot B^\pi$. Let us write out this statement formally. We know:

$$\Pi_\Phi = \Phi \cdot (\Phi^T \cdot D \cdot \Phi)^{-1} \cdot \Phi^T \cdot D$$

$$B^\pi \cdot V = \mathcal{R}^\pi + \gamma \mathcal{P}^\pi \cdot V$$

Therefore, the statement that $\Phi \cdot w_{PBE}$ is a fixed point of operator $\Pi_\Phi \cdot B^\pi$ can be written as follows:

$$\Phi \cdot (\Phi^T \cdot D \cdot \Phi)^{-1} \cdot \Phi^T \cdot D \cdot (\mathcal{R}^\pi + \gamma \mathcal{P}^\pi \cdot \Phi \cdot w_{PBE}) = \Phi \cdot w_{PBE}$$

Since the columns of Φ are assumed to be independent (full rank),

$$(\Phi^T \cdot D \cdot \Phi)^{-1} \cdot \Phi^T \cdot D \cdot (\mathcal{R}^\pi + \gamma \mathcal{P}^\pi \cdot \Phi \cdot w_{PBE}) = w_{PBE}i$$

$$\Phi^T \cdot D \cdot (\mathcal{R}^\pi + \gamma \mathcal{P}^\pi \cdot \Phi \cdot w_{PBE}) = \Phi^T \cdot D \cdot \Phi \cdot w_{PBE}$$

$$\Phi^T \cdot D \cdot (\Phi - \gamma \mathcal{P}^\pi \cdot \Phi) \cdot w_{PBE} = \Phi^T \cdot D \cdot \mathcal{R}^\pi \tag{13.4}$$

This is a square linear system of the form $A \cdot w_{PBE} = b$ whose solution is:

$$w_{PBE} = A^{-1} \cdot b = (\Phi^T \cdot D \cdot (\Phi - \gamma \mathcal{P}^\pi \cdot \Phi))^{-1} \cdot \Phi^T \cdot D \cdot \mathcal{R}^\pi$$

The above formulation can be used to compute w_{PBE} if we know the model probabilities $\mathcal{P}^\pi$ and reward function $\mathcal{R}^\pi$. But often, in practice, we don't know $\mathcal{P}^\pi$ and $\mathcal{R}^\pi$, in which case we seek model-free learning of w_{PBE}, specifically with a TD (bootstrapped) algorithm.

The question is how do we construct matrix

$$A = \Phi^T \cdot D \cdot (\Phi - \gamma \mathcal{P}^\pi \cdot \Phi)$$

and vector

$$b = \Phi^T \cdot D \cdot \mathcal{R}^\pi$$

without a model?

Following policy π, each time we perform an individual transition from s to s' getting reward r, we get a sample estimate of A and b. The sample estimate of A is the outer-product of vectors $\phi(s)$ and $\phi(s) - \gamma \cdot \phi(s')$. The sample estimate of b is scalar r times vector $\phi(s)$. We average these sample estimates across many such individual transitions.

Note that this algorithm is exactly the Least Squares Temporal Difference (LSTD) algorithm we've covered earlier in this chapter. Thus, we now know that LSTD converges to w_{PBE}, i.e., minimizes (in fact, takes down to 0) PBE. If the number of features m is large or if we are doing non-linear function approximation or Off-Policy, then we seek a gradient-based TD algorithm. It turns out that our usual Semi-Gradient TD algorithm converges to w_{PBE} in the case of on-policy linear function approximation. Note that the update for the usual Semi-Gradient TD algorithm in the case of on-policy linear function approximation is as follows:

$$\Delta w = \alpha \cdot (r + \gamma \cdot \phi(s')^T \cdot w - \phi(s)^T \cdot w) \cdot \phi(s)$$

This converges to w_{PBE} because at convergence, we have: $\mathbb{E}_\pi[\Delta w] = 0$, which can be expressed as:

$$\Phi^T \cdot D \cdot (\mathcal{R}^\pi + \gamma \mathcal{P}^\pi \cdot \Phi \cdot w - \Phi \cdot w) = 0$$

$$\Rightarrow \Phi^T \cdot D \cdot (\Phi - \gamma \mathcal{P}^\pi \cdot \Phi) \cdot w = \Phi^T \cdot D \cdot \mathcal{R}^\pi$$

which is satisfied for $w = w_{PBE}$ (as seen from Equation (13.4)).

13.8   GRADIENT TEMPORAL-DIFFERENCE (GRADIENT TD)

For on-policy linear function approximation, the semi-gradient TD algorithm gives us w_{PBE}. But to obtain w_{PBE} in the case of non-linear function approximation or in the case of Off-Policy, we need a different approach. The different approach is Gradient Temporal-Difference (abbreviated, Gradient TD), the subject of this section.

The original Gradient TD algorithm, due to Sutton, Szepesvari, Maei (R. S. Sutton, Szepesvári, and Maei 2008) is typically abbreviated as GTD. Researchers then came up with a second-generation Gradient TD algorithm (R. S. Sutton et al. 2009), which is typically abbreviated as GTD-2. The same researchers also came up with a TD algorithm with Gradient Correction (R. S. Sutton et al. 2009), which is typically abbreviated as TDC.

We now cover the TDC algorithm. For simplicity of articulation and ease of understanding, we restrict to the case of linear function approximation in our coverage of the TDC algorithm below. However, do bear in mind that much of the concepts below extend to non-linear function approximation (which is where we reap the benefits of Gradient TD).

Our first task is to set up the appropriate loss function whose gradient will drive the Stochastic Gradient Descent.

$$w_{PBE} = \arg\min_w d(\Pi_\Phi \cdot B^\pi \cdot V_w, V_w) = \arg\min_w d(\Pi_\Phi \cdot B^\pi \cdot V_w, \Pi_\Phi \cdot V_w)$$

So we define the loss function (denoting $B^\pi \cdot V_w - V_w$ as δ_w) as:

$$\mathcal{L}(w) = (\Pi_\Phi \cdot \delta_w)^T \cdot D \cdot (\Pi_\Phi \cdot \delta_w) = \delta_w^T \cdot \Pi_\Phi^T \cdot D \cdot \Pi_\Phi \cdot \delta_w$$

$$= \delta_w^T \cdot (\Phi \cdot (\Phi^T \cdot D \cdot \Phi)^{-1} \cdot \Phi^T \cdot D)^T \cdot D \cdot (\Phi \cdot (\Phi^T \cdot D \cdot \Phi)^{-1} \cdot \Phi^T \cdot D) \cdot \delta_w$$

$$= \delta_w^T \cdot (D \cdot \Phi \cdot (\Phi^T \cdot D \cdot \Phi)^{-1} \cdot \Phi^T) \cdot D \cdot (\Phi \cdot (\Phi^T \cdot D \cdot \Phi)^{-1} \cdot \Phi^T \cdot D) \cdot \delta_w$$

$$= (\delta_w^T \cdot D \cdot \Phi) \cdot (\Phi^T \cdot D \cdot \Phi)^{-1} \cdot (\Phi^T \cdot D \cdot \Phi) \cdot (\Phi^T \cdot D \cdot \Phi)^{-1} \cdot (\Phi^T \cdot D \cdot \delta_w)$$

$$= (\Phi^T \cdot D \cdot \delta_w)^T \cdot (\Phi^T \cdot D \cdot \Phi)^{-1} \cdot (\Phi^T \cdot D \cdot \delta_w)$$

We derive the TDC Algorithm based on $\nabla_w \mathcal{L}(w)$.

$$\nabla_w \mathcal{L}(w) = 2 \cdot (\nabla_w (\Phi^T \cdot D \cdot \delta_w)^T) \cdot (\Phi^T \cdot D \cdot \Phi)^{-1} \cdot (\Phi^T \cdot D \cdot \delta_w)$$

We want to estimate this gradient from individual transitions data. So we express each of the 3 terms forming the product in the gradient expression above as expectations of functions of individual transitions $s \xrightarrow{\pi} (r, s')$. Denoting $r + \gamma \cdot \phi(s')^T \cdot w - \phi(s)^T \cdot w$ as δ, we get:

$$\mathbf{\Phi}^T \cdot \mathbf{D} \cdot \boldsymbol{\delta_w} = \mathbb{E}[\delta \cdot \boldsymbol{\phi}(s)]$$

$$\nabla_w(\mathbf{\Phi}^T \cdot \mathbf{D} \cdot \boldsymbol{\delta_w})^T = \mathbb{E}[(\nabla_w \delta) \cdot \boldsymbol{\phi}(s)^T] = \mathbb{E}[(\gamma \cdot \boldsymbol{\phi}(s') - \boldsymbol{\phi}(s)) \cdot \boldsymbol{\phi}(s)^T]$$

$$\mathbf{\Phi}^T \cdot \mathbf{D} \cdot \mathbf{\Phi} = \mathbb{E}[\boldsymbol{\phi}(s) \cdot \boldsymbol{\phi}(s)^T]$$

Substituting, we get:

$$\nabla_w \mathcal{L}(w) = 2 \cdot \mathbb{E}[(\gamma \cdot \boldsymbol{\phi}(s') - \boldsymbol{\phi}(s)) \cdot \boldsymbol{\phi}(s)^T] \cdot (\mathbb{E}[\boldsymbol{\phi}(s) \cdot \boldsymbol{\phi}(s)^T])^{-1} \cdot \mathbb{E}[\delta \cdot \boldsymbol{\phi}(s)]$$

$$\Delta w = -\alpha \cdot \frac{1}{2} \cdot \nabla_w \mathcal{L}(w)$$

$$= \alpha \cdot \mathbb{E}[(\boldsymbol{\phi}(s) - \gamma \cdot \boldsymbol{\phi}(s')) \cdot \boldsymbol{\phi}(s)^T] \cdot (\mathbb{E}[\boldsymbol{\phi}(s) \cdot \boldsymbol{\phi}(s)^T])^{-1} \cdot \mathbb{E}[\delta \cdot \boldsymbol{\phi}(s)]$$

$$= \alpha \cdot (\mathbb{E}[\boldsymbol{\phi}(s) \cdot \boldsymbol{\phi}(s)^T] - \gamma \cdot \mathbb{E}[\boldsymbol{\phi}(s') \cdot \boldsymbol{\phi}(s)^T]) \cdot (\mathbb{E}[\boldsymbol{\phi}(s) \cdot \boldsymbol{\phi}(s)^T])^{-1} \cdot \mathbb{E}[\delta \cdot \boldsymbol{\phi}(s)]$$

$$= \alpha \cdot (\mathbb{E}[\delta \cdot \boldsymbol{\phi}(s)] - \gamma \cdot \mathbb{E}[\boldsymbol{\phi}(s') \cdot \boldsymbol{\phi}(s)^T] \cdot (\mathbb{E}[\boldsymbol{\phi}(s) \cdot \boldsymbol{\phi}(s)^T])^{-1} \cdot \mathbb{E}[\delta \cdot \boldsymbol{\phi}(s)])$$

$$= \alpha \cdot (\mathbb{E}[\delta \cdot \boldsymbol{\phi}(s)] - \gamma \cdot \mathbb{E}[\boldsymbol{\phi}(s') \cdot \boldsymbol{\phi}(s)^T] \cdot \boldsymbol{\theta})$$

where $\boldsymbol{\theta} = (\mathbb{E}[\boldsymbol{\phi}(s) \cdot \boldsymbol{\phi}(s)^T])^{-1} \cdot \mathbb{E}[\delta \cdot \boldsymbol{\phi}(s)]$ is the solution to the weighted least-squares linear regression of $\mathbf{B}^\pi \cdot \mathbf{V} - \mathbf{V}$ against $\mathbf{\Phi}$, with weights as μ_π.

We can perform this gradient descent with a technique known as *Cascade Learning*, which involves simultaneously updating both w and $\boldsymbol{\theta}$ (with $\boldsymbol{\theta}$ converging faster). The updates are as follows:

$$\Delta w = \alpha \cdot \delta \cdot \boldsymbol{\phi}(s) - \alpha \cdot \gamma \cdot \boldsymbol{\phi}(s') \cdot (\boldsymbol{\phi}(s)^T \cdot \boldsymbol{\theta})$$

$$\Delta \boldsymbol{\theta} = \beta \cdot (\delta - \boldsymbol{\phi}(s)^T \cdot \boldsymbol{\theta}) \cdot \boldsymbol{\phi}(s)$$

where β is the learning rate for $\boldsymbol{\theta}$. Note that $\boldsymbol{\phi}(s)^T \cdot \boldsymbol{\theta}$ operates as an estimate of the TD error δ for current state s.

Repeating what we had said in Chapter 12, Gradient TD converges reliably for the Prediction problem even when we are faced with the Deadly Triad of [Bootstrapping, Off-Policy, Non-Linear Function Approximation]. The picture is less rosy for Control. Gradient Q-Learning (Gradient TD for Off-Policy Control) converges reliably for both on-policy and off-policy linear function approximations, but there are divergence issues for non-linear function approximations. For Control problems with non-linear function approximations (especially, neural network approximations with off-policy learning), one can leverage the approach of the DQN algorithm (Experience Replay with fixed Target Network helps overcome the Deadly Triad).

13.9   KEY TAKEAWAYS FROM THIS CHAPTER

- Batch RL makes efficient use of data.
- DQN uses Experience-Replay and fixed Q-learning targets, avoiding the pitfalls of time-correlation and varying TD Target.
- LSTD is a direct (gradient-free) solution of Batch TD Prediction.
- LSPI is an Off-Policy, Experience-Replay Control Algorithm using LSTDQ for Policy Evaluation.
- Optimal Exercise of American Options can be tackled with LSPI and Deep Q-Learning algorithms.

- For Prediction, the 4 Value Function vectors of interest in the $\mathbf{\Phi}$ subspace are $w_\pi, w_{BE}, w_{TDE}, w_{PBE}$ with w_{PBE} as the key sought-after function approximation for Value Function V^π.
- For Prediction, Gradient TD solves for w_{PBE} efficiently and robustly in the case of non-linear function approximation and in the case of Off-Policy.

CHAPTER 14

Policy Gradient Algorithms

It's time to take stock of what we have learnt so far to set up context for this chapter. So far, we have covered a range of RL Control algorithms, all of which are based on Generalized Policy Iteration (GPI). All of these algorithms perform GPI by learning the Q-Value Function and improving the policy by identifying the action that fetches the best Q-Value (i.e., action value) for each state. Notice that the way we implemented this *best action identification* is by sweeping through all the actions for each state. This works well only if the set of actions for each state is reasonably small. But if the action space is large/continuous, we have to resort to some sort of optimization method to identify the best action for each state, which is potentially complicated and expensive.

In this chapter, we cover RL Control algorithms that take a vastly different approach. These Control algorithms are still based on GPI, but the Policy Improvement of their GPI is not based on consulting the Q-Value Function, as has been the case with Control algorithms we covered in the previous two chapters. Rather, the approach in the class of algorithms we cover in this chapter is to directly find the Policy that fetches the "Best Expected Returns". Specifically, the algorithms of this chapter perform a Gradient Ascent on "Expected Returns" with the gradient defined with respect to the parameters of a Policy function approximation. We shall work with a stochastic policy of the form $\pi(s, a; \theta)$, with θ denoting the parameters of the policy function approximation π. So we are basically learning this parameterized policy that selects actions without consulting a Value Function. Note that we might still engage a Value Function approximation (call it $Q(s; a; w)$) in our algorithm, but its role is to only help learn the policy parameters θ and not to identify the action with the best action-value for each state. So the two function approximations $\pi(s, a; \theta)$ and $Q(s, a; w)$ collaborate to improve the policy using gradient ascent (based on gradient of "expected returns" with respect to θ). $\pi(s, a; \theta)$ is the primary worker here (known as *Actor*) and $Q(s, a; w)$ is the support worker (known as *Critic*). The Critic parameters w are optimized by minimizing a suitable loss function defined in terms of $Q(s, a; w)$ while the Actor parameters θ are optimized by maximizing a suitable "Expected Returns" function. Note that we still haven't defined what this "Expected Returns" function is (we will do so shortly), but we already see that this idea is appealing for large/continuous action spaces where sweeping through actions is infeasible. We will soon dig into the details of this new approach to RL Control (known as *Policy Gradient*, abbreviated as PG)—for now, it's important to recognize the big picture that PG is basically GPI with Policy Improvement done as a *Policy Gradient Ascent*.

The contrast between the RL Control algorithms covered in the previous two chapters and the algorithms of this chapter actually is part of the following bigger-picture classification of learning algorithms for Control:

- Value Function-based: Here we learn the Value Function (typically with a function approximation for the Value Function) and the Policy is implicit, readily derived from the Value Function (e.g., ϵ-greedy).
- Policy-based: Here we learn the Policy (with a function approximation for the Policy), and there is no need to learn a Value Function.
- Actor-Critic: Here we primarily learn the Policy (with a function approximation for the Policy, known as *Actor*), and secondarily learn the Value Function (with a function approximation for the Value Function, known as *Critic*).

PG Algorithms can be Policy-based or Actor-Critic, whereas the Control algorithms we covered in the previous two chapters are Value Function-based.

In this chapter, we start by enumerating the advantages and disadvantages of Policy Gradient Algorithms, state and prove the Policy Gradient Theorem (which provides the fundamental calculation underpinning Policy Gradient Algorithms), then go on to address how to lower the bias and variance in these algorithms, give an overview of special cases of Policy Gradient algorithms that have found success in practical applications, and finish with a description of Evolutionary Strategies that although technically not RL, resemble Policy Gradient algorithms and are quite effective in solving certain Control problems.

14.1 ADVANTAGES AND DISADVANTAGES OF POLICY GRADIENT ALGORITHMS

Let us start by enumerating the advantages of PG algorithms. We've already said that PG algorithms are effective in large action spaces, especially high-dimensional or continuous action spaces, because in such spaces selecting an action by deriving an improved policy from an updating Q-Value function is intractable. A key advantage of PG is that it naturally *explores* because the policy function approximation is configured as a stochastic policy. Moreover, PG finds the best Stochastic Policy. This is not a factor for MDPs since we know that there exists an optimal Deterministic Policy for any MDP but we often deal with Partially-Observable MDPs (POMDPs) in the real-world, for which the set of optimal policies might all be stochastic policies. We have an advantage in the case of MDPs as well since PG algorithms naturally converge to the deterministic policy (the variance in the policy distribution will automatically converge to 0) whereas in Value Function-based algorithms, we have to reduce the ϵ of the ϵ-greedy policy by-hand and the appropriate declining trajectory of ϵ is typically hard to figure out by manual tuning. In situations where the policy function is a simpler function compared to the Value Function, we naturally benefit from pursuing Policy-based algorithms than Value Function-based algorithms. Perhaps the biggest advantage of PG algorithms is that prior knowledge of the functional form of the Optimal Policy enables us to structure the known functional form in the function approximation for the policy. Lastly, PG offers numerical benefits as small changes in θ yield small changes in π, and consequently small changes in the distribution of occurrences of states. This results in stronger convergence guarantees for PG algorithms relative to Value Function-based algorithms.

Now let's understand the disadvantages of PG Algorithms. The main disadvantage of PG Algorithms is that because they are based on gradient ascent, they typically converge to a local optimum whereas Value Function-based algorithms converge to a global optimum. Furthermore, the Policy Evaluation of PG is typically inefficient and can have high variance. Lastly, the Policy Improvements of PG happen in small steps and so, PG algorithms are slow to converge.

14.2   POLICY GRADIENT THEOREM

In this section, we start by setting up some notation, and then state and prove the Policy Gradient Theorem (abbreviated as PGT). The PGT provides the key calculation for PG Algorithms.

14.2.1   Notation and Definitions

Denoting the discount factor as γ, we shall assume either episodic sequences with $0 \leq \gamma \leq 1$ or non-episodic (continuing) sequences with $0 \leq \gamma < 1$. We shall use our usual notation of discrete-time, countable-spaces, time-homogeneous MDPs although we can indeed extend PGT and PG Algorithms to more general settings as well. We lighten $\mathcal{P}(s, a, s')$ notation to $\mathcal{P}^a_{s,s'}$ and $\mathcal{R}(s, a)$ notation to $\mathcal{R}^a_s$ because we want to save some space in the very long equations in the derivation of PGT.

We denote the probability distribution of the starting state as $p_0 : \mathcal{N} \to [0, 1]$. The policy function approximation is denoted as $\pi(s, a; \boldsymbol{\theta}) = \mathbb{P}[A_t = a | S_t = s; \boldsymbol{\theta}]$.

The PG coverage is quite similar for non-discounted, non-episodic MDPs, by considering the average-reward objective, but we won't cover it in this book.

Now we formalize the "Expected Returns" Objective $J(\boldsymbol{\theta})$.

$$J(\boldsymbol{\theta}) = \mathbb{E}_\pi[\sum_{t=0}^{\infty} \gamma^t \cdot R_{t+1}]$$

Value Function $V^\pi(s)$ and Action Value function $Q^\pi(s, a)$ are defined as:

$$V^\pi(s) = \mathbb{E}_\pi[\sum_{k=t}^{\infty} \gamma^{k-t} \cdot R_{k+1} | S_t = s] \text{ for all } t = 0, 1, 2, \ldots$$

$$Q^\pi(s, a) = \mathbb{E}_\pi[\sum_{k=t}^{\infty} \gamma^{k-t} \cdot R_{k+1} | S_t = s, A_t = a] \text{ for all } t = 0, 1, 2, \ldots$$

$J(\boldsymbol{\theta}), V^\pi, Q^\pi$ are all measures of Expected Returns, so it pays to specify exactly how they differ. $J(\boldsymbol{\theta})$ is the Expected Return when following policy π (that is parameterized by θ), *averaged over all states* $s \in \mathcal{N}$ *and all actions* $a \in \mathcal{A}$. The idea is to perform a gradient ascent with $J(\boldsymbol{\theta})$ as the objective function, with each step in the gradient ascent essentially pushing θ (and hence, π) in a desirable direction, until $J(\boldsymbol{\theta})$ is maximized. $V^\pi(s)$ is the Expected Return for a specific state $s \in \mathcal{N}$ when following policy π. $Q^\pi(s, a)$ is the Expected Return for a specific state $s \in \mathcal{N}$ and specific action $a \in \mathcal{A}$ when following policy π.

We define the *Advantage Function* as:

$$A^\pi(s, a) = Q^\pi(s, a) - V^\pi(s)$$

The advantage function captures how much more value does a particular action provide relative to the average value across actions (for a given state). The advantage function plays an important role in reducing the variance in PG Algorithms.

Also, $p(s \to s', t, \pi)$ will be a key function for us in the PGT proof—it denotes the probability of going from state s to s' in t steps by following policy π.

We express the "Expected Returns" Objective $J(\boldsymbol{\theta})$ as follows:

$$J(\boldsymbol{\theta}) = \mathbb{E}_\pi[\sum_{t=0}^{\infty} \gamma^t \cdot R_{t+1}] = \sum_{t=0}^{\infty} \gamma^t \cdot \mathbb{E}_\pi[R_{t+1}]$$

$$= \sum_{t=0}^{\infty} \gamma^t \cdot \sum_{s \in \mathcal{N}} (\sum_{S_0 \in \mathcal{N}} p_0(S_0) \cdot p(S_0 \to s, t, \pi)) \cdot \sum_{a \in \mathcal{A}} \pi(s, a; \boldsymbol{\theta}) \cdot \mathcal{R}_s^a$$

$$= \sum_{s \in \mathcal{N}} (\sum_{S_0 \in \mathcal{N}} \sum_{t=0}^{\infty} \gamma^t \cdot p_0(S_0) \cdot p(S_0 \to s, t, \pi)) \cdot \sum_{a \in \mathcal{A}} \pi(s, a; \boldsymbol{\theta}) \cdot \mathcal{R}_s^a$$

Definition 14.2.1.

$$J(\boldsymbol{\theta}) = \sum_{s \in \mathcal{N}} \rho^\pi(s) \cdot \sum_{a \in \mathcal{A}} \pi(s, a; \boldsymbol{\theta}) \cdot \mathcal{R}_s^a$$

where

$$\rho^\pi(s) = \sum_{S_0 \in \mathcal{N}} \sum_{t=0}^{\infty} \gamma^t \cdot p_0(S_0) \cdot p(S_0 \to s, t, \pi)$$

is the key function (for PG) that we shall refer to as *Discounted-Aggregate State-Visitation Measure*. Note that $\rho^\pi(s)$ is a measure over the set of non-terminal states, but is not a probability measure. Think of $\rho^\pi(s)$ as weights reflecting the relative likelihood of occurrence of states on a trace experience (adjusted for discounting, i.e, lesser importance to reaching a state later on a trace experience). We can still talk about the distribution of states under the measure ρ^π, but we say that this distribution is *improper* to convey the fact that $\sum_{s \in \mathcal{N}} \rho^\pi(s) \neq 1$ (i.e., the distribution is not normalized). We talk about this improper distribution of states under the measure ρ^π so we can use (as a convenience) the "expected value" notation for any random variable $f : \mathcal{N} \to \mathbb{R}$ under this improper distribution, i.e., we use the notation:

$$\mathbb{E}_{s \sim \rho^\pi}[f(s)] = \sum_{s \in \mathcal{N}} \rho^\pi(s) \cdot f(s)$$

Using this notation, we can re-write the above definition of $J(\boldsymbol{\theta})$ as:

$$J(\boldsymbol{\theta}) = \mathbb{E}_{s \sim \rho^\pi, a \sim \pi}[\mathcal{R}_s^a]$$

14.2.2   Statement of the Policy Gradient Theorem

The Policy Gradient Theorem (PGT) provides a powerful formula for the gradient of $J(\boldsymbol{\theta})$ with respect to $\boldsymbol{\theta}$ so we can perform Gradient Ascent. The key challenge is that $J(\boldsymbol{\theta})$ depends not only on the selection of actions through policy π (parameterized by $\boldsymbol{\theta}$), but also on the probability distribution of occurrence of states (also affected by π, and hence by $\boldsymbol{\theta}$). With knowledge of the functional form of π on $\boldsymbol{\theta}$, it is not difficult to evaluate the dependency of actions selection on $\boldsymbol{\theta}$, but evaluating the dependency of the probability distribution of occurrence of states on $\boldsymbol{\theta}$ is difficult since the environment only provides atomic experiences at a time (and not probabilities of transitions). However, the PGT (below) comes to our rescue because the gradient of $J(\boldsymbol{\theta})$ with respect to $\boldsymbol{\theta}$ involves only the gradient of π with respect to $\boldsymbol{\theta}$, and not the gradient of the probability distribution of occurrence of states with respect to $\boldsymbol{\theta}$. Precisely, we have:

Theorem 14.2.1 (Policy Gradient Theorem).

$$\nabla_{\boldsymbol{\theta}} J(\boldsymbol{\theta}) = \sum_{s \in \mathcal{N}} \rho^\pi(s) \cdot \sum_{a \in \mathcal{A}} \nabla_{\boldsymbol{\theta}} \pi(s, a; \boldsymbol{\theta}) \cdot Q^\pi(s, a)$$

As mentioned above, note that $\rho^\pi(s)$ (representing the discounting-adjusted probability distribution of occurrence of states, ignoring normalizing factor turning the ρ^π measure into a probability measure) depends on θ but there's no $\nabla_\theta \rho^\pi(s)$ term in $\nabla_\theta J(\theta)$.

Also note that:

$$\nabla_\theta \pi(s, a; \theta) = \pi(s, a; \theta) \cdot \nabla_\theta \log \pi(s, a; \theta)$$

$\nabla_\theta \log \pi(s, a; \theta)$ is the Score function (Gradient of log-likelihood) that is commonly used in Statistics.

Since ρ^π is the *Discounted-Aggregate State-Visitation Measure*, we can sample-estimate $\nabla_\theta J(\theta)$ by calculating $\gamma^t \cdot (\nabla_\theta \log \pi(S_t, A_t; \theta)) \cdot Q^\pi(S_t, A_t)$ at each time step in each trace experience (noting that the state occurrence probabilities and action occurrence probabilities are implicit in the trace experiences, and ignoring the probability measure-normalizing factor), and update the parameters θ (according to Stochastic Gradient Ascent) using each atomic experience's $\nabla_\theta J(\theta)$ estimate.

We typically calculate the Score $\nabla_\theta \log \pi(s, a; \theta)$ using an analytically-convenient functional form for the conditional probability distribution $a|s$ (in terms of θ) so that the derivative of the logarithm of this functional form is analytically tractable (this will be clear in the next section when we consider a couple of examples of canonical functional forms for $a|s$). In many PG Algorithms, we estimate $Q^\pi(s, a)$ with a function approximation $Q(s, a; w)$. We will later show how to avoid the estimate bias of $Q(s, a; w)$.

Thus, the PGT enables a numerical estimate of $\nabla_\theta J(\theta)$ which in turn enables *Policy Gradient Ascent*.

14.2.3 Proof of the Policy Gradient Theorem

We begin the proof by noting that:

$$J(\theta) = \sum_{S_0 \in \mathcal{N}} p_0(S_0) \cdot V^\pi(S_0) = \sum_{S_0 \in \mathcal{N}} p_0(S_0) \cdot \sum_{A_0 \in \mathcal{A}} \pi(S_0, A_0; \theta) \cdot Q^\pi(S_0, A_0)$$

Calculate $\nabla_\theta J(\theta)$ by its product parts $\pi(S_0, A_0; \theta)$ and $Q^\pi(S_0, A_0)$.

$$\nabla_\theta J(\theta) = \sum_{S_0 \in \mathcal{N}} p_0(S_0) \cdot \sum_{A_0 \in \mathcal{A}} \nabla_\theta \pi(S_0, A_0; \theta) \cdot Q^\pi(S_0, A_0)$$
$$+ \sum_{S_0 \in \mathcal{N}} p_0(S_0) \cdot \sum_{A_0 \in \mathcal{A}} \pi(S_0, A_0; \theta) \cdot \nabla_\theta Q^\pi(S_0, A_0)$$

Now expand $Q^\pi(S_0, A_0)$ as:

$$\mathcal{R}_{S_0}^{A_0} + \sum_{S_1 \in \mathcal{N}} \gamma \cdot \mathcal{P}_{S_0, S_1}^{A_0} \cdot V^\pi(S_1) \text{ (Bellman Policy Equation)}$$

$$\nabla_\theta J(\theta) = \sum_{S_0 \in \mathcal{N}} p_0(S_0) \cdot \sum_{A_0 \in \mathcal{A}} \nabla_\theta \pi(S_0, A_0; \theta) \cdot Q^\pi(S_0, A_0)$$
$$+ \sum_{S_0 \in \mathcal{N}} p_0(S_0) \cdot \sum_{A_0 \in \mathcal{A}} \pi(S_0, A_0; \theta) \cdot \nabla_\theta (\mathcal{R}_{S_0}^{A_0} + \sum_{S_1 \in \mathcal{N}} \gamma \cdot \mathcal{P}_{S_0, S_1}^{A_0} \cdot V^\pi(S_1))$$

Note: $\nabla_\theta \mathcal{R}_{S_0}^{A_0} = 0$, so remove that term.

$$\nabla_\theta J(\theta) = \sum_{S_0 \in \mathcal{N}} p_0(S_0) \cdot \sum_{A_0 \in \mathcal{A}} \nabla_\theta \pi(S_0, A_0; \theta) \cdot Q^\pi(S_0, A_0)$$
$$+ \sum_{S_0 \in \mathcal{N}} p_0(S_0) \cdot \sum_{A_0 \in \mathcal{A}} \pi(S_0, A_0; \theta) \cdot \nabla_\theta (\sum_{S_1 \in \mathcal{N}} \gamma \cdot \mathcal{P}_{S_0, S_1}^{A_0} \cdot V^\pi(S_1))$$

Now bring the $\nabla_{\boldsymbol{\theta}}$ inside the $\sum_{S_1 \in \mathcal{N}}$ to apply only on $V^\pi(S_1)$.

$$\nabla_{\boldsymbol{\theta}} J(\boldsymbol{\theta}) = \sum_{S_0 \in \mathcal{N}} p_0(S_0) \cdot \sum_{A_0 \in \mathcal{A}} \nabla_{\boldsymbol{\theta}} \pi(S_0, A_0; \boldsymbol{\theta}) \cdot Q^\pi(S_0, A_0)$$

$$+ \sum_{S_0 \in \mathcal{N}} p_0(S_0) \cdot \sum_{A_0 \in \mathcal{A}} \pi(S_0, A_0; \boldsymbol{\theta}) \cdot \sum_{S_1 \in \mathcal{N}} \gamma \cdot \mathcal{P}^{A_0}_{S_0, S_1} \cdot \nabla_{\boldsymbol{\theta}} V^\pi(S_1)$$

Now bring $\sum_{S_0 \in \mathcal{N}}$ and $\sum_{A_0 \in \mathcal{A}}$ inside the $\sum_{S_1 \in \mathcal{N}}$

$$\nabla_{\boldsymbol{\theta}} J(\boldsymbol{\theta}) = \sum_{S_0 \in \mathcal{N}} p_0(S_0) \cdot \sum_{A_0 \in \mathcal{A}} \nabla_{\boldsymbol{\theta}} \pi(S_0, A_0; \boldsymbol{\theta}) \cdot Q^\pi(S_0, A_0)$$

$$+ \sum_{S_1 \in \mathcal{N}} \sum_{S_0 \in \mathcal{N}} \gamma \cdot p_0(S_0) \cdot (\sum_{A_0 \in \mathcal{A}} \pi(S_0, A_0; \boldsymbol{\theta}) \cdot \mathcal{P}^{A_0}_{S_0, S_1}) \cdot \nabla_{\boldsymbol{\theta}} V^\pi(S_1)$$

Note that $\sum_{A_0 \in \mathcal{A}} \pi(S_0, A_0; \boldsymbol{\theta}) \cdot \mathcal{P}^{A_0}_{S_0, S_1} = p(S_0 \to S_1, 1, \pi)$

$$\nabla_{\boldsymbol{\theta}} J(\boldsymbol{\theta}) = \sum_{S_0 \in \mathcal{N}} p_0(S_0) \cdot \sum_{A_0 \in \mathcal{A}} \nabla_{\boldsymbol{\theta}} \pi(S_0, A_0; \boldsymbol{\theta}) \cdot Q^\pi(S_0, A_0)$$

$$+ \sum_{S_1 \in \mathcal{N}} \sum_{S_0 \in \mathcal{N}} \gamma \cdot p_0(S_0) \cdot p(S_0 \to S_1, 1, \pi) \cdot \nabla_{\boldsymbol{\theta}} V^\pi(S_1)$$

Now expand $V^\pi(S_1)$ to $\sum_{A_1 \in \mathcal{A}} \pi(S_1, A_1; \boldsymbol{\theta}) \cdot Q^\pi(S_1, A_1)$

$$\nabla_{\boldsymbol{\theta}} J(\boldsymbol{\theta}) = \sum_{S_0 \in \mathcal{N}} p_0(S_0) \cdot \sum_{A_0 \in \mathcal{A}} \nabla_{\boldsymbol{\theta}} \pi(S_0, A_0; \boldsymbol{\theta}) \cdot Q^\pi(S_0, A_0)$$

$$+ \sum_{S_1 \in \mathcal{N}} \sum_{S_0 \in \mathcal{N}} \gamma \cdot p_0(S_0) \cdot p(S_0 \to S_1, 1, \pi) \cdot \nabla_{\boldsymbol{\theta}} (\sum_{A_1 \in \mathcal{A}} \pi(S_1, A_1; \boldsymbol{\theta}) \cdot Q^\pi(S_1, A_1))$$

We are now back to when we started calculating gradient of $\sum_a \pi \cdot Q^\pi$. Follow the same process of calculating the gradient of $\pi \cdot Q^\pi$ by parts, then Bellman-expanding Q^π (to calculate its gradient), and iterate.

$$\nabla_{\boldsymbol{\theta}} J(\boldsymbol{\theta}) = \sum_{S_0 \in \mathcal{N}} p_0(S_0) \cdot \sum_{A_0 \in \mathcal{A}} \nabla_{\boldsymbol{\theta}} \pi(S_0, A_0; \boldsymbol{\theta}) \cdot Q^\pi(S_0, A_0) +$$

$$\sum_{S_1 \in \mathcal{N}} \sum_{S_0 \in \mathcal{N}} \gamma \cdot p_0(S_0) \cdot p(S_0 \to S_1, 1, \pi) \cdot (\sum_{A_1 \in \mathcal{A}} \nabla_{\boldsymbol{\theta}} \pi(S_1, A_1; \boldsymbol{\theta}) \cdot Q^\pi(S_1, A_1) + \ldots)$$

This iterative process leads us to:

$$\nabla_{\boldsymbol{\theta}} J(\boldsymbol{\theta}) = \sum_{t=0}^{\infty} \sum_{S_t \in \mathcal{N}} \sum_{S_0 \in \mathcal{N}} \gamma^t \cdot p_0(S_0) \cdot p(S_0 \to S_t, t, \pi) \cdot \sum_{A_t \in \mathcal{A}} \nabla_{\boldsymbol{\theta}} \pi(S_t, A_t; \boldsymbol{\theta}) \cdot Q^\pi(S_t, A_t)$$

Bring $\sum_{t=0}^{\infty}$ inside $\sum_{S_t \in \mathcal{N}} \sum_{S_0 \in \mathcal{N}}$ and note that

$$\sum_{A_t \in \mathcal{A}} \nabla_{\boldsymbol{\theta}} \pi(S_t, A_t; \boldsymbol{\theta}) \cdot Q^\pi(S_t, A_t) \text{ is independent of } t$$

$$\nabla_{\boldsymbol{\theta}} J(\boldsymbol{\theta}) = \sum_{s \in \mathcal{N}} \sum_{S_0 \in \mathcal{N}} \sum_{t=0}^{\infty} \gamma^t \cdot p_0(S_0) \cdot p(S_0 \rightarrow s, t, \pi) \cdot \sum_{a \in \mathcal{A}} \nabla_{\boldsymbol{\theta}} \pi(s, a; \boldsymbol{\theta}) \cdot Q^{\pi}(s, a)$$

Remember that $\displaystyle\sum_{S_0 \in \mathcal{N}} \sum_{t=0}^{\infty} \gamma^t \cdot p_0(S_0) \cdot p(S_0 \rightarrow s, t, \pi) \overset{\text{def}}{=} \rho^{\pi}(s)$. So,

$$\nabla_{\boldsymbol{\theta}} J(\boldsymbol{\theta}) = \sum_{s \in \mathcal{N}} \rho^{\pi}(s) \cdot \sum_{a \in \mathcal{A}} \nabla_{\boldsymbol{\theta}} \pi(s, a; \boldsymbol{\theta}) \cdot Q^{\pi}(s, a)$$

$$\mathbb{Q}.\mathbb{E}.\mathbb{D}.$$

This proof is borrowed from the Appendix of the famous paper by Sutton, McAllester, Singh, Mansour on Policy Gradient Methods for Reinforcement Learning with Function Approximation (R. Sutton et al. 2001).

Note that using the "Expected Value" notation under the improper distribution implied by the Discounted-Aggregate State-Visitation Measure ρ^{π}, we can write the statement of PGT as:

$$\nabla_{\boldsymbol{\theta}} J(\boldsymbol{\theta}) = \sum_{s \in \mathcal{N}} \rho^{\pi}(s) \cdot \sum_{a \in \mathcal{A}} \pi(s, a; \boldsymbol{\theta}) \cdot (\nabla_{\boldsymbol{\theta}} \log \pi(s, a; \boldsymbol{\theta})) \cdot Q^{\pi}(s, a)$$

$$= \mathbb{E}_{s \sim \rho^{\pi}, a \sim \pi}[(\nabla_{\boldsymbol{\theta}} \log \pi(s, a; \boldsymbol{\theta})) \cdot Q^{\pi}(s, a)]$$

As explained earlier, since the state occurrence probabilities and action occurrence probabilities are implicit in the trace experiences, we can sample-estimate $\nabla_{\boldsymbol{\theta}} J(\boldsymbol{\theta})$ by calculating $\gamma^t \cdot (\nabla_{\boldsymbol{\theta}} \log \pi(S_t, A_t; \boldsymbol{\theta})) \cdot Q^{\pi}(S_t, A_t)$ at each time step in each trace experience, and update the parameters $\boldsymbol{\theta}$ (according to Stochastic Gradient Ascent) with this calculation.

14.3   SCORE FUNCTION FOR CANONICAL POLICY FUNCTIONS

Now we illustrate how the Score function $\nabla_{\boldsymbol{\theta}} \pi(s, a; \boldsymbol{\theta})$ is calculated using an analytically-convenient functional form for the conditional probability distribution $a|s$ (in terms of $\boldsymbol{\theta}$) so that the derivative of the logarithm of this functional form is analytically tractable. We do this for a couple of canonical functional forms for $a|s$, one for finite action spaces and one for single-dimensional continuous action spaces.

14.3.1   Canonical $\pi(s, a; \boldsymbol{\theta})$ for Finite Action Spaces

For finite action spaces, we often use the Softmax Policy. Assume $\boldsymbol{\theta}$ is an m-vector $(\theta_1, \ldots, \theta_m)$ and assume feature vector $\boldsymbol{\phi}(s, a)$ is given by: $(\phi_1(s, a), \ldots, \phi_m(s, a))$ for all $s \in \mathcal{N}, a \in \mathcal{A}$.

We weight actions using linear combinations of features, i.e., $\boldsymbol{\phi}(s, a)^T \cdot \boldsymbol{\theta}$, and we set the action probabilities to be proportional to exponentiated weights, as follows:

$$\pi(s, a; \boldsymbol{\theta}) = \frac{e^{\boldsymbol{\phi}(s, a)^T \cdot \boldsymbol{\theta}}}{\sum_{b \in \mathcal{A}} e^{\boldsymbol{\phi}(s, b)^T \cdot \boldsymbol{\theta}}} \text{ for all } s \in \mathcal{N}, a \in \mathcal{A}$$

Then the score function is given by:

$$\nabla_{\boldsymbol{\theta}} \log \pi(s, a; \boldsymbol{\theta}) = \boldsymbol{\phi}(s, a) - \sum_{b \in \mathcal{A}} \pi(s, b; \boldsymbol{\theta}) \cdot \boldsymbol{\phi}(s, b) = \boldsymbol{\phi}(s, a) - \mathbb{E}_{\pi}[\boldsymbol{\phi}(s, \cdot)]$$

The intuitive interpretation is that the score function for an action a represents the "advantage" of the feature vector for action a over the mean feature vector (across all actions), for a given state s.

14.3.2 Canonical $\pi(s, a; \boldsymbol{\theta})$ for Single-Dimensional Continuous Action Spaces

For single-dimensional continuous action spaces (i.e., $\mathcal{A} = \mathbb{R}$), we often use a Gaussian distribution for the Policy. Assume $\boldsymbol{\theta}$ is an m-vector $(\theta_1, \ldots, \theta_m)$ and assume the state features vector $\boldsymbol{\phi}(s)$ is given by $(\phi_1(s), \ldots, \phi_m(s))$ for all $s \in \mathcal{N}$.

We set the mean of the Gaussian distribution for the Policy as a linear combination of state features, i.e., $\boldsymbol{\phi}(s)^T \cdot \boldsymbol{\theta}$, and we set the variance to be a fixed value, say σ^2. We could make the variance parameterized as well, but let's work with fixed variance to keep things simple.

The Gaussian policy selects an action a as follows:

$$a \sim \mathcal{N}(\boldsymbol{\phi}(s)^T \cdot \boldsymbol{\theta}, \sigma^2) \text{ for a given } s \in \mathcal{N}$$

Then the score function is given by:

$$\nabla_{\boldsymbol{\theta}} \log \pi(s, a; \boldsymbol{\theta}) = \frac{(a - \boldsymbol{\phi}(s)^T \cdot \boldsymbol{\theta}) \cdot \boldsymbol{\phi}(s)}{\sigma^2}$$

This is easily extensible to multi-dimensional continuous action spaces by considering a multi-dimensional Gaussian distribution for the Policy.

The intuitive interpretation is that the score function for an action a is proportional to the feature vector for given state s scaled by the "advantage" of the action a over the mean action (note: each $a \in \mathbb{R}$).

For each of the above two examples (finite action spaces and continuous action spaces), think of the "features advantage" of an action as the compass for the Gradient Ascent. The gradient estimate for an encountered action is proportional to the action's "features advantage" scaled by the action's Value Function. The intuition is that the Gradient Ascent encourages picking actions that are yielding more favorable outcomes (*Policy Improvement*) so as to ultimately get to a point where the optimal action is selected for each state.

14.4 REINFORCE ALGORITHM (MONTE-CARLO POLICY GRADIENT)

Now we are ready to write our first Policy Gradient algorithm. As ever, the simplest algorithm is a Monte-Carlo algorithm. In the case of Policy Gradient, a simple Monte-Carlo calculation provides us with an algorithm known as REINFORCE, due to R.J.Williams (Williams 1992), which we cover in this section.

We've already explained that we can calculate the Score function using an analytical derivative of a specified functional form for $\pi(S_t, A_t; \boldsymbol{\theta})$ for each atomic experience (S_t, A_t, R_t, S_{t+1}). What remains is to obtain an estimate of $Q^\pi(S_t, A_t)$ for each atomic experience (S_t, A_t, R_t, S_{t+1}). REINFORCE uses the trace experience return G_t for (S_t, A_t), while following policy π, as an unbiased sample of $Q^\pi(S_t, A_t)$. Thus, at every time step (i.e., at every atomic experience) in each episode, we estimate $\nabla_{\boldsymbol{\theta}} J(\boldsymbol{\theta})$ by calculating $\gamma^t \cdot (\nabla_{\boldsymbol{\theta}} \log \pi(S_t, A_t; \boldsymbol{\theta})) \cdot G_t$ (noting that the state occurrence probabilities and action occurrence probabilities are implicit in the trace experiences), and update the parameters $\boldsymbol{\theta}$ at the end of each episode (using each atomic experience's $\nabla_{\boldsymbol{\theta}} J(\boldsymbol{\theta})$ estimate) according to Stochastic Gradient Ascent as follows:

$$\Delta\boldsymbol{\theta} = \alpha \cdot \gamma^t \cdot (\nabla_{\boldsymbol{\theta}} \log \pi(S_t, A_t; \boldsymbol{\theta})) \cdot G_t$$

where α is the learning rate.

This Policy Gradient algorithm is Monte-Carlo because it is not bootstrapped (complete returns are used as an unbiased sample of Q^π, rather than a bootstrapped estimate). In terms of our previously-described classification of RL algorithms as Value Function-based or Policy-based or Actor-Critic, REINFORCE is a Policy-based algorithm since REINFORCE does not involve learning a Value Function.

Now let's write some code to implement the REINFORCE algorithm. In this chapter, we will focus our Python code implementation of Policy Gradient algorithms to continuous action spaces, although it should be clear based on the discussion so far that the Policy Gradient approach applies to arbitrary action spaces (we've already seen an example of the policy function parameterization for discrete action spaces). To keep things simple, the function `reinforce_gaussian` below implements REINFORCE for the simple case of single-dimensional continuous action space (i.e. $\mathcal{A} = \mathbb{R}$), although this can be easily extended to multi-dimensional continuous action spaces. So in the code below, we work with a generic state space given by `TypeVar('S')` and the action space is specialized to `float` (representing $\mathbb{R}$).

As seen earlier in the canonical example for single-dimensional continuous action space, we assume a Gaussian distribution for the policy. Specifically, the policy is represented by an arbitrary parameterized function approximation using the class `FunctionApprox`. As a reminder, an instance of `FunctionApprox` represents a probability distribution function f of the conditional random variable variable $y|x$ where x belongs to an arbitrary domain $\mathcal{X}$ and $y \in \mathbb{R}$ (probability of y conditional on x denoted as $f(x; \boldsymbol{\theta})(y)$ where $\boldsymbol{\theta}$ denotes the parameters of the `FunctionApprox`). Note that the `evaluate` method of `FunctionApprox` takes as input an `Iterable` of x values and calculates $g(x; \boldsymbol{\theta}) = \mathbb{E}_{f(x;\boldsymbol{\theta})}[y]$ for each of the x values. In our case here, x represents non-terminal states in $\mathcal{N}$ and y represents actions in $\mathbb{R}$, so $f(s; \boldsymbol{\theta})$ denotes the probability distribution of actions, conditional on state $s \in \mathcal{N}$, and $g(s; \boldsymbol{\theta})$ represents the *Expected Value* of (real-numbered) actions, conditional on state $s \in \mathcal{N}$. Since we have assumed the policy to be Gaussian,

$$\pi(s, a; \boldsymbol{\theta}) = \frac{1}{\sqrt{2\pi\sigma^2}} \cdot e^{-\frac{(a - g(s;\boldsymbol{\theta}))^2}{2\sigma^2}}$$

To be clear, our code below works with the `@abstractclass` `FunctionApprox` (meaning it is an arbitrary parameterized function approximation) with the assumption that the probability distribution of actions given a state is Gaussian whose variance σ^2 is assumed to be a constant. Assume we have m features for our function approximation, denoted as $\boldsymbol{\phi}(s) = (\phi_1(s), \ldots, \phi_m(s))$ for all $s \in \mathcal{N}$.

σ is specified in the code below as `policy_stdev`. The input `policy_mean_approx0`: `FunctionApprox[NonTerminal[S]]` specifies the function approximation we initialize the algorithm with (it is up to the user of `reinforce_gaussian` to configure `policy_mean_approx0` with the appropriate functional form for the function approximation, the hyper-parameter values, and the initial values of the parameters $\boldsymbol{\theta}$ that we want to solve for).

The Gaussian policy (of the type `GaussianPolicyFromApprox`) selects an action a (given state s) by sampling from the Gaussian distribution defined by mean $g(s; \boldsymbol{\theta})$ and variance σ^2.

The score function is given by:

$$\nabla_{\boldsymbol{\theta}} \log \pi(s, a; \boldsymbol{\theta}) = \frac{(a - g(s; \boldsymbol{\theta})) \cdot \nabla_{\boldsymbol{\theta}} g(s; \boldsymbol{\theta})}{\sigma^2}$$

The outer loop of `while True:` loops over trace experiences produced by the method `simulate_actions` of the input `mdp` for a given input `start_states_distribution` (specifying the initial states distribution $p_0 : \mathcal{N} \to [0, 1]$), and the current policy π (that is parameterized by $\boldsymbol{\theta}$, which updates after each trace experience). The inner loop loops over an `Iterator` of `step: ReturnStep[S, float]` objects produced by the `returns` method for each trace experience.

The variable grad is assigned the value of the negative score for an encountered (S_t, A_t) in a trace experience, i.e., it is assigned the value:

$$-\nabla_\theta \log \pi(S_t, A_t; \theta) = \frac{(g(S_t; \theta) - A_t) \cdot \nabla_\theta g(S_t; \theta)}{\sigma^2}$$

We negate the sign of the score because we are performing Gradient Ascent rather than Gradient Descent (the FunctionApprox class has been written for Gradient Descent). The variable scaled_grad multiplies the negative of score (grad) with γ^t (gamma_prod) and return G_t (step.return_). The rest of the code should be self-explanatory.

reinforce_gaussian returns an Iterable of FunctionApprox representing the stream of updated policies $\pi(s, \cdot; \theta)$, with each of these FunctionApprox being generated (using yield) at the end of each trace experience.

```python
import numpy as np
from rl.distribution import Distribution, Gaussian
from rl.policy import Policy
from rl.markov_process import NonTerminal
from rl.markov_decision_process import MarkovDecisionProcess, TransitionStep
from rl.function_approx import FunctionApprox, Gradient

S = TypeVar('S')

@dataclass(frozen=True)
class GaussianPolicyFromApprox(Policy[S, float]):
    function_approx: FunctionApprox[NonTerminal[S]]
    stdev: float

    def act(self, state: NonTerminal[S]) -> Gaussian:
        return Gaussian(
            mu=self.function_approx(state),
            sigma=self.stdev
        )

def reinforce_gaussian(
    mdp: MarkovDecisionProcess[S, float],
    policy_mean_approx0: FunctionApprox[NonTerminal[S]],
    start_states_distribution: Distribution[NonTerminal[S]],
    policy_stdev: float,
    gamma: float,
    episode_length_tolerance: float
) -> Iterator[FunctionApprox[NonTerminal[S]]]:
    policy_mean_approx: FunctionApprox[NonTerminal[S]] = policy_mean_approx0
    yield policy_mean_approx
    while True:
        policy: Policy[S, float] = GaussianPolicyFromApprox(
            function_approx=policy_mean_approx,
            stdev=policy_stdev
        )
        trace: Iterable[TransitionStep[S, float]] = mdp.simulate_actions(
            start_states=start_states_distribution,
            policy=policy
        )
        gamma_prod: float = 1.0
        for step in returns(trace, gamma, episode_length_tolerance):
            def obj_deriv_out(
                states: Sequence[NonTerminal[S]],
                actions: Sequence[float]
            ) -> np.ndarray:
                return (policy_mean_approx.evaluate(states) -
                        np.array(actions)) / (policy_stdev * policy_stdev)
            grad: Gradient[FunctionApprox[NonTerminal[S]]] = \
                policy_mean_approx.objective_gradient(
                    xy_vals_seq=[(step.state, step.action)],
                    obj_deriv_out_fun=obj_deriv_out
                )
            scaled_grad: Gradient[FunctionApprox[NonTerminal[S]]] = \
```

```
            grad * gamma_prod * step.return_
        policy_mean_approx = \
            policy_mean_approx.update_with_gradient(scaled_grad)
        gamma_prod *= gamma
    yield policy_mean_approx
```

The above code is in the file rl/policy_gradient.py.

14.5   OPTIMAL ASSET ALLOCATION (REVISITED)

In this chapter, we will test the PG algorithms we implement on the Optimal Asset Allocation problem of Chapter 8, specifically the setting of the class AssetAllocDiscrete covered
in Section 8.5. As a reminder, in this setting, we have a single risky asset and at each of a
fixed finite number of time steps, one has to make a choice of the quantity of current wealth
to invest in the risky asset (remainder in the riskless asset) with the goal of maximizing the
expected utility of wealth at the end of the finite horizon. Thus, this finite-horizon MDP's
state at any time t is the pair (t, W_t) where $W_t \in \mathbb{R}$ denotes the wealth at time t, and the
action at time t is the investment $x_t \in \mathbb{R}$ in the risky asset.

We provided an ADP backward-induction solution to this problem in Chapter 6, implemented with AssetAllocDiscrete (code in rl/chapter7/asset_alloc_discrete.py). Now
we want to solve it with PG algorithms, starting with REINFORCE. So we require
a new interface and hence, we implement a new class AssetAllocPG with appropriate tweaks to AssetAllocDiscrete. The key change in the interface is that we have inputs policy_feature_funcs, policy_mean_dnn_spec and policy_stdev (see code below).
policy_feature_funcs represents the sequence of feature functions for the FunctionApprox
representing the mean action for a given state (i.e., $g(s; \boldsymbol{\theta}) = \mathbb{E}_{f(s;\boldsymbol{\theta})}[a]$ where f represents the policy probability distribution of actions for a given state). policy_mean_dnn_spec
specifies the architecture of a deep neural network a user would like to use for the
FunctionApprox. policy_stdev represents the fixed standard deviation σ of the policy probability distribution of actions for any state. Unlike the backward-induction solution of
AssetAllocDiscrete where we had to model a separate MDP for each time step in the finite
horizon (where the state for each time step's MDP is the wealth), here we model a single
MDP across all time steps with the state as the pair of time step index t and the wealth W_t.
AssetAllocState = Tuple[int, float] is the data type for the state (t, W_t).

```
from rl.function_approx import DNNSpec

AssetAllocState = Tuple[int, float]

@dataclass(frozen=True)
class AssetAllocPG:
    risky_return_distributions: Sequence[Distribution[float]]
    riskless_returns: Sequence[float]
    utility_func: Callable[[float], float]
    policy_feature_funcs: Sequence[Callable[[AssetAllocState], float]]
    policy_mean_dnn_spec: DNNSpec
    policy_stdev: float
    policy_mean
    initial_wealth_distribution: Distribution[float]
```

The method get_mdp below sets up this MDP (should be self-explanatory as the construction is very similar to the construction of the single-step MDPs in AssetAllocDiscrete).

```
from rl.distribution import SampledDistribution

    def time_steps(self) -> int:
        return len(self.risky_return_distributions)

    def get_mdp(self) -> MarkovDecisionProcess[AssetAllocState, float]:
```

```
    steps: int = self.time_steps()
    distrs: Sequence[Distribution[float]] = self.risky_return_distributions
    rates: Sequence[float] = self.riskless_returns
    utility_f: Callable[[float], float] = self.utility_func

    class AssetAllocMDP(MarkovDecisionProcess[AssetAllocState, float]):

        def step(
            self,
            state: NonTerminal[AssetAllocState],
            action: float
        ) -> SampledDistribution[Tuple[State[AssetAllocState], float]]:

            def sr_sampler_func(
                state=state,
                action=action
            ) -> Tuple[State[AssetAllocState], float]:
                time, wealth = state.state
                next_wealth: float = action * (1 + distrs[time].sample()) \
                    + (wealth - action) * (1 + rates[time])
                reward: float = utility_f(next_wealth) \
                    if time == steps - 1 else 0.
                next_pair: AssetAllocState = (time + 1, next_wealth)
                next_state: State[AssetAllocState] = \
                    Terminal(next_pair) if time == steps - 1 \
                    else NonTerminal(next_pair)
                return (next_state, reward)

            return SampledDistribution(sampler=sr_sampler_func)

        def actions(self, state: NonTerminal[AssetAllocState]) \
                -> Sequence[float]:
            return []

    return AssetAllocMDP()
```

The methods `start_states_distribution` and `policy_mean_approx` below create the `SampledDistribution` of start states and the `DNNApprox` representing the mean action for a given state, respectively. Finally, the `reinforce` method below simply collects all the ingredients and passes along to `reinforce_gaussian` to solve this asset allocation problem.

```
from rl.function_approx import AdamGradient, DNNApprox

    def start_states_distribution(self) -> \
            SampledDistribution[NonTerminal[AssetAllocState]]:

        def start_states_distribution_func() -> NonTerminal[AssetAllocState]:
            wealth: float = self.initial_wealth_distribution.sample()
            return NonTerminal((0, wealth))

        return SampledDistribution(sampler=start_states_distribution_func)

    def policy_mean_approx(self) -> \
            FunctionApprox[NonTerminal[AssetAllocState]]:
        adam_gradient: AdamGradient = AdamGradient(
            learning_rate=0.003,
            decay1=0.9,
            decay2=0.999
        )
        ffs: List[Callable[[NonTerminal[AssetAllocState]], float]] = []
        for f in self.policy_feature_funcs:
            def this_f(st: NonTerminal[AssetAllocState], f=f) -> float:
                return f(st.state)
            ffs.append(this_f)
        return DNNApprox.create(
            feature_functions=ffs,
            dnn_spec=self.policy_mean_dnn_spec,
            adam_gradient=adam_gradient
        )

    def reinforce(self) -> \
            Iterator[FunctionApprox[NonTerminal[AssetAllocState]]]:
```

```
        return reinforce_gaussian(
            mdp=self.get_mdp(),
            policy_mean_approx0=self.policy_mean_approx(),
            start_states_distribution=self.start_states_distribution(),
            policy_stdev=self.policy_stdev,
            gamma=1.0,
            episode_length_tolerance=1e-5
    )
```

The above code is in the file rl/chapter13/asset_alloc_pg.py.

Let's now test this out on an instance of the problem for which we have a closed-form solution (so we can verify the REINFORCE solution against the closed-form solution). The special instance is the setting covered in Section 8.4 of Chapter 8. From Equation (8.25), we know that the optimal action in state (t, W_t) is linear in a single feature defined as $(1 + r)^t$ where r is the constant riskless rate across time steps. So we need to set up the function approximation policy_mean_dnn_spec: DNNSpec as linear in this single feature (no hidden layers and identity function as the output layer activation function), and check if the optimized weight (coefficient of this single feature) matches up with the closed-form solution of Equation (8.25).

Let us use similar settings that we had used in Chapter 8 to test AssetAllocDiscrete. In the code below, we create an instance of AssetAllocPG with time steps $T = 5$, $\mu = 13\%$, $\sigma = 20\%$, $r = 7\%$, coefficient of CARA $a = 1.0$. We set up risky_return_distributions as a sequence of identical Gaussian distributions, riskless_returns as a sequence of identical riskless rate of returns, and utility_func as a lambda parameterized by the coefficient of CARA a. We set the probability distribution of wealth at time $t = 0$ (start of each trace experience) as $\mathcal{N}(1.0, 0.1)$, and we set the constant standard deviation σ of the policy probability distribution of actions for a given state as 0.5.

```
steps: int = 5
mu: float = 0.13
sigma: float = 0.2
r: float = 0.07
a: float = 1.0
init_wealth: float = 1.0
init_wealth_stdev: float = 0.1
policy_stdev: float = 0.5
```

Next, we print the closed-form solution of the optimal action for states at each time step (note: the closed-form solution for optimal action is independent of wealth W_t, and is only dependent on t).

```
base_alloc: float = (mu - r) / (a * sigma * sigma)
for t in range(steps):
    alloc: float = base_alloc / (1 + r) ** (steps - t - 1)
    print(f"Time {t:d}: Optimal Risky Allocation = {alloc:.3f}")
```

This prints:

```
Time 0: Optimal Risky Allocation = 1.144
Time 1: Optimal Risky Allocation = 1.224
Time 2: Optimal Risky Allocation = 1.310
Time 3: Optimal Risky Allocation = 1.402
Time 4: Optimal Risky Allocation = 1.500
```

Next, we set up an instance of AssetAllocPG with the above parameters. Note that the policy_mean_dnn_spec argument to the constructor of AssetAllocPG is set up as a trivial neural network with no hidden layers and the identity function as the output layer activation function. Note also that the policy_feature_funcs argument to the constructor is set up with the single feature function $(1 + r)^t$.

```
from rl.distribution import Gaussian
from rl.function_approx import
risky_ret: Sequence[Gaussian] = [Gaussian(mu=mu, sigma=sigma)
                                 for _ in range(steps)]
riskless_ret: Sequence[float] = [r for _ in range(steps)]
utility_function: Callable[[float], float] = lambda x: - np.exp(-a * x) / a
policy_feature_funcs: Sequence[Callable[[AssetAllocState], float]] = \
    [
        lambda w_t: (1 + r) ** w_t[1]
    ]
init_wealth_distr: Gaussian = Gaussian(mu=init_wealth, sigma=init_wealth_stdev)
policy_mean_dnn_spec: DNNSpec = DNNSpec(
    neurons=[],
    bias=False,
    hidden_activation=lambda x: x,
    hidden_activation_deriv=lambda y: np.ones_like(y),
    output_activation=lambda x: x,
    output_activation_deriv=lambda y: np.ones_like(y)
)

aad: AssetAllocPG = AssetAllocPG(
    risky_return_distributions=risky_ret,
    riskless_returns=riskless_ret,
    utility_func=utility_function,
    policy_feature_funcs=policy_feature_funcs,
    policy_mean_dnn_spec=policy_mean_dnn_spec,
    policy_stdev=policy_stdev,
    initial_wealth_distribution=init_wealth_distr
)
```

Next, we invoke the method `reinforce` of this `AssetAllocPG` instance. In practice, we'd have parameterized the standard deviation of the policy probability distribution just like we parameterized the mean of the policy probability distribution, and we'd have updated those parameters in a similar manner (the standard deviation would converge to 0, i.e., the policy would converge to the optimal deterministic policy given by the closed-form solution). As an exercise, extend the function `reinforce_gaussian` to include a second `FunctionApprox` for the standard deviation of the policy probability distribution and update this `FunctionApprox` along with the updates to the mean `FunctionApprox`. However, since we set the standard deviation of the policy probability distribution to be a constant σ and since we use a Monte-Carlo method, the variance of the mean estimate of the policy probability distribution is significantly high. So we take the average of the mean estimate over several iterations (below we average the estimate from iteration 10000 to iteration 20000).

```
reinforce_policies: Iterator[FunctionApprox[
    NonTerminal[AssetAllocState]]] = aad.reinforce()

num_episodes: int = 10000
averaging_episodes: int = 10000

policies: Sequence[FunctionApprox[NonTerminal[AssetAllocState]]] = \
    list(itertools.islice(
        reinforce_policies,
        num_episodes,
        num_episodes + averaging_episodes
    ))
for t in range(steps):
    opt_alloc: float = np.mean([p(NonTerminal((init_wealth, t)))
                                for p in policies])
    print(f"Time {t:d}: Optimal Risky Allocation = {opt_alloc:.3f}")
```

This prints:

```
Time 0: Optimal Risky Allocation = 1.215
Time 1: Optimal Risky Allocation = 1.300
```

```
Time 2: Optimal Risky Allocation = 1.392
Time 3: Optimal Risky Allocation = 1.489
Time 4: Optimal Risky Allocation = 1.593
```

So we see that the estimate of the mean action for the 5 time steps from our implementation of the REINFORCE method gets fairly close to the closed-form solution.

The above code is in the file rl/chapter13/asset_alloc_reinforce.py. As ever, we encourage you to tweak the parameters and explore how the results vary.

As an exercise, we encourage you to implement an extension of this problem. Along with the risky asset allocation choice as the action at each time step, also include a consumption quantity (wealth to be extracted at each time step, along the lines of Merton's Dynamic Portfolio Allocation and Consumption problem) as part of the action at each time step. So the action at each time step would be a pair (c, a) where c is the quantity to consume and a is the quantity to allocate to the risky asset. Note that the consumption is constrained to be non-negative and at most the amount of wealth at any time step (a is unconstrained). The reward at each time step is the Utility of Consumption.

14.6   ACTOR-CRITIC AND VARIANCE REDUCTION

As we've mentioned in the previous section, REINFORCE has high variance since it's a Monte-Carlo method. So it can take quite long for REINFORCE to converge. A simple way to reduce the variance is to use a function approximation for the Q-Value Function instead of using the trace experience return as an unbiased sample of the Q-Value Function. Variance reduction happens from the simple fact that a function approximation of the Q-Value Function updates gradually (using gradient descent) and so, does not vary enormously like the trace experience returns would. Let us denote the function approximation of the Q-Value Function as $Q(s, a; w)$ where w denotes the parameters of the function approximation. We refer to $Q(s, a; w)$ as the *Critic* and we refer to the $\pi(s, a; \theta)$ function approximation as the *Actor*. The two function approximations $\pi(s, a; \theta)$ and $Q(s, a; w)$ collaborate to improve the policy using gradient ascent (guided by the PGT, using $Q(s, a; w)$ in place of the true Q-Value Function $Q^\pi(s, a)$). $\pi(s, a; \theta)$ is called *Actor* because it is the primary worker and $Q(s, a; w)$ is called *Critic* because it is the support worker. The intuitive way to think about this is that the Actor updates policy parameters in a direction that is suggested by the Critic.

Bear in mind though that the efficient way to use the Critic is in the spirit of GPI, i.e., we don't take $Q(s, a; w)$ for the current policy (current θ) all the way to convergence (thinking about updates of w for a given Policy as Policy Evaluation phase of GPI). Instead, we switch between Policy Evaluation (updates of w) and Policy Improvement (updates of θ) quite frequently. In fact, with a bootstrapped (TD) approach, we would update both w and θ after each atomic experience. w is updated such that a suitable loss function is minimized. This can be done using any of the usual Value Function approximation methods we have covered previously, including:

- Monte-Carlo, i.e., w updated using trace experience returns G_t.
- Temporal-Difference, i.e., w updated using TD Targets.
- TD(λ), i.e., w updated using targets based on eligibility traces.
- It could even be LSTD if we assume a linear function approximation for the critic $Q(s, a; w)$.

This method of calculating the gradient of $J(\theta)$ can be thought of as *Approximate Policy Gradient* due to the bias of the Critic $Q(s, a; w)$ (serving as an approximation of $Q^\pi(s, a)$), i.e.,

$$\nabla_{\boldsymbol{\theta}} J(\boldsymbol{\theta}) \approx \sum_{s \in \mathcal{N}} \rho^{\pi}(s) \cdot \sum_{a \in \mathcal{A}} \nabla_{\boldsymbol{\theta}} \pi(s, a; \boldsymbol{\theta}) \cdot Q(s, a; \boldsymbol{w})$$

Now let's implement some code to perform Policy Gradient with the Critic updated using Temporal-Difference (again, for the simple case of single-dimensional continuous action space). In the function `actor_critic_gaussian` below, the key changes (from the code in `reinforce_gaussian`) are:

- The Q-Value function approximation parameters $\boldsymbol{w}$ are updated after each atomic experience as:

$$\Delta \boldsymbol{w} = \beta \cdot (R_{t+1} + \gamma \cdot Q(S_{t+1}, A_{t+1}; \boldsymbol{w}) - Q(S_t, A_t; \boldsymbol{w})) \cdot \nabla_{\boldsymbol{w}} Q(S_t, A_t; \boldsymbol{w})$$

where β is the learning rate for the Q-Value function approximation.
- The policy mean parameters $\boldsymbol{\theta}$ are updated after each atomic experience as:

$$\Delta \boldsymbol{\theta} = \alpha \cdot \gamma^t \cdot (\nabla_{\boldsymbol{\theta}} \log \pi(S_t; A_t; \boldsymbol{\theta})) \cdot Q(S_t, A_t; \boldsymbol{w})$$

(instead of $\alpha \cdot \gamma^t \cdot (\nabla_{\boldsymbol{\theta}} \log \pi(S_t, A_t; \boldsymbol{\theta})) \cdot G_t$).

```python
from rl.approximate_dynamic_programming import QValueFunctionApprox
from rl.approximate_dynamic_programming import NTStateDistribution
def actor_critic_gaussian(
    mdp: MarkovDecisionProcess[S, float],
    policy_mean_approx0: FunctionApprox[NonTerminal[S]],
    q_value_func_approx0: QValueFunctionApprox[S, float],
    start_states_distribution: NTStateDistribution[S],
    policy_stdev: float,
    gamma: float,
    max_episode_length: float
) -> Iterator[FunctionApprox[NonTerminal[S]]]:
    policy_mean_approx: FunctionApprox[NonTerminal[S]] = policy_mean_approx0
    yield policy_mean_approx
    q: QValueFunctionApprox[S, float] = q_value_func_approx0
    while True:
        steps: int = 0
        gamma_prod: float = 1.0
        state: NonTerminal[S] = start_states_distribution.sample()
        action: float = Gaussian(
            mu=policy_mean_approx(state),
            sigma=policy_stdev
        ).sample()
        while isinstance(state, NonTerminal) and steps < max_episode_length:
            next_state, reward = mdp.step(state, action).sample()
            if isinstance(next_state, NonTerminal):
                next_action: float = Gaussian(
                    mu=policy_mean_approx(next_state),
                    sigma=policy_stdev
                ).sample()
                q = q.update([(
                    (state, action),
                    reward + gamma * q((next_state, next_action))
                )])
                action = next_action
            else:
                q = q.update([(((state, action), reward)])

        def obj_deriv_out(
            states: Sequence[NonTerminal[S]],
            actions: Sequence[float]
        ) -> np.ndarray:
            return (policy_mean_approx.evaluate(states) -
```

```
                np.array(actions)) / (policy_stdev * policy_stdev)
grad: Gradient[FunctionApprox[NonTerminal[S]]] = \
    policy_mean_approx.objective_gradient(
        xy_vals_seq=[(state, action)],
        obj_deriv_out_fun=obj_deriv_out
)
scaled_grad: Gradient[FunctionApprox[NonTerminal[S]]] = \
    grad * gamma_prod * q((state, action))
policy_mean_approx = \
    policy_mean_approx.update_with_gradient(scaled_grad)
yield policy_mean_approx
gamma_prod *= gamma
steps += 1
state = next_state
```

The above code is in the file rl/policy_gradient.py. We leave it to you as an exercise to implement the update of $Q(s, a; w)$ with $TD(\lambda)$, i.e., with eligibility traces.

We can reduce the variance of this Actor-Critic method by subtracting a Baseline Function $B(s)$ from $Q(s, a; w)$ in the Policy Gradient estimate. This means we update the parameters θ as:

$$\Delta\theta = \alpha \cdot \gamma^t \cdot \nabla_\theta \log \pi(S_t, A_t; \theta) \cdot (Q(S_t, A_t; w) - B(S_t))$$

Note that the Baseline Function $B(s)$ is only a function of state s (and not of action a). This ensures that subtracting the Baseline Function $B(s)$ does not add bias. This is because:

$$\sum_{s \in \mathcal{N}} \rho^\pi(s) \sum_{a \in \mathcal{A}} \nabla_\theta \pi(s, a; \theta) \cdot B(s)$$

$$= \sum_{s \in \mathcal{N}} \rho^\pi(s) \cdot B(s) \cdot \nabla_\theta (\sum_{a \in \mathcal{A}} \pi(s, a; \theta))$$

$$= \sum_{s \in \mathcal{N}} \rho^\pi(s) \cdot B(s) \cdot \nabla_\theta 1$$

$$= 0$$

A good Baseline Function $B(s)$ is a function approximation $V(s; v)$ of the State-Value Function $V^\pi(s)$. So then we can rewrite the Actor-Critic Policy Gradient algorithm using an estimate of the Advantage Function, as follows:

$$A(s, a; w, v) = Q(s, a; w) - V(s; v)$$

With this, the approximation for $\nabla_\theta J(\theta)$ is given by:

$$\nabla_\theta J(\theta) \approx \sum_{s \in \mathcal{N}} \rho^\pi(s) \cdot \sum_{a \in \mathcal{A}} \nabla_\theta \pi(s, a; \theta) \cdot A(s, a; w, v)$$

The function actor_critic_advantage_gaussian in the file rl/policy_gradient.py implements this algorithm, i.e., Policy Gradient with two Critics $Q(s, a; w)$ and $V(s; v)$, each updated using Temporal-Difference (again, for the simple case of single-dimensional continuous action space). Specifically, in the code of actor_critic_advantage_gaussian:

- The Q-Value function approximation parameters w are updated after each atomic experience as:

$$\Delta w = \beta_w \cdot (R_{t+1} + \gamma \cdot Q(S_{t+1}, A_{t+1}; w) - Q(S_t, A_t; w)) \cdot \nabla_w Q(S_t, A_t; w)$$

where β_w is the learning rate for the Q-Value function approximation.

- The State-Value function approximation parameters v are updated after each atomic experience as:

$$\Delta v = \beta_v \cdot (R_{t+1} + \gamma \cdot V(S_{t+1}; v) - V(S_t; v)) \cdot \nabla_v V(S_t; v)$$

  where β_v is the learning rate for the State-Value function approximation.
- The policy mean parameters θ are updated after each atomic experience as:

$$\Delta \theta = \alpha \cdot \gamma^t \cdot (\nabla_\theta \log \pi(S_t; A_t; \theta)) \cdot (Q(S_t, A_t; w) - V(S_t; v))$$

A simpler way is to use the TD Error of the State-Value Function as an estimate of the Advantage Function. To understand this idea, let δ^π denote the TD Error for the *true* State-Value Function $V^\pi(s)$. Then,

$$\delta^\pi = r + \gamma \cdot V^\pi(s') - V^\pi(s)$$

Note that δ^π is an unbiased estimate of the Advantage function $A^\pi(s, a)$. This is because

$$\mathbb{E}_\pi[\delta^\pi | s, a] = \mathbb{E}_\pi[r + \gamma \cdot V^\pi(s') | s, a] - V^\pi(s) = Q^\pi(s, a) - V^\pi(s) = A^\pi(s, a)$$

So we can write Policy Gradient in terms of $\mathbb{E}_\pi[\delta^\pi | s, a]$:

$$\nabla_\theta J(\theta) = \sum_{s \in \mathcal{N}} \rho^\pi(s) \cdot \sum_{a \in \mathcal{A}} \nabla_\theta \pi(s, a; \theta) \cdot \mathbb{E}_\pi[\delta^\pi | s, a]$$

In practice, we use a function approximation for the TD error, and sample:

$$\delta(s, r, s'; v) = r + \gamma \cdot V(s'; v) - V(s; v)$$

This approach requires only one set of critic parameters v, and we don't have to worry about the Action-Value Function Q.

Now let's implement some code for this TD Error-based PG Algorithm (again, for the simple case of single-dimensional continuous action space). In the function `actor_critic_td_error_gaussian` below:

- The State-Value function approximation parameters v are updated after each atomic experience as:

$$\Delta v = \alpha_v \cdot (R_{t+1} + \gamma \cdot V(S_{t+1}; v) - V(S_t; v)) \cdot \nabla_v V(S_t; v)$$

  where α_v is the learning rate for the State-Value function approximation.
- The policy mean parameters θ are updated after each atomic experience as:

$$\Delta \theta = \alpha_\theta \cdot \gamma^t \cdot (\nabla_\theta \log \pi(S_t; A_t; \theta)) \cdot (R_{t+1} + \gamma \cdot V(S_{t+1}; v) - V(S_t; v))$$

  where α_θ is the learning rate for the Policy Mean function approximation.

```
from rl.approximate_dynamic_programming import ValueFunctionApprox
def actor_critic_td_error_gaussian(
    mdp: MarkovDecisionProcess[S, float],
    policy_mean_approx0: FunctionApprox[NonTerminal[S]],
    value_func_approx0: ValueFunctionApprox[S],
    start_states_distribution: NTStateDistribution[S],
    policy_stdev: float,
    gamma: float,
    max_episode_length: float
```

```
) -> Iterator[FunctionApprox[NonTerminal[S]]]:
    policy_mean_approx: FunctionApprox[NonTerminal[S]] = policy_mean_approx0
    yield policy_mean_approx
    vf: ValueFunctionApprox[S] = value_func_approx0
    while True:
        steps: int = 0
        gamma_prod: float = 1.0
        state: NonTerminal[S] = start_states_distribution.sample()
        while isinstance(state, NonTerminal) and steps < max_episode_length:
            action: float = Gaussian(
                mu=policy_mean_approx(state),
                sigma=policy_stdev
            ).sample()
            next_state, reward = mdp.step(state, action).sample()
            if isinstance(next_state, NonTerminal):
                td_target: float = reward + gamma * vf(next_state)
            else:
                td_target = reward
            td_error: float = td_target - vf(state)
            vf = vf.update([(state, td_target)])

            def obj_deriv_out(
                states: Sequence[NonTerminal[S]],
                actions: Sequence[float]
            ) -> np.ndarray:
                return (policy_mean_approx.evaluate(states) -
                        np.array(actions)) / (policy_stdev * policy_stdev)

            grad: Gradient[FunctionApprox[NonTerminal[S]]] = \
                policy_mean_approx.objective_gradient(
                    xy_vals_seq=[(state, action)],
                    obj_deriv_out_fun=obj_deriv_out
                )
            scaled_grad: Gradient[FunctionApprox[NonTerminal[S]]] = \
                grad * gamma_prod * td_error
            policy_mean_approx = \
                policy_mean_approx.update_with_gradient(scaled_grad)
            yield policy_mean_approx
            gamma_prod *= gamma
            steps += 1
            state = next_state
```

The above code is in the file rl/policy_gradient.py.

Likewise, we can implement an Actor-Critic algorithm using Eligibility Traces (i.e., TD(λ)) for the State-Value Function Approximation and also for the Policy Mean Function Approximation. The updates after each atomic experience to parameters v of the State-Value function approximation and parameters θ of the policy mean function approximation are given by:

$$\Delta v = \alpha_v \cdot (R_{t+1} + \gamma \cdot V(S_{t+1}; v) - V(S_t; v)) \cdot E_v$$

$$\Delta \theta = \alpha_\theta \cdot (R_{t+1} + \gamma \cdot V(S_{t+1}; v) - V(S_t; v)) \cdot E_\theta$$

where the Eligibility Traces E_v and E_θ are updated after each atomic experience as follows:

$$E_v \leftarrow \gamma \cdot \lambda_v \cdot E_v + \nabla_v V(S_t; v)$$

$$E_\theta \leftarrow \gamma \cdot \lambda_\theta \cdot E_\theta + \gamma^t \cdot \nabla_\theta \log \pi(S_t, A_t; \theta)$$

where λ_v and λ_θ are the TD(λ) parameters respectively for the State-Value Function Approximation and the Policy Mean Function Approximation.

We encourage you to implement in code this Actor-Critic algorithm using Eligibility Traces.

Now let's compare these methods on the `AssetAllocPG` instance we had created earlier to test REINFORCE, i.e., for time steps $T = 5, \mu = 13\%, \sigma = 20\%, r = 7\%$, coefficient of CARA $a = 1.0$, probability distribution of wealth at the start of each trace experience as $\mathcal{N}(1.0, 0.1)$, and constant standard deviation σ of the policy probability distribution of actions for a given state as 0.5. The `__main__` code in rl/chapter13/asset_alloc_pg.py evaluates the mean action for the start state of $(t = 0, W_0 = 1.0)$ after each episode (over 50,000 episodes) for each of the above-implemented PG algorithms' function approximation for the policy mean. It then plots the progress of the evaluated mean action for the start state over the 50,000 episodes (each point plotted as an average over a batch of 200 episodes), along with the benchmark of the optimal action for the start state from the known closed-form solution. Figure 14.1 shows the graph, validating the points we have made above on bias and variance of these algorithms.

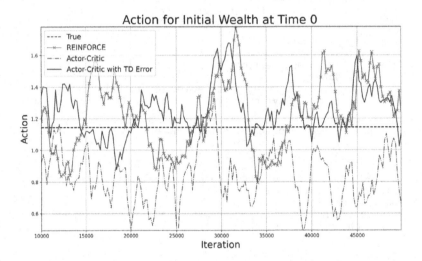

Figure 14.1 **Progress of PG Algorithms**

Actor-Critic methods were developed in the late 1970s and 1980s, but not paid attention to in the 1990s. In the past two decades, there has been a revival of Actor-Critic methods. For a more detailed coverage of Actor-Critic methods, see the paper by Degris, White, Sutton (Degris, White, and Sutton 2012).

14.7 OVERCOMING BIAS WITH COMPATIBLE FUNCTION APPROXIMATION

We've talked a lot about reducing variance for faster convergence of PG Algorithms. Specifically, we've talked about the following proxies for $Q^\pi(s, a)$ in the form of Actor-Critic algorithms in order to reduce variance.

- $Q(s, a; \boldsymbol{w})$
- $A(s, a; \boldsymbol{w}, \boldsymbol{v}) = Q(s, a; \boldsymbol{w}) - V(s; \boldsymbol{v})$
- $\delta(s, s', r; \boldsymbol{v}) = r + \gamma \cdot V(s'; \boldsymbol{v}) - V(s; \boldsymbol{v})$

However, each of the above proxies for $Q^\pi(s, a)$ in PG algorithms have a bias. In this section, we talk about how to overcome bias. The basis for overcoming bias is an important

Theorem known as the *Compatible Function Approximation Theorem*. We state and prove this theorem, and then explain how we could use it in a PG algorithm.

Theorem 14.7.1 (Compatible Function Approximation Theorem). *Let $\boldsymbol{w}_{\boldsymbol{\theta}}^*$ denote the Critic parameters $\boldsymbol{w}$ that minimize the following mean-squared-error for given policy parameters $\boldsymbol{\theta}$:*

$$\sum_{s \in \mathcal{N}} \rho^\pi(s) \cdot \sum_{a \in \mathcal{A}} \pi(s, a; \boldsymbol{\theta}) \cdot (Q^\pi(s, a) - Q(s, a; \boldsymbol{w}))^2$$

Assume that the data type of $\boldsymbol{\theta}$ is the same as the data type of $\boldsymbol{w}$ and furthermore, assume that for any policy parameters $\boldsymbol{\theta}$, the Critic gradient at $\boldsymbol{w}_{\boldsymbol{\theta}}^*$ is compatible with the Actor score function, i.e.,

$$\nabla_{\boldsymbol{w}} Q(s, a; \boldsymbol{w}_{\boldsymbol{\theta}}^*) = \nabla_{\boldsymbol{\theta}} \log \pi(s, a; \boldsymbol{\theta}) \text{ for all } s \in \mathcal{N}, \text{ for all } a \in \mathcal{A}$$

Then the Policy Gradient using critic $Q(s, a; \boldsymbol{w}_{\boldsymbol{\theta}}^*)$ is exact:

$$\nabla_{\boldsymbol{\theta}} J(\boldsymbol{\theta}) = \sum_{s \in \mathcal{N}} \rho^\pi(s) \cdot \sum_{a \in \mathcal{A}} \nabla_{\boldsymbol{\theta}} \pi(s, a; \boldsymbol{\theta}) \cdot Q(s, a; \boldsymbol{w}_{\boldsymbol{\theta}}^*)$$

Proof. For a given $\boldsymbol{\theta}$, since $\boldsymbol{w}_{\boldsymbol{\theta}}^*$ minimizes the mean-squared-error as defined above, we have:

$$\sum_{s \in \mathcal{N}} \rho^\pi(s) \cdot \sum_{a \in \mathcal{A}} \pi(s, a; \boldsymbol{\theta}) \cdot (Q^\pi(s, a) - Q(s, a; \boldsymbol{w}_{\boldsymbol{\theta}}^*)) \cdot \nabla_{\boldsymbol{w}} Q(s, a; \boldsymbol{w}_{\boldsymbol{\theta}}^*) = 0$$

But since $\nabla_{\boldsymbol{w}} Q(s, a; \boldsymbol{w}_{\boldsymbol{\theta}}^*) = \nabla_{\boldsymbol{\theta}} \log \pi(s, a; \boldsymbol{\theta})$, we have:

$$\sum_{s \in \mathcal{N}} \rho^\pi(s) \cdot \sum_{a \in \mathcal{A}} \pi(s, a; \boldsymbol{\theta}) \cdot (Q^\pi(s, a) - Q(s, a; \boldsymbol{w}_{\boldsymbol{\theta}}^*)) \cdot \nabla_{\boldsymbol{\theta}} \log \pi(s, a; \boldsymbol{\theta}) = 0$$

Therefore,

$$\sum_{s \in \mathcal{N}} \rho^\pi(s) \cdot \sum_{a \in \mathcal{A}} \pi(s, a; \boldsymbol{\theta}) \cdot Q^\pi(s, a) \cdot \nabla_{\boldsymbol{\theta}} \log \pi(s, a; \boldsymbol{\theta})$$
$$= \sum_{s \in \mathcal{N}} \rho^\pi(s) \cdot \sum_{a \in \mathcal{A}} \pi(s, a; \boldsymbol{\theta}) \cdot Q(s, a; \boldsymbol{w}_{\boldsymbol{\theta}}^*) \cdot \nabla_{\boldsymbol{\theta}} \log \pi(s, a; \boldsymbol{\theta})$$

But $\nabla_{\boldsymbol{\theta}} J(\boldsymbol{\theta}) = \sum_{s \in \mathcal{N}} \rho^\pi(s) \cdot \sum_{a \in \mathcal{A}} \pi(s, a; \boldsymbol{\theta}) \cdot Q^\pi(s, a) \cdot \nabla_{\boldsymbol{\theta}} \log \pi(s, a; \boldsymbol{\theta})$

So, $\nabla_{\boldsymbol{\theta}} J(\boldsymbol{\theta}) = \sum_{s \in \mathcal{N}} \rho^\pi(s) \cdot \sum_{a \in \mathcal{A}} \pi(s, a; \boldsymbol{\theta}) \cdot Q(s, a; \boldsymbol{w}_{\boldsymbol{\theta}}^*) \cdot \nabla_{\boldsymbol{\theta}} \log \pi(s, a; \boldsymbol{\theta})$

$$= \sum_{s \in \mathcal{N}} \rho^\pi(s) \cdot \sum_{a \in \mathcal{A}} \nabla_{\boldsymbol{\theta}} \pi(s, a; \boldsymbol{\theta}) \cdot Q(s, a; \boldsymbol{w}_{\boldsymbol{\theta}}^*)$$

Q.E.D.

□

This proof originally appeared in the famous paper by Sutton, McAllester, Singh, Mansour on Policy Gradient Methods for Reinforcement Learning with Function Approximation (R. Sutton et al. 2001).

This means if the compatibility assumption of the Theorem is satisfied, we can use the critic function approximation $Q(s, a; w^*_{\theta})$ and still have exact Policy Gradient (i.e., no bias due to using a function approximation for the Q-Value Function). However, note that in practice, we invoke the spirit of GPI and don't take $Q(s, a; w)$ to convergence for the current θ. Rather, we update both w and θ frequently, and this turns out to be good enough in terms of lowering the bias.

A simple way to enable Compatible Function Approximation is to make $Q(s, a; w)$ a linear function approximation, with the features of the linear function approximation equal to the Score of the policy function approximation, as follows:

$$Q(s, a; w) = \sum_{i=1}^{m} \frac{\partial \log \pi(s, a; \theta)}{\partial \theta_i} \cdot w_i \text{ for all } s \in \mathcal{N}, \text{ for all } a \in \mathcal{A}$$

which means the feature functions $\boldsymbol{\eta}(s, a) = (\eta_1(s, a), \eta_2(s, a), \ldots, \eta_m(s, a))$ of the linear function approximation are given by:

$$\eta_i(s, a) = \frac{\partial \log \pi(s, a; \theta)}{\partial \theta_i} \text{ for all } s \in \mathcal{N}, \text{ for all } a \in \mathcal{A}, \text{ for all } i = 1, \ldots, m$$

This means the feature functions $\boldsymbol{\eta}$ is identically equal to the *Score*. Note that although here we assume $Q(s, a; w)$ to be a linear function approximation, the policy function approximation $\pi(s, a; \theta)$ can be more flexible. All that is required is that θ consists of exactly m parameters (matching the number of number of parameters m of w) and that each of the partial derivatives $\frac{\partial \log \pi(s, a; \theta)}{\partial \theta_i}$ lines up with a corresponding feature $\eta_i(s, a)$ of the linear function approximation $Q(s, a; w)$. This means that as θ updates (as a consequence of Stochastic Gradient Ascent), $\pi(s, a; \theta)$ updates, and consequently the feature functions $\boldsymbol{\eta}(s, a) = \nabla_\theta \log \pi(s, a; \theta)$ update. This means the feature vector $\boldsymbol{\eta}(s, a)$ is not constant for a given (s, a) pair. Rather, the feature vector $\boldsymbol{\eta}(s, a)$ for a given (s, a) pair varies in accordance with θ varying.

If we assume the canonical function approximation for $\pi(s, a; \theta)$ for finite action spaces that we had described in Section 14.3, then:

$$\boldsymbol{\eta}(s, a) = \boldsymbol{\phi}(s, a) - \sum_{b \in \mathcal{A}} \pi(s, b; \theta) \cdot \boldsymbol{\phi}(s, b) \text{ for all } s \in \mathcal{N} \text{ for all } a \in \mathcal{A}$$

Note the dependency of feature vector $\boldsymbol{\eta}(s, a)$ on θ.

If we assume the canonical function approximation for $\pi(s, a; \theta)$ for single-dimensional continuous action spaces that we had described in Section 14.3, then:

$$\boldsymbol{\eta}(s, a) = \frac{(a - \boldsymbol{\phi}(s)^T \cdot \theta) \cdot \boldsymbol{\phi}(s)}{\sigma^2} \text{ for all } s \in \mathcal{N} \text{ for all } a \in \mathcal{A}$$

Note the dependency of feature vector $\boldsymbol{\eta}(s, a)$ on θ.

We note that any compatible linear function approximation $Q(s, a; w)$ serves as an approximation of the advantage function because:

$$\sum_{a \in \mathcal{A}} \pi(s, a; \theta) \cdot Q(s, a; w) = \sum_{a \in \mathcal{A}} \pi(s, a; \theta) \cdot \left(\sum_{i=1}^{m} \frac{\partial \log \pi(s, a; \theta)}{\partial \theta_i} \cdot w_i \right)$$

$$= \sum_{a \in \mathcal{A}} (\sum_{i=1}^{m} \frac{\partial \pi(s, a; \boldsymbol{\theta})}{\partial \theta_i} \cdot w_i) = \sum_{i=1}^{m} (\sum_{a \in \mathcal{A}} \frac{\partial \pi(s, a; \boldsymbol{\theta})}{\partial \theta_i}) \cdot w_i$$

$$= \sum_{i=1}^{m} \frac{\partial}{\partial \theta_i} (\sum_{a \in \mathcal{A}} \pi(s, a; \boldsymbol{\theta})) \cdot w_i = \sum_{i=1}^{m} \frac{\partial 1}{\partial \theta_i} \cdot w_i = 0$$

Denoting $\nabla_{\boldsymbol{\theta}} \log \pi(s, a; \boldsymbol{\theta})$ as the score column vector $\boldsymbol{SC}(s, a; \boldsymbol{\theta})$ and assuming compatible linear-approximation critic:

$$\nabla_{\boldsymbol{\theta}} J(\boldsymbol{\theta}) = \sum_{s \in \mathcal{N}} \rho^{\pi}(s) \cdot \sum_{a \in \mathcal{A}} \pi(s, a; \boldsymbol{\theta}) \cdot \boldsymbol{SC}(s, a; \boldsymbol{\theta}) \cdot (\boldsymbol{SC}(s, a; \boldsymbol{\theta})^T \cdot \boldsymbol{w}_{\boldsymbol{\theta}}^*)$$

$$= \sum_{s \in \mathcal{N}} \rho^{\pi}(s) \cdot \sum_{a \in \mathcal{A}} \pi(s, a; \boldsymbol{\theta}) \cdot (\boldsymbol{SC}(s, a; \boldsymbol{\theta}) \cdot \boldsymbol{SC}(s, a; \boldsymbol{\theta})^T) \cdot \boldsymbol{w}_{\boldsymbol{\theta}}^*$$

$$= \mathbb{E}_{s \sim \rho^{\pi}, a \sim \pi}[\boldsymbol{SC}(s, a; \boldsymbol{\theta}) \cdot \boldsymbol{SC}(s, a; \boldsymbol{\theta})^T] \cdot \boldsymbol{w}_{\boldsymbol{\theta}}^*$$

Note that $\mathbb{E}_{s \sim \rho^{\pi}, a \sim \pi}[\boldsymbol{SC}(s, a; \boldsymbol{\theta}) \cdot \boldsymbol{SC}(s, a; \boldsymbol{\theta})^T]$ is the Fisher Information Matrix $\boldsymbol{FIM}_{\rho^{\pi}, \pi}(\boldsymbol{\theta})$ with respect to $s \sim \rho^{\pi}, a \sim \pi$. Therefore, we can write $\nabla_{\boldsymbol{\theta}} J(\boldsymbol{\theta})$ more succinctly as:

$$\nabla_{\boldsymbol{\theta}} J(\boldsymbol{\theta}) = \boldsymbol{FIM}_{\rho^{\pi}, \pi}(\boldsymbol{\theta}) \cdot \boldsymbol{w}_{\boldsymbol{\theta}}^* \tag{14.1}$$

Thus, we can update $\boldsymbol{\theta}$ after each atomic experience at time step t by calculating the gradient of $J(\boldsymbol{\theta})$ for the atomic experience as the outer product of $\boldsymbol{SC}(S_t, A_t; \boldsymbol{\theta})$ with itself (which gives a $m \times m$ matrix), then multiply this matrix with the vector $\boldsymbol{w}$, and then scale by γ^t, i.e.

$$\Delta \boldsymbol{\theta} = \alpha_{\boldsymbol{\theta}} \cdot \gamma^t \cdot \boldsymbol{SC}(S_t, A_t; \boldsymbol{w}) \cdot \boldsymbol{SC}(S_t, A_t; \boldsymbol{w})^T \cdot \boldsymbol{w}$$

The update for $\boldsymbol{w}$ after each atomic experience is the usual Q-Value Function Approximation update with Q-Value loss function gradient for the atomic experience calculated as:

$$\Delta \boldsymbol{w} = \alpha_{\boldsymbol{w}} \cdot (R_{t+1} + \gamma \cdot \boldsymbol{SC}(S_{t+1}, A_{t+1}; \boldsymbol{\theta})^T \cdot \boldsymbol{w} - \boldsymbol{SC}(S_t, A_t; \boldsymbol{\theta})^T \cdot \boldsymbol{w}) \cdot \boldsymbol{SC}(S_t, A_t; \boldsymbol{\theta})$$

This completes our coverage of the basic Policy Gradient Methods. Next, we cover a couple of special Policy Gradient Methods that have worked well in practice—Natural Policy Gradient and Deterministic Policy Gradient.

14.8   POLICY GRADIENT METHODS IN PRACTICE

14.8.1   Natural Policy Gradient

Natural Policy Gradient (abbreviated NPG) is due to a paper by Kakade (Kakade 2001) that utilizes the idea of Natural Gradient first introduced by Amari (Amari 1998). We won't cover the theory of Natural Gradient in detail here, and refer you to the above two papers instead. Here we give a high-level overview of the concepts, and describe the algorithm.

The core motivation for Natural Gradient is that when the parameters space has a certain underlying structure (as is the case with the parameters space of $\boldsymbol{\theta}$ in the context of maximizing $J(\boldsymbol{\theta})$), the usual gradient does not represent its steepest descent direction, but the Natural Gradient does. The steepest descent direction of an arbitrary function $f(\boldsymbol{\theta})$ to be minimized is defined as the vector $\Delta \boldsymbol{\theta}$ that minimizes $f(\boldsymbol{\theta} + \Delta \boldsymbol{\theta})$ under the constraint that the length $|\Delta \boldsymbol{\theta}|$ is a constant. In general, the length $|\Delta \boldsymbol{\theta}|$ is defined with respect to some positive-definite matrix $\boldsymbol{G}(\boldsymbol{\theta})$ governed by the underlying structure of the $\boldsymbol{\theta}$ parameters space, i.e.,

$$|\Delta \boldsymbol{\theta}|^2 = (\Delta \boldsymbol{\theta})^T \cdot \boldsymbol{G}(\boldsymbol{\theta}) \cdot \Delta \boldsymbol{\theta}$$

We can show that under the length metric defined by the matrix G, the steepest descent direction is:

$$\nabla_{\theta}^{nat} f(\theta) = G^{-1}(\theta) \cdot \nabla_{\theta} f(\theta)$$

We refer to this steepest descent direction $\nabla_{\theta}^{nat} f(\theta)$ as the Natural Gradient. We can update the parameters θ in this Natural Gradient direction in order to achieve steepest descent (according to the matrix G), as follows:

$$\Delta\theta = -\alpha \cdot \nabla_{\theta}^{nat} f(\theta)$$

Amari showed that for a supervised learning problem of estimating the conditional probability distribution of $y|x$ with a function approximation (i.e., where the loss function is defined as the KL divergence between the data distribution and the model distribution), the matrix G is the Fisher Information Matrix for $y|x$.

Kakade specialized this idea of Natural Gradient to the case of Policy Gradient (naming it Natural Policy Gradient) with the objective function $f(\theta)$ equal to the negative of the Expected Returns $J(\theta)$. This gives the Natural Policy Gradient $\nabla_{\theta}^{nat} J(\theta)$ defined as:

$$\nabla_{\theta}^{nat} J(\theta) = FIM_{\rho^{\pi}, \pi}^{-1}(\theta) \cdot \nabla_{\theta} J(\theta)$$

where $FIM_{\rho_{\pi}, \pi}$ denotes the Fisher Information Matrix with respect to $s \sim \rho^{\pi}, a \sim \pi$.

We've noted in the previous section that if we enable Compatible Function Approximation with a linear function approximation for $Q^{\pi}(s, a)$, then we have Equation (14.1), i.e.,

$$\nabla_{\theta} J(\theta) = FIM_{\rho^{\pi}, \pi}(\theta) \cdot w_{\theta}^{*}$$

This means:

$$\nabla_{\theta}^{nat} J(\theta) = w_{\theta}^{*}$$

This compact result enables a simple algorithm for Natural Policy Gradient (NPG):

- After each atomic experience, update Critic parameters w with the critic loss gradient as:

$$\Delta w = \alpha_{w} \cdot (R_{t+1} + \gamma \cdot SC(S_{t+1}, A_{t+1}; \theta)^{T} \cdot w - SC(S_t, A_t; \theta)^{T} \cdot w) \cdot SC(S_t, A_t; \theta)$$

- After each atomic experience, update Actor parameters θ in the direction of w:

$$\Delta\theta = \alpha_{\theta} \cdot w$$

14.8.2 Deterministic Policy Gradient

Deterministic Policy Gradient (abbreviated DPG) is a creative adaptation of Policy Gradient wherein instead of a parameterized function approximation for a stochastic policy, we have a parameterized function approximation for a deterministic policy for the case of continuous action spaces. DPG is due to a paper by Silver, Lever, Heess, Degris, Wiestra, Riedmiller (Silver et al. 2014). DPG is expressed in terms of the Expected Gradient of the Q-Value Function and can be estimated much more efficiently than the usual (stochastic) PG. (Stochastic) PG integrates over both the state and action spaces, whereas DPG integrates over only the state space. As a result, computing (stochastic) PG would require more samples if the action space has many dimensions.

In Actor-Critic DPG, the Actor is the function approximation for the deterministic policy and the Critic is the function approximation for the Q-Value Function. The paper by Silver et al. provides a Compatible Function Approximation Theorem for DPG to overcome Critic approximation bias. The paper also shows that DPG is the limiting case of (Stochastic) PG, as policy variance tends to 0. This means the usual machinery of PG (such as Actor-Critic, Compatible Function Approximation, Natural Policy Gradient etc.) is also applicable to DPG.

We use the notation $a = \pi_D(s; \boldsymbol{\theta})$ to represent (a potentially multi-dimensional) continuous-valued action a equal to the value of a deterministic policy function approximation π_D (parameterized by $\boldsymbol{\theta}$), when evaluated for a state s.

The core idea of DPG can be understood intuitively by orienting on the basics of GPI and specifically, on Policy Improvement in GPI. For continuous action spaces, greedy policy improvement (with an argmax over actions, for each state) is problematic. So a simple and attractive alternative is to move the policy in the direction of the gradient of the Q-Value Function (rather than globally maximizing the Q-Value Function, at each step). Specifically, for each state s that is encountered, the policy approximation parameters $\boldsymbol{\theta}$ are updated in proportion to $\nabla_{\boldsymbol{\theta}} Q(s, \pi_D(s; \boldsymbol{\theta}))$. Note that the direction of policy improvement is different for each state, and so the average direction of policy improvements is given by:

$$\mathbb{E}_{s \sim \rho^{\mu}}[\nabla_{\boldsymbol{\theta}} Q(s, \pi_D(s; \boldsymbol{\theta}))]$$

where ρ^{μ} is the same Discounted-Aggregate State-Visitation Measure we had defined for PG (now for exploratory behavior policy μ).

Using chain-rule, the above expression can be written as:

$$\mathbb{E}_{s \sim \rho^{\mu}}[\nabla_{\boldsymbol{\theta}} \pi_D(s; \boldsymbol{\theta}) \cdot \nabla_a Q^{\pi_D}(s, a)\Big|_{a = \pi_D(s; \boldsymbol{\theta})}]$$

Note that $\nabla_{\boldsymbol{\theta}} \pi_D(s; \boldsymbol{\theta})$ is a Jacobian matrix as it takes the partial derivatives of a potentially multi-dimensional action $a = \pi_D(s; \boldsymbol{\theta})$ with respect to each parameter in $\boldsymbol{\theta}$. As we've pointed out during the coverage of (stochastic) PG, when $\boldsymbol{\theta}$ changes, policy π_D changes, which changes the state distribution ρ^{π_D}. So it's not clear that this calculation indeed guarantees improvement—it doesn't take into account the effect of changing $\boldsymbol{\theta}$ on ρ^{π_D}. However, as was the case in PGT, Deterministic Policy Gradient Theorem (abbreviated DPGT) ensures that there is no need to compute the gradient of ρ^{π_D} with respect to $\boldsymbol{\theta}$, and that the update described above indeed follows the gradient of the Expected Return objective function. We formalize this now by stating the DPGT.

Analogous to the Expected Returns Objective defined for (stochastic) PG, we define the Expected Returns Objective $J(\boldsymbol{\theta})$ for DPG as:

$$J(\boldsymbol{\theta}) = \mathbb{E}_{\pi_D}\left[\sum_{t=0}^{\infty} \gamma^t \cdot R_{t+1}\right]$$

$$= \sum_{s \in \mathcal{N}} \rho^{\pi_D}(s) \cdot \mathcal{R}_s^{\pi_D(s;\boldsymbol{\theta})}$$

$$= \mathbb{E}_{s \sim \rho^{\pi_D}}\left[\mathcal{R}_s^{\pi_D(s;\boldsymbol{\theta})}\right]$$

where

$$\rho^{\pi_D}(s) = \sum_{S_0 \in \mathcal{N}} \sum_{t=0}^{\infty} \gamma^t \cdot p_0(S_0) \cdot p(S_0 \to s, t, \pi_D)$$

is the Discounted-Aggregate State-Visitation Measure when following deterministic policy $\pi_D(s; \boldsymbol{\theta})$.

With a derivation similar to the proof of the PGT, we have the DPGT, as follows:

Theorem 14.8.1 (Deterministic Policy Gradient Theorem). *Given an MDP with action space* $\mathbb{R}^k$, *with appropriate gradient existence conditions,*

$$\nabla_{\boldsymbol{\theta}} J(\boldsymbol{\theta}) = \sum_{s \in \mathcal{N}} \rho^{\pi_D}(s) \cdot \nabla_{\boldsymbol{\theta}} \pi_D(s; \boldsymbol{\theta}) \cdot \nabla_a Q^{\pi_D}(s, a) \Big|_{a = \pi_D(s; \boldsymbol{\theta})}$$

$$= \mathbb{E}_{s \sim \rho^{\pi_D}} \left[\nabla_{\boldsymbol{\theta}} \pi_D(s; \boldsymbol{\theta}) \cdot \nabla_a Q^{\pi_D}(s, a) \Big|_{a = \pi_D(s; \boldsymbol{\theta})} \right]$$

In practice, we use an Actor-Critic algorithm with a function approximation $Q(s, a; \boldsymbol{w})$ for the Q-Value Function as the Critic. Since the policy approximated is Deterministic, we need to address the issue of exploration—this is typically done with Off-Policy Control wherein we employ an exploratory (stochastic) behavior policy, while the policy being approximated (and learnt with DPG) is the target (deterministic) policy. The Expected Return Objective is a bit different in the case of Off-Policy—it is the Expected Q-Value for the target policy under state-occurrence probabilities while following the behavior policy, and the Off-Policy Deterministic Policy Gradient is an approximate (not exact) formula. We avoid importance sampling in the Actor because DPG doesn't involve an integral over actions, and we avoid importance sampling in the Critic by employing Q-Learning. As a result, for Off-Policy Actor-Critic DPG, we update the Critic parameters $\boldsymbol{w}$ and the Actor parameters $\boldsymbol{\theta}$ after each atomic experience in a trace experience generated by the behavior policy.

$$\Delta \boldsymbol{w} = \alpha_{\boldsymbol{w}} \cdot (R_{t+1} + \gamma \cdot Q(S_{t+1}, \pi_D(S_{t+1}; \boldsymbol{\theta}); \boldsymbol{w}) - Q(S_t, A_t; \boldsymbol{w})) \cdot \nabla_{\boldsymbol{w}} Q(S_t, A_t; \boldsymbol{w})$$

$$\Delta \boldsymbol{\theta} = \alpha_{\boldsymbol{\theta}} \cdot \nabla_{\boldsymbol{\theta}} \pi_D(S_t; \boldsymbol{\theta}) \cdot \nabla_a Q(S_t, a; \boldsymbol{w}) \Big|_{a = \pi_D(S_t; \boldsymbol{\theta})}$$

Critic Bias can be resolved with a Compatible Function Approximation Theorem for DPG (see Silver et al. paper for details). Instabilities caused by Bootstrapped Off-Policy Learning with Function Approximation can be resolved with Gradient Temporal Difference (GTD).

14.9   EVOLUTIONARY STRATEGIES

We conclude this chapter with a section on Evolutionary Strategies—a class of algorithms to solve MDP Control problems. We want to highlight right upfront that Evolutionary Strategies are technically not RL algorithms (for reasons we shall illuminate once we explain the technique of Evolutionary Strategies). However, Evolutionary Strategies can sometimes be quite effective in solving MDP Control problems and so, we give them appropriate coverage as part of a wide range of approaches to solve MDP Control. We cover them in this chapter because of their superficial resemblance to Policy Gradient Algorithms (again, they are not RL algorithms and hence, not Policy Gradient algorithms).

Evolutionary Strategies (abbreviated as ES) actually refers to a technique/approach that is best understood as a type of Black-Box Optimization. It was popularized in the 1970s as *Heuristic Search Methods*. It is loosely inspired by natural evolution of living beings. We focus on a subclass of ES known as Natural Evolutionary Strategies (abbreviated as NES).

The original setting for this approach was quite generic and not at all specific to solving MDPs. Let us understand this generic setting first. Given an objective function $F(\psi)$, where ψ refers to parameters, we consider a probability distribution $p_{\theta}(\psi)$ over ψ, where

θ refers to the parameters of the probability distribution. The goal in this generic setting is to maximize the average objective $\mathbb{E}_{\psi \sim p_\theta}[F(\psi)]$.

We search for optimal θ with stochastic gradient ascent as follows:

$$\nabla_\theta(\mathbb{E}_{\psi \sim p_\theta}[F(\psi)]) = \nabla_\theta(\int_\psi p_\theta(\psi) \cdot F(\psi) \cdot d\psi)$$

$$= \int_\psi \nabla_\theta(p_\theta(\psi)) \cdot F(\psi) \cdot d\psi$$

$$= \int_\psi p_\theta(\psi) \cdot \nabla_\theta(\log p_\theta(\psi)) \cdot F(\psi) \cdot d\psi$$

$$= \mathbb{E}_{\psi \sim p_\theta}[\nabla_\theta(\log p_\theta(\psi)) \cdot F(\psi)] \qquad (14.2)$$

Now let's see how NES can be applied to solving MDP Control. We set $F(\cdot)$ to be the (stochastic) *Return* of an MDP. ψ corresponds to the parameters of a deterministic policy $\pi_\psi : \mathcal{N} \to \mathcal{A}$. $\psi \in \mathbb{R}^m$ is drawn from an isotropic m-variate Gaussian distribution, i.e., Gaussian with mean vector $\theta \in \mathbb{R}^m$ and fixed diagonal covariance matrix $\sigma^2 I_m$ where $\sigma \in \mathbb{R}$ is kept fixed and I_m is the $m \times m$ identity matrix. The average objective (*Expected Return*) can then be written as:

$$\mathbb{E}_{\psi \sim p_\theta}[F(\psi)] = \mathbb{E}_{\epsilon \sim \mathcal{N}(0, I_m)}[F(\theta + \sigma \cdot \epsilon)]$$

where $\epsilon \in \mathbb{R}^m$ is the standard normal random variable generating ψ. Hence, from Equation (14.2), the gradient (∇_θ) of *Expected Return* can be written as:

$$\mathbb{E}_{\psi \sim p_\theta}[\nabla_\theta(\log p_\theta(\psi)) \cdot F(\psi)]$$

$$= \mathbb{E}_{\psi \sim \mathcal{N}(\theta, \sigma^2 I_m)}[\nabla_\theta(\frac{-(\psi - \theta)^T \cdot (\psi - \theta)}{2\sigma^2}) \cdot F(\psi)]$$

$$= \frac{1}{\sigma} \cdot \mathbb{E}_{\epsilon \sim \mathcal{N}(0, I_m)}[\epsilon \cdot F(\theta + \sigma \cdot \epsilon)]$$

Now we come up with a sampling-based algorithm to solve the MDP. The above formula helps estimate the gradient of *Expected Return* by sampling several ϵ (each ϵ represents a *Policy* $\pi_{\theta + \sigma \cdot \epsilon}$), and averaging $\epsilon \cdot F(\theta + \sigma \cdot \epsilon)$ across a large set (n) of ϵ samples.

Note that evaluating $F(\theta + \sigma \cdot \epsilon)$ involves playing an episode for a given sampled ϵ, and obtaining that episode's *Return* $F(\theta + \sigma \cdot \epsilon)$. Hence, we have n values of ϵ, n Policies $\pi_{\theta + \sigma \cdot \epsilon}$, and n *Returns* $F(\theta + \sigma \cdot \epsilon)$.

Given the gradient estimate, we update θ in this gradient direction, which in turn leads to new samples of ϵ (new set of *Policies* $\pi_{\theta + \sigma \cdot \epsilon}$), and the process repeats until $\mathbb{E}_{\epsilon \sim \mathcal{N}(0, I_m)}[F(\theta + \sigma \cdot \epsilon)]$ is maximized.

The key inputs to the algorithm are:

- Learning rate (SGD Step Size) α
- Standard Deviation σ
- Initial value of parameter vector θ_0

With these inputs, for each iteration $t = 0, 1, 2, \ldots$, the algorithm performs the following steps:

- Sample $\epsilon_1, \epsilon_2, \ldots \epsilon_n \sim \mathcal{N}(0, I_m)$.
- Compute Returns $F_i \leftarrow F(\theta_t + \sigma \cdot \epsilon_i)$ for $i = 1, 2, \ldots, n$.

- $\theta_{t+1} \leftarrow \theta_t + \frac{\alpha}{n\sigma} \sum_{i=1}^{n} \epsilon_i \cdot F_i$

On the surface, this NES algorithm looks like PG because it's not Value Function-based (it's Policy-based, like PG). Also, similar to PG, it uses a gradient to move the policy towards optimality. But, ES does not interact with the environment (like PG/RL does). ES operates at a high-level, ignoring the (state, action, reward) interplay. Specifically, it does not aim to assign credit to actions in specific states. Hence, ES doesn't have the core essence of RL: *Estimating the Q-Value Function for a Policy and using it to Improve the Policy*. Therefore, we don't classify ES as Reinforcement Learning. Rather, we consider ES to be an alternative approach to RL Algorithms.

What is the effectiveness of ES compared to RL? The traditional view has been that ES won't work on high-dimensional problems. Specifically, ES has been shown to be data-inefficient relative to RL. This is because ES resembles simple hill-climbing based only on finite differences along a few random directions at each step. However, ES is very simple to implement (no Value Function approximation or back-propagation needed), and is highly parallelizable. ES has the benefits of being indifferent to distribution of rewards and to action frequency, and is tolerant of long horizons. A paper from OpenAI Research (Salimans et al. 2017) shows techniques to make NES more robust and more data-efficient, and they demonstrate that NES has more exploratory behavior than advanced PG algorithms.

14.10 KEY TAKEAWAYS FROM THIS CHAPTER

- Policy Gradient Algorithms are based on GPI with Policy Improvement as a Stochastic Gradient Ascent for "Expected Returns" Objective $J(\theta)$ where θ are the parameters of the function approximation for the Policy.
- Policy Gradient Theorem gives us a simple formula for $\nabla_\theta J(\theta)$ in terms of the score of the policy function approximation (i.e., gradient of the log of the policy with respect to the policy parameters θ).
- We can reduce variance in PG algorithms by using a critic and by using an estimate of the advantage function for the Q-Value Function.
- Compatible Function Approximation Theorem enables us to overcome bias in PG Algorithms.
- Natural Policy Gradient and Deterministic Policy Gradient are specialized PG algorithms that have worked well in practice.
- Evolutionary Strategies are technically not RL, but they resemble PG Algorithms and can sometimes be quite effective in solving MDP Control problems.

IV

Finishing Touches

CHAPTER 15

Multi-Armed Bandits: Exploration versus Exploitation

We learnt in Chapter 12 that balancing exploration and exploitation is vital in RL Control algorithms. While we want to exploit actions that seem to be fetching good returns, we also want to adequately explore all possible actions so we can obtain an accurate-enough estimate of their Q-Values. We had mentioned that this is essentially the Explore-Exploit dilemma of the famous Multi-Armed Bandit Problem. The Multi-Armed Bandit problem provides a simple setting to understand the explore-exploit tradeoff and to develop explore-exploit balancing algorithms. The approaches followed by the Multi-Armed Bandit algorithms are then well-transportable to the more complex setting of RL Control.

In this chapter, we start by specifying the Multi-Armed Bandit problem, followed by coverage of a variety of techniques to solve the Multi-Armed Bandit problem (i.e., effectively balancing exploration against exploitation). We've actually seen one of these algorithms already for RL Control—following an ϵ-greedy policy, which naturally is applicable to the simpler setting of Multi-Armed Bandits. We had mentioned in Chapter 12 that we can simply replace the ϵ-greedy approach with any other algorithm for explore-exploit tradeoff. In this chapter, we consider a variety of such algorithms, many of which are far more sophisticated compared to the simple ϵ-greedy approach. However, we cover these algorithms for the simple setting of Multi-Armed Bandits as it promotes understanding and development of intuition. After covering a range of algorithms for Multi-Armed Bandits, we consider an extended problem known as Contextual Bandits, that is a step between the Multi-Armed Bandits problem and the RL Control problem (in terms of problem complexity). Finally, we explain how the algorithms for Multi-Armed Bandits can be easily transported to the more nuanced/extended setting of Contextual Bandits, and further extended to RL Control.

15.1 INTRODUCTION TO THE MULTI-ARMED BANDIT PROBLEM

At various points in past chapters, we've emphasized the importance of the Explore-Exploit tradeoff in the context of RL Control—selecting actions for any given state that balances the notions of exploration and exploitation. If you think about it, you will realize that many situations in business (and in our lives!) present this explore-exploit dilemma on choices one has to make. *Exploitation* involves making choices that *seem to be best* based on past outcomes, while *Exploration* involves making choices one hasn't yet tried (or not tried sufficiently enough).

Exploitation has intuitive notions of "being greedy" and of being "short-sighted", and too much exploitation could lead to some regret of having missing out on unexplored "gems". Exploration has intuitive notions of "gaining information" and of being "long-sighted", and too much exploration could lead to some regret of having wasting time on "duds". This naturally leads to the idea of balancing exploration and exploitation so we can combine *information-gains* and *greedy-gains* in the most optimal manner. The natural question then is whether we can set up this problem of explore-exploit dilemma in a mathematically disciplined manner. Before we do that, let's look at a few common examples of the explore-exploit dilemma.

15.1.1 Some Examples of Explore-Exploit Dilemma

- Restaurant Selection: We like to go to our favorite restaurant (Exploitation) but we also like to try out a new restaurant (Exploration).
- Online Banner Advertisements: We like to repeat the most successful advertisement (Exploitation) but we also like to show a new advertisement (Exploration).
- Oil Drilling: We like to drill at the best known location (Exploitation) but we also like to drill at a new location (Exploration).
- Learning to play a game: We like to play the move that has worked well for us so far (Exploitation) but we also like to play a new experimental move (Exploration).

The term *Multi-Armed Bandit* (abbreviated as MAB) is a spoof name that stands for "Many One-Armed Bandits" and the term *One-Armed Bandit* refers to playing a slot-machine in a casino (that has a single lever to be pulled, that presumably addicts us and eventually takes away all our money, hence the term "bandit"). Multi-Armed Bandit refers to the problem of playing several slot machines (each of which has an unknown fixed pay-out probability distribution) in a manner that we can make the maximum cumulative gains by playing over multiple rounds (by selecting a single slot machine in a single round). The core idea is that to achieve maximum cumulative gains, one would need to balance the notions of exploration and exploitation, no matter which selection strategy one would pursue.

15.1.2 Problem Definition

Definition 15.1.1. A *Multi-Armed Bandit* (MAB) comprises of:

- A finite set of *Actions* $\mathcal{A}$ (known as the "arms").

- Each action ("arm") $a \in \mathcal{A}$ is associated with a probability distribution over $\mathbb{R}$ (unknown to the AI Agent) denoted as $\mathcal{R}^a$, defined as:

$$\mathcal{R}^a(r) = \mathbb{P}[r|a] \text{ for all } r \in \mathbb{R}$$

- A time-indexed sequence of AI Agent-selected actions $A_t \in \mathcal{A}$ for time steps $t = 1, 2, \ldots,$ and a time-indexed sequence of Environment-generated *Reward* random variables $R_t \in \mathbb{R}$ for time steps $t = 1, 2, \ldots,$ with R_t randomly drawn from the probability distribution $\mathcal{R}^{A_t}$.

The AI Agent's goal is to maximize the following *Expected Cumulative Rewards* over a certain number of time steps T:

$$\mathbb{E}[\sum_{t=1}^{T} R_t]$$

So the AI Agent has T selections of actions to make (in sequence), basing each of those selections only on the rewards it has observed before that time step (specifically, the AI Agent does not have knowledge of the probability distributions $\mathcal{R}^a$). Any selection strategy to maximize the Expected Cumulative Rewards risks wasting time on "duds" while exploring and also risks missing untapped "gems" while exploiting.

It is immediately observable that the Environment doesn't have a notion of *State*. When the AI Agent selects an arm, the Environment simply samples from the probability distribution for that arm. However, the AI Agent might maintain relevant features of the history (of actions taken and rewards obtained) as its *State*, which would help the AI Agent in making the arm-selection (action) decision. The arm-selection action is then based on a (*Policy*) function of the agent's *State*. So, the agent's arm-selection strategy is basically this *Policy*. Thus, even though a MAB is not posed as an MDP, the agent could model it as an MDP and solve it with an appropriate Planning or Learning algorithm. However, many MAB algorithms don't take this formal MDP approach. Instead, they rely on heuristic methods that don't aim to *optimize*—they simply strive for *good* Cumulative Rewards (in Expectation). Note that even in a simple heuristic algorithm, A_t is a random variable simply because it is a function of past (random) rewards.

15.1.3 Regret

The idea of *Regret* is quite fundamental in designing algorithms for MAB. In this section, we illuminate this idea.

We define the *Action Value* $Q(a)$ as the (unknown) mean reward of action a, i.e.,

$$Q(a) = \mathbb{E}[r|a]$$

We define the *Optimal Value* V^* and *Optimal Action* a^* (noting that there could be multiple optimal actions) as:

$$V^* = \max_{a \in \mathcal{A}} Q(a) = Q(a^*)$$

We define *Regret* l_t as the opportunity loss at a single time step t, as follows:

$$l_t = \mathbb{E}[V^* - Q(A_t)]$$

We define the *Total Regret* L_T as the total opportunity loss, as follows:

$$L_T = \sum_{t=1}^{T} l_t = \sum_{t=1}^{T} \mathbb{E}[V^* - Q(A_t)]$$

Maximizing the *Expected Cumulative Rewards* is the same as Minimizing *Total Regret*.

15.1.4 Counts and Gaps

Let $N_t(a)$ be the (random) number of selections of an action a across the first t time steps. Let us refer to $\mathbb{E}[N_t(a)]$ for a given action-selection strategy as the *Count* of an action a over the first t steps, denoted as $Count_t(a)$. Let us refer to the Value difference between an action a and the optimal action a^* as the *Gap* for a, denoted as Δ_a, i.e,

$$\Delta_a = V^* - Q(a)$$

We define Total Regret as the sum-product (over actions) of *Counts* and *Gaps*, as follows:

$$L_T = \sum_{t=1}^{T} \mathbb{E}[V^* - Q(A_t)] \qquad = \sum_{a \in \mathcal{A}} \mathbb{E}[N_T(a)] \cdot (V^* - Q(a)) = \sum_{a \in \mathcal{A}} Count_T(a) \cdot \Delta_a$$

A good algorithm ensures small *Counts* for large *Gaps*. The core challenge though is that *we don't know the Gaps*.

In this chapter, we implement (in code) a few different algorithms for the MAB problem. So let's invest in an abstract base class whose interface can be implemented by each of the algorithms we develop. The code for this abstract base class `MABBase` is shown below. Its constructor takes 3 inputs:

- `arm_distributions` which is a Sequence of `Distribution[float]`s, one for each arm.
- `time_steps` which represents the number of time steps T
- `num_episodes` which represents the number of episodes we can run the algorithm on (each episode having T time steps), in order to produce metrics to evaluate how well the algorithm does in expectation (averaged across the episodes).

Each of the algorithms we'd like to write simply needs to implement the `@abstractmethod` `get_episode_rewards_actions` which is meant to return a 1-D `ndarray` of actions taken by the algorithm across the T time steps (for a single episode), and a 1-D `ndarray` of rewards produced in response to those actions.

```python
from rl.distribution import Distribution
from numpy import ndarray

class MABBase(ABC):
    def __init__(
        self,
        arm_distributions: Sequence[Distribution[float]],
        time_steps: int,
        num_episodes: int
    ) -> None:
        self.arm_distributions: Sequence[Distribution[float]] = \
            arm_distributions
        self.num_arms: int = len(arm_distributions)
        self.time_steps: int = time_steps
        self.num_episodes: int = num_episodes

    @abstractmethod
    def get_episode_rewards_actions(self) -> Tuple[ndarray, ndarray]:
        pass
```

We write the following self-explanatory methods for the abstract base class `MABBase`:

```python
from numpy import mean, vstack, cumsum, full, bincount
    def get_all_rewards_actions(self) -> Sequence[Tuple[ndarray, ndarray]]:
        return [self.get_episode_rewards_actions()
                for _ in range(self.num_episodes)]

    def get_rewards_matrix(self) -> ndarray:
        return vstack([x for x, _ in self.get_all_rewards_actions()])

    def get_actions_matrix(self) -> ndarray:
        return vstack([y for _, y in self.get_all_rewards_actions()])

    def get_expected_rewards(self) -> ndarray:
        return mean(self.get_rewards_matrix(), axis=0)

    def get_expected_cum_rewards(self) -> ndarray:
        return cumsum(self.get_expected_rewards())

    def get_expected_regret(self, best_mean) -> ndarray:
        return full(self.time_steps, best_mean) - self.get_expected_rewards()

    def get_expected_cum_regret(self, best_mean) -> ndarray:
        return cumsum(self.get_expected_regret(best_mean))

    def get_action_counts(self) -> ndarray:
        return vstack([bincount(ep, minlength=self.num_arms)
                       for ep in self.get_actions_matrix()])
```

```
def get_expected_action_counts(self) -> ndarray:
    return mean(self.get_action_counts(), axis=0)
```

The above code is in the file rl/chapter14/mab_base.py.
Next, we cover some simple heuristic algorithms.

15.2 SIMPLE ALGORITHMS

We consider algorithms that estimate a Q-Value $\hat{Q}_t(a)$ for each $a \in \mathcal{A}$, as an approximation to the true Q-Value $Q(a)$. The subscript t in $\hat{Q}_t$ refers to the fact that this is an estimate after t time steps that takes into account all of the information available up to t time steps.

A natural way of estimating $\hat{Q}_t(a)$ is by *rewards-averaging*, i.e.,

$$\hat{Q}_t(a) = \frac{1}{N_t(a)} \sum_{s=1}^{t} R_s \cdot \mathbb{I}_{A_s=a}$$

where $\mathbb{I}$ refers to the indicator function.

15.2.1 Greedy and ϵ-Greedy

First, consider an algorithm that *never* explores (i.e., *always* exploits). This is known as the *Greedy Algorithm* which selects the action with highest estimated value, i.e.,

$$A_t = \arg\max_{a \in \mathcal{A}} \hat{Q}_{t-1}(a)$$

As ever, arg max ties are broken with an arbitrary rule in prioritizing actions. We've noted in Chapter 12 that such an algorithm can lock into a suboptimal action forever (suboptimal a is an action for which $\Delta_a > 0$). This results in $Count_T(a)$ being a linear function of T for some suboptimal a, which means the Total Regret is a linear function of T (we refer to this as *Linear Total Regret*).

Now let's consider the ϵ-greedy algorithm, which explores forever. At each time-step t:

- With probability $1 - \epsilon$, select action equal to $\arg\max_{a \in \mathcal{A}} \hat{Q}_{t-1}(a)$
- With probability ϵ, select a random action (uniformly) from $\mathcal{A}$

A constant value of ϵ ensures a minimum regret proportional to the mean gap, i.e.,

$$l_t \geq \frac{\epsilon}{|\mathcal{A}|} \sum_{a \in \mathcal{A}} \Delta_a$$

Hence, the ϵ-Greedy algorithm also has Linear Total Regret.

15.2.2 Optimistic Initialization

Next, we consider a simple and practical idea: Initialize $\hat{Q}_0(a)$ to a high value for all $a \in \mathcal{A}$ and update $\hat{Q}_t$ by incremental-averaging. Starting with $N_0(a) \geq 0$ for all $a \in \mathcal{A}$, the updates at each time step t are as follows:

$$N_t(A_t) = N_{t-1}(A_t) + 1$$

$$\hat{Q}_t(A_t) = \hat{Q}_{t-1}(A_t) + \frac{R_t - \hat{Q}_{t-1}(A_t)}{N_t(A_t)}$$

The idea here is that by setting a high initial value for the estimate of Q-Values (which we refer to as *Optimistic Initialization*), we encourage systematic exploration early on. Another way of doing optimistic initialization is to set a high value for $N_0(a)$ for all $a \in \mathcal{A}$, which likewise encourages systematic exploration early on. However, these optimistic initialization ideas only serve to promote exploration early on and eventually, one can still lock into a suboptimal action. Specifically, the Greedy algorithm together with optimistic initialization cannot be prevented from having Linear Total Regret in the general case. Likewise, the ϵ-Greedy algorithm together with optimistic initialization cannot be prevented from having Linear Total Regret in the general case. But in practice, these simple ideas of doing optimistic initialization work quite well.

15.2.3 Decaying ϵ_t-Greedy Algorithm

The natural question that emerges is whether it is possible to construct an algorithm with Sublinear Total Regret in the general case. Along these lines, we consider an ϵ-Greedy algorithm with ϵ decaying as time progresses. We call such an algorithm Decaying ϵ_t-Greedy.

For any fixed $c > 0$, consider a decay schedule for $\epsilon_1, \epsilon_2, \ldots$ as follows:

$$d = \min_{a | \Delta_a > 0} \Delta_a$$

$$\epsilon_t = \min(1, \frac{c|\mathcal{A}|}{d^2(t+1)})$$

It can be shown that this decay schedule achieves *Logarithmic* Total Regret. However, note that the above schedule requires advance knowledge of the gaps Δ_a (which by definition, is not known to the AI Agent). In practice, implementing *some* decay schedule helps considerably. Let's now write some code to implement Decaying ϵ_t-Greedy algorithm along with Optimistic Initialization.

The class `EpsilonGreedy` shown below implements the interface of the abstract base class `MABBase`. Its constructor inputs `arm_distributions`, `time_steps` and `num_episodes` are the inputs we have seen before (used to pass to the constructor of the abstract base class `MABBase`). `epsilon` and `epsilon_half_life` are the inputs used to specify the declining trajectory of ϵ_t. `epsilon` refers to ϵ_0 (initial value of ϵ) and `epsilon_half_life` refers to the half life of an exponentially-decaying ϵ_t (used in the @staticmethod `get_epsilon_decay_func`). `count_init` and `mean_init` refer to values of N_0 and $\hat{Q}_0$, respectively. `get_episode_rewards_actions` implements `MABBase`'s @abstracmethod interface, and its code below should be self-explanatory.

```
from operator import itemgetter
from rl.distribution import Distribution, Range, Bernoulli
from numpy import ndarray, empty

class EpsilonGreedy(MABBase):
    def __init__(
        self,
        arm_distributions: Sequence[Distribution[float]],
        time_steps: int,
        num_episodes: int,
        epsilon: float,
        epsilon_half_life: float = 1e8,
        count_init: int = 0,
        mean_init: float = 0.,
    ) -> None:
        if epsilon < 0 or epsilon > 1 or \
                epsilon_half_life <= 1 or count_init < 0:
            raise ValueError
```

```python
        super().__init__(
            arm_distributions=arm_distributions,
            time_steps=time_steps,
            num_episodes=num_episodes
        )
        self.epsilon_func: Callable[[int], float] = \
            EpsilonGreedy.get_epsilon_decay_func(epsilon, epsilon_half_life)
        self.count_init: int = count_init
        self.mean_init: float = mean_init

    @staticmethod
    def get_epsilon_decay_func(
        epsilon,
        epsilon_half_life
    ) -> Callable[[int], float]:
        def epsilon_decay(
            t: int,
            epsilon=epsilon,
            epsilon_half_life=epsilon_half_life
        ) -> float:
            return epsilon * 2 ** -(t / epsilon_half_life)

        return epsilon_decay

    def get_episode_rewards_actions(self) -> Tuple[ndarray, ndarray]:
        counts: List[int] = [self.count_init] * self.num_arms
        means: List[float] = [self.mean_init] * self.num_arms
        ep_rewards: ndarray = empty(self.time_steps)
        ep_actions: ndarray = empty(self.time_steps, dtype=int)
        for i in range(self.time_steps):
            max_action: int = max(enumerate(means), key=itemgetter(1))[0]
            epsl: float = self.epsilon_func(i)
            action: int = max_action if Bernoulli(1 - epsl).sample() else \
                Range(self.num_arms).sample()
            reward: float = self.arm_distributions[action].sample()
            counts[action] += 1
            means[action] += (reward - means[action]) / counts[action]
            ep_rewards[i] = reward
            ep_actions[i] = action
        return ep_rewards, ep_actions
```

The above code is in the file rl/chapter14/epsilon_greedy.py.

Figure 15.1 shows the results of running the above code for 1000 time steps over 500 episodes, with N_0 and $\hat{Q}_0$ both set to 0. This graph was generated (see __main__ in rl/chapter14/epsilon_greedy.py) by creating 3 instances of EpsilonGreedy—the first with epsilon set to 0 (i.e., Greedy), the second with epsilon set to 0.12 and epsilon_half_life set to a very high value (i.e, ϵ-Greedy, with no decay for ϵ), and the third with epsilon set to 0.12 and epsilon_half_life set to 150 (i.e., Decaying ϵ_t-Greedy). We can see that Greedy produces Linear Total Regret since it locks to a suboptimal value. We can also see that ϵ-Greedy has higher total regret than Greedy initially because of exploration, and then settles in with Linear Total Regret, commensurate with the constant amount of exploration ($\epsilon = 0.12$ in this case). Lastly, we can see that Decaying ϵ_t-Greedy produces Sublinear Total Regret as the initial effort spent in exploration helps identify the best action and as time elapses, the exploration keeps reducing so as to keep reducing the single-step regret.

In the __main__ code in rl/chapter14/epsilon_greedy.py, we encourage you to experiment with different arm_distributions, epsilon, epsilon_half_life, count_init (N_0) and mean_init ($\hat{Q}_0$), observe how the graphs change, and develop better intuition for these simple algorithms.

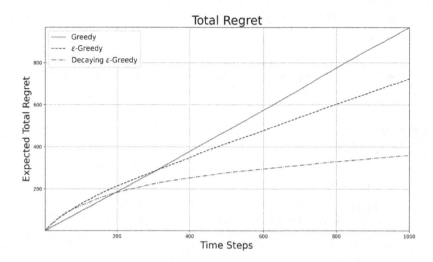

Figure 15.1 **Total Regret Curves**

15.3 LOWER BOUND

It should be clear by now that we strive for algorithms with Sublinear Total Regret for any MAB problem (i.e., without any prior knowledge of the arm-reward distributions $\mathcal{R}^a$). Intuitively, the performance of any algorithm is determined by the similarity between the optimal arm's reward-distribution and the other arms's reward-distributions. Hard MAB problems are those with similar-distribution arms with different means $Q(a)$. This can be formally described in terms of the KL Divergence $KL(\mathcal{R}^a||\mathcal{R}^{a^*})$ and gaps Δ_a. Indeed, Lai and Robbins (Lai and Robbins 1985) established a logarithmic lower bound for the Asymptotic Total Regret, with a factor expressed in terms of the KL Divergence $KL(\mathcal{R}^a||\mathcal{R}^{a^*})$ and gaps Δ_a. Specifically,

Theorem 15.3.1 (Lai and Robbins Lower-Bound). *Asymptotic Total Regret is at least logarithmic in the number of time steps, i.e., as $T \to \infty$,*

$$L_T \geq logT \sum_{a|\Delta_a>0} \frac{1}{\Delta_a} \geq \log T \sum_{a|\Delta_a>0} \frac{\Delta_a}{KL(\mathcal{R}^a||\mathcal{R}^{a^*})}$$

This makes intuitive sense because it would be hard for an algorithm to have low total regret if the KL Divergence of arm reward-distributions (relative to the optimal arm's reward-distribution) are low (i.e., arms that look distributionally-similar to the optimal arm) but the Gaps (Expected Rewards of Arms relative to Optimal Arm) are not small—these are the MAB problem instances where the algorithm will have a hard time isolating the optimal arm simply from reward samples (we'd get similar sampling reward-distributions of arms), and suboptimal arm selections would inflate the Total Regret.

15.4 UPPER CONFIDENCE BOUND ALGORITHMS

Now we come to an important idea that is central to many algorithms for MAB. This idea goes by the catchy name of *Optimism in the Face of Uncertainty*. As ever, this idea is best understood with intuition first, followed by mathematical rigor. To develop intuition, imagine you are given 3 arms. You'd like to develop an estimate of $Q(a) = \mathbb{E}[r|a]$ for each of the 3 arms a. After playing the arms a few times, you start forming beliefs in your mind of what the $Q(a)$ might be for each arm. Unlike the simple algorithms we've seen so far where one averaged the sample rewards for each arm to maintain a $\hat{Q}(a)$ estimate for each a, here we maintain the sampling distribution of the mean rewards (for each a) that represents our (probabilistic) beliefs of what $Q(a)$ might be for each arm a.

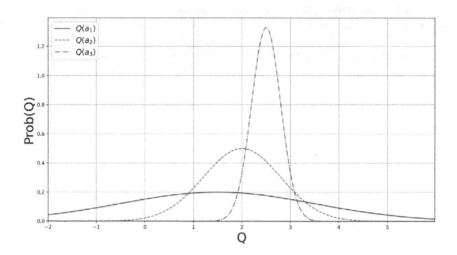

Figure 15.2 **Q-Value Distributions**

To keep things simple, let's assume the sampling distribution of the mean reward is a Gaussian distribution (for each a), and so we maintain an estimate of μ_a and σ_a for each arm a to represent the mean and standard deviation of the sampling distribution of mean reward for a. μ_a would be calculated as the average of the sample rewards seen so far for an arm a. σ_a would be calculated as the standard error of the mean reward estimate, i.e., the sample standard deviation of the rewards seen so far, divided by the square root of the number of samples (for a given arm a). Let us say that after playing the arms a few times, we arrive at the Gaussian sampling distribution of mean reward for each of the 3 arms, as illustrated in Figure 15.2. Let's refer to the three arms as red, blue and green. The normal distributions in Figure 15.2 show the red arm as the solid curve, the blue arm as the dashed curve and the green arm as the dotted-and-dashed curve. The blue arm has the highest σ_a. This could be either because the sample standard deviation is high or it could be because we have played the blue arm just a few times (remember the square root of number of samples appears in the denominator of the standard error calculation). Now looking at this figure, we have to decide which arm to select next. The intuition behind *Optimism in the Face of Uncertainty* is that the more uncertain we are about the $Q(a)$ for an arm a, the more important it is to play that arm. This is because more uncertainty on $Q(a)$ makes it more likely to be the best arm (all else being equal on the arms). The rough heuristic then would be to select the arm with the highest value of $\mu_a + c \cdot \sigma_a$ across the arms (for some fixed $c \in \mathbb{R}^+$). Thus, we are comparing (across actions) c standard errors higher than the mean reward estimate (i.e., the upper-end of an appropriate confidence interval for the mean reward). In this figure, let's say $\mu_a + c \cdot \sigma_a$ is highest for the blue arm. So we play the blue arm, and let's say we get a somewhat low reward for the blue arm. This might do two things to the blue arm's sampling distribution—it can move blue's μ_a lower and it can also also lower blue's σ_a (simply due to the fact that the number of blue arm samples has grown).

With the new μ_a and σ_a for the blue arm, let's say the updated sampling distributions are as shown in Figure 15.3. With the blue arm's sampling distribution of the mean reward narrower, let's say the red arm now has the highest $\mu_a + c \cdot \sigma_a$, and so we play the red arm. This process goes on until the sampling distributions get narrow enough to give us adequate confidence in the mean rewards for the actions (i.e., obtain confident estimates of $Q(a)$) so we can home in on the action with highest $Q(a)$.

It pays to emphasize that *Optimism in the Face of Uncertainty* is a great approach to resolve the Explore-Exploit dilemma because you gain regardless of whether the exploration due to Optimism produces large rewards or not. If it does produce large rewards, you gain immediately by collecting the large rewards. If it does not produce large rewards, you still gain by acquiring the knowledge that certain actions (that you have explored) might not be the best actions, which helps you in the long-run by focusing your attention on other actions.

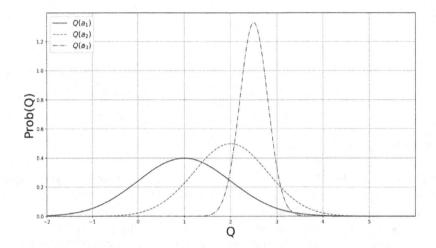

Figure 15.3 **Q-Value Distributions**

A formalization of the above intuition on *Optimism in the Face of Uncertainty* is the idea of *Upper Confidence Bounds* (abbreviated as UCB). The idea of UCB is that along with an estimate $\hat{Q}_t(a)$ (for each a after t time steps), we also maintain an estimate $\hat{U}_t(a)$ representing the upper confidence interval width for the mean reward of a (after t time steps) such that $Q(a) < \hat{Q}_t(a) + \hat{U}_t(a)$ with high probability. This naturally depends on the number of times that a has been selected so far (call it $N_t(a)$). A small value of $N_t(a)$ would imply a large value of $\hat{U}_t(a)$ since the estimate of the mean reward would be fairly uncertain. On the other hand, a large value of $N_t(a)$ would imply a small value of $\hat{U}_t(a)$ since the estimate of the mean reward would be fairly certain. We refer to $\hat{Q}_t(a) + \hat{U}_t(a)$ as the *Upper Confidence Bound* (or simply UCB). The idea is to select the action that maximizes the UCB. Formally, the action A_{t+1} selected for the next $(t + 1)$ time step is as follows:

$$A_{t+1} = \arg\max_{a \in \mathcal{A}}\{\hat{Q}_t(a) + \hat{U}_t(a)\}$$

Next, we develop the famous UCB1 Algorithm. In order to do that, we tap into an important result from Statistics known as Hoeffding's Inequality.

15.4.1   Hoeffding's Inequality

We state Hoeffding's Inequality without proof.

Theorem 15.4.1 (Hoeffding's Inequality). *Let* $X_1, \ldots, X_n$ *be independent and identically distributed random variables in the range* $[0, 1]$*, and let*

$$\bar{X}_n = \frac{1}{n} \sum_{i=1}^{n} X_i$$

be the sample mean. Then for any $u \geq 0$,

$$\mathbb{P}[\mathbb{E}[\bar{X}_n] > \bar{X}_n + u] \leq e^{-2nu^2}$$

We can apply Hoeffding's Inequality to MAB problem instances whose rewards have probability distributions with $[0, 1]$-support. Conditioned on selecting action a at time step t, sample mean $\bar{X}_n$ specializes to $\hat{Q}_t(a)$, and we set $n = N_t(a)$ and $u = \hat{U}_t(a)$. Therefore,

$$\mathbb{P}[Q(a) > \hat{Q}_t(a) + \hat{U}_t(a)] \leq e^{-2N_t(a) \cdot \hat{U}_t(a)^2}$$

Next, we pick a small probability p for $Q(a)$ exceeding UCB $\hat{Q}_t(a) + \hat{U}_t(a)$. Now solve for $\hat{U}_t(a)$, as follows:

$$e^{-2N_t(a) \cdot \hat{U}_t(a)^2} = p \Rightarrow \hat{U}_t(a) = \sqrt{\frac{-\log p}{2N_t(a)}}$$

We reduce p as we observe more rewards, e.g., $p = t^{-\alpha}$ (for some fixed $\alpha > 0$). This ensures we select the optimal action as $t \to \infty$. Thus,

$$\hat{U}_t(a) = \sqrt{\frac{\alpha \log t}{2N_t(a)}}$$

15.4.2   UCB1 Algorithm

This yields the UCB1 algorithm by Auer, Cesa-Bianchi, Fischer (Auer, Cesa-Bianchi, and Fischer 2002) for arbitrary-distribution arms bounded in $[0, 1]$:

$$A_{t+1} = \arg\max_{a \in \mathcal{A}} \{\hat{Q}_t(a) + \sqrt{\frac{\alpha \log t}{2N_t(a)}}\}$$

It has been shown that the UCB1 Algorithm achieves logarithmic total regret asymptotically. Specifically,

Theorem 15.4.2 (UCB1 Logarithmic Total Regret). *As* $T \to \infty$,

$$L_T \leq \sum_{a|\Delta_a > 0} \frac{4\alpha \cdot \log T}{\Delta_a} + \frac{2\alpha \cdot \Delta_a}{\alpha - 1}$$

Now let's implement the UCB1 Algorithm in code. The class UCB1 below implements the interface of the abstract base class MABBase. We've implemented the below code for rewards range $[0, B]$ (adjusting the above UCB1 formula apropriately from $[0, 1]$ range to $[0, B]$ range). B is specified as the constructor input bounds_range. The constructor input alpha corresponds to the parameter α specified above. get_episode_rewards_actions implements MABBase's @abstracmethod interface, and its code below should be self-explanatory.

```
from numpy import ndarray, empty, sqrt, log
from operator import itemgetter
class UCB1(MABBase):
    def __init__(
        self,
        arm_distributions: Sequence[Distribution[float]],
        time_steps: int,
        num_episodes: int,
        bounds_range: float,
        alpha: float
    ) -> None:
        if bounds_range < 0 or alpha <= 0:
            raise ValueError
        super().__init__(
            arm_distributions=arm_distributions,
            time_steps=time_steps,
            num_episodes=num_episodes
        )
        self.bounds_range: float = bounds_range
        self.alpha: float = alpha

    def get_episode_rewards_actions(self) -> Tuple[ndarray, ndarray]:
        ep_rewards: ndarray = empty(self.time_steps)
        ep_actions: ndarray = empty(self.time_steps, dtype=int)
        for i in range(self.num_arms):
            ep_rewards[i] = self.arm_distributions[i].sample()
            ep_actions[i] = i
        counts: List[int] = [1] * self.num_arms
        means: List[float] = [ep_rewards[j] for j in range(self.num_arms)]
        for i in range(self.num_arms, self.time_steps):
            ucbs: Sequence[float] = [means[j] + self.bounds_range *
                                     sqrt(0.5 * self.alpha * log(i) /
                                          counts[j])
                                     for j in range(self.num_arms)]
            action: int = max(enumerate(ucbs), key=itemgetter(1))[0]
            reward: float = self.arm_distributions[action].sample()
            counts[action] += 1
            means[action] += (reward - means[action]) / counts[action]
            ep_rewards[i] = reward
            ep_actions[i] = action
        return ep_rewards, ep_actions
```

The above code is in the file rl/chapter14/ucb1.py. The code in __main__ sets up a UCB1 instance with 6 arms, each having a binomial distribution with $n = 10$ and $p = \{0.4, 0.8, 0.1, 0.5, 0.9, 0.2\}$ for the 6 arms. When run with 1000 time steps, 500 episodes and $\alpha = 4$, we get the Total Regret Curve as shown in Figure 15.4.

We encourage you to modify the code in __main__ to model other distributions for the arms, examine the results obtained, and develop more intuition for the UCB1 Algorithm.

15.4.3 Bayesian UCB

The algorithms we have covered so far have not made any assumptions about the rewards distributions $\mathcal{R}^a$ (except for the range of the rewards). Now we assume that the rewards distributions are restricted to a family of analytically-tractable probability distributions, which enables us to make analytically-favorable inferences about the rewards distributions. Let us refer to the sequence of distributions $[\mathcal{R}^a | a \in \mathcal{A}]$ as $\mathcal{R}$. To be clear, the AI Agent (algorithm) does not have knowledge of $\mathcal{R}$ and aims to estimate $\mathcal{R}$ from the rewards data obtained upon performing actions. Bayesian Bandit Algorithms (abbreviated as *Bayesian Bandits*) achieve this by maintaining an estimate of the probability distribution over $\mathcal{R}$ based on rewards data seen for each of the selected arms. The idea is to compute the posterior distribution $\mathbb{P}[\mathcal{R}|H_t]$ by exploiting prior knowledge of $\mathbb{P}[\mathcal{R}]$, where

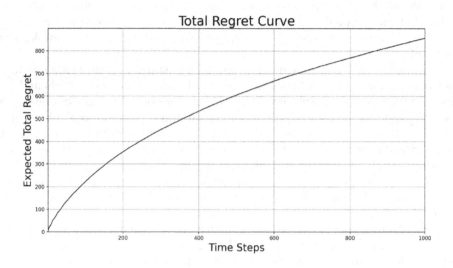

Figure 15.4 **UCB1 Total Regret Curve**

$H_t = A_1, R_1, A_1, R_1, \ldots, A_t, R_t$ is the history. Note that the prior distribution $\mathbb{P}[\mathcal{R}]$ and the posterior distribution $\mathbb{P}[\mathcal{R}|H_t]$ are probability distributions over probability distributions (since each $\mathcal{R}^a$ in $\mathcal{R}$ is a probability distribution). This posterior distribution is then used to guide exploration. This leads to two types of algorithms:

- Upper Confidence Bounds (Bayesian UCB), which we give an example of below.
- Probability Matching, which we cover in the next section in the form of Thompson Sampling.

We get a better performance if our prior knowledge of $\mathbb{P}[\mathcal{R}]$ is accurate. A simple example of Bayesian UCB is to model independent Gaussian distributions. Assume the reward distribution is Gaussian: $\mathcal{R}^a(r) = \mathcal{N}(r; \mu_a, \sigma_a^2)$ for all $a \in \mathcal{A}$, where μ_a and σ_a^2 denote the mean and variance respectively of the Gaussian reward distribution of a. The idea is to compute a Gaussian posterior over μ_a, σ_a^2, as follows:

$$\mathbb{P}[\mu_a, \sigma_a^2 | H_t] \propto \mathbb{P}[\mu_a, \sigma_a^2] \cdot \prod_{t | A_t = a} \mathcal{N}(R_t; \mu_a, \sigma_a^2)$$

This posterior calculation can be performed in an incremental manner by updating $\mathbb{P}[\mu_{A_t}, \sigma_{A_t}^2 | H_t]$ after each time step t (observing R_t after selecting action A_t). This incremental calculation with Bayesian updates to hyperparameters (parameters controlling the probability distributions of μ_a and σ_a^2) is described in detail in Section G.1 in Appendix G.

Given this posterior distribution for μ_a and σ_a^2 for all $a \in \mathcal{A}$ after each time step t, we select the action that maximizes the Expectation of "c standard-errors above mean", i.e.,

$$A_{t+1} = \arg\max_{a \in \mathcal{A}} \mathbb{E}_{\mathbb{P}[\mu_a, \sigma_a^2 | H_t]}[\mu_a + \frac{c \cdot \sigma_a}{\sqrt{N_t(a)}}]$$

15.5 PROBABILITY MATCHING

As mentioned in the previous section, calculating the posterior distribution $\mathbb{P}[\mathcal{R}|H_t]$ after each time step t also enables a different approach known as *Probability Matching*. The idea behind Probability Matching is to select an action a probabilistically in proportion to the probability that a might be the optimal action (based on the rewards data seen so far). Before describing Probability Matching formally, we illustrate the idea with a simple example to develop intuition.

Let us say we have only two actions a_1 and a_2. For simplicity, let us assume that the posterior distribution $\mathbb{P}[\mathcal{R}^{a_1}|H_t]$ has only two distribution outcomes (call them $\mathcal{R}_1^{a_1}$ and $\mathcal{R}_2^{a_1}$) and that the posterior distribution $\mathbb{P}[\mathcal{R}^{a_2}|H_t]$ also has only two distribution outcomes (call them $\mathcal{R}_1^{a_2}$ and $\mathcal{R}_2^{a_2}$). Typically, there will be an infinite (continuum) of distribution outcomes for $\mathbb{P}[\mathcal{R}|H_t]$—here we assume only two distribution outcomes for each of the actions' estimated conditional probability of rewards purely for simplicity so as to convey the intuition behind Probability Matching. Assume that $\mathbb{P}[\mathcal{R}^{a_1} = \mathcal{R}_1^{a_1}|H_t] = 0.7$ and $\mathbb{P}[\mathcal{R}^{a_1} = \mathcal{R}_2^{a_1}|H_t] = 0.3$, and that $\mathcal{R}_1^{a_1}$ has mean 5.0 and $\mathcal{R}_2^{a_1}$ has mean 10.0. Assume that $\mathbb{P}[\mathcal{R}^{a_2} = \mathcal{R}_1^{a_2}|H_t] = 0.2$ and $\mathbb{P}[\mathcal{R}^{a_2} = \mathcal{R}_2^{a_2}|H_t] = 0.8$, and that $\mathcal{R}_1^{a_2}$ has mean 2.0 and $\mathcal{R}_2^{a_2}$ has mean 7.0.

Probability Matching calculates at each time step t how often does each action a have the maximum $\mathbb{E}[r|a]$ among all actions, across all the probabilistic outcomes for the posterior distribution $\mathbb{P}[\mathcal{R}|H_t]$, and then selects that action a probabilistically in proportion to this calculation. Let's do this probability calculation for our simple case of two actions and two probabilistic outcomes each for the posterior distribution for each action. So here, we have 4 probabilistic outcomes when considering the two actions jointly, as follows:

- Outcome 1: $\mathcal{R}_1^{a_1}$ (with probability 0.7) and $\mathcal{R}_1^{a_2}$ (with probability 0.2). Thus, Outcome 1 has probability 0.7 * 0.2 = 0.14. In Outcome 1, a_1 has the maximum $\mathbb{E}[r|a]$ among all actions since $\mathcal{R}_1^{a_1}$ has mean 5.0 and $\mathcal{R}_1^{a_2}$ has mean 2.0.
- Outcome 2: $\mathcal{R}_1^{a_1}$ (with probability 0.7) and $\mathcal{R}_2^{a_2}$ (with probability 0.8). Thus, Outcome 2 has probability 0.7 * 0.8 = 0.56. In Outcome 2, a_2 has the maximum $\mathbb{E}[r|a]$ among all actions since $\mathcal{R}_1^{a_1}$ has mean 5.0 and $\mathcal{R}_2^{a_2}$ has mean 7.0.
- Outcome 3: $\mathcal{R}_2^{a_1}$ (with probability 0.3) and $\mathcal{R}_1^{a_2}$ (with probability 0.2). Thus, Outcome 3 has probability 0.3 * 0.2 = 0.06. In Outcome 3, a_1 has the maximum $\mathbb{E}[r|a]$ among all actions since $\mathcal{R}_2^{a_1}$ has mean 10.0 and $\mathcal{R}_1^{a_2}$ has mean 2.0.
- Outcome 4: $\mathcal{R}_2^{a_1}$ (with probability 0.3) and $\mathcal{R}_2^{a_2}$ (with probability 0.8). Thus, Outcome 4 has probability 0.3 * 0.8 = 0.24. In Outcome 4, a_1 has the maximum $\mathbb{E}[r|a]$ among all actions since $\mathcal{R}_2^{a_1}$ has mean 10.0 and $\mathcal{R}_2^{a_2}$ has mean 7.0.

Thus, a_1 has the maximum $\mathbb{E}[r|a]$ among the two actions in Outcomes 1, 3 and 4, amounting to a total outcomes probability of 0.14 + 0.06 + 0.24 = 0.44, and a_2 has the maximum $\mathbb{E}[r|a]$ among the two actions only in Outcome 2, which has an outcome probability of 0.56. Therefore, in the next time step $(t+1)$, the Probability Matching method will select action a_1 with probability 0.44 and a_2 with probability 0.56.

Generalizing this Probability Matching method to an arbitrary number of actions and to an arbitrary number of probabilistic outcomes for the conditional reward distributions for each action, we can write the probabilistic selection of actions at time step $t+1$ as:

$$\mathbb{P}[A_{t+1}|H_t] = \mathbb{P}_{\mathcal{D}_t \sim \mathbb{P}[\mathcal{R}|H_t]}[\mathbb{E}_{\mathcal{D}_t}[r|A_{t+1}] > \mathbb{E}_{\mathcal{D}_t}[r|a] \text{ for all } a \neq A_{t+1}] \quad (15.1)$$

where $\mathcal{D}_t$ refers to a particular random outcome of a distribution of rewards for each action, drawn from the posterior distribution $\mathbb{P}[\mathcal{R}|H_t]$. As ever, ties between actions are broken with an arbitrary rule prioritizing actions.

Note that the Probability Matching method is also based on the principle of *Optimism in the Face of Uncertainty* because an action with more uncertainty in its mean reward is more likely to have the highest mean reward among all actions (all else being equal), and hence deserves to be selected more frequently.

We see that the Probability Matching approach is mathematically disciplined in driving towards cumulative reward maximization while balancing exploration and exploitation. However, the right-hand-side of Equation 15.1 can be difficult to compute analytically from the posterior distributions. We resolve this difficulty with a sampling approach to Probability Matching known as *Thompson Sampling*.

15.5.1 Thompson Sampling

We can reformulate the right-hand-side of Equation 15.1 as follows:

$$\mathbb{P}[A_{t+1}|H_t] = \mathbb{P}_{\mathcal{D}_t \sim \mathbb{P}[\mathcal{R}|H_t]}[\mathbb{E}_{\mathcal{D}_t}[r|A_{t+1}] > \mathbb{E}_{\mathcal{D}_t}[r|a] \text{for all } a \neq A_{t+1}]$$
$$= \mathbb{E}_{\mathcal{D}_t \sim \mathbb{P}[\mathcal{R}|H_t]}[\mathbb{I}_{A_{t+1}=\arg\max_{a \in \mathcal{A}} \mathbb{E}_{\mathcal{D}_t}[r|a]}]$$

where $\mathbb{I}$ refers to the indicator function. This reformulation in terms of an *Expectation* is convenient because we can estimate the Expectation by sampling various $\mathcal{D}_t$ probability distributions and for each sample of $\mathcal{D}_t$, we simply check if an action has the best mean reward (compared to other actions) under the distribution $\mathcal{D}_t$. This sampling-based approach to Probability Matching is known as *Thompson Sampling*. Specifically, Thompson Sampling performs the following calculations for time step $t + 1$:

- Compute the posterior distribution $\mathbb{P}[\mathcal{R}|H_t]$ by performing Bayesian updates of the hyperparameters that govern the estimated probability distributions of the parameters of the reward distributions for each action.
- *Sample* a joint (across actions) rewards distribution $\mathcal{D}_t$ from the posterior distribution $\mathbb{P}[\mathcal{R}|H_t]$.
- Calculate a sample Action-Value function with sample $\mathcal{D}_t$ as:

$$\hat{Q}_t(a) = \mathbb{E}_{\mathcal{D}_t}[r|a]$$

- Select the action (for time step $t + 1$) that maximizes this sample Action-Value function:

$$A_{t+1} = \arg\max_{a \in \mathcal{A}} \hat{Q}_t(a)$$

It turns out that Thompson Sampling achieves the Lai-Robbins lower bound for Logarithmic Total Regret. To learn more about Thompson Sampling, we refer you to the excellent tutorial on Thompson Sampling by Russo, Roy, Kazerouni, Osband, Wen (Russo et al. 2018).

Now we implement Thompson Sampling by assuming a Gaussian distribution of rewards for each action. The posterior distributions for each action are produced by performing Bayesian updates of the hyperparameters that govern the estimated Gaussian-Inverse-Gamma Probability Distributions of the parameters of the Gaussian reward distributions for each action. Section G.1 of Appendix G describes the Bayesian updates of the hyperparameters θ, α, β, and the code below implements this update in the variable bayes in method get_episode_rewards_actions (this method implements the @abstractmethod interface of abstract base class MABBase). The sample mean rewards are obtained by invoking the sample method of Gaussian and Gamma classes, and assigned to the variable mean_draws. The variable theta refers to the hyperparameter θ, the variable alpha refers to the hyperparameter α, and the variable beta refers to the hyperparameter β. The rest of the code in the method get_episode_rewards_actions should be self-explanatory.

```
from rl.distribution import Gaussian, Gamma
from operator import itemgetter
```

```python
from numpy import ndarray, empty, sqrt
class ThompsonSamplingGaussian(MABBase):
    def __init__(
        self,
        arm_distributions: Sequence[Gaussian],
        time_steps: int,
        num_episodes: int,
        init_mean: float,
        init_stdev: float
    ) -> None:
        super().__init__(
            arm_distributions=arm_distributions,
            time_steps=time_steps,
            num_episodes=num_episodes
        )
        self.theta0: float = init_mean
        self.n0: int = 1
        self.alpha0: float = 1
        self.beta0: float = init_stdev * init_stdev

    def get_episode_rewards_actions(self) -> Tuple[ndarray, ndarray]:
        ep_rewards: ndarray = empty(self.time_steps)
        ep_actions: ndarray = empty(self.time_steps, dtype=int)
        bayes: List[Tuple[float, int, float, float]] =\
            [(self.theta0, self.n0, self.alpha0, self.beta0)] * self.num_arms

        for i in range(self.time_steps):
            mean_draws: Sequence[float] = [Gaussian(
                mu=theta,
                sigma=1 / sqrt(n * Gamma(alpha=alpha, beta=beta).sample())
            ).sample() for theta, n, alpha, beta in bayes]
            action: int = max(enumerate(mean_draws), key=itemgetter(1))[0]
            reward: float = self.arm_distributions[action].sample()
            theta, n, alpha, beta = bayes[action]
            bayes[action] = (
                (reward + n * theta) / (n + 1),
                n + 1,
                alpha + 0.5,
                beta + 0.5 * n / (n + 1) * (reward - theta) * (reward - theta)
            )
            ep_rewards[i] = reward
            ep_actions[i] = action
        return ep_rewards, ep_actions
```

The above code is in the file rl/chapter14/ts_gaussian.py. The code in __main__ sets up a ThompsonSamplingGaussian instance with 6 arms, each having a Gaussian rewards distribution. When run with 1000 time steps and 500 episodes, we get the Total Regret Curve as shown in Figure 15.5.

We encourage you to modify the code in __main__ to try other mean and variance settings for the Gaussian reward distributions of the arms, examine the results obtained, and develop more intuition for Thompson Sampling for Gaussians.

Now we implement Thompson Sampling by assuming a Bernoulli distribution of rewards for each action. The posterior distributions for each action are produced by performing Bayesian updates of the hyperparameters that govern the estimated Beta Probability Distributions of the parameters of the Bernoulli reward distributions for each action. Section G.2 of Appendix G describes the Bayesian updates of the hyperparameters α and β, and the code below implements this update in the variable bayes in method get_episode_rewards_actions (this method implements the @abstractmethod interface of abstract base class MABBase). The sample mean rewards are obtained by invoking the sample method of the Beta class, and assigned to the variable mean_draws. The variable alpha refers to the hyperparameter α and the variable beta refers to the hyperparameter β. The rest of the code in the method get_episode_rewards_actions should be self-explanatory.

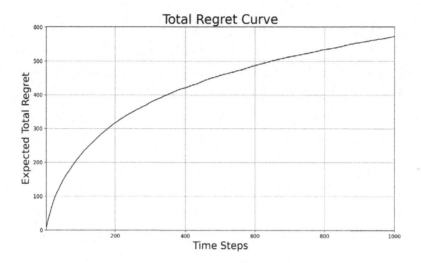

Figure 15.5 Thompson Sampling (Gaussian) Total Regret Curve

```
from rl.distribution import Bernoulli, Beta
from operator import itemgetter
from numpy import ndarray, empty

class ThompsonSamplingBernoulli(MABBase):
    def __init__(
        self,
        arm_distributions: Sequence[Bernoulli],
        time_steps: int,
        num_episodes: int
    ) -> None:
        super().__init__(
            arm_distributions=arm_distributions,
            time_steps=time_steps,
            num_episodes=num_episodes
        )

    def get_episode_rewards_actions(self) -> Tuple[ndarray, ndarray]:
        ep_rewards: ndarray = empty(self.time_steps)
        ep_actions: ndarray = empty(self.time_steps, dtype=int)
        bayes: List[Tuple[int, int]] = [(1, 1)] * self.num_arms

        for i in range(self.time_steps):
            mean_draws: Sequence[float] = \
                [Beta(alpha=alpha, beta=beta).sample() for alpha, beta in bayes]
            action: int = max(enumerate(mean_draws), key=itemgetter(1))[0]
            reward: float = float(self.arm_distributions[action].sample())
            alpha, beta = bayes[action]
            bayes[action] = (alpha + int(reward), beta + int(1 - reward))
            ep_rewards[i] = reward
            ep_actions[i] = action
        return ep_rewards, ep_actions
```

The above code is in the file rl/chapter14/ts_bernoulli.py. The code in __main__ sets up a ThompsonSamplingBernoulli instance with 6 arms, each having a Bernoulli rewards distribution. When run with 1000 time steps and 500 episodes, we get the Total Regret Curve as shown in Figure 15.6.

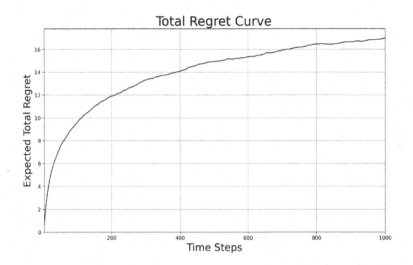

Figure 15.6 Thompson Sampling (Bernoulli) Total Regret Curve

We encourage you to modify the code in __main__ to try other mean settings for the Bernoulli reward distributions of the arms, examine the results obtained, and develop more intuition for Thompson Sampling for Bernoullis.

15.6   GRADIENT BANDITS

Now we cover a MAB algorithm that is similar to Policy Gradient for MDPs. This MAB algorithm's action selection is randomized and the action selection probabilities are constructed through Gradient Ascent (much like Stochastic Policy Gradient for MDPs). This MAB Algorithm and its variants are cheekily referred to as *Gradient Bandits*. Our coverage below follows the coverage of Gradient Bandit algorithm in the RL book by Sutton and Barto (Richard S. Sutton and Barto 2018).

The basic idea is that we have m *Score* parameters (to be optimized), one for each action, denoted as $\{s_a | a \in \mathcal{A}\}$ that define the action-selection probabilities, which in turn defines an *Expected Reward* Objective function to be maximized, as follows:

$$J(s_{a_1}, \ldots, s_{a_m}) = \sum_{a \in \mathcal{A}} \pi(a) \cdot \mathbb{E}[r|a]$$

where $\pi : \mathcal{A} \to [0, 1]$ refers to the function for action-selection probabilities, that is defined as follows:

$$\pi(a) = \frac{e^{s_a}}{\sum_{b \in \mathcal{A}} e^{s_b}} \text{ for all } a \in \mathcal{A}$$

The *Score* parameters are meant to represent the relative value of actions based on the rewards seen until a certain time step, and are adjusted appropriately after each time step (using Gradient Ascent). Note that $\pi(\cdot)$ is a Softmax function of the *Score* parameters.

Gradient Ascent moves the *Score* parameters s_a (and hence, action probabilities $\pi(a)$) in the direction of the gradient of the objective function $J(s_{a_1}, \ldots, s_{a_m})$ with respect to

$(s_{a_1}, \ldots, s_{a_m})$. To construct this gradient of $J(\cdot)$, we calculate $\frac{\partial J}{\partial s_a}$ for each $a \in \mathcal{A}$, as follows:

$$
\begin{aligned}
\frac{\partial J}{\partial s_a} &= \frac{\partial(\sum_{a' \in \mathcal{A}} \pi(a') \cdot \mathbb{E}[r|a'])}{\partial s_a} \\
&= \sum_{a' \in \mathcal{A}} \mathbb{E}[r|a'] \cdot \frac{\partial \pi(a')}{\partial s_a} \\
&= \sum_{a' \in \mathcal{A}} \pi(a') \cdot \mathbb{E}[r|a'] \cdot \frac{\partial \log \pi(a')}{\partial s_a} \\
&= \mathbb{E}_{a' \sim \pi, r \sim \mathcal{R}^{a'}}[r \cdot \frac{\partial \log \pi(a')}{\partial s_a}]
\end{aligned}
$$

We know from standard softmax-function calculus that:

$$
\frac{\partial \log \pi(a')}{\partial s_a} = \frac{\partial(\log \frac{e^{s_{a'}}}{\sum_{b \in \mathcal{A}} e^{s_b}})}{\partial s_a} = \mathbb{I}_{a=a'} - \pi(a)
$$

Therefore, $\frac{\partial J}{\partial s_a}$ can be re-written as:

$$
\frac{\partial J}{\partial s_a} = \mathbb{E}_{a' \sim \pi, r \sim \mathcal{R}^{a'}}[r \cdot (\mathbb{I}_{a=a'} - \pi(a))]
$$

At each time step t, we approximate the gradient with the (A_t, R_t) sample as:

$$
R_t \cdot (\mathbb{I}_{a=A_t} - \pi_t(a)) \text{ for all } a \in \mathcal{A}
$$

$\pi_t(a)$ is the probability of selecting action a at time step t, derived from the *Score* $s_t(a)$ at time step t.

We can reduce the variance of this estimate with a baseline B that is independent of a, as follows:

$$
(R_t - B) \cdot (\mathbb{I}_{a=A_t} - \pi_t(a)) \text{ for all } a \in \mathcal{A}
$$

This doesn't introduce any bias in the estimate of the gradient of $J(\cdot)$ because:

$$
\begin{aligned}
\mathbb{E}_{a' \sim \pi}[B \cdot (\mathbb{I}_{a=a'} - \pi(a))] &= \mathbb{E}_{a' \sim \pi}[B \cdot \frac{\partial \log \pi(a')}{\partial s_a}] \\
&= B \cdot \sum_{a' \in \mathcal{A}} \pi(a') \cdot \frac{\partial \log \pi(a')}{\partial s_a} \\
&= B \cdot \sum_{a' \in \mathcal{A}} \frac{\partial \pi(a')}{\partial s_a} \\
&= B \cdot \frac{\partial(\sum_{a' \in \mathcal{A}} \pi(a'))}{\partial s_a} \\
&= B \cdot \frac{\partial 1}{\partial s_a} \\
&= 0
\end{aligned}
$$

We can use $B = \bar{R}_t = \frac{1}{t} \sum_{s=1}^{t} R_s$ (average of all rewards obtained until time step t). So, the update to scores $s_t(a)$ for all $a \in \mathcal{A}$ is:

$$
s_{t+1}(a) = s_t(a) + \alpha \cdot (R_t - \bar{R}_t) \cdot (\mathbb{I}_{a=A_t} - \pi_t(a))
$$

It should be noted that this Gradient Bandit algorithm and its variant Gradient Bandit algorithms are simply a special case of policy gradient-based RL algorithms.

Now let's write some code to implement this Gradient Algorithm. Apart from the usual constructor inputs `arm_distributions`, `time_steps` and `num_episodes` that are passed along to the constructor of the abstract base class `MABBase`, `GradientBandits`' constructor also takes as input `learning_rate` (specifying the initial learning rate) and `learning_rate_decay` (specifying the speed at which the learning rate decays), which influence how the variable `step_size` is set at every time step. The variable `scores` represents $s_t(a)$ for all $a \in \mathcal{A}$ and the variable `probs` represents $\pi_t(a)$ for all $a \in \mathcal{A}$. The rest of the code below should be self-explanatory, based on the above description of the calculations.

```python
from rl.distribution import Distribution, Categorical
from operator import itemgetter
from numpy import ndarray, empty, exp

class GradientBandits(MABBase):
    def __init__(
        self,
        arm_distributions: Sequence[Distribution[float]],
        time_steps: int,
        num_episodes: int,
        learning_rate: float,
        learning_rate_decay: float
    ) -> None:
        super().__init__(
            arm_distributions=arm_distributions,
            time_steps=time_steps,
            num_episodes=num_episodes
        )
        self.learning_rate: float = learning_rate
        self.learning_rate_decay: float = learning_rate_decay

    def get_episode_rewards_actions(self) -> Tuple[ndarray, ndarray]:
        ep_rewards: ndarray = empty(self.time_steps)
        ep_actions: ndarray = empty(self.time_steps, dtype=int)
        scores: List[float] = [0.] * self.num_arms
        avg_reward: float = 0.

        for i in range(self.time_steps):
            max_score: float = max(scores)
            exp_scores: Sequence[float] = [exp(s - max_score) for s in scores]
            sum_exp_scores = sum(exp_scores)
            probs: Sequence[float] = [s / sum_exp_scores for s in exp_scores]
            action: int = Categorical(
                {i: p for i, p in enumerate(probs)}
            ).sample()
            reward: float = self.arm_distributions[action].sample()
            avg_reward += (reward - avg_reward) / (i + 1)
            step_size: float = self.learning_rate *\
                (i / self.learning_rate_decay + 1) ** -0.5
            for j in range(self.num_arms):
                scores[j] += step_size * (reward - avg_reward) *\
                    ((1 if j == action else 0) - probs[j])

            ep_rewards[i] = reward
            ep_actions[i] = action
        return ep_rewards, ep_actions
```

The above code is in the file rl/chapter14/gradient_bandits.py. The code in __main__ sets up a `GradientBandits` instance with 6 arms, each having a Gaussian reward distribution. When run with 1000 time steps and 500 episodes, we get the Total Regret Curve as shown in Figure 15.7.

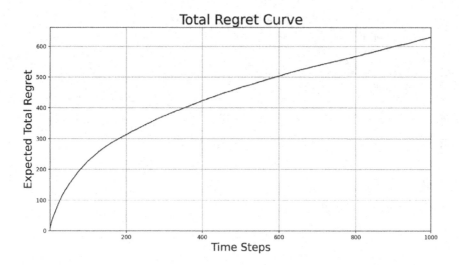

Figure 15.7 Gradient Algorithm Total Regret Curve

We encourage you to modify the code in __main__ to try other mean and standard deviation settings for the Gaussian reward distributions of the arms, examine the results obtained, and develop more intuition for this Gradient Algorithm.

15.7 HORSE RACE

We've implemented several algorithms for the MAB problem. Now it's time for a competition between them, that we will call a *Horse Race*. In this Horse Race, we will compare the *Total Regret* across the algorithms, and we will also examine the number of times the different arms get pulled by the various algorithms. We expect a good algorithm to have small total regret and we expect a good algorithm to pull the arms with high *Gaps* few number of times and pull the arms with low (and zero) gaps large number of times.

The code in the file rl/chapter14/plot_mab_graphs.py has a function to run a horse race for Gaussian arms with the following algorithms:

- Greedy with Optimistic Initialization
- ϵ-Greedy
- Decaying ϵ_t-Greedy
- Thompson Sampling
- Gradient Bandit

Running this horse race for 7 Gaussian arms with 500 time steps, 500 episodes and the settings as specified in the file rl/chapter14/plot_mab_graphs.py, we obtain Figure 15.8 for the Total Regret Curves for each of these algorithms.

Figure 15.9 shows the number of times each arm is pulled (for each of these algorithms). The X-axis is sorted by the mean of the reward distributions of the arms. For each arm, the left-to-right order of the arm-pulls count is the order in which the 5 MAB algorithms are listed above. As we can see, the arms with low means are pulled only a few times and the arms with high means are pulled often.

The file rl/chapter14/plot_mab_graphs.py also has a function to run a horse race for Bernoulli arms with the following algorithms:

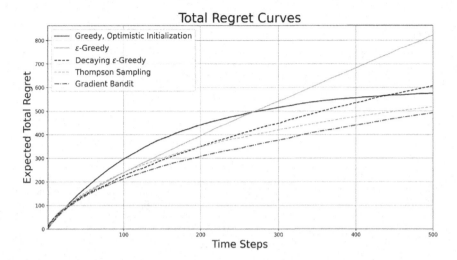

Figure 15.8 **Gaussian Horse Race—Total Regret Curves**

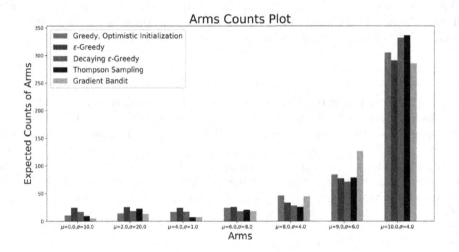

Figure 15.9 **Gaussian Horse Race—Arms Count**

- Greedy with Optimistic Initialization
- ϵ-Greedy
- Decaying ϵ_t-Greedy
- UCB1
- Thompson Sampling
- Gradient Bandit

Running this horse race for 9 Bernoulli arms with 500 time steps, 500 episodes and the settings as specified in the file rl/chapter14/plot_mab_graphs.py, we obtain Figure 15.10 for the Total Regret Curves for each of these algorithms.

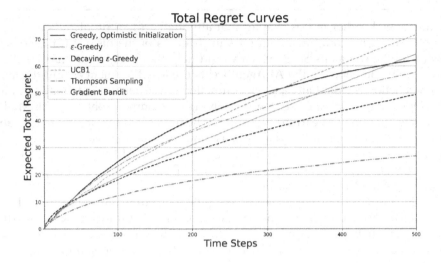

Figure 15.10 Bernoulli Horse Race—Total Regret Curves

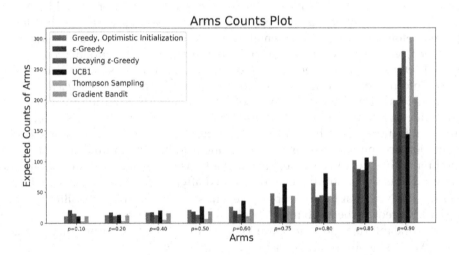

Figure 15.11 Bernoulli Horse Race—Arms Count

Figure 15.11 shows the number of times each arm is pulled (for each of the algorithms). The X-axis is sorted by the mean of the reward distributions of the arms. For each arm, the left-to-right order of the arm-pulls count is the order in which the 6 MAB algorithms are listed above. As we can see, the arms with low means are pulled only a few times and the arms with high means are pulled often.

We encourage you to experiment with the code in rl/chapter14/plot_mab_graphs.py: try different arm distributions, try different input parameters for each of the algorithms, plot the graphs, and try to explain the relative performance of the algorithms (perhaps by writing some more diagnostics code). This will help build tremendous intuition on the pros and cons of these algorithms.

15.8 INFORMATION STATE SPACE MDP

We had mentioned earlier in this chapter that although a MAB problem is not posed as an MDP, the AI Agent could maintain relevant features of the history (of actions taken and rewards obtained) as its *State*, which would help the AI Agent in making the arm-selection (action) decision. So the AI Agent treats the MAB problem as an MDP and the arm-selection action is essentially a (*Policy*) function of the agent's *State*. One can then arrive at the Optimal arm-selection strategy by solving the Control problem of this MDP with an appropriate Planning or Learning algorithm. The representation of *State* as relevant features of history is known as *Information State* (to indicate that the agent captures all of the relevant information known so far in the *State* of the modeled MDP). Before we explain this *Information State Space MDP* approach in more detail, it pays to develop an intuitive understanding of the *Value of Information*.

The key idea is that *Exploration* enables the agent to acquire information, which in turn enables the agent to make more informed decisions as far as its future arm-selection strategy is concerned. The natural question to ask then is whether we can quantify the value of this information that can be acquired by *Exploration*. In other words, how much would a decision-maker be willing to pay to acquire information (through exploration), prior to making a decision? Vaguely speaking, the decision-maker should be paying an amount equal to the gains in long-term (accumulated) reward that can be obtained upon getting the information, less the sacrifice of excess immediate reward one would have obtained had one exploited rather than explored. We can see that this approach aims to settle the explore-exploit trade-off in a mathematically rigorous manner by establishing the *Value of Information*. Note that information gain is higher in a more uncertain situation (all else being equal). Therefore, it makes sense to explore uncertain situations more. By formalizing the value of information, we can trade-off exploration and exploitation *optimally*.

Now let us formalize the approach of treating a MAB as an Information State Space MDP. After each time step of a MAB, we construct an *Information State* $\tilde{s}$, which comprises of relevant features of the history until that time step. Essentially, $\tilde{s}$ summarizes all of the information accumulated so far that is pertinent to be able to predict the reward distribution for each action. Each action a causes a transition to a new information state $\tilde{s}'$ (by adding information about the reward obtained after performing action a), with probability $\tilde{\mathcal{P}}(\tilde{s}, a, \tilde{s}')$. Note that this probability depends on the reward probability function $\mathcal{R}^a$ of the MAB. Moreover, the MAB reward r obtained upon performing action a constitutes the Reward of the Information State Space MDP for that time step. Putting all this together, we have an MDP $\tilde{M}$ in information state space as follows:

- Denote the Information State Space of $\tilde{M}$ as $\tilde{\mathcal{S}}$.
- The Action Space of $\tilde{M}$ is the action space of the given MAB: $\mathcal{A}$.
- The State Transition Probability function of $\tilde{M}$ is $\tilde{\mathcal{P}}$.
- The Reward function of $\tilde{M}$ is given by the Reward probability function $\mathcal{R}^a$ of the MAB.
- Discount Factor $\gamma = 1$.

The key point to note is that since $\mathcal{R}^a$ is unknown to the AI Agent in the MAB problem, the State Transition Probability function and the Reward function of the Information State Space MDP $\tilde{M}$ are unknown to the AI Agent. However, at any given time step, the AI Agent can utilize the information within $\tilde{s}$ to form an estimate of $\mathcal{R}^a$, which in turn gives estimates of the State Transition Probability function and the Reward function of the Information State Space MDP $\tilde{M}$.

Note that $\tilde{M}$ will typically be a fairly complex MDP over an infinite number of information states, and hence is not easy to solve. However, since it is after all an MDP, we can use Dynamic Programming or Reinforcement Learning algorithms to arrive at the Optimal Policy, which prescribes the optimal MAB action to take at that time step. If a Dynamic Programming approach is taken, then after each time step, as new information arrives (in the form of the MAB reward in response to the action taken), the estimates of the State Transition probability function and the Reward function change, meaning the Information State Space MDP to be solved changes, and consequently the Action-Selection strategy for the MAB problem (prescribed by the Optimal Policy of the Information State Space MDP) changes. A common approach is to treat the Information State Space MDP as a *Bayes-Adaptive MDP*. Specifically, if we have m arms $a_1, \ldots, a_m$, the state $\tilde{s}$ is modeled as $(\tilde{s_{a_1}}, \ldots, \tilde{s_{a_m}})$ such that $\tilde{s_a}$ for any $a \in \mathcal{A}$ represents a posterior probability distribution over $\mathcal{R}^a$, which is Bayes-updated after observing the reward upon each pull of the arm a. This Bayes-Adaptive MDP can be tackled with the highly-celebrated Dynamic Programming method known as Gittins Index, which was introduced in a 1979 paper by Gittins (Gittins 1979). The Gittins Index approach finds the Bayes-optimal explore-exploit trade-off with respect to the prior distribution.

To grasp the concept of Information State Space MDP, let us consider a Bernoulli Bandit problem with m arms with arm a's reward probability distribution $\mathcal{R}^a$ given by the Bernoulli distribution $\mathcal{B}(\mu_a)$, where $\mu_a \in [0, 1]$ (i.e., reward = 1 with probability μ_a, and reward = 0 with probability $1 - \mu_a$). If we denote the m arms by $a_1, a_2, \ldots, a_m$, then the information state is $\tilde{s} = (\alpha_{a_1}, \beta_{a_1}, \alpha_{a_2}, \beta_{a_2} \ldots, \alpha_{a_m}, \beta_{a_m})$, where α_a is the number of pulls of arm a (so far) for which the reward was 1 and β_a is the number of pulls of arm a (so far) for which the reward was 0. Note that by the Law of Large Numbers, in the long-run, $\frac{\alpha_a}{\alpha_a + \beta_a} \to \mu_a$.

We can treat this as a Bayes-adaptive MDP as follows: We model the prior distribution over $\mathcal{R}^a$ as the Beta Distribution $Beta(\alpha_a, \beta_a)$ over the unknown parameter μ_a. Each time arm a is pulled, we update the posterior for $\mathcal{R}^a$ as:

- $Beta(\alpha_a + 1, \beta_a)$ if $r = 1$
- $Beta(\alpha_a, \beta_a + 1)$ if $r = 0$

Note that the component (α_a, β_a) within the information state provides the model $Beta(\alpha_a, \beta_a)$ as the probability distribution over μ_a. Moreover, note that each state transition (updating either α_a or β_a by 1) is essentially a Bayesian model update (Section G.2 in Appendix G provides details of Bayesian updates to a Beta distribution over a Bernoulli parameter).

Note that in general, an exact solution to a Bayes-adaptive MDP is typically intractable. In 2014, Guez, Heess, Silver, Dayan (Guez et al. 2014) came up with a Simulation-based Search method, which involves a forward search in information state space using simulations from current information state, to solve a Bayes-adaptive MDP.

15.9 EXTENDING TO CONTEXTUAL BANDITS AND RL CONTROL

A Contextual Bandit problem is a natural extension of the MAB problem, by introducing the concept of *Context* that has an influence on the rewards probability distribution for each arm. Before we provide a formal definition of a Contextual Bandit problem, we will provide an intuitive explanation with a canonical example. Consider the problem of showing a banner advertisement on a web site where there is a choice of displaying one among m different advertisements at a time. If the user clicks on the advertisement, there is a reward of 1 (if the user doesn't click, the reward is 0). The selection of the advertisement to display is the arm-selection (out of m arms, i.e., advertisements). This seems like a standard MAB problem, except that on a web site, we don't have a single user. In each round, a random user (among typically millions of users) appears. Each user will have their own characteristics of how they would respond to advertisements, meaning the rewards probability distribution for each arm would depend on the user. We refer to the user characteristics (as relevant to their likelihood to respond to specific advertisements) as the *Context*. This means, the *Context* influences the rewards probability distribution for each arm. This is known as the *Contextual Bandit* problem, which we formalize below:

Definition 15.9.1. A *Contextual Bandit* comprises of:

- A finite set of *Actions* $\mathcal{A}$ (known as the "arms").

- A probability distribution $\mathcal{C}$ over *Contexts*, defined as:

$$\mathcal{C}(c) = \mathbb{P}[c] \text{ for all Contexts } c$$

- Each pair of a context c and an action ("arm") $a \in \mathcal{A}$ is associated with a probability distribution over $\mathbb{R}$ (unknown to the AI Agent) denoted as $\mathcal{R}_c^a$, defined as:

$$\mathcal{R}_c^a(r) = \mathbb{P}[r|c, a] \text{ for all } r \in \mathbb{R}$$

- A time-indexed sequence of Environment-generated random Contexts C_t for time steps $t = 1, 2, \ldots$, a time-indexed sequence of AI Agent-selected actions $A_t \in \mathcal{A}$ for time steps $t = 1, 2, \ldots$, and a time-indexed sequence of Environment-generated *Reward* random variables $R_t \in \mathbb{R}$ for time steps $t = 1, 2, \ldots$, such that for each time step t, C_t is first randomly drawn from the probability distribution $\mathcal{C}$, after which the AI Agent selects the action A_t, after which R_t is randomly drawn from the probability distribution $\mathcal{R}_{C_t}^{A_t}$.

The AI Agent's goal is to maximize the following *Expected Cumulative Rewards* over a certain number of time steps T:

$$\mathbb{E}[\sum_{t=1}^{T} R_t]$$

Each of the algorithms we've covered for the MAB problem can be easily extended to the Contextual Bandit problem. The key idea in the extension of the MAB algorithms is that we have to take into account the Context, when dealing with the rewards probability distribution. In the MAB problem, the algorithms deal with a finite set of reward distributions, one for each of the actions. Here in the Contextual Bandit problem, the algorithms work with function approximations for the rewards probability distributions where each function approximation takes as input a pair of (Context, Action).

We won't cover the details of the extensions of all MAB Algorithms to Contextual Bandit algorithms. Rather, we simply sketch a simple Upper-Confidence-Bound algorithm for the Contextual Bandit problem to convey a sense of how to extend the MAB algorithms to the Contextual Bandit problem. Assume that the sampling distribution of the mean reward for each (Context, Action) pair is a Gaussian distribution, and so we maintain two function approximations $\mu(c, a; w)$ and $\sigma(c, a; v)$ to represent the mean and standard deviation of the sampling distribution of mean reward for any context c and any action a. It's important to note that for MAB, we simply maintained a finite set of estimates μ_a and σ_a, i.e., two parameters for each action a. Here we replace μ_a with function approximation $\mu(c, a; w)$ and we replace σ_a with function approximation $\sigma(c, a; v)$. After the receipt of a reward from the Environment, the parameters w and v are appropriately updated. We essentially perform supervised learning in an incremental manner when updating these parameters of the function approximations. Note that $\sigma(c, a; v)$ represents a function approximation for the standard error of the mean reward for a given context c and given action a. A simple Upper-Confidence-Bound algorithm would then select the action for a given context C_t at time step t that maximizes $\mu(C_t, a; w) + \alpha \cdot \sigma(C_t, a; v)$ over all choices of $a \in \mathcal{A}$, for some fixed α. Thus, we are comparing (across actions) α standard errors higher than the mean reward estimate (i.e., the upper-end of an appropriate confidence interval for the mean reward) for Context C_t.

We want to highlight that many authors refer to the *Context* in Contextual Bandits as *State*. We desist from using the term *State* in Contextual Bandits since we want to reserve the term *State* to refer to the concept of "transitions" (as is the case in MDPs). Note that the Context does not "transition" to the next Context in the next time step in Contextual Bandits problems. Rather, the Context is drawn at random independently at each time step from the Context probability distribution $\mathcal{C}$. This is in contrast to the *State* in MDPs which transitions to the next state at the next time step based on the State Transition probability function of the MDP.

We finish this chapter by simply pointing out that the approaches of the MAB algorithms can be further extended to resolve the Explore-Exploit dilemma in RL Control. From the perspective of this extension, it pays to emphasize that MAB algorithms that fall under the category of *Optimism in the Face of Uncertainty* can be roughly split into:

- Those that estimate the Q-Values (i.e., estimate $\mathbb{E}[r|a]$ from observed data) and the uncertainty of the Q-Values estimate. When extending to RL Control, we estimate the Q-Value Function for the (unknown) MDP and the uncertainty of the Q-Value Function estimate. Note that when moving from MAB to RL Control, the Q-Values are no longer simply the Expected Reward for a given action—rather, they are the Expected Return (i.e., accumulated rewards) from a given state and a given action. This extension from Expected Reward to Expected Return introduces significant complexity in the calculation of the uncertainty of the Q-Value Function estimate.
- Those that estimate the Model of the MDP, i.e., estimate of the State-Reward Transition Probability function $\mathcal{P}_R$ of the MDP, and the uncertainty of the $\mathcal{P}_R$ estimate. This includes extension of Bayesian Bandits, Thompson Sampling and Bayes-Adaptive MDP (for Information State Space MDP) where we replace $\mathbb{P}[\mathcal{R}|H_t]$ in the case of Bandits with $\mathbb{P}[\mathcal{P}_R|H_t]$ in the case of RL Control. Some of these algorithms sample from the estimated $\mathcal{P}_R$, and learn the Optimal Value Function/Optimal Policy from the samples. Some other algorithms are Planning-oriented. Specifically, the Planning-oriented approach is to run a Planning method (e.g., Policy Iteration, Value Iteration) using the estimated $\mathcal{P}_R$, then generate more data using the Optimal Policy (produced by the Planning method), use the generated data to improve the $\mathcal{P}_R$ estimate, then run the Planning method again to come up with the Optimal Policy (for the MDP based on the improved $\mathcal{P}_R$ estimate), and loop on in this manner until convergence. As an example of this Planning-oriented approach, we refer you to the paper on RMax Algorithm (Brafman and Tennenholtz 2001) to learn more.

15.10   KEY TAKEAWAYS FROM THIS CHAPTER

- The Multi-Armed Bandit problem provides a simple setting to understand and appreciate the nuances of the Explore-Exploit dilemma that we typically need to resolve within RL Control algorithms.
- In this chapter, we covered the following broad approaches to resolve the Explore-Exploit dilemma:

- Naive Exploration, e.g., ϵ-greedy
- Optimistic Initialization
- Optimism in the Face of Uncertainty, e.g., UCB, Bayesian UCB
- Probability Matching, e.g., Thompson Sampling
- Gradient Bandit Algorithms
- Information State Space MDPs (incorporating value of Information), typically solved by treating as Bayes-Adaptive MDPs

• The above MAB algorithms are well-extensible to Contextual Bandits and RL Control.

CHAPTER 16

Blending Learning and Planning

After coverage of the issue of Exploration versus Exploitation in the last chapter, in this chapter, we cover the topic of Planning versus Learning (and how to blend the two approaches) in the context of solving MDP Prediction and Control problems. In this chapter, we also provide some coverage of the much-celebrated Monte-Carlo Tree-Search (abbreviated as MCTS) algorithm and it's spiritual origin—the Adaptive Multi-Stage Sampling (abbreviated as AMS) algorithm. MCTS and AMS are examples of Planning algorithms tackled with sampling/RL-based techniques.

16.1 PLANNING VERSUS LEARNING

In the language of AI, we use the terms *Planning* and *Learning* to refer to two different approaches to solve an AI problem. Let us understand these terms from the perspective of solving MDP Prediction and Control. Let us zoom out and look at the big picture. The AI Agent has access to an MDP Environment E. In the process of *interacting* with the MDP Environment E, the AI Agent receives experiences data. Each unit of experience data is in the form of a (next state, reward) pair for the current state and action. The AI Agent's goal is to estimate the requisite Value Function/Policy through this process of interaction with the MDP Environment E (for Prediction, the AI Agent estimates the Value Function for a given policy and for Control, the AI Agent estimates the Optimal Value Function and the Optimal Policy). The AI Agent can go about this in one of two ways:

1. By interacting with the MDP Environment E, the AI Agent can build a *Model of the Environment* (call it M) and then use that model to estimate the requisite Value Function/Policy. We refer to this as the *Model-Based* approach. Solving Prediction/Control using a Model of the Environment (i.e., *Model-Based* approach) is known as *Planning* the solution. The term *Planning* comes from the fact that the AI Agent projects (with the help of the model M) probabilistic scenarios of future states/rewards for various choices of actions from specific states, and solves for the requisite Value Function/Policy based on the model-projected future outcomes.

2. By interacting with the MDP Environment E, the AI Agent can directly estimate the requisite Value Function/Policy, without bothering to build a Model of the Environment. We refer to this as the *Model-Free* approach. Solving Prediction/Control without using a model (i.e., *Model-Free* approach) is known as *Learning* the solution. The term *Learning* comes from the fact that the AI Agent "learns" the requisite Value

Function/Policy directly from experiences data obtained by interacting with the MDP Environment E (without requiring any model).

Let us now dive a bit deeper into both these approaches to understand them better.

16.1.1 Planning the Solution of Prediction/Control

In the first approach (*Planning* the solution of Prediction/Control), we first need to "build a model". By "model", we refer to the State-Reward Transition Probability Function $\mathcal{P}_R$. By "building a model", we mean estimating $\mathcal{P}_R$ from experiences data obtained by interacting with the MDP Environment E. How does the AI Agent do this? Well, this is a matter of estimating the conditional probability density function of pairs of (next state, reward), conditioned on a particular pair of (state, action). This is an exercise in Supervised Learning, where the y-values are (next state, reward) pairs and the x-values are (state, action) pairs. We covered how to do Supervised Learning in Chapter 6. Also, note that Equation (11.12) in Chapter 11 provides a simple tabular calculation to estimate the $\mathcal{P}_R$ function for an MRP from a fixed, finite set of atomic experiences of (state, reward, next state) triples. Following this Equation, we had written the function `finite_mrp` to construct a `FiniteMarkovRewardProcess` (which includes a tabular $\mathcal{P}_R$ function of explicit probabilities of transitions), given as input a `Sequence[TransitionStep[S]]` (i.e., fixed, finite set of MRP atomic experiences). This approach can be easily extended to estimate the $\mathcal{P}_R$ function for an MDP. Ok—now we have a model M in the form of an estimated $\mathcal{P}_R$. The next thing to do in this approach of *Planning* the solution of Prediction/Control is to use the model M to estimate the requisite Value Function/Policy. There are two broad approaches to do this:

1. By constructing $\mathcal{P}_R$ as an explicit representation of probabilities of transitions, the AI Agent can utilize one of the Dynamic Programming Algorithms (e.g., Policy Evaluation, Policy Iteration, Value Iteration) or a Tree-search method (by growing out a tree of future states/rewards/actions from a given state/action, e.g., the MCTS/AMS algorithms we will cover later in this chapter). Note that in this approach, there is *no need to interact with an MDP Environment* since a model of transition probabilities are available that can be used to project any (probabilistic) future outcome (for any choice of action) that is desired to estimate the requisite Value Function/Policy.

2. By treating $\mathcal{P}_R$ as a *sampling model*, by which we mean that the AI agent uses $\mathcal{P}_R$ as simply an (on-demand) interface to sample an individual pair of (next state, reward) from a given (state, action) pair. This means the AI Agent treats this *sampling model* view of $\mathcal{P}_R$ as a *Simulated MDP Environment* (let us refer to this Simulated MDP Environment as S). Note that S serves as a proxy interaction-interface to the real MDP Environment E. A significant advantage of interacting with S instead of E is that we can sample infinitely many times without any of the real-world interaction constraints that a real MDP Environment E poses. Think about a robot learning to walk on an actual street versus learning to walk on a simulator of the street's activities. Furthermore, the user could augment his/her views on top of an experiences-data-learnt simulator. For example, the user might say that the experiences data obtained by interacting with E doesn't include certain types of scenarios but the user might have knowledge of how those scenarios would play out, thus creating a "human-knowledge-augmented simulator" (more on this in Chapter 17). By interacting with the simulated MDP Environment S (instead of the real MDP Environment E), the AI Agent can use any of the RL Algorithms we covered in Module III of this book to estimate the requisite Value Function/Policy. Since this approach uses a model M (albeit a sampling model) and since this approach uses RL, we refer to this approach as *Model-Based RL*. To summarize this approach, the AI Agent first learns (supervised

learning) a model M as an approximation of the real MDP Environment E, and then the AI Agent plans the solution to Prediction/Control by using the model M in the form of a simulated MDP Environment S which an RL algorithm interacts with. Here the Planning/Learning terminology often gets confusing to new students of this topic since this approach is supervised learning followed by planning (the planning being done with a Reinforcement Learning algorithm interacting with the learnt simulator).

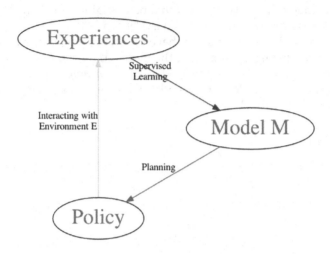

Figure 16.1 **Planning with a Supervised-Learnt Model**

Figure 16.1 depicts the above-described approach of *Planning* the solution of Prediction/Control. We start with an arbitrary Policy that is used to interact with the Environment E (upward-pointing arrow in the figure). These interactions generate Experiences, which are used to perform Supervised Learning (rightward-pointing arrow in the figure) to learn a model M. This model M is used to plan the requisite Value Function/Policy (leftward-pointing arrow in the figure). The Policy produced through this process of Planning is then used to further interact with the Environment E, which in turn generates a fresh set of Experiences, which in turn are used to update the Model M (incremental supervised learning), which in turn is used to plan an updated Value Function/Policy, and so the cycle repeats.

16.1.2 Learning the Solution of Prediction/Control

In the second approach (*Learning* the solution of Prediction/Control), we don't bother to build a model. Rather, the AI Agent directly estimates the requisite Value Function/Policy from the experiences data obtained by interacting with the real MDP Environment E. The AI Agent does this by using any of the RL algorithms we covered in Module III of this book. Since this approach is "model-free", we refer to this approach as *Model-Free RL*.

16.1.3 Advantages and Disadvantages of Planning versus Learning

In the previous two subsections, we covered the two different approaches to solving Prediction/Control, either by *Planning* (subsection 16.1.1) or by *Learning* (subsection 16.1.2). Let us now talk about their advantages and disadvantages.

Planning involves constructing a Model, so its natural advantage is to be able to construct a model (from experiences data) with efficient and robust supervised learning methods.

The other key advantage of *Planning* is that we can reason about Model Uncertainty. Specifically, when we learn the Model M using supervised learning, we typically obtain the standard errors for estimation of model parameters, which can then be used to create confidence intervals for the Value Function and Policy planned using the model. Furthermore, since modeling real-world problems tends to be rather difficult, it is valuable to create a family of models with differing assumptions, with different functional forms, with differing parameterizations etc., and reason about how the Value Function/Policy would disperse as a function of this range of models. This is quite beneficial in typical real-world problems since it enables us to do Prediction/Control in a *robust* manner.

The disadvantage of *Planning* is that we have two sources of approximation error—the first from supervised learning in estimating the model M, and the second from constructing the Value Function/Policy (given the model). The *Learning* approach (without resorting to a model, i.e., Model-Free RL) is thus advantageous is not having the first source of approximation error (i.e., Model Error).

16.1.4 Blending Planning and Learning

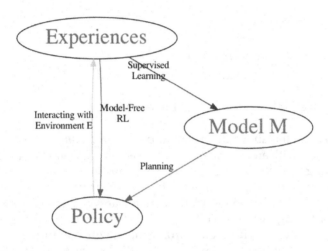

Figure 16.2 **Blending Planning and Learning**

In this subsection, we show a rather creative and practically powerful approach to solve real-world Prediction and Control problems. We basically extend Figure 16.1 to Figure 16.2. As you can see in Figure 16.2, the change is that there is a downward-pointing arrow from the *Experiences* node to the *Policy* node. This downward-pointing arrow refers to *Model-Free Reinforcement Learning*, i.e., learning the Value Function/Policy directly from experiences obtained by interacting with Environment E, i.e., Model-Free RL. This means we obtain the requisite Value Function/Policy through the collaborative approach of *Planning* (using the model M) and *Learning* (using Model-Free RL).

Note that when Planning is based on RL using experiences obtained by interacting with the Simulated Environment S (based on Model M), then we obtain the requisite Value Function/Policy from two sources of experiences (from E and S) that are combined and provided to an RL Algorithm. This means we simultaneously do Model-Based RL and Model-Free RL. This is creative and powerful because it blends the best of both worlds—Planning (with Model-Based RL) and Learning (with Model-Free RL). Apart from Model-Free RL and Model-Based RL being blended here to obtain a more accurate Value Function/Policy, the Model is simultaneously being updated with incremental supervised learning (rightward-pointing arrow in Figure 16.2) as new experiences are being generated as a result of the Policy interacting with the Environment E (upward-pointing arrow in Figure 16.2).

This framework of blending Planning and Learning was created by Richard Sutton which he named as Dyna (Richard S. Sutton 1991).

16.2   DECISION-TIME PLANNING

In the next two sections of this chapter, we cover a couple of Planning methods that are sampling-based (experiences obtained by interacting with a sampling model) and use RL techniques to solve for the requisite Value Function/Policy from the model-sampled experiences. We cover the famous Monte-Carlo Tree-Search (MCTS) algorithm, followed by an algorithm which is MCTS' spiritual origin—the Adaptive Multi-Stage Sampling (AMS) algorithm.

Both these algorithms are examples of *Decision-Time Planning*. The term *Decision-Time Planning* requires some explanation. When it comes to Planning (with a model), there are two possibilities:

- Background Planning: This refers to a planning method where the AI Agent precomputes the requisite Value Function/Policy *for all states*, and when it is time for the AI Agent to perform the requisite action for a given state, it simply has to refer to the pre-calculated policy and apply that policy to the given state. Essentially, in the *background*, the AI Agent is constantly improving the requisite Value Function/Policy, irrespective of which state the AI Agent is currently required to act on. Hence, the term *Background Planning*.
- Decision-Time Planning: This approach contrasts with Background Planning. In this approach, when the AI Agent has to identify the best action to take for a specific state that the AI Agent currently encounters, the calculations for that best-action-identification happens only when the AI Agent *reaches that state*. This is appropriate in situations when there are such a large number of states in the state space that Background Planning is infeasible. However, for Decision-Time Planning to be effective, the AI Agent needs to have sufficient time to be able to perform the calculations to identify the action to take *upon reaching a given state*. This is feasible in games like Chess where there is indeed some time for the AI Agent to make its move upon encountering a specific state of the chessboard (the move response doesn't need to be immediate). However, this is not feasible for a self-driving car, where the decision to accelerate/brake or to steer must be immediate (this requires *Background Planning*).

Hence, with Decision-Time Planning, the AI Agent focuses all of the available computation and memory resources for the sole purpose of identifying the best action for *a particular state* (the state that has just been reached by the AI Agent). Decision-Time Planning is typically successful because of this focus on a single state and consequently, on the states that are most likely to be reached within the next few time steps (essentially, avoiding any wasteful computation on states that are unlikely to be reached from the given state).

Decision-Time Planning typically looks much deeper than just a single time step ahead (DP algorithms only look a single time step ahead) and evaluates action choices leading to many different state and reward possibilities over the next several time steps. Searching deeper than a single time step ahead is required because these Decision-Time Planning algorithms typically work with imperfect Q-Values.

Decision-Time Planning methods sometimes go by the name *Heuristic Search*. Heuristic Search refers to the method of growing out a tree of future states/actions/rewards from the given state (which serves as the root of the tree). In classical Heuristic Search, an approximate Value Function is calculated at the leaves of the tree and the Value Function is then backed up to the root of the tree. Knowing the backed-up Q-Values at the root of the tree enables the calculation of the best action for the root state. Modern methods of Heuristic Search are very efficient in how the Value Function is approximated and backed up. Monte-Carlo Tree-Search (MCTS) in one such efficient method that we cover in the next section.

16.3 MONTE-CARLO TREE-SEARCH (MCTS)

Monte-Carlo Tree-Search (abbreviated as MCTS) is a Heuristic Search method that involves growing out a Search Tree from the state for which we seek the best action (hence, it is a Decision-Time Planning algorithm). MCTS was popularized in 2016 by Deep Mind's AlphaGo algorithm (Silver et al. 2016). MCTS was first introduced by Remi Coulom for game trees (Coulom 2006).

For every state in the Search Tree, we maintain the Q-Values for all actions from that state. The basic idea is to form several sampling traces from the root of the tree (i.e., from the given state) to terminal states. Each such sampling trace threads through the Search Tree to a leaf node of the tree, and then extends beyond the tree from the leaf node to a terminal state. This separation of the two pieces of each sampling trace is important—the first piece within the tree, and the second piece outside the tree. Of particular importance is the fact that the first piece of various sampling traces will pass through states (within the Search Tree) that are quite likely to be reached from the given state (at the root node). MCTS benefits from states in the tree being revisited several times, as it enables more accurate Q-Values for those states (and consequently, a more accurate Q-Value at the root of the tree, from backing-up of Q-Values). Moreover, these sampling traces prioritize actions with good Q-Values. Prioritizing actions with good Q-Values has to be balanced against actions that haven't been tried sufficiently, and this is essentially the explore-exploit tradeoff that we covered in detail in Chapter 15.

Each sampling trace round of MCTS consists of four steps:

- Selection: Starting from the root node R (given state), we successively select children nodes all the way until a leaf node L. This involves selecting actions based on a *tree policy*, and selecting next states by sampling from the model of state transitions. The trees in Figure 16.3 show states colored as white and actions colored as gray. This figure shows the Q-Values for a 2-player game (e.g., Chess) where the reward is 1 at termination for a win, 0 at termination for a loss, and 0 throughout the time the game is in play. So the Q-Values in the figure are displayed at each node in the form of Wins as a fractions of Games Played that passed through the node (Games through a node means the number of sampling traces that have run through the node). So the label "1/6" for one of the State nodes (under "Selection", the first image in the figure) means that we've had 6 sampling traces from the root node that have passed through this State node labeled "1/6", and 1 of those games was won by us. For Actions nodes (gray nodes), the labels correspond to *Opponent Wins* as a fraction of Games through the Action node. So the label "2/3" for one of the Action leaf nodes means that we've had 3 sampling traces from the root node that have passed through this Action leaf node, and 2 of those resulted in wins *for the opponent* (i.e., 1 win for us).
- Expansion: On some rounds, the tree is expanded from L by adding a child node C to it. In the figure, we see that L is the Action leaf node labeled as "3/3" and we add a child node C (state) to it labeled "0/0" (because we don't yet have any sampling traces running through this added state C).

- Simulation: From L (or from C if this round involved adding C), we complete the sampling trace (that started from R and ran through L) all the way to a terminal state T. This entire sampling trace from R to T is known as a single Monte-Carlo Simulation Trace, in which actions are selected according to the *tree policy* when within the tree, and according to a *rollout policy* beyond the tree (the term "rollout" refers to "rolling out" a simulation from the leaf node to termination). The tree policy is based on an Explore-Exploit tradeoff using estimated Q-Values, and the rollout policy is typically a simple policy such as a uniform policy.
- Backpropagation: The return generated by the sampling trace is backed up ("back-propagated") to update (or initialize) the Q-Values at the nodes in the tree that are part of the sampling trace. Note that in the figure, the rolled out simulation trace resulted in a win for the opponent (loss for us). So the backed up Q-Values reflect an extra win for the opponent (on the gray nodes, i.e., action nodes) and an extra loss for us (on the white nodes, i.e., state nodes).

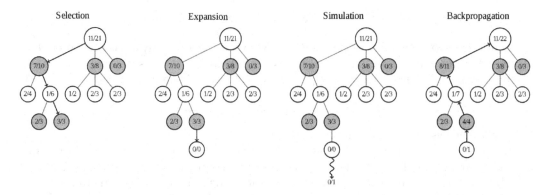

Figure 16.3 Monte-Carlo Tree-Search (This wikipedia image is being used under the creative commons license CC BY-SA 4.0)

The Selection Step in MCTS involves picking a child node (action) with "most promise", for each state in the sampling trace of the Selection Step. This means prioritizing actions with higher Q-Value estimates. However, this needs to be balanced against actions that haven't been tried sufficiently (i.e., those actions whose Q-Value estimates have considerable uncertainty). This is our usual *Explore v/s Exploit* tradeoff that we covered in detail in Chapter 15. The Explore v/s Exploit formula for games was first provided by Kocsis and Szepesvari (Kocsis and Szepesvári 2006). This formula is known as *Upper Confidence Bound 1 for Trees* (abbreviated as UCT). Most current MCTS Algorithms are based on some variant of UCT. UCT is based on the UCB1 formula of Auer, Cesa-Bianchi, Fischer (Auer, Cesa-Bianchi, and Fischer 2002).

16.4   ADAPTIVE MULTI-STAGE SAMPLING

Its not well known that MCTS and UCT concepts first appeared in the Adaptive Multi-Stage Sampling algorithm by Chang, Fu, Hu, Marcus (Chang et al. 2005). Adaptive Multi-Stage Sampling (abbreviated as AMS) is a generic sampling-based algorithm to solve finite-horizon Markov Decision Processes (although the paper describes how to extend this algorithm for infinite-horizon MDPs). We consider AMS to be the "spiritual origin" of MCTS/UCT, and hence we dedicate this section to coverage of AMS.

AMS is a planning algorithm—a sampling model is provided for the next state (conditional on given state and action), and a model of Expected Reward (conditional on given state and action) is also provided. AMS overcomes the curse of dimensionality by sampling the next state. The key idea in AMS is to adaptively select actions based on a suitable tradeoff between Exploration and Exploitation. AMS was the first algorithm to apply the theory of Multi-Armed Bandits to derive a provably convergent algorithm for solving finite-horizon MDPs. Moreover, it performs far better than the typical backward-induction approach to solving finite-horizon MDPs, in cases where the state space is very large and the action space is fairly small.

We use the same notation we used in Section 5.13 of Chapter 5 for Finite-Horizon MDPs (time steps $t = 0, 1, \ldots T$). We assume that the state space $\mathcal{S}_t$ for time step t is very large for all $t = 0, 1, \ldots, T - 1$ (the state space $\mathcal{S}_T$ for time step T consists of all terminal states). We assume that the action space $\mathcal{A}_t$ for time step t is fairly small for all $t = 0, 1, \ldots, T - 1$. We denote the probability distribution for the next state, conditional on the current state and action (for time step t) as the function $\mathcal{P}_t : (\mathcal{S}_t \times \mathcal{A}_t) \rightarrow (\mathcal{S}_{t+1} \rightarrow [0, 1])$, defined as:

$$\mathcal{P}_t(s_t, a_t)(s_{t+1}) = \mathbb{P}[S_{t+1} = s_{t+1} | (S_t = s_t, A_t = a_t)]$$

As mentioned above, for all $t = 0, 1, \ldots, T - 1$, AMS has access to only a sampling model of $\mathcal{P}_t$, that can be used to fetch a sample of the next state from $\mathcal{S}_{t+1}$. We also assume that we are given the Expected Reward function $\mathcal{R}_t : \mathcal{S}_t \times \mathcal{A}_t \rightarrow \mathbb{R}$ for each time step $t = 0, 1, \ldots, T - 1$ defined as:

$$\mathcal{R}_t(s_t, a_t) = \mathbb{E}[R_{t+1} | (S_t = s_t, A_t = a_t)]$$

We denote the Discount Factor as γ.

The problem is to calculate an approximation to the Optimal Value Function $V_t^*(s_t)$ for all $s_t \in \mathcal{S}_t$ for all $t = 0, 1, \ldots, T - 1$. Using only samples from the state-transition probability distribution functions $\mathcal{P}_t$ and the Expected Reward functions $\mathcal{R}_t$, AMS aims to do better than backward induction for the case where $\mathcal{S}_t$ is very large and $\mathcal{A}_t$ is small for all $t = 0, 1, \ldots T - 1$.

The AMS algorithm is based on a fixed allocation of the number of action selections for each state at each time step. Denote the number of action selections for each state at time step t as N_t. We ensure that each action $a_t \in \mathcal{A}_t$ is selected at least once, hence $N_t \geq |\mathcal{A}_t|$. While the algorithm is running, we denote $N_t^{s_t, a_t}$ to be the number of selections of a particular action a_t (for a given state s_t) *until that point in the algorithm*.

Denote $\hat{V}_t^{N_t}(s_t)$ as the AMS Algorithm's approximation of $V_t^*(s_t)$, utilizing all of the N_t action selections. For a given state s_t, for each selection of an action a_t, *one* next state is sampled from the probability distribution $\mathcal{P}_t(s_t, a_t)$ (over the state space $\mathcal{S}_{t+1}$). For a fixed s_t and fixed a_t, let us denote the j-th sample of the next state (for $j = 1, \ldots, N_t^{s_t, a_t}$) as $s_{t+1}^{(s_t, a_t, j)}$. Each such next state sample $s_{t+1}^{(s_t, a_t, j)} \sim \mathcal{P}_t(s_t, a_t)$ leads to a recursive call to $\hat{V}_{t+1}^{N_{t+1}}(s_{t+1}^{(s_t, a_t, j)})$ in order to calculate the approximation $\hat{Q}_t(s_t, a_t)$ of the Optimal Action Value Function $Q_t^*(s_t, a_t)$ as:

$$\hat{Q}_t(s_t, a_t) = \mathcal{R}_t(s_t, a_t) + \gamma \cdot \frac{\sum_{j=1}^{N_t^{s_t, a_t}} \hat{V}_{t+1}^{N_{t+1}}(s_{t+1}^{(s_t, a_t, j)})}{N_t^{s_t, a_t}}$$

Now let us understand how the N_t action selections are done for a given state s_t. First, we select each of the actions in $\mathcal{A}_t$ exactly once. This is a total of $|\mathcal{A}_t|$ action selections. Each of the remaining $N_t - |\mathcal{A}_t|$ action selections (indexed as i ranging from $|\mathcal{A}_t|$ to $N_t - 1$) is made based on $|\mathcal{A}_t|$ that maximizes the following UCT formula (thus balancing exploration and exploitation):

$$\hat{Q}_t(s_t, a_t) + \sqrt{\frac{2 \log i}{N_t^{s_t, a_t}}} \tag{16.1}$$

When all N_t action selections are made for a given state s_t, $V_t^*(s_t) = \max_{a_t \in \mathcal{A}_t} Q_t^*(s_t, a_t)$ is approximated as:

$$\hat{V}_t^{N_t}(s_t) = \sum_{a_t \in \mathcal{A}_t} \frac{N_t^{s_t, a_t}}{N_t} \cdot \hat{Q}_t(s_t, a_t) \tag{16.2}$$

Now let's write a Python class to implement AMS. We start by writing its constructor. For convenience, we assume each of the state spaces $\mathcal{S}_t$ (for $t = 0, 1, \ldots, T$) is the same (denoted as $\mathcal{S}$) and the allowable actions are the same across all time steps (denoted as $\mathcal{A}$).

```
from rl.distribution import Distribution
A = TypeVar('A')
S = TypeVar('S')
class AMS(Generic[S, A]):
    def __init__(
        self,
        actions_funcs: Sequence[Callable[[S], Set[A]]],
        state_distr_funcs: Sequence[Callable[[S, A], Distribution[S]]],
        expected_reward_funcs: Sequence[Callable[[S, A], float]],
        num_samples: Sequence[int],
        gamma: float
    ) -> None:
        self.num_steps: int = len(actions_funcs)
        self.actions_funcs: Sequence[Callable[[S], Set[A]]] = \
            actions_funcs
        self.state_distr_funcs: Sequence[Callable[[S, A], Distribution[S]]] = \
            state_distr_funcs
        self.expected_reward_funcs: Sequence[Callable[[S, A], float]] = \
            expected_reward_funcs
        self.num_samples: Sequence[int] = num_samples
        self.gamma: float = gamma
```

Let us understand the inputs to the constructor __init__.

- actions_funcs consists of a Sequence (for all of $t = 0, 1, \ldots, T - 1$) of functions, each mapping a state in $\mathcal{S}_t$ to a set of actions within $\mathcal{A}$, that we denote as $\mathcal{A}_t(s_t)$ (i.e., Callable[[S], Set[A]]).
- state_distr_funcs represents $\mathcal{P}_t$ for all $t = 0, 1, \ldots, T - 1$ (accessible only as a sampling model since the return type of each Callable is Distribution).
- expected_reward_funcs represents $\mathcal{R}_t$ for all $t = 0, 1, \ldots, T - 1$.
- num_samples represents N_t (the number of actions selections for each state s_t) for all $t = 0, 1, \ldots, T - 1$.
- gamma represents the discount factor γ.

self.num_steps represents the number of time steps T.

Next, we write the method optimal_vf_and_policy to compute $\hat{V}_t^{N_t}(s_t)$ and the associated recommended action for state s_t (note the type of the output, representing this pair as Tuple[float, A]).

In the code below, vals_sum builds up the sum $\sum_{j=1}^{N_t^{s_t, a_t}} \hat{V}_{t+1}^{N_{t+1}}(s_{t+1}^{(s_t, a_t, j)})$, and counts represents $N_t^{s_t, a_t}$. Before the for loop, we initialize vals_sum by selecting each action $a_t \in \mathcal{A}_t(s_t)$ exactly once. Then, for each iteration i of the for loop (for i ranging from $|\mathcal{A}_t(s_t)|$ to $N_t - 1$), we calculate the Upper-Confidence Value (ucb_vals in the code below) for each of the actions $a_t \in \mathcal{A}_t(s_t)$ using the UCT formula of Equation (16.1), and pick an action a_t^* that maximizes ucb_vals. After the termination of the for loop, optimal_vf_and_policy returns the Optimal Value Function approximation for s_t based on Equation (16.2) and the recommended action for s_t as the action that maximizes $\hat{Q}_t(s_t, a_t)$

```python
import numpy as np
from operator import itemgetter
    def optimal_vf_and_policy(self, t: int, s: S) -> \
            Tuple[float, A]:
        actions: Set[A] = self.actions_funcs[t](s)
        state_distr_func: Callable[[S, A], Distribution[S]] = \
            self.state_distr_funcs[t]
        expected_reward_func: Callable[[S, A], float] = \
            self.expected_reward_funcs[t]
        rewards: Mapping[A, float] = {a: expected_reward_func(s, a)
                                for a in actions}
        val_sums: Dict[A, float] = {a: (self.optimal_vf_and_policy(
            t + 1,
            state_distr_func(s, a).sample()
        )[0] if t < self.num_steps - 1 else 0.) for a in actions}
        counts: Dict[A, int] = {a: 1 for a in actions}
        for i in range(len(actions), self.num_samples[t]):
            ucb_vals: Mapping[A, float] = \
                {a: rewards[a] + self.gamma * val_sums[a] / counts[a] +
                    np.sqrt(2 * np.log(i) / counts[a]) for a in actions}
            max_actions: Sequence[A] = [a for a, u in ucb_vals.items()
                                if u == max(ucb_vals.values())]
            a_star: A = np.random.default_rng().choice(max_actions)
            val_sums[a_star] += (self.optimal_vf_and_policy(
                t + 1,
                state_distr_func(s, a_star).sample()
            )[0] if t < self.num_steps - 1 else 0.)
            counts[a_star] += 1
        return (
            sum(counts[a] / self.num_samples[t] *
                (rewards[a] + self.gamma * val_sums[a] / counts[a])
                for a in actions),
            max(
                [(a, rewards[a] + self.gamma * val_sums[a] / counts[a])
                    for a in actions],
                key=itemgetter(1)
            )[0]
        )
```

The above code is in the file rl/chapter15/ams.py. The __main__ in this file tests the AMS algorithm for the simple case of the Dynamic Pricing problem that we had covered in Section 5.14 of Chapter 5, although the Dynamic Pricing problem itself is not a problem where AMS would do better than backward induction (since its state space is not very large). We encourage you to play with our implementation of AMS by constructing a finite-horizon MDP with a large state space (and small-enough action space). An example of such a problem is Optimal Stopping (in particular, pricing of American Options) that we had covered in Chapter 9.

Now let's analyze the running-time complexity of AMS. Let $N = \max(N_0, N_1, \ldots, N_{T-1})$. At each time step t, the algorithm makes at most N recursive calls, and so the running-time complexity is $O(N^T)$. Note that since we need to select every action at least once for every state at every time step, $N \geq |\mathcal{A}|$, meaning the running-time complexity is at least $|\mathcal{A}|^T$. Compare this against the running-time complexity of backward induction, which is $O(|\mathcal{S}|^2 \cdot |\mathcal{A}| \cdot T)$. So, AMS is more efficient when $\mathcal{S}$ is very large (which is typical in many real-world problems). In their paper, Chang, Fu, Hu, Marcus proved that the Value Function approximation $\hat{V}_0^{N_0}$ is asymptotically unbiased, i.e.,

$$\lim_{N_0 \to \infty} \lim_{N_1 \to \infty} \cdots \lim_{N_{T-1} \to \infty} \mathbb{E}[\hat{V}_0^{N_0}(s_0)] = V_0^*(s_0) \text{ for all } s_0 \in \mathcal{S}$$

They also proved that the worst-possible bias is bounded by a quantity that converges to zero at the rate of $O(\sum_{t=0}^{T-1} \frac{\ln N_t}{N_t})$. Specifically,

$$0 \leq V_0^*(s_0) - \mathbb{E}[\hat{V}_0^{N_0}(s_0)] \leq O(\sum_{t=0}^{T-1} \frac{\ln N_t}{N_t}) \text{ for all } s_0 \in \mathcal{S}$$

16.5   SUMMARY OF KEY LEARNINGS FROM THIS CHAPTER

- Planning versus Learning, and how to blend Planning and Learning.
- Monte-Carlo Tree-Search (MCTS): An example of a Planning algorithm based on Tree-Search and based on sampling/RL techniques.
- Adaptive Multi-Stage Sampling (AMS): The spiritual origin of MCTS—it is an efficient algorithm for finite-horizon MDPs with very large state space and fairly small action space.

CHAPTER 17

Summary and Real-World Considerations

The purpose of this chapter is two-fold: Firstly to summarize the key learnings from this book, and secondly to provide some commentary on how to take the learnings from this book into practice (to solve real-world problems). On the latter, we specifically focus on the challenges one faces in the real-world—modeling difficulties, problem-size difficulties, operational challenges, data challenges (access, cleaning, organization), product management challenges (e.g., addressing the gap between the technical problem being solved and the business problem to be solved), and also change-management challenges as one shifts an enterprise from legacy systems to an AI system.

17.1 SUMMARY OF KEY LEARNINGS FROM THIS BOOK

In Module I, we covered the Markov Decision Process framework, the Bellman Equations, Dynamic Programming algorithms, Function Approximation, and Approximate Dynamic Programming.

Module I started with Chapter 3, where we first introduced the very important *Markov Property*, a concept that enables us to reason effectively and compute efficiently in practical systems involving sequential uncertainty. Such systems are best approached through the very simple framework of *Markov Processes*, involving probabilistic state transitions. Next, we developed the framework of *Markov Reward Processes* (MRP), the *MRP Value Function*, and the *MRP Bellman Equation*, which expresses the MRP Value Function recursively. We showed how this MRP Bellman Equation can be solved with simple linear-algebraic-calculations when the state space is finite and not too large.

In Chapter 4, we developed the framework of *Markov Decision Processes* (MDP). A key learning from this chapter is that an MDP evaluated with a fixed Policy is equivalent to an MRP. Calculating the Value Function of an MDP evaluated with a fixed Policy (i.e. calculating the Value Function of an MRP) is known as the *Prediction* problem. We developed the 4 forms of the MDP Bellman Policy Equations (which are essentially equivalent to the MRP Bellman Equation). Next, we defined the *Control* problem as the calculation of the Optimal Value Function (and an associated Optimal Policy) of an MDP. Correspondingly, we developed the 4 forms of the MDP Bellman Optimality Equation. We stated and proved an important theorem on the existence of an Optimal Policy, and of each Optimal Policy achieving the Optimal Value Function. We finished this chapter with some commentary on variants and extensions of MDPs. Here we introduced the two Curses in the context of solving MDP Prediction and Control—the Curse of Dimensionality and the Curse of

DOI: 10.1201/9781003229193-17

Modeling, which can be battled with appropriate approximation of the Value Function and with appropriate sampling from the state-reward transition probability function. Here, we also covered Partially-Observable Markov Decision Processes (POMDP), which refers to situations where all components of the *State* are not observable (quite typical in the real-world). Often, we pretend a POMDP is an MDP as the MDP framework fetches us computational tractability. Modeling a POMDP as an MDP is indeed a very challenging endeavor in the real-world. However, sometimes partial state-observability cannot be ignored, and in such situations, we have to employ (computationally expensive) algorithms to solve the POMDP.

In Chapter 5, we first covered the foundation of the classical Dynamic Programming (DP) algorithms—the Banach Fixed-Point Theorem, which gives us a simple method for iteratively solving for a fixed-point of a contraction function. Next, we constructed the Bellman Policy Operator and showed that it's a contraction function, meaning we can take advantage of the Banach Fixed-Point Theorem, yielding a DP algorithm to solve the Prediction problem, referred to as the Policy Evaluation algorithm. Next, we introduced the notions of a Greedy Policy and Policy Improvement, which yields a DP algorithm known as Policy Iteration to solve the Control problem. Next, we constructed the Bellman Optimality Operator and showed that it's a contraction function, meaning we can take advantage of the Banach Fixed-Point Theorem, yielding a DP algorithm to solve the Control problem, referred to as the Value Iteration algorithm. Next, we introduced the all-important concept of *Generalized Policy Iteration* (GPI)—the powerful idea of alternating between *any* method for Policy Evaluation and *any* method for Policy Improvement, including methods that are partial applications of Policy Evaluation or Policy Improvement. This generalized perspective unifies almost all of the algorithms that solve MDP Control problems (including Reinforcement Learning algorithms). We finished this chapter with coverage of Backward Induction algorithms to solve Prediction and Control problems for finite-horizon MDPs—Backward Induction is a simple technique to backpropagate the Value Function from horizon-end to the start. It is important to note that the DP algorithms in this chapter apply to MDPs with a finite number of states and that these algorithms are computationally feasible only if the state space is not too large (the next chapter extends these DP algorithms to handle large state spaces, including infinite state spaces).

In Chapter 6, we first covered a refresher on Function Approximation by developing the calculations first for linear function approximation and then for feed-forward fully-connected deep neural networks. We also explained that a Tabular prediction can be viewed as a special form of function approximation (since it satisfies the interface we designed for Function Approximation). With this apparatus for Function Approximation, we extended the DP algorithms of the previous chapter to Approximate Dynamic Programming (ADP) algorithms in a rather straightforward manner. In fact, DP algorithms can be viewed as special cases of ADP algorithms by setting the function approximation to be Tabular. Essentially, we replace tabular Value Function updates with updates to Function Approximation parameters (where the Function Approximation represents the Value Function). The sweeps over all states in the tabular (DP) algorithms are replaced by sampling states in the ADP algorithms, and expectation calculations in Bellman Operators are handled in ADP as averages of the corresponding calculations over transition samples (versus calculations using explicit transition probabilities in the DP algorithms).

Module II was about Modeling Financial Applications as MDPs. We started Module II with a basic coverage of Utility Theory in Chapter 7. The concept of Utility is vital since Utility of cashflows is the appropriate *Reward* in the MDP for many financial applications. In this chapter, we explained that an individual's financial risk-aversion is represented by the concave nature of the individual's Utility as a function of financial outcomes. We showed that the Risk-Premium (compensation an individual seeks for taking financial risk) is roughly proportional to the individual's financial risk-aversion and also proportional to the measure of uncertainty in financial outcomes. Risk-Adjusted-Return in finance should be thought of as the Certainty-Equivalent-Value, whose Utility is the Expected Utility across uncertain (risky) financial outcomes. We finished this chapter by covering the Constant Absolute Risk-Aversion (CARA) and the Constant Relative Risk-Aversion (CRRA) Utility functions, along with simple asset allocation examples for each of CARA and CRRA Utility functions.

In Chapter 8, we covered the problem of Dynamic Asset-Allocation and Consumption. This is a fundamental problem in Mathematical Finance of jointly deciding on A) optimal investment allocation (among risky and riskless investment assets) and B) optimal consumption, over a finite horizon. We first covered Merton's landmark paper from 1969 that provided an elegant closed-form solution under assumptions of continuous-time, normal distribution of returns on the assets, CRRA utility, and frictionless transactions. In a more general setting of this problem, we need to model it as an MDP. If the MDP is not too large and if the asset return distributions are known, we can employ finite-horizon ADP algorithms to solve it. However, in typical real-world situations, the action space can be quite large and the asset return distributions are unknown. This points to RL, and specifically RL algorithms that are well suited to tackle large action spaces (such as Policy Gradient Algorithms).

In Chapter 9, we covered the problem of pricing and hedging of derivative securities. We started with the fundamental concepts of Arbitrage, Market-Completeness and Risk-Neutral Probability Measure. Based on these concepts, we stated and proved the two fundamental theorems of Asset Pricing for the simple case of a single discrete time-step. These theorems imply that the pricing of derivatives in an arbitrage-free and complete market can be done in two equivalent ways: A) Based on construction of a replicating portfolio and B) Based on riskless rate-discounted expectation in the risk-neutral probability measure. Finally, we covered two financial trading problems that can be cast as MDPs. The first problem is the Optimal Exercise of American Options (and its generalization to Optimal Stopping problems). The second problem is the Pricing and Hedging of Derivatives in an Incomplete (real-world) Market.

In Chapter 10, we covered problems involving trading optimally on an Order Book. We started with developing an understanding of the core ingredients of an Order Book: Limit Orders, Market Orders, Order Book Dynamics, and Price Impact. The rest of the chapter covered two important problems that can be cast as MDPs. These are the problems of Optimal Order Execution and Optimal Market-Making. For each of these two problems, we derived closed-form solutions under highly simplified assumptions (e.g., Bertsimas-Lo, Avellaneda-Stoikov formulations), which helps develop intuition. Since these problems are modeled as finite-horizon MDPs, we can implement backward-induction ADP algorithms to solve them. However, in practice, we need to develop Reinforcement Learning algorithms (and associated market simulators) to solve these problems in real-world settings to overcome the Curse of Dimensionality and Curse of Modeling.

Module III covered Reinforcement Learning algorithms. Module III starts by motivating the case for Reinforcement Learning (RL). In the real-world, we typically do not have access to a model of state-reward transition probabilities. Typically, we simply have access to an environment, that serves up the next state and reward, given current state and action, at each step in the AI Agent's interaction with the environment. The environment could be the real environment or could be a simulated environment (the latter from a learnt model of the environment). RL algorithms for Prediction/Control learn the requisite Value Function/Policy by obtaining sufficient data (*atomic experiences*) from interaction with the environment. This is a sort of "trial and error" learning, through a process of prioritizing actions that seems to fetch good rewards, and deprioritizing actions that seem to fetch poor rewards. Specifically, RL algorithms are in the business of learning an approximate Q-Value Function, an estimate of the Expected Return for any given action in any given state. The success of RL algorithms depends not only on their ability to learn the Q-Value Function in an incremental manner through interactions with the environment, but also on their ability to perform good generalization of the Q-Value Function with appropriate function approximation (often using deep neural networks, in which case we term it as Deep RL). Most RL algorithms are founded on the Bellman Equations and all RL Control algorithms are based on the fundamental idea of *Generalized Policy Iteration*.

In Chapter 11, we covered RL Prediction algorithms. Specifically, we covered Monte-Carlo (MC) and Temporal-Difference (TD) algorithms for Prediction. A key learning from this chapter was the Bias-Variance tradeoff in MC versus TD. Another key learning was that while MC Prediction learns the statistical mean of the observed returns, TD Prediction learns something "deeper"—TD implicitly estimates an MRP from the observed data and produces the Value Function of the implicitly-estimated MRP. We emphasized viewing TD versus MC versus DP from the perspectives of "bootstrapping" and "experiencing". We finished this chapter by covering λ-Return Prediction and TD(λ) Prediction algorithms, which give us a way to tradeoff bias versus variance (along the spectrum of MC to TD) by tuning the λ parameter. TD is equivalent to TD(0) and MC is "equivalent" to TD(1).

In Chapter 12, we covered RL Control algorithms. We re-emphasized that RL Control is based on the idea of Generalized Policy Iteration (GPI). We explained that Policy Evaluation is done for the Q-Value Function (instead of the State-Value Function), and that the Improved Policy needs to be exploratory, e.g., ϵ-greedy. Next, we described an important concept—*Greedy in the Limit with Infinite Exploration* (GLIE). Our first RL Control algorithm was GLIE Monte-Carlo Control. Next, we covered two important TD Control algorithms: SARSA (which is On-Policy) and Q-Learning (which is Off-Policy). We briefly covered Importance Sampling, which is a different way of doing Off-Policy algorithms. We wrapped up this chapter with some commentary on the convergence of RL Prediction and RL Control algorithms. We highlighted a strong pattern of situations when we run into convergence issues—it is when all three of [Bootstrapping, Function Approximation, Off-Policy] are done together. We've seen how each of these three is individually beneficial, but when the three come together, it's "too much of a good thing", bringing about convergence issues. The confluence of these three is known as the *Deadly Triad* (an example of this would be Q-Learning with Function Approximation).

In Chapter 13, we covered the more nuanced RL Algorithms, going beyond the plain-vanilla MC and TD algorithms we covered in chapters 11 and 12. We started this chapter by introducing the novel ideas of *Batch RL* and *Experience-Replay*. Next, we covered the Least-Squares Monte-Carlo (LSMC) Prediction algorithm and the Least-Squares Temporal-Difference (LSTD) algorithm, which is a direct (gradient-free) solution of Batch TD. Next, we covered the very important Deep Q-Networks (DQN) algorithm, which uses Experience-Replay and fixed Q-learning targets, in order to avoid the pitfalls of time-correlation and varying TD Target. Next, we covered the Least-Squares Policy Iteration (LSPI) algorithm, which is an Off-Policy, Experience-Replay Control Algorithm using LSTDQ for Policy Evaluation. Then we showed how Optimal Exercise of American Options can be tackled with LSPI and Deep Q-Learning algorithms. In the second half of this chapter, we looked deeper into the issue of the *Deadly Triad* by viewing Value Functions as Vectors so as to understand Value Function Vector transformations with a balance of geometric intuition and mathematical rigor, providing insights into convergence issues for a variety of traditional loss functions used to develop RL algorithms. Finally, this treatment of Value Functions as Vectors led us in the direction of overcoming the Deadly Triad by defining an appropriate loss function, calculating whose gradient provides a more robust set of RL algorithms known as Gradient Temporal Difference (Gradient TD).

In Chapter 14, we covered Policy Gradient (PG) algorithms, which are based on GPI with Policy Improvement as a Stochastic Gradient Ascent for an *Expected Returns Objective* using a policy function approximation. We started with the Policy Gradient Theorem that gives us a simple formula for the gradient of the Expected Returns Objective in terms of the score of the policy function approximation. Our first PG algorithm was the REINFORCE algorithm, a Monte-Carlo Policy Gradient algorithm with no bias but high variance. We showed how to tackle the Optimal Asset Allocation problem with REINFORCE. Next, we showed how we can reduce variance in PG algorithms by using a critic and by using an estimate of the advantage function in place of the Q-Value Function. Next, we showed how to overcome bias in PG Algorithms based on the *Compatible Function Approximation Theorem*. Finally, we covered two specialized PG algorithms that have worked well in practice—Natural Policy Gradient and Deterministic Policy Gradient. We also provided some coverage of Evolutionary Strategies, which are technically not RL algorithms, but they resemble PG Algorithms and can sometimes be quite effective in solving MDP Control problems.

In Module IV, we provided some finishing touches by covering the topic of Exploration versus Exploitation and the topic of Blending Learning and Planning in some detail. In Chapter 15, we provided significant coverage of algorithms for the Multi-Armed Bandit (MAB) problem, which provides a simple setting to understand and appreciate the nuances of the Explore versus Exploit dilemma that we typically need to resolve within RL Control algorithms. We started with simple methods such as Naive Exploration (e.g., ϵ-greedy) and Optimistic Initialization. Next, we covered methods based on the broad approach of *Optimism in the Face of Uncertainty* (e.g., Upper-Confidence Bounds). Next, we covered the powerful and practically effective method of Probability Matching (e.g., Thompson Sampling). Then we also covered Gradient Bandit Algorithms and a disciplined approach to balancing exploration and exploitation by forming Information State Space MDPs (incorporating value of Information), typically solved by treating as Bayes-Adaptive MDPs. Finally, we noted that the above MAB algorithms are well-extensible to Contextual Bandits and RL Control.

In Chapter 16, we covered the issue of Planning versus Learning, and showed how to blend Planning and Learning. Next, we covered Monte-Carlo Tree-Search (MCTS), which is a Planning algorithm based on Tree-Search and based on sampling/RL techniques. Lastly, we covered Adaptive Multi-Stage Sampling (AMS), that we consider to be the spiritual origin of MCTS—it is an efficient algorithm for finite-horizon MDPs with very large state space and fairly small action space.

17.2   RL IN THE REAL-WORLD

Although this is an academic book on the Foundations of RL, we (the authors of this book) have significant experience in leveraging the power of Applied Mathematics to solve problems in the real-world. So we devote this subsection to the various nuances of applying RL in the real-world.

The most important point we'd like to make is that in order to develop models and algorithms that will be effective in the real-world, one should not only have a deep technical understanding but to also have a deep understanding of the business domain. If the business domain is Financial Trading, one needs to be well-versed in the practical details of the specific market one is working in, and the operational details and transactional frictions involved in trading. These details need to be carefully captured in the MDP one is constructing. These details have ramifications on the choices made in defining the state space and the action space. More importantly, defining the reward function is typically not an obvious choice at all—it requires considerable thought and typically one would need to consult with the business head to identify what exactly is the objective function in running the business, e.g., the precise definition of the Utility function. One should also bear in mind that a typical real-world problem is actually a Partially Observable Markov Decision Process (POMDP) rather than an MDP. In the pursuit of computational tractability, one might approximate the POMDP as an MDP but in order to do so, one requires strong understanding of the business domain. However, sometimes partial state-observability cannot be ignored, and in such situations, we have to employ (computationally expensive) algorithms to solve the POMDP. Indeed, controlling state space explosion is one of the biggest challenges in the real-world. Much of the effort in modeling an MDP is to define a state space that finds the appropriate balance between capturing the key aspects of the real-world problem and attaining computational tractability.

Now we'd like to share the approach we usually take when encountering a new problem, like one of the Financial Applications we covered in Module II. Our first stab at the problem is to create a simpler version of the problem that lends itself to analytical tractability, exploring ways to develop a closed-form solution (like we obtained for some of the Financial Applications in Module II). This typically requires removing some of the frictions and constraints of the real-world problem. For Financial Applications, this might involve assuming no transaction costs, perhaps assuming continuous trading, perhaps assuming no liquidity constraints. There are multiple advantages of deriving a closed-form solution with simplified assumptions. Firstly, the closed-form solution immediately provides tremendous intuition as it shows the analytical dependency of the Optimal Value Function/Optimal Policy on the inputs and parameters of the problem. Secondly, when we eventually obtain the solution to the full-fledged model, we can test the solution by creating a special case of the full-fledged model that reduces to the simplified model for which we have a closed-form solution. Thirdly, the expressions within the closed-form solution provide us with some guidance on constructing appropriate features for function approximation when solving the full-fledged model.

The next stage would be to bring in some of the real-world frictions and constraints, and attempt to solve the problem with Dynamic Programming (or Approximate Dynamic Programming). This means we need to construct a model of state-reward transition probabilities. Such a model would be estimated from real-world data obtained from interaction with the real environment. However, often we find that Dynamic Programming (or Approximate Dynamic Programming) is not an option due to the Curse of Modeling (i.e., hard to build a model of transition probabilities). This leaves us with the eventual go-to option of pursuing a Reinforcement Learning technique. In most real-world problems, we'd employ RL not with real environment interactions, but with simulated environment interactions. This means we need to build a sampling model estimated from real-world data obtained from interactions with the real environment. In fact, in many real-world problems, we'd want to augment the data-learnt simulator with human knowledge/assumptions (specifically information that might not be readily obtained from electronic data that a human expert might be knowledgeable about). Having a simulator of the environment is very valuable because we can run it indefinitely and also because we can create a variety of scenarios (with different settings/assumptions) to run the simulator in. Deep Learning-based function approximations have been quite successful in the context of Reinforcement Learning algorithms (we refer to this as Deep Reinforcement Learning). Lastly, it pays to re-emphasize that the learnings from Chapter 16 are very important for real-world problems. In particular, the idea of blending model-based RL with model-free RL (Figure 16.2) is an attractive option for real-world applications because the real-world is typically not stationary and hence, models need to be updated continuously.

Given the plethora of choices for different types of RL algorithms, it is indeed difficult to figure out which RL algorithm would be most suitable for a given real-world problem. As ever, we recommend starting with a simple algorithm such as the MC and TD methods we used in Chapters 11 and 12. Although the simple algorithms may not be powerful enough for many real-world applications, they are a good place to start to try out on a smaller size of the actual problem—these simple RL algorithms are very easy to implement, reason about and debug. However, the most important advice we can give you is that after having understood the various nuances of the specific real-world problem you want to solve, you should aim to construct an RL algorithm that is customized for your problem. One must recognize that the set of RL algorithms is not a fixed menu to choose from. Rather, there are various pieces of RL algorithms that are open to modification. In fact, we can combine different aspects of different algorithms to suit our specific needs for a given real-world problem. We not only make choices on features in function approximations and on hyper-parameters, we also make choices on the exact design of the algorithm method, e.g., how exactly we'd like to do Off-Policy Learning, or how exactly we'd like to do the Policy Evaluation component of Generalized Policy Iteration in our Control algorithm. In practice, we've found that we often end up with the more advanced algorithms due to the typical real-world problem complexity or state-space/action-space size. There is no silver bullet here, and one has to try various algorithms to see which one works best for the given problem. However, it pays to share that the algorithms that have worked well for us in real-world problems are Least-Squares Policy Iteration, Gradient Temporal-Difference, Deep Q-Networks and Natural Policy Gradient. We have always paid attention to Richard Sutton's mantra of avoiding the Deadly Triad. We recommend the excellent paper by Hasselt, Doron, Strub, Hessel, Sonnerat, Modayil (Hasselt et al. 2018) to understand the nuances of the Deadly Triad in the context of Deep Reinforcement Learning.

It's important to recognize that the code we developed in this book is for educational purposes and we barely made an attempt to make the code performant. In practice, this type of educational code won't suffice—we need to develop highly performant code and make the code parallelizable wherever possible. This requires an investment in a suitable distributed system for storage and compute, so the RL algorithms can be trained in an efficient manner.

When it comes to making an RL algorithm successful in a real-world application, the design and implementation of the model and the algorithm is only a small piece of the overall puzzle. Indeed, one needs to build an entire ecosystem of data management, software engineering, model training infrastructure, model deployment platform, tools for easy debugging, measurements/instrumentation and explainability of results. Moreover, it is vital to have a strong Product Management practice in order to ensure that the algorithm is serving the needs of the overall product being built. Indeed, the goal is to build a successful product, not just a model and an algorithm. A key challenge in many organizations is to replace a legacy system or a manual system with a modern solution (e.g., with an RL-based solution). This requires investment in a culture change in the organization so that all stakeholders are supportive, otherwise the change management will be very challenging.

When the product carrying the RL algorithm runs in production, it is vital to evaluate whether the real-world problem is actually being solved effectively by defining, evaluating and reporting the appropriate success metrics. If those metrics are found to be inadequate, we need the appropriate feedback system in the organization to investigate why the product (and perhaps the model) is not delivering the requisite results. It could be that we have designed a model which is not quite the right fit for the real-world problem, in which case we improve the model in the next iteration of this feedback system. It often takes several iterations of evaluating the success metrics, providing feedback, and improving the model (and sometimes the algorithm) in order to achieve adequate results. An important point to note is that typically in practice, we rarely need to solve all the way to an optima—typically, being close to optimum is good enough to achieve the requisite success metrics. A Product Manager must constantly question whether we are solving the right problem, and whether we are investing our efforts in the most important aspects of the problem (e.g., ask if it suffices to be reasonably close to optimum).

Lastly, one must recognize that typically in the real-world, we are plagued with noisy data, incomplete data and sometimes plain wrong data. The design of the model needs to take this into account. Also, there is no such thing as the "perfect model"—in practice, a model is simply a crude approximation of reality. It should be assumed by default that we have bad data and that we have an imperfect model. Hence, it is important to build a system that can reason about uncertainties in data and about uncertainties with the model.

A book can simply not do justice to explaining the various nuances and complications that arise in developing and deploying an RL-based solution in the real-world. Here we have simply scratched the surface of the various issues that arise. You would truly understand and appreciate these nuances and complications only by stepping into the real-world and experiencing it for yourself. However, it is important to first be grounded in the foundations of RL, which is what we hope you got from this book.

Moment Generating Function and Its Applications

The purpose of this Appendix is to introduce the *Moment Generating Function (MGF)* and demonstrate its utility in several applications in Applied Mathematics.

A.1 THE MOMENT GENERATING FUNCTION (MGF)

The Moment Generating Function (MGF) of a random variable x (discrete or continuous) is defined as a function $f_x : \mathbb{R} \to \mathbb{R}^+$ such that:

$$f_x(t) = \mathbb{E}_x[e^{tx}] \text{ for all } t \in \mathbb{R} \tag{A.1}$$

Let us denote the nth-derivative of f_x as $f_x^{(n)} : \mathbb{R} \to \mathbb{R}$ for all $n \in \mathbb{Z}_{\geq 0}$ ($f_x^{(0)}$ is defined to be simply the MGF f_x).

$$f_x^{(n)}(t) = \mathbb{E}_x[x^n \cdot e^{tx}] \text{ for all } n \in \mathbb{Z}_{\geq 0} \text{ for all } t \in \mathbb{R} \tag{A.2}$$

$$f_x^{(n)}(0) = \mathbb{E}_x[x^n] \tag{A.3}$$

$$f_x^{(n)}(1) = \mathbb{E}_x[x^n \cdot e^x] \tag{A.4}$$

Equation (A.3) tells us that $f_x^{(n)}(0)$ gives us the nth moment of x. In particular, $f_x^{(1)}(0) = f_x'(0)$ gives us the mean and $f_x^{(2)}(0) - (f_x^{(1)}(0))^2 = f_x''(0) - (f_x'(0))^2$ gives us the variance. Note that this holds true for any distribution for x. This is rather convenient since all we need is the functional form for the distribution of x. This would lead us to the expression for the MGF (in terms of t). Then, we take derivatives of this MGF and evaluate those derivatives at 0 to obtain the moments of x.

Equation (A.4) helps us calculate the often-appearing expectation $\mathbb{E}_x[x^n \cdot e^x]$. In fact, $\mathbb{E}_x[e^x]$ and $\mathbb{E}_x[x \cdot e^x]$ are very common in several areas of Applied Mathematics. Again, note that this holds true for any distribution for x.

The MGF should be thought of as an alternative specification of a random variable (alternative to specifying its Probability Distribution). This alternative specification is very valuable because it can sometimes provide better analytical tractability than working with the Probability Density Function or Cumulative Distribution Function (as an example, see the below section on the MGF for linear functions of independent random variables).

DOI: 10.1201/9781003229193-A

A.2   MGF FOR LINEAR FUNCTIONS OF RANDOM VARIABLES

Consider m independent random variables $x_1, x_2, \ldots, x_m$. Let $\alpha_0, \alpha_1, \ldots, \alpha_m \in \mathbb{R}$. Now consider the random variable

$$x = \alpha_0 + \sum_{i=1}^{m} \alpha_i x_i$$

The Probability Density Function of x is complicated to calculate as it involves convolutions. However, observe that the MGF f_x of x is given by:

$$f_x(t) = \mathbb{E}[e^{t(\alpha_0 + \sum_{i=1}^{m} \alpha_i x_i)}] = e^{\alpha_0 t} \cdot \prod_{i=1}^{m} \mathbb{E}[e^{t\alpha_i x_i}] = e^{\alpha_0 t} \cdot \prod_{i=1}^{m} f_{\alpha_i x_i}(t) = e^{\alpha_0 t} \cdot \prod_{i=1}^{m} f_{x_i}(\alpha_i t)$$

This means the MGF of x can be calculated as $e^{\alpha_0 t}$ times the product of the MGFs of $\alpha_i x_i$ (or of α_i-scaled MGFs of x_i) for all $i = 1, 2, \ldots, m$. This gives us a much better way to analytically tract the probability distribution of x (compared to the convolution approach).

A.3   MGF FOR THE NORMAL DISTRIBUTION

Here we assume that the random variables x follows a normal distribution. Let $x \sim \mathcal{N}(\mu, \sigma^2)$.

$$
\begin{aligned}
f_{x \sim \mathcal{N}(\mu, \sigma^2)}(t) &= \mathbb{E}_{x \sim \mathcal{N}(\mu, \sigma^2)}[e^{tx}] \\
&= \int_{-\infty}^{+\infty} \frac{1}{\sqrt{2\pi}\sigma} \cdot e^{-\frac{(x-\mu)^2}{2\sigma^2}} \cdot e^{tx} \cdot dx \\
&= \int_{-\infty}^{+\infty} \frac{1}{\sqrt{2\pi}\sigma} \cdot e^{-\frac{(x-(\mu+t\sigma^2))^2}{2\sigma^2}} \cdot e^{\mu t + \frac{\sigma^2 t^2}{2}} \cdot dx \\
&= e^{\mu t + \frac{\sigma^2 t^2}{2}} \cdot \mathbb{E}_{x \sim \mathcal{N}(\mu + t\sigma^2, \sigma^2)}[1] \\
&= e^{\mu t + \frac{\sigma^2 t^2}{2}}
\end{aligned}
\tag{A.5}
$$

$$f'_{x \sim \mathcal{N}(\mu, \sigma^2)}(t) = \mathbb{E}_{x \sim \mathcal{N}(\mu, \sigma^2)}[x \cdot e^{tx}] = (\mu + \sigma^2 t) \cdot e^{\mu t + \frac{\sigma^2 t^2}{2}} \tag{A.6}$$

$$f''_{x \sim \mathcal{N}(\mu, \sigma^2)}(t) = \mathbb{E}_{x \sim \mathcal{N}(\mu, \sigma^2)}[x^2 \cdot e^{tx}] = ((\mu + \sigma^2 t)^2 + \sigma^2) \cdot e^{\mu t + \frac{\sigma^2 t^2}{2}} \tag{A.7}$$

$$f'_{x \sim \mathcal{N}(\mu, \sigma^2)}(0) = \mathbb{E}_{x \sim \mathcal{N}(\mu, \sigma^2)}[x] = \mu$$

$$f''_{x \sim \mathcal{N}(\mu, \sigma^2)}(0) = \mathbb{E}_{x \sim \mathcal{N}(\mu, \sigma^2)}[x^2] = \mu^2 + \sigma^2$$

$$f'_{x \sim \mathcal{N}(\mu, \sigma^2)}(1) = \mathbb{E}_{x \sim \mathcal{N}(\mu, \sigma^2)}[x \cdot e^x] = (\mu + \sigma^2) e^{\mu + \frac{\sigma^2}{2}}$$

$$f''_{x \sim \mathcal{N}(\mu, \sigma^2)}(1) = \mathbb{E}_{x \sim \mathcal{N}(\mu, \sigma^2)}[x^2 \cdot e^x] = ((\mu + \sigma^2)^2 + \sigma^2) e^{\mu + \frac{\sigma^2}{2}}$$

A.4   MINIMIZING THE MGF

Now let us consider the problem of minimizing the MGF. The problem is to:

$$\min_{t \in \mathbb{R}} f_x(t) = \min_{t \in \mathbb{R}} \mathbb{E}_x[e^{tx}]$$

This problem of minimizing $\mathbb{E}_x[e^{tx}]$ shows up a lot in various places in Applied Mathematics when dealing with exponential functions (e.g., when optimizing the Expectation of a Constant Absolute Risk-Aversion (CARA) Utility function $U(y) = \frac{1-e^{-\gamma y}}{\gamma}$ where γ is the coefficient of risk-aversion and where y is a parameterized function of a random variable x).

Let us denote t^* as the value of t that minimizes the MGF. Specifically,

$$t^* = \arg\min_{t \in \mathbb{R}} f_x(t) = \arg\min_{t \in \mathbb{R}} \mathbb{E}_x[e^{tx}]$$

A.4.1   Minimizing the MGF When x Follows a Normal Distribution

Here we consider the fairly typical case where x follows a normal distribution. Let $x \sim \mathcal{N}(\mu, \sigma^2)$. Then we have to solve the problem:

$$\min_{t \in \mathbb{R}} f_{x \sim \mathcal{N}(\mu, \sigma^2)}(t) = \min_{t \in \mathbb{R}} \mathbb{E}_{x \sim \mathcal{N}(\mu, \sigma^2)}[e^{tx}] = \min_{t \in \mathbb{R}} e^{\mu t + \frac{\sigma^2 t^2}{2}}$$

From Equation (A.6) above, we have:

$$f'_{x \sim \mathcal{N}(\mu, \sigma^2)}(t) = (\mu + \sigma^2 t) \cdot e^{\mu t + \frac{\sigma^2 t^2}{2}}$$

Setting this to 0 yields:

$$(\mu + \sigma^2 t^*) \cdot e^{\mu t^* + \frac{\sigma^2 t^{*2}}{2}} = 0$$

which leads to:

$$t^* = \frac{-\mu}{\sigma^2} \tag{A.8}$$

From Equation (A.7) above, we have:

$$f''_{x \sim \mathcal{N}(\mu, \sigma^2)}(t) = ((\mu + \sigma^2 t)^2 + \sigma^2) \cdot e^{\mu t + \frac{\sigma^2 t^2}{2}} > 0 \text{ for all } t \in \mathbb{R}$$

which confirms that t^* is a minima.

Substituting $t = t^*$ in $f_{x \sim \mathcal{N}(\mu, \sigma^2)}(t) = e^{\mu t + \frac{\sigma^2 t^2}{2}}$ yields:

$$\min_{t \in \mathbb{R}} f_{x \sim \mathcal{N}(\mu, \sigma^2)}(t) = e^{\mu t^* + \frac{\sigma^2 t^{*2}}{2}} = e^{\frac{-\mu^2}{2\sigma^2}} \tag{A.9}$$

A.4.2   Minimizing the MGF When x Follows a Symmetric Binary Distribution

Here we consider the case where x follows a binary distribution: x takes values $\mu + \sigma$ and $\mu - \sigma$ with probability 0.5 each. Let us refer to this distribution as $x \sim \mathcal{B}(\mu + \sigma, \mu - \sigma)$. Note that the mean and variance of x under $\mathcal{B}(\mu + \sigma, \mu - \sigma)$ are μ and σ^2, respectively. So we have to solve the problem:

$$\min_{t \in \mathbb{R}} f_{x \sim \mathcal{B}(\mu + \sigma, \mu - \sigma)}(t) = \min_{t \in \mathbb{R}} \mathbb{E}_{x \sim \mathcal{B}(\mu + \sigma, \mu - \sigma)}[e^{tx}] = \min_{t \in \mathbb{R}} 0.5(e^{(\mu + \sigma)t} + e^{(\mu - \sigma)t})$$

$$f'_{x \sim \mathcal{B}(\mu + \sigma, \mu - \sigma)}(t) = 0.5((\mu + \sigma) \cdot e^{(\mu + \sigma)t} + (\mu - \sigma) \cdot e^{(\mu - \sigma)t})$$

Note that unless $\mu \in$ open interval $(-\sigma, \sigma)$ (i.e., absolute value of mean is less than standard deviation), $f'_{x \sim \mathcal{B}(\mu + \sigma, \mu - \sigma)}(t)$ will not be 0 for any value of t. Therefore, for this minimization to be non-trivial, we will henceforth assume $\mu \in (-\sigma, \sigma)$. With this assumption in place, setting $f'_{x \sim \mathcal{B}(\mu + \sigma, \mu - \sigma)}(t)$ to 0 yields:

$$(\mu + \sigma) \cdot e^{(\mu + \sigma)t^*} + (\mu - \sigma) \cdot e^{(\mu - \sigma)t^*} = 0$$

which leads to:

$$t^* = \frac{1}{2\sigma} \ln\left(\frac{\sigma - \mu}{\mu + \sigma}\right)$$

Note that

$$f''_{x \sim \mathcal{B}(\mu+\sigma, \mu-\sigma)}(t) = 0.5((\mu + \sigma)^2 \cdot e^{(\mu+\sigma)t} + (\mu - \sigma)^2 \cdot e^{(\mu-\sigma)t}) > 0 \text{ for all } t \in \mathbb{R}$$

which confirms that t^* is a minima.

Substituting $t = t^*$ in $f_{x \sim \mathcal{B}(\mu+\sigma, \mu-\sigma)}(t) = 0.5(e^{(\mu+\sigma)t} + e^{(\mu-\sigma)t})$ yields:

$$\min_{t \in \mathbb{R}} f_{x \sim \mathcal{B}(\mu+\sigma, \mu-\sigma)}(t) = 0.5(e^{(\mu+\sigma)t^*} + e^{(\mu-\sigma)t^*}) = 0.5\left(\left(\frac{\sigma - \mu}{\mu + \sigma}\right)^{\frac{\mu+\sigma}{2\sigma}} + \left(\frac{\sigma - \mu}{\mu + \sigma}\right)^{\frac{\mu-\sigma}{2\sigma}}\right)$$

Portfolio Theory

In this Appendix, we provide a quick and terse introduction to *Portfolio Theory*. While this topic is not a direct pre-requisite for the topics we cover in the chapters, we believe one should have some familiarity with the risk versus reward considerations when constructing portfolios of financial assets, and know of the important results. To keep this Appendix brief, we will provide the minimal content required to understand the *essence* of the key concepts. We won't be doing rigorous proofs. We will also ignore details pertaining to edge-case/irregular-case conditions so as to focus on the core concepts.

B.1   SETTING AND NOTATION

In this section, we go over the core setting of Portfolio Theory, along with the requisite notation.

Assume there are n assets in the economy and that their mean returns are represented in a column vector $R \in \mathbb{R}^n$. We denote the covariance of returns of the n assets by an $n \times n$ non-singular matrix V.

We consider arbitrary portfolios p comprised of investment quantities in these n assets that are normalized to sum up to 1. Denoting column vector $X_p \in \mathbb{R}^n$ as the investment quantities in the n assets for portfolio p, we can write the normality of the investment quantities in vector notation as:

$$X_p^T \cdot 1_n = 1$$

where $1_n \in \mathbb{R}^n$ is a column vector comprising of all 1's.

We shall drop the subscript p in X_p whenever the reference to portfolio p is clear.

B.2   PORTFOLIO RETURNS

- A single portfolio's mean return is $X^T \cdot R \in \mathbb{R}$.
- A single portfolio's variance of return is the quadratic form $X^T \cdot V \cdot X \in \mathbb{R}$.
- Covariance between portfolios p and q is the bilinear form $X_p^T \cdot V \cdot X_q \in \mathbb{R}$.
- Covariance of the n assets with a single portfolio is the vector $V \cdot X \in \mathbb{R}^n$.

B.3   DERIVATION OF EFFICIENT FRONTIER CURVE

An asset which has no variance in terms of how its value evolves in time is known as a riskless asset. The Efficient Frontier is defined for a world with no riskless assets. The Efficient Frontier is the set of portfolios with minimum variance of return for each level of

DOI: 10.1201/9781003229193-B

portfolio mean return (we refer to a portfolio in the Efficient Frontier as an *Efficient Portfolio*). Hence, to determine the Efficient Frontier, we solve for X so as to minimize portfolio variance $X^T \cdot V \cdot X$ subject to constraints:

$$X^T \cdot 1_n = 1$$

$$X^T \cdot R = r_p$$

where r_p is the mean return for Efficient Portfolio p. We set up the Lagrangian and solve to express X in terms of R, V, r_p. Substituting for X gives us the efficient frontier parabola of Efficient Portfolio Variance σ_p^2 as a function of its mean r_p:

$$\sigma_p^2 = \frac{a - 2br_p + cr_p^2}{ac - b^2}$$

where

- $a = R^T \cdot V^{-1} \cdot R$
- $b = R^T \cdot V^{-1} \cdot 1_n$
- $c = 1_n^T \cdot V^{-1} \cdot 1_n$

B.4 GLOBAL MINIMUM VARIANCE PORTFOLIO (GMVP)

The global minimum variance portfolio (GMVP) is the portfolio at the tip of the efficient frontier parabola, i.e., the portfolio with the lowest possible variance among all portfolios on the Efficient Frontier. Here are the relevant characteristics for the GMVP:

- It has mean $r_0 = \frac{b}{c}$.
- It has variance $\sigma_0^2 = \frac{1}{c}$.
- It has investment proportions $X_0 = \frac{V^{-1} \cdot 1_n}{c}$.

GMVP is positively correlated with all portfolios and with all assets. GMVP's covariance with all portfolios and with all assets is a constant value equal to $\sigma_0^2 = \frac{1}{c}$ (which is also equal to its own variance).

B.5 ORTHOGONAL EFFICIENT PORTFOLIOS

For every efficient portfolio p (other than GMVP), there exists a unique orthogonal efficient portfolio z (i.e. $Covariance(p, z) = 0$) with finite mean

$$r_z = \frac{a - br_p}{b - cr_p}$$

z always lies on the opposite side of p on the (efficient frontier) parabola. If we treat the Efficient Frontier as a curve of mean (y-axis) versus variance (x-axis), the straight line from p to GMVP intersects the mean axis (y-axis) at r_z. If we treat the Efficient Frontier as a curve of mean (y-axis) versus standard deviation (x-axis), the tangent to the efficient frontier at p intersects the mean axis (y-axis) at r_z. Moreover, all portfolios on one side of the efficient frontier are positively correlated with each other.

B.6 TWO-FUND THEOREM

The X vector (normalized investment quantities in assets) of any efficient portfolio is a linear combination of the X vectors of two other efficient portfolios. Notationally,

$$X_p = \alpha X_{p_1} + (1 - \alpha) X_{p_2} \text{ for some scalar } \alpha$$

Varying α from $-\infty$ to $+\infty$ basically traces the entire efficient frontier. So to construct all efficient portfolios, we just need to identify two canonical efficient portfolios. One of them is GMVP. The other is a portfolio we call Special Efficient Portfolio (SEP) with:

* Mean $r_1 = \frac{a}{b}$.
* Variance $\sigma_1^2 = \frac{a}{b^2}$.
* Investment proportions $X_1 = \frac{V^{-1} \cdot R}{b}$.

The orthogonal portfolio to SEP has mean $r_z = \frac{a - b\frac{a}{b}}{b - c\frac{a}{b}} = 0$

B.7 AN EXAMPLE OF THE EFFICIENT FRONTIER FOR 16 ASSETS

Figure B.1 shows a plot of the mean daily returns versus the standard deviation of daily returns collected over a 3-year period for 16 assets. The curve is the Efficient Frontier for these 16 assets. Note the special portfolios GMVP and SEP on the Efficient Frontier. This curve was generated from the code at rl/appendix2/efficient_frontier.py. We encourage you to play with different choices (and count) of assets, and to also experiment with different time ranges as well as to try weekly and monthly returns.

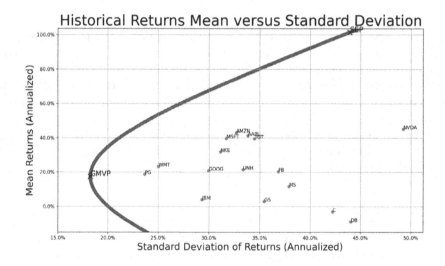

Figure B.1 **Efficient Frontier for 16 Assets**

B.8 CAPM: LINEARITY OF COVARIANCE VECTOR W.R.T. MEAN RETURNS

Important Theorem: The covariance vector of individual assets with a portfolio (note: covariance vector $= V \cdot X \in \mathbb{R}^n$) can be expressed as an exact linear function of the individual assets' mean returns vector if and only if the portfolio is efficient. If the efficient portfolio is p (and its orthogonal portfolio z), then:

$$R = r_z 1_n + \frac{r_p - r_z}{\sigma_p^2}(V \cdot X_p) = r_z 1_n + (r_p - r_z)\beta_p$$

where $\beta_p = \frac{V \cdot X_p}{\sigma_p^2} \in \mathbb{R}^n$ is the vector of slope coefficients of regressions where the explanatory variable is the portfolio mean return $r_p \in \mathbb{R}$ and the n dependent variables are the asset mean returns $R \in \mathbb{R}^n$.

The linearity of β_p w.r.t. mean returns R is famously known as the Capital Asset Pricing Model (CAPM).

B.9   USEFUL COROLLARIES OF CAPM

- If p is SEP, $r_z = 0$ which would mean:

$$R = r_p \beta_p = \frac{r_p}{\sigma_p^2} \cdot V \cdot X_p$$

- So, in this case, covariance vector $V \cdot X_p$ and β_p are just scalar multiples of asset mean vector.
- The investment proportion X in a given individual asset changes monotonically along the efficient frontier.
- Covariance $V \cdot X$ is also monotonic along the efficient frontier.
- But β is not monotonic, which means that for every individual asset, there is a unique pair of efficient portfolios that result in maximum and minimum βs for that asset.

B.10   CROSS-SECTIONAL VARIANCE

- The cross-sectional variance in βs (variance in βs across assets for a fixed efficient portfolio) is zero when the efficient portfolio is GMVP and is also zero when the efficient portfolio has infinite mean.
- The cross-sectional variance in βs is maximum for the two efficient portfolios with means: $r_0 + \sigma_0^2 \sqrt{|A|}$ and $r_0 - \sigma_0^2 \sqrt{|A|}$ where A is the 2×2 symmetric matrix consisting of a, b, b, c.
- These two portfolios lie symmetrically on opposite sides of the efficient frontier (their βs are equal and of opposite signs), and are the only two orthogonal efficient portfolios with the same variance ($= 2\sigma_0^2$).

B.11   EFFICIENT SET WITH A RISK-FREE ASSET

If we have a riskless asset with return r_F, then V is singular. So we first form the Efficient Frontier without the riskless asset. The Efficient Set (including the riskless asset) is defined as the tangent to this Efficient Frontier (without the riskless asset) from the point $(0, r_F)$ when the Efficient Frontier is considered to be a curve of mean returns (y-axis) against standard deviation of returns (x-axis).

Let's say the tangent touches the Efficient Frontier at the point (Portfolio) T and let its return be r_T. Then:

- If $r_F < r_0, r_T > r_F$.
- If $r_F > r_0, r_T < r_F$.
- All portfolios on this efficient set are perfectly correlated.

Introduction to and Overview of Stochastic Calculus Basics

In this Appendix, we provide a quick introduction to the *Basics of Stochastic Calculus*. To be clear, Stochastic Calculus is a vast topic requiring an entire graduate-level course to develop a good understanding. We shall only be scratching the surface of Stochastic Calculus and even with the very basics of this subject, we will focus more on intuition than rigor, and familiarize you with just the most important results relevant to this book. For an adequate treatment of Stochastic Calculus relevant to Finance, we recommend Steven Shreve's two-volume discourse Stochastic Calculus for Finance I (Shreve 2003) and Stochastic Calculus for Finance II (Shreve 2004). For a broader treatment of Stochastic Calculus, we recommend Bernt Oksendal's book on Stochastic Differential Equations (Øksendal 2003).

C.1 SIMPLE RANDOM WALK

The best way to get started with Stochastic Calculus is to first get familiar with key properties of a *simple random walk* viewed as a discrete-time, countable state-space, time-homogeneous Markov Process. The state space is the set of integers $\mathbb{Z}$. Denoting the random state at time t as Z_t, the state transitions are defined in terms of the independent and identically distributed (i.i.d.) random variables Y_t for all $t = 0, 1, \ldots$

$$Z_{t+1} = Z_t + Y_t \text{ and } \mathbb{P}[Y_t = 1] = \mathbb{P}[Y_t = -1] = 0.5 \text{ for all } t = 0, 1, \ldots$$

A quick point on notation: We refer to the random state at time t as Z_t (i.e., as a random variable at time t), whereas we refer to the Markov Process for this simple random walk as Z (i.e., without any subscript).

Since the random variables $\{Y_t | t = 0, 1, \ldots\}$ are i.i.d, the *increments* $Z_{t_{i+1}} - Z_{t_i}$ (for $i = 0, 1, \ldots n-1$) in the random walk states for any set of time steps $t_0 < t_1 < \ldots < t_n$ have the following properties:

- **Independent Increments**: Increments $Z_{t_1} - Z_{t_0}, Z_{t_2} - Z_{t_1}, \ldots, Z_{t_n} - Z_{t_{n-1}}$ are independent of each other.
- **Martingale (i.e., Zero-Drift) Property**: Expected Value of any Increment is 0.

$$\mathbb{E}[Z_{t_{i+1}} - Z_{t_i}] = \sum_{j=t_i}^{t_{i+1}-1} \mathbb{E}[Z_{j+1} - Z_j] = 0 \text{ for all } i = 0, 1, \ldots, n-1$$

DOI: 10.1201/9781003229193-C

- **Variance of any Increment equals Time Steps of the Increment:**

$$\mathbb{E}[(Z_{t_{i+1}} - Z_{t_i})^2] = \mathbb{E}[(\sum_{j=t_i}^{t_{i+1}-1} Y_j)^2] = \sum_{j=t_i}^{t_{i+1}-1} \mathbb{E}[Y_j^2] + 2\sum_{j=t_i}^{t_{i+1}-1}\sum_{k=j+1}^{t_{i+1}} \mathbb{E}[Y_j]\cdot\mathbb{E}[Y_k] = t_{i+1} - t_i$$

for all $i = 0, 1, \ldots, n - 1$.

Moreover, we have an important property that **Quadratic Variation equals Time Steps**. Quadratic Variation over the time interval $[t_i, t_{i+1}]$ for all $i = 0, 1, \ldots, n - 1$ is defined as:

$$\sum_{j=t_i}^{t_{i+1}-1} (Z_{j+1} - Z_j)^2$$

Since $(Z_{j+1} - Z_j)^2 = Y_j^2 = 1$ for all $j = t_i, t_i + 1, \ldots, t_{i+1} - 1$, Quadratic Variation

$$\sum_{j=t_i}^{t_{i+1}-1} (Z_{j+1} - Z_j)^2 = t_{i+1} - t_i \text{ for all } i = 0, 1, \ldots n - 1$$

It pays to emphasize the important conceptual difference between the Variance of Increment property and the Quadratic Variation property. The Variance of Increment property is a statement about the *expectation* of the square of the $Z_{t_{i+1}} - Z_{t_i}$ increment whereas the Quadratic Variation property is a statement of certainty (note: there is no $\mathbb{E}[\cdots]$ in this statement) about the sum of squares of *atomic* increments Y_j over the discrete-steps time-interval $[t_i, t_{i+1}]$. The Quadratic Variation property owes to the fact that $\mathbb{P}[Y_t^2 = 1] = 1$ for all $t = 0, 1, \ldots$.

We can view the Quadratic Variations of a Process X over all discrete-step time intervals $[0, t]$ as a Process denoted $[X]$, defined as:

$$[X]_t = \sum_{j=0}^{t} (X_{j+1} - X_j)^2$$

Thus, for the simple random walk Markov Process Z, we have the succinct formula: $[Z]_t = t$ for all t (i.e., this Quadratic Variation process is a deterministic process).

C.2   BROWNIAN MOTION AS SCALED RANDOM WALK

Now let us take our simple random walk process Z, and simultaneously A) speed up time and B) scale down the size of the atomic increments Y_t. Specifically, define for any fixed positive integer n:

$$z_t^{(n)} = \frac{1}{\sqrt{n}} \cdot Z_{nt} \text{ for all } t = 0, \frac{1}{n}, \frac{2}{n}, \ldots$$

It's easy to show that the above properties of the simple random walk process hold for the $z^{(n)}$ process as well. Now consider the continuous-time process z defined as:

$$z_t = \lim_{n \to \infty} z_t^{(n)} \text{ for all } t \in \mathbb{R}_{\geq 0}$$

This continuous-time process z with $z_0 = 0$ is known as standard Brownian Motion. z retains the same properties as those of the simple random walk process that we have listed above (independent increments, martingale, increment variance equal to time interval, and quadratic variation equal to time interval). Also, by Central Limit Theorem,

$$z_t | z_s \sim \mathcal{N}(z_s, t - s) \text{ for any } 0 \leq s < t$$

We denote dz_t as the increment in z over the infinitesimal time interval $[t, t + dt]$.

$$dz_t \sim \mathcal{N}(0, dt)$$

C.3 CONTINUOUS-TIME STOCHASTIC PROCESSES

Brownian motion z is our first example of a Continuous-Time Stochastic Process. Now let us define a general continuous-time stochastic process, although for the sake of simplicity, we shall restrict ourselves to one-dimensional real-valued continuous-time stochastic processes.

Definition C.3.1. A *One-dimensional Real-Valued Continuous-Time Stochastic Process* denoted X is defined as a collection of real-valued random variables $\{X_t | t \in [0, T]\}$ (for some fixed $T \in \mathbb{R}$, with index t interpreted as continuous-time) defined on a common probability space $(\Omega, \mathcal{F}, \mathbb{P})$, where Ω is a sample space, $\mathcal{F}$ is a σ-algebra and $\mathbb{P}$ is a probability measure (so, $X_t : \Omega \to \mathbb{R}$ for each $t \in [0, T]$).

We can view a stochastic process X as an $\mathbb{R}$-valued function of two variables:

- $t \in [0, T]$
- $\omega \in \Omega$

As a two-variable function, if we fix t, then we get the random variable $X_t : \Omega \to \mathbb{R}$ for time t and if we fix ω, then we get a single $\mathbb{R}$-valued outcome for each random variable across time (giving us a *sample trace* across time, denoted $X(\omega)$).

Now let us come back to Brownian motion, viewed as a Continuous-Time Stochastic Process.

C.4 PROPERTIES OF BROWNIAN MOTION SAMPLE TRACES

- Sample traces $z(\omega)$ of Brownian motion z are continuous.
- Sample traces $z(\omega)$ are almost always non-differentiable, meaning:

$$\text{Random variable } \lim_{h \to 0} \frac{z_{t+h} - z_t}{h} \text{ is almost always infinite}$$

The intuition is that $\frac{z_{t+h} - z_t}{h}$ has standard deviation of $\frac{1}{\sqrt{h}}$, which goes to ∞ as h goes to 0.

- Sample traces $z(\omega)$ have infinite total variation, meaning:

$$\text{Random variable } \int_S^T |dz_t| = \infty \text{ (almost always)}$$

The quadratic variation property can be expressed as:

$$\int_S^T (dz_t)^2 = T - S$$

This means each sample random trace of Brownian motion has quadratic variation equal to the time interval of the trace. The quadratic variation of z expressed as a process $[z]$ has the deterministic value of t at time t. Expressed in infinitesimal terms, we say that:

$$(dz_t)^2 = dt$$

This formula generalizes to:

$$(dz_t^{(1)}) \cdot (dz_t^{(2)}) = \rho \cdot dt$$

where $z^{(1)}$ and $z^{(2)}$ are two different Brownian motions with correlation between the random variables $z_t^{(1)}$ and $z_t^{(2)}$ equal to ρ for all $t > 0$.

You should intuitively interpret the formula $(dz_t)^2 = dt$ (and its generalization) as a deterministic statement, and in fact, this statement is used as an algebraic convenience in Brownian motion-based stochastic calculus, forming the core of *Ito Isometry* and *Ito's Lemma* (which we cover shortly, but first we need to define the Ito Integral).

C.5   ITO INTEGRAL

We want to define a stochastic process Y from a stochastic process X as follows:

$$Y_t = \int_0^t X_s \cdot dz_s$$

In the interest of focusing on intuition rather than rigor, we skip the technical details of filtrations and adaptive processes that make the above integral sensible. Instead, we simply say that this integral makes sense only if random variable X_s for any time s is disallowed from depending on $z_{s'}$ for any $s' > s$ (i.e., the stochastic process X cannot peek into the future) and that the time-integral $\int_0^t X_s^2 \cdot ds$ is finite for all $t \geq 0$. So we shall roll forward with the assumption that the stochastic process Y is defined as the above-specified integral (known as the *Ito Integral*) of a stochastic process X with respect to Brownian motion. The equivalent notation is:

$$dY_t = X_t \cdot dz_t$$

We state without proof the following properties of the Ito Integral stochastic process Y:

- Y is a martingale, i.e., $\mathbb{E}[(Y_t - Y_s)|Y_s] = 0$ (i.e., $\mathbb{E}[Y_t|Y_s] = Y_s$) for all $0 \leq s < t$
- **Ito Isometry**: $\mathbb{E}[Y_t^2] = \int_0^t \mathbb{E}[X_s^2] \cdot ds$.
- Quadratic Variation formula: $[Y]_t = \int_0^t X_s^2 \cdot ds$

Note that we have generalized the notation $[X]$ for discrete-time processes to continuous-time processes, defined as $[X]_t = \int_0^t (dX_s)^2$ for any continuous-time stochastic process.

Ito Isometry generalizes to:

$$\mathbb{E}[(\int_S^T X_t^{(1)} \cdot dz_t^{(1)})(\int_S^T X_t^{(2)} \cdot dz_t^{(2)})] = \int_S^T \mathbb{E}[X_t^{(1)} \cdot X_t^{(2)} \cdot \rho \cdot dt]$$

where $X^{(1)}$ and $X^{(2)}$ are two different stochastic processes, and $z^{(1)}$ and $z^{(2)}$ are two different Brownian motions with correlation between the random variables $z_t^{(1)}$ and $z_t^{(2)}$ equal to ρ for all $t > 0$.

Likewise, the Quadratic Variation formula generalizes to:

$$\int_S^T (X_t^{(1)} \cdot dz_t^{(1)})(X_t^{(2)} \cdot dz_t^{(2)}) = \int_S^T X_t^{(1)} \cdot X_t^{(2)} \cdot \rho \cdot dt$$

C.6 ITO'S LEMMA

We can extend the above Ito Integral to an Ito process Y as defined below:

$$dY_t = \mu_t \cdot dt + \sigma_t \cdot dz_t$$

We require the same conditions for the stochastic process σ as we required above for X in the definition of the Ito Integral. Moreover, we require that: $\int_0^t |\mu_s| \cdot ds$ is finite for all $t \geq 0$.

In the context of this Ito process Y described above, we refer to μ as the *drift* process and we refer to σ as the *dispersion* process.

Now, consider a twice-differentiable function $f : [0, T] \times \mathbb{R} \to \mathbb{R}$. We define a stochastic process whose (random) value at time t is $f(t, Y_t)$. Let's write its Taylor series with respect to the variables t and Y_t.

$$df(t, Y_t) = \frac{\partial f(t, Y_t)}{\partial t} \cdot dt + \frac{\partial f(t, Y_t)}{\partial Y_t} \cdot dY_t + \frac{1}{2} \cdot \frac{\partial^2 f(t, Y_t)}{\partial Y_t^2} \cdot (dY_t)^2 + \dots$$

Substituting for dY_t and lightening notation, we get:

$$df(t, Y_t) = \frac{\partial f}{\partial t} \cdot dt + \frac{\partial f}{\partial Y_t} \cdot (\mu_t \cdot dt + \sigma_t \cdot dz_t) + \frac{1}{2} \cdot \frac{\partial^2 f}{\partial Y_t^2} \cdot (\mu_t \cdot dt + \sigma_t \cdot dz_t)^2 + \dots$$

Next, we use the rules: $(dt)^2 = 0, dt \cdot dz_t = 0, (dz_t)^2 = dt$ to get **Ito's Lemma**:

$$df(t, Y_t) = \left(\frac{\partial f}{\partial t} + \mu_t \cdot \frac{\partial f}{\partial Y_t} + \frac{\sigma_t^2}{2} \cdot \frac{\partial^2 f}{\partial Y_t^2} \right) \cdot dt + \sigma_t \cdot \frac{\partial f}{\partial Y_t} \cdot dz_t \qquad \text{(C.1)}$$

Ito's Lemma describes the stochastic process of a function (f) of an Ito Process (Y) in terms of the partial derivatives of f, and in terms of the drift (μ) and dispersion (σ) processes that define Y.

If we generalize Y to be an n-dimensional stochastic process (as a column vector) with $\boldsymbol{\mu}_t$ as an n-dimensional (stochastic) column vector, $\boldsymbol{\sigma}_t$ as an $n \times m$ (stochastic) matrix, and $\boldsymbol{z}_t$ as an m-dimensional vector of m independent standard Brownian motions (as follows):

$$d\boldsymbol{Y}_t = \boldsymbol{\mu}_t \cdot dt + \boldsymbol{\sigma}_t \cdot d\boldsymbol{z}_t$$

then we get the multi-variate version of Ito's Lemma, as follows:

$$df(t, \boldsymbol{Y}_t) = \left(\frac{\partial f}{\partial t} + (\nabla_{\boldsymbol{Y}} f)^T \cdot \boldsymbol{\mu}_t + \frac{1}{2} Tr[\boldsymbol{\sigma}_t^T \cdot (\Delta_{\boldsymbol{Y}} f) \cdot \boldsymbol{\sigma}_t] \right) \cdot dt + (\nabla_{\boldsymbol{Y}} f)^T \cdot \boldsymbol{\sigma}_t \cdot d\boldsymbol{z}_t \qquad \text{(C.2)}$$

where the symbol ∇ represents the gradient of a function, the symbol Δ represents the Hessian of a function, and the symbol Tr represents the Trace of a matrix.

Next, we cover two common Ito processes, and use Ito's Lemma to solve the Stochastic Differential Equation represented by these Ito Processes:

C.7 A LOGNORMAL PROCESS

Consider a stochastic process x described in the form of the following Ito process:

$$dx_t = \mu(t) \cdot x_t \cdot dt + \sigma(t) \cdot x_t \cdot dz_t$$

Note that here z is standard (one-dimensional) Brownian motion, and μ, σ are deterministic functions of time t. This is solved easily by defining an appropriate function of x_t and applying Ito's Lemma, as follows:

$$y_t = \log(x_t)$$

Applying Ito's Lemma on y_t with respect to x_t, we get:

$$dy_t = (\mu(t) \cdot x_t \cdot \frac{1}{x_t} - \frac{\sigma^2(t) \cdot x_t^2}{2} \cdot \frac{1}{x_t^2}) \cdot dt + \sigma(t) \cdot x_t \cdot \frac{1}{x_t} \cdot dz_t$$

$$= (\mu(t) - \frac{\sigma^2(t)}{2}) \cdot dt + \sigma(t) \cdot dz_t$$

So,

$$y_T = y_S + \int_S^T (\mu(t) - \frac{\sigma^2(t)}{2}) \cdot dt + \int_S^T \sigma(t) \cdot dz_t$$

$$x_T = x_S \cdot e^{\int_S^T (\mu(t) - \frac{\sigma^2(t)}{2}) \cdot dt + \int_S^T \sigma(t) \cdot dz_t}$$

$x_T | x_S$ follows a lognormal distribution, i.e.,

$$y_T = \log(x_T) \sim \mathcal{N}(\log(x_S) + \int_S^T (\mu(t) - \frac{\sigma^2(t)}{2}) \cdot dt, \int_S^T \sigma^2(t) \cdot dt)$$

$$E[x_T | x_S] = x_S \cdot e^{\int_S^T \mu(t) \cdot dt}$$

$$E[x_T^2 | x_S] = x_S^2 \cdot e^{\int_S^T (2\mu(t) + \sigma^2(t)) \cdot dt}$$

$$Variance[x_T | x_S] = E[x_T^2 | x_S] - (E[x_T | x_S])^2 = x_S^2 \cdot e^{\int_S^T 2\mu(t) \cdot dt} \cdot (e^{\int_S^T \sigma^2(t) \cdot dt} - 1)$$

The special case of $\mu(t) = \mu$ (constant) and $\sigma(t) = \sigma$ (constant) is a very common Ito process used all over Finance/Economics (for its simplicity, tractability as well as practicality), and is known as Geometric Brownian Motion, to reflect the fact that the stochastic increment of the process ($\sigma \cdot x_t \cdot dz_t$) is multiplicative to the level of the process x_t. If we consider this special case, we get:

$$y_T = \log(x_T) \sim \mathcal{N}(\log(x_S) + (\mu - \frac{\sigma^2}{2})(T - S), \sigma^2(T - S))$$

$$E[x_T | x_S] = x_S \cdot e^{\mu(T-S)}$$

$$Variance[x_T | x_S] = x_S^2 \cdot e^{2\mu(T-S)} \cdot (e^{\sigma^2(T-S)} - 1)$$

C.8 A MEAN-REVERTING PROCESS

Now we consider a stochastic process x described in the form of the following Ito process:

$$dx_t = \mu(t) \cdot x_t \cdot dt + \sigma(t) \cdot dz_t$$

As in the process of the previous section, z is standard (one-dimensional) Brownian motion, and μ, σ are deterministic functions of time t. This is solved easily by defining an appropriate function of x_t and applying Ito's Lemma, as follows:

$$y_t = x_t \cdot e^{-\int_0^t \mu(u) \cdot du}$$

Applying Ito's Lemma on y_t with respect to x_t, we get:

$$dy_t = (-x_t \cdot \mu(t) \cdot e^{-\int_0^t \mu(u) \cdot du} + \mu(t) \cdot x_t \cdot e^{-\int_0^t \mu(u) \cdot du}) \cdot dt + \sigma(t) \cdot e^{-\int_0^t \mu(u) \cdot du} \cdot dz_t$$
$$= \sigma(t) \cdot e^{-\int_0^t \mu(u) \cdot du} \cdot dz_t$$

So the process y is a martingale. Using Ito Isometry, we get:

$$y_T \sim \mathcal{N}(y_S, \int_S^T \sigma^2(t) \cdot e^{-\int_0^t 2\mu(u) \cdot du} \cdot dt)$$

Therefore,

$$x_T \sim \mathcal{N}(x_S \cdot e^{\int_S^T \mu(t) \cdot dt}, e^{\int_0^T 2\mu(t) \cdot dt} \cdot \int_S^T \sigma^2(t) \cdot e^{-\int_0^t 2\mu(u) \cdot du} \cdot dt)$$

We call this process "mean-reverting" because with negative $\mu(t)$, the process is "pulled" to a baseline level of 0, at a speed whose expectation is proportional to $-\mu(t)$ and proportional to the distance from the baseline (so we say the process reverts to a baseline of 0 and the strength of mean-reversion is greater if the distance from the baseline is greater). If $\mu(t)$ is positive, then we say that the process is "mean-diverting" to signify that it gets pulled away from the baseline level of 0.

The special case of $\mu(t) = \mu$ (constant) and $\sigma(t) = \sigma$ (constant) is a fairly common Ito process (again for its simplicity, tractability as well as practicality), and is known as the Ornstein-Uhlenbeck Process with the mean (baseline) level set to 0. If we consider this special case, we get:

$$x_T \sim \mathcal{N}(x_S \cdot e^{\mu(T-S)}, \frac{\sigma^2}{2\mu} \cdot (e^{2\mu(T-S)} - 1))$$

The Hamilton-Jacobi-Bellman (HJB) Equation

In this Appendix, we provide a quick coverage of the Hamilton-Jacobi-Bellman (HJB) Equation, which is the continuous-time version of the Bellman Optimality Equation. Although much of this book covers Markov Decision Processes in a discrete-time setting, we do cover some classical Mathematical Finance Stochastic Control formulations in continuous-time. To understand these formulations, one must first understand the HJB Equation, which is the purpose of this Appendix. As is the norm in the Appendices in this book, we will compromise on some of the rigor and emphasize the intuition to develop basic familiarity with HJB.

D.1 HJB AS A CONTINUOUS-TIME VERSION OF BELLMAN OPTIMALITY EQUATION

In order to develop the continuous-time setting, we shall consider a (not necessarily time-homogeneous) process where the set of states at time t are denoted as $\mathcal{S}_t$ and the set of allowable actions for each state at time t are denoted as $\mathcal{A}_t$. Since time is continuous, Rewards are represented as a *Reward Rate* function $\mathcal{R}$ such that for any state $s_t \in \mathcal{S}_t$ and for any action $a_t \in \mathcal{A}_t$, $\mathcal{R}(t, s_t, a_t) \cdot dt$ is the *Expected Reward* in the time interval $(t, t + dt]$, conditional on state s_t and action a_t (note the functional dependency of $\mathcal{R}$ on t since we will be integrating $\mathcal{R}$ over time). Instead of the discount factor γ as in the case of discrete-time MDPs, here we employ a *discount rate* (akin to interest-rate discounting) $\rho \in \mathbb{R}_{\geq 0}$ so that the discount factor over any time interval $(t, t + dt]$ is $e^{-\rho \cdot dt}$.

We denote the Optimal Value Function as V^* such that the Optimal Value for state $s_t \in \mathcal{S}_t$ at time t is $V^*(t, s_t)$. Note that unlike Section 5.13 in Chapter 5 where we denoted the Optimal Value Function as a time-indexed sequence $V_t^*(s_t)$, here we make t an explicit functional argument of V^*. This is because in the continuous-time setting, we are interested in the time-differential of the Optimal Value Function.

Now let us write the Bellman Optimality Equation in its continuous-time version, i.e, let us consider the process V^* over the time interval $(t, t + dt]$ as follows:

$$V^*(t, s_t) = \max_{a_t \in \mathcal{A}_t} \{\mathcal{R}(t, s_t, a_t) \cdot dt + \mathbb{E}_{(t, s_t, a_t)}[e^{-\rho \cdot dt} \cdot V^*(t + dt, s_{t+dt})]\}$$

Multiplying throughout by $e^{-\rho t}$ and re-arranging, we get:

$$\max_{a_t \in \mathcal{A}_t} \{e^{-\rho t} \cdot \mathcal{R}(t, s_t, a_t) \cdot dt + \mathbb{E}_{(t, s_t, a_t)}[e^{-\rho(t+dt)} \cdot V^*(t + dt, s_{t+dt}) - e^{-\rho t} \cdot V^*(t, s_t)]\} = 0$$

$$\Rightarrow \max_{a_t \in \mathcal{A}_t} \{e^{-\rho t} \cdot \mathcal{R}(t, s_t, a_t) \cdot dt + \mathbb{E}_{(t, s_t, a_t)}[d\{e^{-\rho t} \cdot V^*(t, s_t)\}]\} = 0$$

$$\Rightarrow \max_{a_t \in \mathcal{A}_t} \{e^{-\rho t} \cdot \mathcal{R}(t, s_t, a_t) \cdot dt + \mathbb{E}_{(t, s_t, a_t)}[e^{-\rho t} \cdot (dV^*(t, s_t) - \rho \cdot V^*(t, s_t) \cdot dt)]\} = 0$$

Multiplying throughout by $e^{\rho t}$ and re-arranging, we get:

$$\rho \cdot V^*(t, s_t) \cdot dt = \max_{a_t \in \mathcal{A}_t} \{\mathbb{E}_{(t, s_t, a_t)}[dV^*(t, s_t)] + \mathcal{R}(t, s_t, a_t) \cdot dt\} \tag{D.1}$$

For a finite-horizon problem terminating at time T, the above equation is subject to terminal condition:

$$V^*(T, s_T) = \mathcal{T}(s_T)$$

for some terminal reward function $\mathcal{T}(\cdot)$.

Equation (D.1) is known as the Hamilton-Jacobi-Bellman Equation—the continuous-time analog of the Bellman Optimality Equation. In the literature, it is often written in a more compact form that essentially takes the above form and "divides throughout by dt". This requires a few technical details involving the stochastic differentiation operator. To keep things simple, we shall stick to the HJB formulation of Equation (D.1).

D.2   HJB WITH STATE TRANSITIONS AS AN ITO PROCESS

Although we have expressed the HJB Equation for V^*, we cannot do anything useful with it unless we know the state transition probabilities (all of which are buried inside the calculation of $\mathbb{E}_{(t, s_t, a_t)}[\cdot]$ in the HJB Equation). In continuous-time, the state transition probabilities are modeled as a stochastic process for states (or of its features). Let us assume that states are real-valued vectors, i.e, state $s_t \in \mathbb{R}^n$ at any time $t \geq 0$ and that the transitions for s are given by an Ito process, as follows:

$$ds_t = \mu(t, s_t, a_t) \cdot dt + \sigma(t, s_t, a_t) \cdot dz_t$$

where the function μ (drift function) gives an $\mathbb{R}^n$ valued process, the function σ (dispersion function) gives an $\mathbb{R}^{n \times m}$-valued process and z is an m-dimensional process consisting of m independent standard Brownian motions.

Now we can apply multivariate Ito's Lemma (Equation (C.2) from Appendix C) for V^* as a function of t and s_t (we lighten notation by writing μ_t and σ_t instead of $\mu(t, s_t, a_t)$ and $\sigma(t, s_t, a_t)$):

$$dV^*(t, s_t) = (\frac{\partial V^*}{\partial t} + (\nabla_s V^*)^T \cdot \mu_t + \frac{1}{2} Tr[\sigma_t^T \cdot (\Delta_s V^*) \cdot \sigma_t]) \cdot dt + (\nabla_s V^*)^T \cdot \sigma_t \cdot dz_t$$

Substituting this expression for $dV^*(t, s_t)$ in Equation (D.1), noting that

$$\mathbb{E}_{(t, s_t, a_t)}[(\nabla_s V^*)^T \cdot \sigma_t \cdot dz_t] = 0$$

and dividing throughout by dt, we get:

$$\rho \cdot V^*(t, s_t) = \max_{a_t \in \mathcal{A}_t} \{\frac{\partial V^*}{\partial t} + (\nabla_s V^*)^T \cdot \mu_t + \frac{1}{2} Tr[\sigma_t^T \cdot (\Delta_s V^*) \cdot \sigma_t] + \mathcal{R}(t, s_t, a_t)\} \tag{D.2}$$

For a finite-horizon problem terminating at time T, the above equation is subject to terminal condition:

$$V^*(T, s_T) = \mathcal{T}(s_T)$$

for some terminal reward function $\mathcal{T}(\cdot)$.

Black-Scholes Equation and Its Solution for Call/Put Options

In this Appendix, we sketch the derivation of the much-celebrated Black-Scholes equation and its solution for Call and Put Options (Black and Scholes 1973). As is the norm in the Appendices in this book, we will compromise on some of the rigor and emphasize the intuition to develop basic familiarity with concepts in continuous-time derivatives pricing and hedging.

E.1 ASSUMPTIONS

The Black-Scholes Model is about pricing and hedging of a derivative on a single underlying asset (henceforth, simply known as "underlying"). The model makes several simplifying assumptions for analytical convenience. Here are the assumptions:

- The underlying (whose price we denote as S_t as time t) follows a special case of the lognormal process we covered in Section C.7 of Appendix C, where the drift $\mu(t)$ is a constant (call it $\mu \in \mathbb{R}$) and the dispersion $\sigma(t)$ is also a constant (call it $\sigma \in \mathbb{R}^+$):

$$dS_t = \mu \cdot S_t \cdot dt + \sigma \cdot S_t \cdot dz_t \qquad (E.1)$$

  This process is often referred to as *Geometric Brownian Motion* to reflect the fact that the stochastic increment of the process ($\sigma \cdot S_t \cdot dz_t$) is multiplicative to the level of the process S_t.
- The derivative has a known payoff at time $t = T$, as a function $f : \mathbb{R}^+ \to \mathbb{R}$ of the underlying price S_T at time T.
- Apart from the underlying, the market also includes a riskless asset (which should be thought of as lending/borrowing money at a constant infinitesimal rate of annual return equal to r). The riskless asset (denote its price as R_t at time t) movements can thus be described as:

$$dR_t = r \cdot R_t \cdot dt$$

- Assume that we can trade in any real-number quantity in the underlying as well as in the riskless asset, in continuous-time, without any transaction costs (i.e., the typical "frictionless" market assumption).

DOI: 10.1201/9781003229193-E

E.2   DERIVATION OF THE BLACK-SCHOLES EQUATION

We denote the price of the derivative at any time t for any price S_t of the underlying as $V(t, S_t)$. Thus, $V(T, S_T)$ is equal to the payoff $f(S_T)$. Applying Ito's Lemma on $V(t, S_t)$ (see Equation (C.1) in Appendix C), we get:

$$dV(t, S_t) = \left(\frac{\partial V}{\partial t} + \mu \cdot S_t \cdot \frac{\partial V}{\partial S_t} + \frac{\sigma^2}{2} \cdot S_t^2 \cdot \frac{\partial^2 V}{\partial S_t^2}\right) \cdot dt + \sigma \cdot S_t \cdot \frac{\partial V}{\partial S_t} \cdot dz_t \qquad (E.2)$$

Now here comes the key idea: create a portfolio comprising of the derivative and the underlying so as to eliminate the incremental uncertainty arising from the Brownian motion increment dz_t. It's clear from the coefficients of dz_t in Equation (E.1) and (E.2) that this can be accomplished with a portfolio comprising of $\frac{\partial V}{\partial S_t}$ units of the underlying and -1 units of the derivative (i.e., by selling a derivative contract written on a single unit of the underlying). Let us refer to the value of this portfolio as Π_t at time t. Thus,

$$\Pi_t = -V(t, S_t) + \frac{\partial V}{\partial S_t} \cdot S_t \qquad (E.3)$$

Over an infinitesimal time-period $[t, t+dt]$, the change in the portfolio value Π_t is given by:

$$d\Pi_t = -dV(t, S_t) + \frac{\partial V}{\partial S_t} \cdot dS_t$$

Substituting for dS_t and $dV(t, S_t)$ from Equations (E.1) and (E.2), we get:

$$d\Pi_t = \left(-\frac{\partial V}{\partial t} - \frac{\sigma^2}{2} \cdot S_t^2 \cdot \frac{\partial^2 V}{\partial S_t^2}\right) \cdot dt \qquad (E.4)$$

Thus, we have eliminated the incremental uncertainty arising from dz_t and hence, this is a riskless portfolio. To ensure the market remains free of arbitrage, the infinitesimal rate of annual return for this riskless portfolio must be the same as that for the riskless asset, i.e., must be equal to r. Therefore,

$$d\Pi_t = r \cdot \Pi_t \cdot dt \qquad (E.5)$$

From Equations (E.4) and (E.5), we infer that:

$$-\frac{\partial V}{\partial t} - \frac{\sigma^2}{2} \cdot S_t^2 \cdot \frac{\partial^2 V}{\partial S_t^2} = r \cdot \Pi_t$$

Substituting for Π_t from Equation (E.3), we get:

$$-\frac{\partial V}{\partial t} - \frac{\sigma^2}{2} \cdot S_t^2 \cdot \frac{\partial^2 V}{\partial S_t^2} = r \cdot \left(-V(t, S_t) + \frac{\partial V}{\partial S_t} \cdot S_t\right)$$

Re-arranging, we arrive at the famous Black-Scholes equation:

$$\frac{\partial V}{\partial t} + \frac{\sigma^2}{2} \cdot S_t^2 \cdot \frac{\partial^2 V}{\partial S_t^2} + r \cdot S_t \cdot \frac{\partial V}{\partial S_t} + r \cdot V(t, S_t) = 0 \qquad (E.6)$$

A few key points to note here:

1. The Black-Scholes equation is a partial differential equation (PDE) in t and S_t, and it is valid for any derivative with arbitary payoff $f(S_T)$ at a fixed time $t = T$, and the derivative price function $V(t, S_t)$ needs to be twice differentiable with respect to S_t and once differentiable with respect to t.

2. The infinitesimal change in the portfolio value ($= d\Pi_t$) incorporates only the infinitesimal changes in the prices of the underlying and the derivative, and not the changes in the units held in the underlying and the derivative (meaning the portfolio is assumed to be self-financing). The portfolio composition does change continuously though since the units held in the underlying at time t needs to be $\frac{\partial V}{\partial S_t}$, which in general would change as time evolves and as the price S_t of the underlying changes. Note that $-\frac{\partial V}{\partial S_t}$ represents the hedge units in the underlying at any time t for any underlying price S_t, which nullifies the risk of changes to the derivative price $V(t, S_t)$.

3. The drift μ of the underlying price movement (interpreted as expected annual rate of return of the underlying) does not appear in the Black-Scholes Equation and hence, the price of any derivative will be independent of the expected rate of return of the underlying. Note though the prominent appearance of σ (referred to as the underlying volatility) and the riskless rate of return r in the Black-Scholes equation.

E.3 SOLUTION OF THE BLACK-SCHOLES EQUATION FOR CALL/PUT OPTIONS

The Black–Scholes PDE can be solved numerically using standard methods such as finite-differences. It turns out we can solve this PDE as an exact formula (closed-form solution) for the case of European call and put options, whose payoff functions are $\max(S_T - K, 0)$ and $\max(K - S_T, 0)$ respectively, where K is the option strike. We shall denote the call and put option prices at time t for underlying price of S_t as $C(t, S_t)$ and $P(t, S_t)$ respectively (as specializations of $V(t, S_t)$). We derive the solution below for call option pricing, with put option pricing derived similarly. Note that we could simply use the put-call parity: $C(t, S_t) - P(t, S_t) = S_t - K \cdot e^{-r \cdot (T-t)}$ to obtain the put option price from the call option price. The put-call parity holds because buying a call option and selling a put option is a combined payoff of $S_T - K$—this means owning the underlying and borrowing $K \cdot e^{-r(T-t)}$ at time t, whose value is $S_t - K \cdot e^{-r \cdot (T-t)}$.

To derive the formula for $C(t, S_t)$, we perform the following change-of-variables transformation:

$$\tau = T - t$$

$$x = \log\left(\frac{S_t}{K}\right) + (r - \frac{\sigma^2}{2}) \cdot \tau$$

$$u(\tau, x) = C(t, S_t) \cdot e^{r\tau}$$

This reduces the Black-Scholes PDE into the *Heat Equation*:

$$\frac{\partial u}{\partial \tau} = \frac{\sigma^2}{2} \cdot \frac{\partial^2 u}{\partial x^2}$$

The terminal condition $C(T, S_T) = \max(S_T - K, 0)$ transforms into the Heat Equation's initial condition:

$$u(0, x) = K \cdot (e^{\max(x,0)} - 1)$$

Using the standard convolution method for solving this Heat Equation with initial condition $u(0, x)$, we obtain the Green's Function Solution:

$$u(\tau, x) = \frac{1}{\sigma\sqrt{2\pi\tau}} \cdot \int_{-\infty}^{+\infty} u(0, y) \cdot e^{-\frac{(x-y)^2}{2\sigma^2\tau}} \cdot dy$$

With some manipulations, this yields:

$$u(\tau, x) = K \cdot e^{x + \frac{\sigma^2 \tau}{2}} \cdot N(d_1) - K \cdot N(d_2)$$

where $N(\cdot)$ is the standard normal cumulative distribution function:

$$N(z) = \frac{1}{\sigma \sqrt{2\pi}} \int_{-\infty}^{z} e^{-\frac{y^2}{2}} \cdot dy$$

and d_1, d_2 are the quantities:

$$d_1 = \frac{x + \sigma^2 \tau}{\sigma \sqrt{\tau}}$$

$$d_2 = d_1 - \sigma \sqrt{\tau}$$

Substituting for $\tau, x, u(\tau, x)$ with $t, S_t, C(t, S_t)$, we get:

$$C(t, S_t) = S_t \cdot N(d_1) - K \cdot e^{-r \cdot (T-t)} \cdot N(d_2) \tag{E.7}$$

where

$$d_1 = \frac{\log\left(\frac{S_t}{K}\right) + (r + \frac{\sigma^2}{2}) \cdot (T - t)}{\sigma \cdot \sqrt{T - t}}$$

$$d_2 = d_1 - \sigma \sqrt{T - t}$$

The put option price is:

$$P(t, S_t) = K \cdot e^{-r \cdot (T-t)} \cdot N(-d_2) - S_t \cdot N(-d_1) \tag{E.8}$$

Function Approximations as Affine Spaces

F.1   VECTOR SPACE

A Vector space is defined as a commutative group $\mathcal{V}$ under an addition operation (written as $+$), together with multiplication of elements of $\mathcal{V}$ with elements of a field $\mathcal{K}$ (known as scalars field), expressed as a binary in-fix operation $* : \mathcal{K} \times \mathcal{V} \to \mathcal{V}$, with the following properties:

- $a * (b * v) = (a * b) * v$, for all $a, b \in \mathcal{K}$, for all $v \in \mathcal{V}$.
- $1 * v = v$ for all $v \in \mathcal{V}$ where 1 denotes the multiplicative identity of $\mathcal{K}$.
- $a * (v_1 + v_2) = a * v_1 + a * v_2$ for all $a \in \mathcal{K}$, for all $v_1, v_2 \in \mathcal{V}$.
- $(a + b) * v = a * v + b * v$ for all $a, b \in \mathcal{K}$, for all $v \in \mathcal{V}$.

F.2   FUNCTION SPACE

The set $\mathcal{F}$ of all functions from an arbitrary generic domain $\mathcal{X}$ to a vector space co-domain $\mathcal{V}$ (over scalars field $\mathcal{K}$) constitutes a vector space (known as function space) over the scalars field $\mathcal{K}$ with addition operation $(+)$ defined as:

$$(f + g)(x) = f(x) + g(x) \text{ for all } f, g \in \mathcal{F}, \text{ for all } x \in \mathcal{X}$$

and scalar multiplication operation $(*)$ defined as:

$$(a * f)(x) = a * f(x) \text{ for all } f \in \mathcal{F}, \text{ for all } a \in \mathcal{K}, \text{ for all } x \in \mathcal{X}$$

Hence, addition and scalar multiplication for a function space are defined point-wise.

F.3   LINEAR MAP OF VECTOR SPACES

A linear map of Vector Spaces is a function $h : \mathcal{V} \to \mathcal{W}$ where $\mathcal{V}$ is a vector space over a scalars field $\mathcal{K}$ and $\mathcal{W}$ is a vector space over the same scalars field $\mathcal{K}$, having the following two properties:

- $h(v_1 + v_2) = h(v_1) + h(v_2)$ for all $v_1, v_2 \in \mathcal{V}$ (i.e., application of h commutes with the addition operation).
- $h(a * v) = a * h(v)$ for all $v \in \mathcal{V}$, for all $a \in \mathcal{K}$ (i.e., application of h commutes with the scalar multiplication operation).

DOI: 10.1201/9781003229193-F

Then the set of all linear maps with domain V and co-domain W constitute a function space (restricted to just this subspace of all linear maps, rather than the space of all $V \to W$ functions) that we denote as $\mathcal{L}(V, W)$.

The specialization of the function space of linear maps to the space $\mathcal{L}(V, \mathcal{K})$ (i.e., specializing the vector space W to the scalars field $\mathcal{K}$) is known as the dual vector space and is denoted as V^*.

F.4 AFFINE SPACE

An Affine Space is defined as a set $\mathcal{A}$ associated with a vector space V and a binary in-fix operation $\oplus : \mathcal{A} \times V \to \mathcal{A}$, with the following properties:

- For all $a \in \mathcal{A}, a \oplus 0 = a$, where 0 is the zero vector in V (this is known as the right identity property).
- For all $v_1, v_2 \in V$, for all $a \in \mathcal{A}, (a \oplus v_1) \oplus v_2 = a \oplus (v_1 + v_2)$ (this is known as the associativity property).
- For each $a \in \mathcal{A}$, the mapping $f_a : V \to \mathcal{A}$ defined as $f_a(v) = a \oplus v$ for all $v \in V$ is a bijection (i.e., one-to-one and onto mapping).

The elements of an affine space are called *points* and the elements of the vector space associated with an affine space are called *translations*. The idea behind affine spaces is that unlike a vector space, an affine space doesn't have a notion of a zero element and one cannot add two *points* in the affine space. Instead one adds a *translation* (from the associated vector space) to a *point* (from the affine space) to yield another *point* (in the affine space). The term *translation* is used to signify that we "translate" (i.e. shift) a point to another point in the affine space with the shift being effected by a *translation* in the associated vector space. The bijection property defined above implies that there is a notion of "subtracting" one *point* of the affine space from another *point* of the affine space (denoted with the operation $\ominus$), yielding a *translation* in the associated vector space. Formally, $\ominus$ is defined as:

For each $a_1, a_2 \in \mathcal{A}$, there exists a unique $v \in V$, denoted $a_2 \ominus a_1$, such that $a_2 = a_1 \oplus v$

A simple way to visualize an affine space is by considering the simple example of the affine space of all 3-D points on the plane defined by the equation $z = 1$, i.e., the set of all points $(x, y, 1)$ for all $x \in \mathbb{R}, y \in \mathbb{R}$. The associated vector space is the set of all 3-D points on the plane defined by the equation $z = 0$, i.e., the set of all points $(x, y, 0)$ for all $x \in \mathbb{R}, y \in \mathbb{R}$ (with element-wise addition and scalar multiplication operations). The $\oplus$ operation is element-wise addition. We see that any point $(x, y, 1)$ on the affine space is *translated* to the point $(x + x', y + y', 1)$ by the translation $(x', y', 0)$ in the vector space. Note that the translation $(0, 0, 0)$ (zero vector) results in the point $(x, y, 1)$ remaining unchanged. Note that translations $(x', y', 0)$ and $(x'', y'', 0)$ applied one after the other is the same as the single translation $(x' + x'', y' + y'', 0)$. Finally, note that for any fixed point $(x, y, 1)$, we have a bijective mapping from the vector space $z = 0$ to the affine space $z = 1$ that maps any translation $(x', y', 0)$ to the point $(x + x', y + y', 1)$.

F.5 AFFINE MAP

An Affine Map is a function $h : \mathcal{A} \to \mathcal{B}$, associated with a linear map $l : V \to W$, where $\mathcal{A}$ is an affine space associated with vector space V and $\mathcal{B}$ is an affine space associated with vector space W, having the following property:

$$h(a_1) \ominus h(a_2) = l(a_1 \ominus a_2) \text{ for all } a_1, a_2 \in \mathcal{A}$$

This implies:
$$h(a \oplus v) = h(a) \oplus l(v) \text{ for all } a \in \mathcal{A}, \text{ for all } v \in V$$

The intuitive way of thinking about an affine map h is that it's completely defined by the image $h(a)$ of *any single point* $a \in \mathcal{A}$ and by it's associated linear map l.

Later in this appendix, we consider a specialization of affine maps—when $V = W$ and l is the identity function. For this specialization, we have:

$$h(a \oplus v) = h(a) \oplus v \text{ for all } a \in \mathcal{A}, \text{ for all } v \in V$$

The way to think about this is that $\oplus : \mathcal{A} \times V \to \mathcal{A}$ simply delegates to $\oplus : \mathcal{B} \times V \to \mathcal{B}$. So we shall refer to such a specialization of affine maps as *Delegating Map* and the corresponding affine spaces $\mathcal{A}$ and $\mathcal{B}$ as *Delegator Space* and *Delegate Space*, respectively.

F.6   FUNCTION APPROXIMATIONS

We represent function approximations by parameterized functions $f : \mathcal{X} \times D[\mathbb{R}] \to \mathbb{R}$ where $\mathcal{X}$ is the input domain and $D[\mathbb{R}]$ is the parameters domain. The notation $D[Y]$ refers to a generic container data type D over a component generic data type Y. The data type D is specified as a generic container data type because we consider generic function approximations here. A specific family of function approximations will customize to a specific container data type for D (e.g., linear function approximations will customize D to a Sequence data type, a feed-forward deep neural network will customize D to a Sequence of 2-dimensional arrays). We are interested in viewing Function Approximations as *point*s in an appropriate Affine Space. To explain this, we start by viewing parameters as *point*s in an Affine Space.

F.6.1 $D[\mathbb{R}]$ as an Affine Space $\mathcal{P}$

When performing Stochastic Gradient Descent or Batch Gradient Descent, parameters $p \in D[\mathbb{R}]$ of a function approximation $f : \mathcal{X} \times D[\mathbb{R}] \to \mathbb{R}$ are updated using an appropriate linear combination of gradients of f with respect to p (at specific values of $x \in \mathcal{X}$). Hence, the parameters domain $D[\mathbb{R}]$ can be treated as an affine space (call it $\mathcal{P}$) whose associated vector space (over scalars field $\mathbb{R}$) is the set of gradients of f with respect to parameters $p \in D[\mathbb{R}]$ (denoted as $\nabla_p f(x, p)$), evaluated at specific values of $x \in \mathcal{X}$, with addition operation defined as element-wise real-numbered addition and scalar multiplication operation defined as element-wise multiplication with real-numbered scalars. We refer to this Affine Space $\mathcal{P}$ as the *Parameters Space* and we refer to its associated vector space (of gradients) as the *Gradient Space* $\mathcal{G}$. Since each *point* in $\mathcal{P}$ and each *translation* in $\mathcal{G}$ is an element in $D[\mathbb{R}]$, the $\oplus$ operation is element-wise real-numbered addition.

We define the gradient function

$$G : \mathcal{X} \to (\mathcal{P} \to \mathcal{G})$$

as:

$$G(x)(p) = \nabla_p f(x, p)$$

for all $x \in \mathcal{X}$, for all $p \in \mathcal{P}$.

F.6.2   Delegator Space $\mathcal{R}$

We consider a function $I : \mathcal{P} \to (\mathcal{X} \to \mathbb{R})$ defined as $I(p) = g : \mathcal{X} \to \mathbb{R}$ for all $p \in \mathcal{P}$ such that $g(x) = f(x, p)$ for all $x \in \mathcal{X}$. The *Range* of this function I forms an affine space $\mathcal{R}$ whose associated vector space is the Gradient Space $\mathcal{G}$, with the $\oplus$ operation defined as:

$$I(p) \oplus v = I(p \oplus v) \text{ for all } p \in \mathcal{P}, v \in \mathcal{G}$$

We refer to this affine space $\mathcal{R}$ as the *Delegator Space* to signify the fact that the $\oplus$ operation for $\mathcal{R}$ simply "delegates" to the $\oplus$ operation for $\mathcal{P}$ and so, the parameters $p \in \mathcal{P}$ basically serve as the internal representation of the function approximation $I(p) : \mathcal{X} \to \mathbb{R}$. This "delegation" from $\mathcal{R}$ to $\mathcal{P}$ implies that I is a *Delegating Map* (as defined earlier) from Parameters Space $\mathcal{P}$ to Delegator Space $\mathcal{R}$.

Notice that the __add__ method of the Gradient class in rl/function_approx.py is overloaded. One of the __add__ methods corresponds to vector addition of two gradients in the Gradient Space $\mathcal{G}$. The other __add__ method corresponds to the $\oplus$ operation adding a gradient (treated as a *translation* in the vector space of gradients) to a function approximation (treated as a *point* in the affine space of function approximations).

F.7 STOCHASTIC GRADIENT DESCENT

Stochastic Gradient Descent is a function

$$SGD : \mathcal{X} \times \mathbb{R} \to (\mathcal{P} \to \mathcal{P})$$

representing a mapping from (predictor, response) data to a "parameters-update" function (in order to improve the function approximation), defined as:

$$SGD(x, y)(p) = p \oplus (\alpha * ((y - f(x, p)) * G(x)(p)))$$

for all $x \in \mathcal{X}, y \in \mathbb{R}, p \in \mathcal{P}$, where $\alpha \in \mathbb{R}^+$ represents the learning rate (step size of SGD).

For a fixed data pair $(x, y) \in \mathcal{X} \times \mathbb{R}$, with prediction error function $e : \mathcal{P} \to \mathbb{R}$ defined as $e(p) = y - f(x, p)$, the (SGD-based) parameters change function

$$U : \mathcal{P} \to \mathcal{G}$$

is defined as:

$$U(p) = SGD(x, y)(p) \ominus p = \alpha * (e(p) * G(x)(p))$$

for all $p \in \mathcal{P}$.

So, we can conceptualize the parameters change function U as the product of:

- Learning rate $\alpha \in \mathbb{R}^+$
- Prediction error function $e : \mathcal{P} \to \mathbb{R}$
- Gradient operator $G(x) : \mathcal{P} \to \mathcal{G}$

Note that the product of functions e and $G(x)$ above is point-wise in their common domain $\mathcal{P} = D[\mathbb{R}]$, resulting in the scalar ($\mathbb{R}$) multiplication of vectors in $\mathcal{G}$.

Updating vector p to vector $p \oplus U(p)$ in the Parameters Space $\mathcal{P}$ results in updating function $I(p) : \mathcal{X} \to \mathbb{R}$ to function $I(p \oplus U(p)) : \mathcal{X} \to \mathbb{R}$ in the Delegator Space $\mathcal{R}$. This is rather convenient since we can view the $\oplus$ operation for the Parameters Space $\mathcal{P}$ as effectively the $\oplus$ operation in the Delegator Space $\mathcal{R}$.

F.8   SGD UPDATE FOR LINEAR FUNCTION APPROXIMATIONS

In this section, we restrict to linear function approximations, i.e., for all $x \in \mathcal{X}$,

$$f(x, \boldsymbol{p}) = \boldsymbol{\Phi}(x)^T \cdot \boldsymbol{p}$$

where $\boldsymbol{p} \in \mathbb{R}^m = \mathcal{P}$ and $\boldsymbol{\Phi} : \mathcal{X} \to \mathbb{R}^m$ represents the feature functions (note: $\boldsymbol{\Phi}(x)^T \cdot \boldsymbol{p}$ is the usual inner-product in $\mathbb{R}^m$).

Then the gradient function $G : \mathcal{X} \to (\mathbb{R}^m \to \mathbb{R}^m)$ can be written as:

$$G(x)(\boldsymbol{p}) = \nabla_{\boldsymbol{p}}(\boldsymbol{\Phi}(x)^T \cdot \boldsymbol{p}) = \boldsymbol{\Phi}(x)$$

for all $x \in \mathcal{X}$, for all $\boldsymbol{p} \in \mathbb{R}^m$.

When SGD-updating vector $\boldsymbol{p}$ to vector $\boldsymbol{p} \oplus (\alpha * ((y - \boldsymbol{\Phi}(x)^T \cdot \boldsymbol{p}) * \boldsymbol{\Phi}(x)))$ in the Parameters Space $\mathcal{P} = \mathbb{R}^m$, applying the affine map $I : \mathbb{R}^m \to \mathcal{R}$ correspondingly updates functions in $\mathcal{R}$. Concretely, a linear function approximation $g : \mathcal{X} \to \mathbb{R}$ defined as $g(z) = \boldsymbol{\Phi}(z)^T \cdot \boldsymbol{p}$ for all $z \in \mathcal{X}$ updates correspondingly to the function $g^{(x,y)} : \mathcal{X} \to \mathbb{R}$ defined as $g^{(x,y)}(z) = \boldsymbol{\Phi}(z)^T \cdot \boldsymbol{p} + \alpha \cdot (y - \boldsymbol{\Phi}(x)^T \cdot \boldsymbol{p}) \cdot (\boldsymbol{\Phi}(z)^T \cdot \boldsymbol{\Phi}(x))$ for all $z \in \mathcal{X}$.

It's useful to note that the change in the evaluation at $z \in \mathcal{X}$, i.e., $g^{(x,y)}(z) - g(z)$, is simply the product of:

- Learning rate $\alpha \in \mathbb{R}^+$
- Prediction Error $y - \boldsymbol{\Phi}(x)^T \cdot \boldsymbol{p} \in \mathbb{R}$ for the updating data $(x, y) \in \mathcal{X} \times \mathbb{R}$
- Inner-product of the feature vector $\boldsymbol{\Phi}(x) \in \mathbb{R}^m$ of the updating input value $x \in \mathcal{X}$ and the feature vector $\boldsymbol{\Phi}(z) \in \mathbb{R}^m$ of the evaluation input value $z \in \mathcal{X}$.

Conjugate Priors for Gaussian and Bernoulli Distributions

The setting for this Appendix is that we receive data incrementally as $x_1, x_2, \ldots$ and we assume a certain probability distribution (e.g., Gaussian, Bernoulli) for each $x_i, i = 1, 2, \ldots$. We utilize an appropriate conjugate prior for the assumed data distribution so that we can derive the posterior distribution for the parameters of the assumed data distribution. We can then say that for any $n \in \mathbb{Z}^+$, the conjugate prior is the probability distribution for the parameters of the assumed data distribution, conditional on the first n data points $(x_1, x_2, \ldots x_n)$ and the posterior is the probability distribution for the parameters of the assumed distribution, conditional on the first $n + 1$ data points $(x_1, x_2, \ldots, x_{n+1})$. This amounts to performing Bayesian updates on the hyperparameters upon receipt of each incremental data x_i (hyperparameters refer to the parameters of the prior and posterior distributions). In this appendix, we shall not cover the derivations of the posterior distribution from the prior distribution and the data distribution. We shall simply state the results (references for derivations can be found on the Conjugate Prior Wikipedia Page).

G.1   CONJUGATE PRIOR FOR GAUSSIAN DISTRIBUTION

Here we assume that each data point is Gaussian-distributed in $\mathbb{R}$. So when we receive the n-th data point x_n, we assume:

$$x_n \sim \mathcal{N}(\mu, \sigma^2)$$

and we assume both μ and σ^2 are unknown random variables with Gaussian-Inverse-Gamma Probability Distribution Conjugate Prior for μ and σ^2, i.e.,

$$\mu | x_1, \ldots, x_n \sim \mathcal{N}(\theta_n, \frac{\sigma^2}{n})$$

$$\sigma^2 | x_1, \ldots, x_n \sim IG(\alpha_n, \beta_n)$$

where $IG(\alpha_n, \beta_n)$ refers to the Inverse Gamma distribution with parameters α_n and β_n. This means $\frac{1}{\sigma^2} | x_1, \ldots, x_n$ follows a Gamma distribution with parameters α_n and β_n, i.e., the probability of $\frac{1}{\sigma^2}$ having a value $y \in \mathbb{R}^+$ is:

$$\frac{\beta^\alpha \cdot y^{\alpha-1} \cdot e^{-\beta y}}{\Gamma(\alpha)}$$

where $\Gamma(\cdot)$ is the Gamma Function.

DOI: 10.1201/9781003229193-G

$\theta_n, \alpha_n, \beta_n$ are hyperparameters determining the probability distributions of μ and σ^2, conditional on data $x_1, \ldots, x_n$.

Then, the posterior distribution is given by:

$$\mu | x_1, \ldots, x_{n+1} \sim \mathcal{N}\left(\frac{n\theta_n + x_{n+1}}{n+1}, \frac{\sigma^2}{n+1}\right)$$

$$\sigma^2 | x_1, \ldots, x_{n+1} \sim IG\left(\alpha_n + \frac{1}{2}, \beta_n + \frac{n(x_{n+1} - \theta_n)^2}{2(n+1)}\right)$$

This means upon receipt of the data point x_{n+1}, the hyperparameters can be updated as:

$$\theta_{n+1} = \frac{n\theta_n + x_{n+1}}{n+1}$$

$$\alpha_{n+1} = \alpha_n + \frac{1}{2}$$

$$\beta_{n+1} = \beta_n + \frac{n(x_{n+1} - \theta_n)^2}{2(n+1)}$$

G.2 CONJUGATE PRIOR FOR BERNOULLI DISTRIBUTION

Here we assume that each data point is Bernoulli-distributed. So when we receive the n-th data point x_n, we assume $x_n = 1$ with probability p and $x_n = 0$ with probability $1 - p$. We assume p is an unknown random variable with Beta Distribution Conjugate Prior for p, i.e,

$$p | x_1, \ldots, x_n \sim Beta(\alpha_n, \beta_n)$$

where $Beta(\alpha_n, \beta_n)$ refers to the Beta distribution with parameters α_n and β_n, i.e., the probability of p having a value $y \in [0, 1]$ is:

$$\frac{\Gamma(\alpha + \beta)}{\Gamma(\alpha) \cdot \Gamma(\beta)} \cdot y^{\alpha - 1} \cdot (1 - y)^{\beta - 1}$$

where $\Gamma(\cdot)$ is the Gamma Function.

α_n, β_n are hyperparameters determining the probability distribution of p, conditional on data $x_1, \ldots, x_n$.

Then, the posterior distribution is given by:

$$p | x_1, \ldots, x_{n+1} \sim Beta(\alpha_n + \mathbb{I}_{x_{n+1}=1}, \beta_n + \mathbb{I}_{x_{n+1}=0})$$

where $\mathbb{I}$ refers to the indicator function.

This means upon receipt of the data point x_{n+1}, the hyperparameters can be updated as:

$$\alpha_{n+1} = \alpha_n + \mathbb{I}_{x_{n+1}=1}$$

$$\beta_{n+1} = \beta_n + \mathbb{I}_{x_{n+1}=0}$$

Bibliography

Almgren, Robert, and Neil Chriss. 2000. 'Optimal Execution of Portfolio Transactions'. *Journal of Risk* 3 (2): 5–39.

Amari, S. 1998. 'Natural Gradient Works Efficiently in Learning'. *Neural Computation* 10 (2): 251–76.

Åström, K. J. 1965. 'Optimal Control of Markov Processes with Incomplete State Information'. *Journal of Mathematical Analysis and Applications* 10 (1): 174–205. https://doi.org/10.1016/0022-247X(65)90154-X.

Auer, Peter, Nicolò Cesa-Bianchi, and Paul Fischer. 2002. 'Finite-Time Analysis of the Multiarmed Bandit Problem'. *Machine Learning* 47 (2): 235–56. https://doi.org/10.1023/A:1013689704352.

Avellaneda, Marco, and Sasha Stoikov. 2008. 'High-Frequency Trading in a Limit Order Book'. *Quantitative Finance* 8 (3): 217–24. http://www.informaworld.com/10.1080/14697680701381228.

Baird, Leemon. 1995. 'Residual Algorithms: Reinforcement Learning with Function Approximation'. In *Machine Learning Proceedings 1995*, edited by Armand Prieditis and Stuart Russell, 30–37. San Francisco, CA: Morgan Kaufmann. https://doi.org/https://doi.org/10.1016/B978-1-55860-377-6.50013-X.

Barbera, S., P. Hammond, and C. Seidl. 1998. *Handbook of Utility Theory: Volume 1: Principles*. Handbook of Utility Theory. Springer US. https://books.google.com/books?id=ksRgriIymBUC.

Bellman, Richard. 1957a. 'A Markovian Decision Process'. *Journal of Mathematics and Mechanics* 6 (5): 679–84. http://www.jstor.org/stable/24900506.

———. 1957b. *Dynamic Programming*. 1st ed. Princeton, NJ: Princeton University Press.

Bertsekas, D. P., and J. N. Tsitsiklis. 1996. *Neuro-Dynamic Programming*. Belmont, MA: Athena Scientific.

Bertsekas, Dimitri P. 1981. 'Distributed Dynamic Programming'. In *1981 20th IEEE Conference on Decision and Control Including the Symposium on Adaptive Processes*, 774–79. https://doi.org/10.1109/CDC.1981.269319.

———. 1983. 'Distributed Asynchronous Computation of Fixed Points'. *Mathematical Programming* 27: 107–20.

———. 2005. *Dynamic Programming and Optimal Control, Volume 1, 3rd Edition*. Athena Scientific.

———. 2012. *Dynamic Programming and Optimal Control, Volume 2: Approximate Dynamic Programming*. Athena Scientific.

Bertsimas, Dimitris, and Andrew W. Lo. 1998. 'Optimal Control of Execution Costs'. *Journal of Financial Markets* 1 (1): 1–50.

Björk, Tomas. 2005. *Arbitrage Theory in Continuous Time*. 2. ed., reprint. Oxford [u.a.]: Oxford Univ. Press. http://gso.gbv.de/DB=2.1/CMD?ACT=SRCHA\&SRT=YOP\&IKT=1016\&TRM=ppn+505893878\&sourceid=fbw_bibsonomy.

Black, Fisher, and Myron S. Scholes. 1973. 'The Pricing of Options and Corporate Liabilities'. *Journal of Political Economy* 81 (3): 637–54.

Bradtke, Steven J., and Andrew G. Barto. 1996. 'Linear Least-Squares Algorithms for Temporal Difference Learning.' *Machine Learning* 22 (1-3): 33–57. http://dblp.uni-trier.de/db/journals/ml/ml22.html\#BradtkeB96.

Brafman, Ronen I., and Moshe Tennenholtz. 2001. 'R-MAX - a General Polynomial Time Algorithm for Near-Optimal Reinforcement Learning.' In *IJCAI*, edited by Bernhard Nebel, 953–58. Morgan Kaufmann. http://dblp.uni-trier.de/db/conf/ijcai/ijcai2001.html\#BrafmanT01.

Bühler, Hans, Lukas Gonon, Josef Teichmann, and Ben Wood. 2018. 'Deep Hedging'. http://arxiv.org/abs/1802.03042.

Chang, Hyeong Soo, Michael C. Fu, Jiaqiao Hu, and Steven I. Marcus. 2005. 'An Adaptive Sampling Algorithm for Solving Markov Decision Processes.' *Operations Research* 53 (1): 126–39. http://dblp.uni-trier.de/db/journals/ior/ior53.html\#ChangFHM05.

Coulom, Rémi. 2006. 'Efficient Selectivity and Backup Operators in Monte-Carlo Tree Search.' In *Computers and Games*, edited by H. Jaap van den Herik, Paolo Ciancarini, and H. H. L. M. Donkers, 4630:72–83. Lecture Notes in Computer Science. Springer. http://dblp.uni-trier.de/db/conf/cg/cg2006.html\#Coulom06.

Cox, J., S. Ross, and M. Rubinstein. 1979. 'Option Pricing: A Simplified Approach'. *Journal of Financial Economics* 7: 229–63.

Degris, Thomas, Martha White, and Richard S. Sutton. 2012. 'Off-Policy Actor-Critic'. In *Proceedings of the 29th International Coference on International Conference on Machine Learning*, 179–86. ICML'12. Madison, WI: Omnipress.

Gagniuc, Paul A. 2017. *Markov Chains: From Theory to Implementation and Experimentation*. John Wiley & Sons.

Ganesh, Sumitra, Nelson Vadori, Mengda Xu, Hua Zheng, Prashant P. Reddy, and Manuela Veloso. 2019. 'Reinforcement Learning for Market Making in a Multi-Agent Dealer Market.' *CoRR* abs/1911.05892. http://dblp.uni-trier.de/db/journals/corr/corr1911.html\#abs-1911-05892.

Gittins, John C. 1979. 'Bandit Processes and Dynamic Allocation Indices'. *Journal of the Royal Statistical Society. Series B (Methodological)*, 148–77.

Goodfellow, Ian, Yoshua Bengio, and Aaron Courville. 2016. *Deep Learning*. Cambridge, MA: MIT Press.

Gueant, Olivier. 2016. *The Financial Mathematics of Market Liquidity: From Optimal Execution to Market Making*. Chapman; Hall/CRC Financial Mathematics Series.

Guez, Arthur, Nicolas Heess, David Silver, and Peter Dayan. 2014. 'Bayes-Adaptive Simulation-Based Search with Value Function Approximation.' In *NIPS*, edited by Zoubin Ghahramani, Max Welling, Corinna Cortes, Neil D. Lawrence, and Kilian Q. Weinberger, 451–59. http://dblp.uni-trier.de/db/conf/nips/nips2014.html\#GuezHSD14.

Hasselt, Hado van, Yotam Doron, Florian Strub, Matteo Hessel, Nicolas Sonnerat, and Joseph Modayil. 2018. 'Deep Reinforcement Learning and the Deadly Triad'. *CoRR* abs/1812.02648. http://arxiv.org/abs/1812.02648.

Howard, R. A. 1960. *Dynamic Programming and Markov Processes*. Cambridge, MA: MIT Press.

Hull, John C. 2010. *Options, Futures, and Other Derivatives*. Seventh. Pearson.

Kakade, Sham M. 2001. 'A Natural Policy Gradient.' In *NIPS*, edited by Thomas G. Dietterich, Suzanna Becker, and Zoubin Ghahramani, 1531–38. Cambridge, MA: MIT Press. http://dblp.uni-trier.de/db/conf/nips/nips2001.html\#Kakade01.

Kingma, Diederik P., and Jimmy Ba. 2015. 'Adam: A Method for Stochastic Optimization'. In *3rd International Conference on Learning Representations, ICLR 2015, San Diego, CA, USA, May 7-9, 2015, Conference Track Proceedings*, edited by Yoshua Bengio and Yann LeCun. http://arxiv.org/abs/1412.6980.

Klopf, A. H., and Air Force Cambridge Research Laboratories (U.S.). Data Sciences Laboratory. 1972. *Brain Function and Adaptive Systems–a Heterostatic Theory*. Special Reports. Data Sciences Laboratory, Air Force Cambridge Research Laboratories, Air Force Systems Command, United States Air Force. `https://books.google.com/books?id=C2hztwEACAAJ`.

Kocsis, L., and Cs. Szepesvári. 2006. 'Bandit Based Monte-Carlo Planning'. In *ECML*, 282–93.

Krishnamurthy, Vikram. 2016. *Partially Observed Markov Decision Processes: From Filtering to Controlled Sensing*. Cambridge University Press. `https://doi.org/10.1017/CBO9781316471104`.

Lagoudakis, Michail G., and Ronald Parr. 2003. 'Least-Squares Policy Iteration.' *Journal of Machine Learning Research* 4: 1107–49. `http://dblp.uni-trier.de/db/journals/jmlr/jmlr4.html\#LagoudakisP03`.

Lai, T. L., and H. Robbins. 1985. 'Asymptotically Efficient Adaptive Allocation Rules'. *Advances in Applied Mathematics* 6: 4–22.

Li, Y., Cs. Szepesvári, and D. Schuurmans. 2009. 'Learning Exercise Policies for American Options'. In *AISTATS*, 5:352–59. `http://www.ics.uci.edu/~aistats/`.

Lin, Long J. 1993. 'Reinforcement Learning for Robots Using Neural Networks'. PhD thesis, Pittsburg: CMU.

Longstaff, Francis A., and Eduardo S. Schwartz. 2001. 'Valuing American Options by Simulation: A Simple Least-Squares Approach'. *Review of Financial Studies* 14 (1): 113–47. `https://doi.org/10.1093/rfs/14.1.113`.

Merton, Robert C. 1969. 'Lifetime Portfolio Selection Under Uncertainty: The Continuous-Time Case'. *The Review of Economics and Statistics* 51 (3): 247–57. `https://doi.org/10.2307/1926560`.

Mnih, Volodymyr, Koray Kavukcuoglu, David Silver, Alex Graves, Ioannis Antonoglou, Daan Wierstra, and Martin Riedmiller. 2013. 'Playing Atari with Deep Reinforcement Learning'. `http://arxiv.org/abs/1312.5602`.

Mnih, Volodymyr, Koray Kavukcuoglu, David Silver, Andrei A. Rusu, Joel Veness, Marc G. Bellemare, Alex Graves, et al. 2015. 'Human-Level Control Through Deep Reinforcement Learning'. *Nature* 518 (7540): 529–33. `https://doi.org/10.1038/nature14236`.

Nevmyvaka, Yuriy, Yi Feng, and Michael J. Kearns. 2006. 'Reinforcement Learning for Optimized Trade Execution.' In *ICML*, edited by William W. Cohen and Andrew W. Moore, 148:673–80. ACM International Conference Proceeding Series. ACM. `http://dblp.uni-trier.de/db/conf/icml/icml2006.html\#NevmyvakaFK06`.

Øksendal, Bernt. 2003. *Stochastic Differential Equations*. 6. ed. Universitext. Berlin ; Heidelberg [u.a.]: Springer. `http://aleph.bib.uni-mannheim.de/F/?func=find-b\&request=106106503\&find_code=020\&adjacent=N\&local_base=MAN01PUBLIC\&x=0\&y=0`.

Puterman, Martin L. 2014. *Markov Decision Processes: Discrete Stochastic Dynamic Programming*. John Wiley & Sons.

Rummery, G. A., and M. Niranjan. 1994. 'On-Line Q-Learning Using Connectionist Systems'. CUED/F-INFENG/TR-166. Engineering Department, Cambridge University.

Russo, Daniel J., Benjamin Van Roy, Abbas Kazerouni, Ian Osband, and Zheng Wen. 2018. 'A Tutorial on Thompson Sampling'. *Foundations and Trends® in Machine Learning* 11 (1): 1–96. `https://doi.org/10.1561/2200000070`.

Salimans, Tim, Jonathan Ho, Xi Chen, Szymon Sidor, and Ilya Sutskever. 2017. 'Evolution Strategies as a Scalable Alternative to Reinforcement Learning'. *arXiv Preprint arXiv:1703.03864*.

Sherman, Jack, and Winifred J. Morrison. 1950. 'Adjustment of an Inverse Matrix Corresponding to a Change in One Element of a Given Matrix'. *The Annals of Mathematical Statistics* 21 (1): 124–27. `https://doi.org/10.1214/aoms/1177729893`.

Shreve, Steven E. 2003. *Stochastic Calculus for Finance I: The Binomial Asset Pricing Model: Binomial Asset Pricing Model*. New York, NY: Springer-Verlag.

———. 2004. *Stochastic Calculus for Finance II: Continuous-Time Models*. New York: Springer.

Silver, David, Aja Huang, Chris J. Maddison, Arthur Guez, Laurent Sifre, George van den Driessche, Julian Schrittwieser, et al. 2016. 'Mastering the Game of Go with Deep Neural Networks and Tree Search'. *Nature* 529 (7587): 484–89. https://doi.org/10.1038/nature16961.

Silver, David, Guy Lever, Nicolas Heess, Thomas Degris, Daan Wierstra, and Martin A. Riedmiller. 2014. 'Deterministic Policy Gradient Algorithms.' In *ICML*, 32:387–95. JMLR Workshop and Conference Proceedings. JMLR.org. http://dblp.uni-trier.de/db/conf/icml/icml2014.html\#SilverLHDWR14.

Spooner, Thomas, John Fearnley, Rahul Savani, and Andreas Koukorinis. 2018. 'Market Making via Reinforcement Learning.' *CoRR* abs/1804.04216. http://dblp.uni-trier.de/db/journals/corr/corr1804.html\#abs-1804-04216.

Sutton, R. S., H. R. Maei, D. Precup, S. Bhatnagar, D. Silver, Cs. Szepesvári, and E. Wiewiora. 2009. 'Fast Gradient-Descent Methods for Temporal-Difference Learning with Linear Function Approximation'. In *ICML*, 993–1000.

Sutton, R. S., Cs. Szepesvári, and H. R. Maei. 2008. 'A Convergent O(n) Algorithm for Off-Policy Temporal-Difference Learning with Linear Function Approximation'. In *NIPS*, 1609–16.

Sutton, Richard S. 1991. 'Dyna, an Integrated Architecture for Learning, Planning, and Reacting.' *SIGART Bull.* 2 (4): 160–63. http://dblp.uni-trier.de/db/journals/sigart/sigart2.html\#Sutton91.

Sutton, Richard S., and Andrew G. Barto. 2018. *Reinforcement Learning: An Introduction*. Second. Cambridge, MA: MIT Press. http://incompleteideas.net/book/the-book-2nd.html.

Sutton, R., D. Mcallester, S. Singh, and Y. Mansour. 2001. 'Policy Gradient Methods for Reinforcement Learning with Function Approximation'. Cambridge, MA: MIT Press.

Vyetrenko, Svitlana, and Shaojie Xu. 2019. 'Risk-Sensitive Compact Decision Trees for Autonomous Execution in Presence of Simulated Market Response.' *CoRR* abs/1906.02312. http://dblp.uni-trier.de/db/journals/corr/corr1906.html\#abs-1906-02312.

Watkins, C. J. C. H. 1989. 'Learning from Delayed Rewards'. PhD thesis, King's College, Oxford.

Williams, R. J. 1992. 'Simple Statistical Gradient-Following Algorithms for Connectionist Reinforcement Learning'. *Machine Learning* 8: 229–56.

Index

Printed in the United States
by Baker & Taylor Publisher Services